SOCIOBIOLOGY OF INDIGENOUS FOODS

AND FOOD SYSTEMS

BY MATTHEW OGWU

SOCIOBIOLOGY OF INDIGENOUS FOODS AND FOOD SYSTEMS

BY MATTHEW OGWU

 COMMON GROUND

First published in 2025 as part of Food in **Society Book Imprint**

Common Ground Research Networks
University of Illinois Research Park
2001 South First St, Suite 201 L
Champaign, IL 61820 USA

Library of Congress Cataloging-in-Publication Data

Cover Image & Design:

TABLE OF CONTENTS

TABLE OF CONTENTS

CHAPTER 1

Sociobiology of Food Systems–An Integrated Perspective

Abstract

This chapter examines the application of sociobiology to Indigenous food systems, offering a novel perspective on how dietary practices, nutritional choices, and food-related rituals evolve in response to ecological and biological imperatives. Sociobiology, which studies the biological foundations of social behavior, is extended to human food systems, emphasizing the interplay between evolutionary pressures and cultural practices. This framework recognizes Indigenous food systems as coevolved constructs shaped by long-standing interactions between humans and their environments. These systems, characterized by deep ecological knowledge, sustainability, and resilience, serve as models for understanding how food practices adapt to local environmental conditions and promote survival. The chapter highlights how food-related behaviors—such as dietary preferences, culinary techniques, and taboos—are adaptive strategies that ensure nutritional outcomes, cultural continuity, and environmental sustainability. By applying a sociobiological lens, this work underscores the importance of Indigenous food systems in addressing contemporary challenges such as food insecurity, biodiversity loss, and climate change. Furthermore, the chapter proposes a sociobiological model for transforming global food systems, focusing on sustainability, cultural resilience, and the revitalization of traditional food practices in response to modern ecological and health challenges.

Keywords: sociobiology, Indigenous food systems, coevolution, ecological adaptation, nutritional ecology, food taboos, cultural resilience, sustainability

1.1 Introduction

Sociobiology is a field traditionally rooted in evolutionary biology and behavioral science that explores the biological basis of social behavior across species. When applied to human food systems, sociobiology provides a robust framework for understanding how dietary practices, nutritional choices, and food-related rituals are not merely cultural artifacts but are also shaped by long-term biological and ecological interactions. This chapter introduces the core thesis of this book: that Indigenous food systems are best understood as coevolved sociobiological constructs, reflecting a deep integration between ecological processes and cultural development.

The concept of sociobiology was popularized by E.O. Wilson in the 1970s, focusing on the evolutionary mechanics behind social behavior. In recent decades, scholars have broadened its application to include human cultural systems, such as language, religion, and, increasingly, food (Gonzalez et al., 2017; Grimm & Steinle, 2011; Hertzler & Owen, 1976; Lewontin, 1980). Sociobiology in this context does not reduce culture to biology but highlights how biological imperatives and ecological conditions shape cultural adaptations over time. Within food systems, sociobiological thinking encourages exploration of how humans adapt to their environment through subsistence strategies that maximize nutritional outcomes, ensure survival, and promote reproductive success (Obongodot & Ogwu, 2023; Ogwu et al., 2017). These adaptations are not random; they are influenced by generations of knowledge, trial-and-error experimentation, and environmental feedback. Over time, food-related practices become ingrained in cultural norms, rituals, taboos, and culinary identities.

Indigenous food systems offer a vibrant arena for sociobiological analysis. Unlike industrial food systems driven by mass production and economic efficiency, Indigenous systems reflect a close, reciprocal relationship between communities and their surrounding ecosystems. These systems are characterized by profound temporal continuity, spiritual and symbolic dimensions, and an ethic of sustainability. For example, the coevolution of humans with drought-resistant grains such as fonio or millet (*Digitaria exilis*) in the Sahel region illustrates a sociobiological relationship where cultural practices like selective breeding, communal harvesting, and ritual use have reinforced the ecological viability and nutritional value of these crops. Likewise, fermentation practices in various cultures often reflect microbial–human symbioses that enhance food preservation and digestion, revealing intricate environmental relationships at the microbial level.

One of the central insights of sociobiology is that environmental pressures profoundly influence human behavior. This is evident in how Indigenous food practices align with local ecological conditions. In arid climates, culinary techniques that minimize water usage or preserve moisture, such as steaming in banana leaves or sun-drying, are common. In tropical regions, diets rich in fermented foods and plant-based fibers correspond with environmental abundance and microbiome health. These adaptations are often formalized through cultural norms. Food taboos, for instance, may reflect ecological constraints, such as prohibiting the consumption of scarce or ecologically important species. Festive foods may coincide with seasonal abundance, reinforcing agricultural cycles and collective memory. These cultural expressions serve both functional and symbolic purposes, connecting communities to their environments while transmitting adaptive knowledge.

1.1.1 Why View Food Systems from a Sociobiological Lens?

The rationale for applying a sociobiological lens to Indigenous food systems may be threefold:

1. Integration of Ecology and Culture: It allows us to view food not merely as a product of culture or environment, but as the result of their dynamic interplay (Berry et al., 2023; Fanzo et al., 2021).
2. Recognition of Coevolution: It emphasizes the reciprocal influence of human biology, behavior, and food resources over millennia (Wells & Stock, 2020).
3. Framework for Food System Sustainability: It provides a model for understanding sustainable food practices aligned with long-term ecological and community health (Çakmakçı et al., 2023).

The aforementioned perspective is especially pertinent given the global challenges of food insecurity, biodiversity loss, and climate change. When examined through a sociobiological lens, Indigenous food systems offer valuable insights into how human communities can adaptively respond to environmental changes while preserving cultural integrity.

This chapter aims to build a foundational understanding of sociobiology as it relates to human food systems. Specifically, it introduces the idea that food behaviors are cultural constructs strongly rooted in evolutionary and ecological processes. The chapter explores how dietary practices, food preferences, and culinary traditions reflect the coevolved relationships between humans and their environments. It takes an interdisciplinary approach to food sociobiology—bridging biology, anthropology, ecology, and food studies—to highlight the adaptive nature of Indigenous food systems. As global food systems face increasing challenges from climate change, biodiversity loss, and industrial agriculture, a sociobiological perspective offers critical insights into sustainable and resilient foodways. By underscoring the deep connections between biology and culture, this chapter lays the groundwork for reconsidering food as both a commodity and a dynamic, coevolved aspect of human and environmental health.

1.2　Historical Foundations of Sociobiology

1.2.1　Overview of E.O. Wilson's Contribution and the Early Framework

Edward O. Wilson, an esteemed biologist, is widely recognized as the architect of sociobiology. His seminal work, *Sociobiology: The New Synthesis*, published in 1975, introduced a new scientific rigor to the study of social behavior through an evolutionary lens. The book integrates concepts from various disciplines, including ethology, population genetics, and anthropology, to clarify the biological basis of social behaviors observed across numerous species, including humans.

In the first half of *Sociobiology*, Wilson elucidates fundamental principles of social behavior within the animal kingdom, examining territoriality, mating strategies, and altruism. He applies evolutionary concepts, particularly natural selection and kin selection, to explain altruism as an adaptive trait that enhances the reproductive success of genetic relatives (Cronk, 1988). For instance, Wilson investigates the role of inclusive fitness, a concept heavily drawn from the works of W.D. Hamilton, which posits that organisms can increase their genetic success through acts that benefit their relatives, thereby improving the survival of shared genes. Wilson's early framework laid the foundation for understanding that certain aspects of behavior could be inherited or influenced by genetics, thereby establishing a connection between biology and social sciences. His assertion that human behavior could be rooted in our evolutionary past sparked conversation about the biological underpinnings of culture and social structures (Foster, 2009). This assertion was particularly controversial, as researchers sought to determine the extent to which biology influences behaviors traditionally regarded as products of culture.

1.2.2 Extension of Sociobiological Principles to Human Behavior and Culture

The principles articulated by Wilson were not merely intended to explain social behaviors in animals; they were also extrapolated to humans. Scholars in both anthropology and psychology began to examine how evolutionary principles could elucidate aspects of human behavior, including mating systems, social hierarchies, and even cultural practices. Applying the sociobiological tenets led to various theories concerning human nature, such as mate selection, parental investment, and the evolutionary origins of social norms (Harpending et al., 1987). For example, the theory of sexual selection posited by Darwin was extended by Wilson and his followers to explain human mating behaviors as a strategic competition for reproductive success. This perspective suggested that male and female mating preferences could be traced back to evolutionary pressures—males might be predisposed to seek youth and beauty, indicators of fertility. In contrast, females might prefer traits indicating resource acquisition potential (Lieberman et al., 1992).

Furthermore, sociobiologists explored the concept of "reciprocal altruism," where individuals act to benefit others with the expectation of future reciprocation. This idea has been applied to human social interactions, suggesting evolutionary advantages in fostering cooperation and building social bonds (Leavitt, 2007). Social behaviors such as cooperation, altruism, and even aggression can be examined through this lens to understand how they may confer survival advantages throughout human history.

1.2.3 Limitations and Critiques of Early Sociobiology

While Wilson's contributions to sociobiology were profound, the field has faced significant critiques since its inception. Critics such as Richard Lewontin and Philip Kitcher pointed out that the sociobiological approach often leans heavily toward genetic determinism (Kitcher, 1987; Lewontin, 1979). This reductionist tendency simplifies the complex tapestry of human behavior by attributing a large portion to biological imperatives, frequently overlooking the roles of sociocultural influences and individual experiences. Additionally, the deterministic narratives present in early sociobiological discourse have led to accusations of promoting socially regressive ideologies, including justifications for discriminatory practices and the reinforcement of gender stereotypes (Lerner & Eye, 1992). For example, critics have argued that suggesting that certain behaviors, such as aggression or nurturing, are biologically rooted can undermine societal efforts toward gender equality and social reform. Consequently, sociobiology faced scrutiny from feminist scholars who contended against the biological essentialism that many early sociobiological claims seemed to endorse (Liesen, 1995). Furthermore, critiques of sociobiological principles have also included discussions of environmental context. The interplay between genetics and environment is tremendously complex, challenging the notion that genetics alone dictates behavior (Ogwu et al., 2024). The rise of cultural anthropology and sociology further suggested that human behaviors cannot be isolated from cultural contexts; they are often shaped by environmental variables, societal structures, and cohort-specific norms (Caporael & Brewer, 1991). This acknowledgment of complexity highlights the risk of oversimplification in a strictly biologically driven narrative.

Wilson's framework of sociobiology effectively introduced biological principles to the study of behavior and culture; however, its ambitious attempts to unify biology with the social sciences have been met with intrigue and skepticism. The extension of sociobiological principles to human behavior not only opened new avenues for research but also required a critical evaluation of the limits of biological determinism. Emphasizing an integrated approach that recognizes the valuable interplay between biological predispositions and sociocultural contexts represents a crucial progression in evolving scholarly discussions on human behavior.

1.3 Indigenous Food as a Sociobiological System

Indigenous food systems represent multifaceted networks of practices, knowledge, and innovations deeply rooted in the natural environment and human evolutionary history. They provide an exemplary case study for understanding how dietary practices fulfill nutritional needs and function as adaptive strategies that reinforce social organization and survival. Analyzing Indigenous food as a sociobiological system draws on evidence from evolutionary biology, cultural anthropology, neurocognitive theory, and social sciences.

1.3.1 Food Behaviors as Adaptive Biological Strategies

Food behaviors, defined as the selection, processing, sharing, and consumption of dietary resources, have evolved as critical adaptive mechanisms in response to ecological variability. The formative period of human evolution was characterized by diverse environments where resource availability was highly seasonal and unpredictable. In this context, the preference for energy-dense foods can be seen as an adaptive strategy to maximize caloric intake from limited resources. For instance, research on adaptive nutritional ecology posits that hominids developed preferences for foods with high energy content, thus sustaining the high metabolic demands associated with increased brain size and encephalization (Ulijaszek, 2002).

The development of cooking and food processing technologies represents a pivotal evolutionary advancement. These technological innovations improved the digestibility and caloric yield of raw ingredients and contributed to reducing pathogen loads, which is an advancement particularly beneficial in environments where microbial threats were abundant (Ogwu, 2023; Soulier, 2021). Indigenous practices of fermentation, drying, and using natural preservatives illustrate how traditional knowledge leads to optimized energy extraction from diverse food sources, thereby buffering communities against bouts of famine or nutritional instability.

1.3.2 The Evolutionary Role of Diet in Human Development

One of the central evolutionary hypotheses concerning human dietary adaptation involves the relationship between energy-rich diets and brain expansion. The consumption of high-quality proteins and fats—often derived from animal sources—has been linked to the metabolic support necessary for developing a large and complex brain. Additionally, the reduction in the size of the digestive tract observed in modern humans, in contrast to other primates, suggests a trade-off that favors cognitive function over extended digestive processes (Ulijaszek, 2002). As an exogenous method to pre-digest food, cooking further highlights this adaptive trend by enhancing nutrient availability and minimizing caloric loss.

Recent advances in neurocognitive research have introduced the concept of a "theory of food," analogous to the theory of mind. This theory suggests that the evolution of complex neural networks related to dietary choices is supported by cultural and familial transmission of food practices (Allen, 2012). These cognitive structures allow individuals to learn, internalize, and adapt traditional food behaviors, indicating that the cognitive architecture behind dietary decision-making results from evolutionary pressures that merged nutritional ecology with cultural innovation.

1.3.3 Cultural Norms Around Food as Extensions of Survival Strategies

Within Indigenous communities, cultural norms surrounding food practices are crucial for ensuring group survival. Ethnographic studies have long shown that food sharing is not merely an economic exchange but a socially regulated behavior that promotes interpersonal trust, group cohesion, and reciprocity. Models of fairness and cooperation in resource distribution have been empirically linked to environmental conditions, where interdependence in resource procurement requires norms that encourage equitable sharing (Mesoudi & Danielson, 2008). This not only secures immediate nutritional needs but also strengthens social bonds essential for community resilience under environmental stress.

Food holds significant symbolic value in many Indigenous cultures, serving diverse functions beyond energy provision. Rituals, feasts, and ceremonial preparations symbolize a deep connection to the land and ancestral heritage (Ogwu et al., 2024). Integrating spiritual beliefs with subsistence practices often functions as an adaptive mechanism, motivating sustainable resource use and encouraging behaviors that protect critical environmental resources. Carrus et al. (2018) illustrate how symbolic food practices enable communities to spawn a collective identity, reinforcing culturally embedded sustainable practices. The persistent transmission of these symbols and rituals underlines the adaptive significance of food as a cultural marker that supports both environmental stewardship and social solidarity.

1.3.4 Social Norms and Child Ontogeny

The early acquisition of cultural norms related to food consumption is central to the sustainability of Indigenous food systems. Normative behaviors, including food sharing and the observance of taboos, begin to be internalized during childhood, highlighting the extensive role of social learning in cultural evolution (Rakoczy & Schmidt, 2012). Through their participation in communal food procurement and processing activities, children develop survival skills and an inherent understanding of group-oriented ethical behavior. This process is crucial in perpetuating food-related traditions and ensuring that adaptive strategies are continually refined across generations.

1.4 Coevolution of Humans and Food Resources

The coevolution of humans and food resources—encompassing plants, animals, and microbes—represents a critical intersection in our evolutionary history.

Domestication practices have fostered mutual adaptation between humans and their food sources, which have been shaped over millennia to optimize agricultural yield and nutritional intake. This relationship is characterized by the domestication of various species, resulting in changes in genetic traits that enhance their compatibility with human needs.

In particular, crops such as maize and rice, along with animals like goats and chickens, have undergone extensive selection processes, resulting in traits that are more suitable for human consumption and cultivation, as shown in Figure 1.1.

Figure 1.1: Different Sources for Plant Selection and Breeding

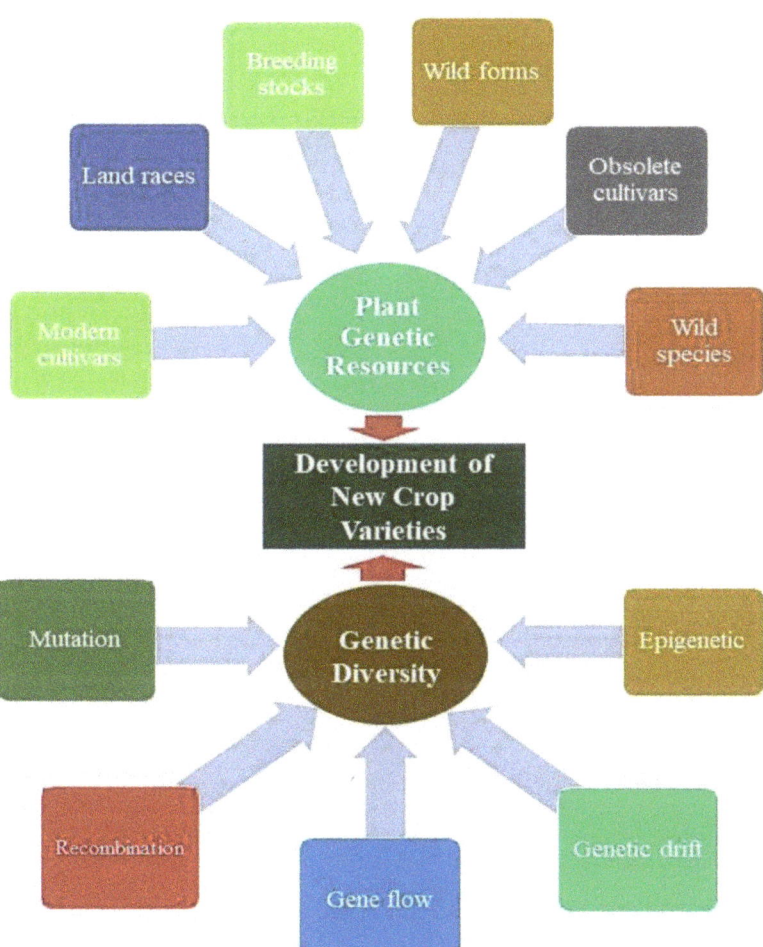

Source: Salgotra and Chauhan (2023)

For instance, maize has been selectively bred for larger cobs and higher kernel counts, becoming a staple food that underpins many diets worldwide (Jones, 2017). Similarly, rice has undergone adaptations that enhance its resistance to environmental stressors, enabling it to thrive in diverse ecological conditions while providing essential nutrients for millions (Remans et al., 2011). Additionally, the domestication of livestock has influenced human diets by promoting the evolution of breeds that are better adapted to the agricultural practices of specific regions. This mutual adaptation extends to the microbial communities associated with these food sources. The fermentation processes employed in traditional food production—such as those observed in yogurt, kimchi, and sourdough—demonstrate how humans have harnessed microbial life for nutritional gain (Gibbons & Rinker, 2015). In these practices, wild microbial populations adapt over time, influencing not only the flavors and textures of foods but also their nutritional profiles, thereby benefiting the human hosts consuming these products (Georgala, 2012).

1.4.1 Examples of Coevolution: Maize, Rice, and Fermented Foods

The specific examples of maize, rice, and fermented foods illustrate the complex dynamics of coevolution. Maize domestication, which can be traced back to ancient Mesoamerican civilizations, involved not only the selection of favorable traits but also cultivation practices that altered maize's genetic landscape, promoting its diversity and adaptability to various climates (Jones, 2017). Maize is not just a food source; it also plays a significant role in cultural practices and nutrition, influencing food security in numerous communities (Remans et al., 2011). Rice serves as another cornerstone of agrarian societies, particularly in Asian countries where it is interwoven with cultural identity and cuisine. The ongoing selection for varieties that yield better under flooded conditions adds another layer to this coevolutionary process, linking human agricultural practices to environmental characteristics (Remans et al., 2011). The relationship between rice crops and local customs emphasizes the blend of ecological knowledge and agricultural practice, demonstrating how food sources evolve alongside human cultivation methods (Anyanwu et al., 2022). Fermented foods embody a unique aspect of this relationship. Fermentation utilizes specific microbes to transform raw ingredients into nutrient-rich foods, enhancing the bioavailability of essential nutrients. For instance, the fermentation of legumes leads to the production of protein-rich foods like natto or tempeh, which have profound implications for the dietary practices of various cultures (Georgala, 2012). As fermentation techniques vary worldwide, they reflect localized microbial adaptations, resulting in diverse taste profiles and health benefits (Bodinaku et al., 2019).

1.4.2 Nutritional Ecology and Bioavailability in Traditional Diets

Nutritional ecology examines how diet impacts health and ecological systems, particularly the interactions between dietary diversity and nutritional outcomes. Traditional diets typically emphasize a wide variety of locally sourced foods, which tend to be more nutritionally balanced and rich in essential micronutrients compared to more modern diets that rely on fewer staple crops (Jones, 2017). Research indicates that diverse diets promote better health outcomes, including lower incidences of chronic diseases, which can be partially attributed to the synergistic effects of consuming various food types that are rich in phytochemicals and bioactive compounds (Luo et al., 2019). Bioavailability is the proportion of nutrients absorbed and utilized by the body, and it also plays a crucial role in traditional diets. Foods like leafy greens and tubers, common in various cultural diets, provide macronutrients and serve as rich sources of essential minerals crucial for maintaining health. However, factors such as preparation methods can significantly influence the bioavailability of these nutrients (Mariat et al., 2009). For example, traditional cooking techniques involving fermentation often enhance nutrient absorption by breaking down antinutritional factors (Georgala, 2012). The interplay between dietary practices and nutritional outcomes illustrates that traditional food systems, through their diversity and methods of preparation, have evolved to optimize human health within specific ecological contexts. Sustainability in food systems also emerges through shared cultural practices and knowledge of local ecosystems, which inform dietary choices and adaptations over time (Maroyi, 2018; Remans et al., 2011).

1.5 Indigenous Food Systems Through a Sociobiological Lens

A hallmark of Indigenous food systems is their emphasis on biodiversity. Indigenous communities tend to cultivate a variety of crop species and landraces. This diversity is maintained through traditional breeding practices that select for traits such as drought tolerance, pest resistance, and nutritional variability. In many parts of West Africa, millet cultivation (Pennisetum glaucum) demonstrates how genetic diversity is preserved through a deep understanding of local environmental conditions (Brown et al., 2006; Manning et al., 2011). Such genetic resilience secures nutritional supply during ecological stress and contributes to overall agroecosystem stability.

Sustainability in Indigenous food systems is achieved through practices that are in harmony with the local ecology. These practices often involve minimal external inputs, reliance on organic fertilizers, and crop rotations that restore soil fertility. The integrated design of these systems allows for the efficient use of water and nutrients, rendering them adaptable to variable climatic conditions. For instance, traditional agroforestry and intercropping regimes help maintain eco-system balance and reduce the likelihood of soil degradation. These sustainability practices are crucial in preserving food security under changing environmental conditions, enabling Indigenous communities to thrive for centuries.

Food in Indigenous societies often transcends its nutritional role, assuming a central role in cultural rituals and spiritual practices. Many Indigenous communities imbue specific food items with symbolic meanings that reflect their cosmology and relationship with nature. Rituals related to food, such as harvest festivals, sacred ceremonies surrounding planting or gleaning, and taboos against certain forms of consumption, reinforce group identity and continuity with ancestral traditions. For example, quinoa in the Andean region has long been venerated as the "mother grain" with deep ritual significance, linking agricultural practices with spiritual beliefs (Bazile et al., 2016; Mizuno et al., 2020).

Some sociobiological mechanisms in Indigenous food systems include the following.

1.5.1 Taboos as Regulatory Mechanisms

Sociobiologically, food taboos are powerful regulatory mechanisms that govern resource use and social behavior. They ensure that food resources are conserved and used consistently with communal welfare. In many Andean societies, the sacred status of quinoa was institutionalized by taboos that restricted its con-sumption and prescribed specific ritual practices. Such taboos regulated dietary behavior and reinforced social hierarchies and ecological conservation, preventing overexploitation of vital crops and maintaining a balance between human needs and environmental sustainability.

1.5.2 Seasonality and Ecological Synchronization

Seasonality is a critical aspect of Indigenous food systems, where biological cycles and ecological rhythms dictate agricultural practices. Indigenous calendars are often intricately tied to seasonal variations, providing practical guidance for sowing, cultivation, and harvesting activities. In West Africa, the cultivation of millet is closely aligned with seasonal rainfall patterns. Traditional ecological knowledge enables communities to predict climatic changes and adjust planting schedules accordingly, optimizing crop productivity even in arid environments and unpredictable rainfall (Brown et al., 2006).

1.5.3 Knowledge Transfer and Cultural Continuity

Knowledge transfer is a cornerstone of Indigenous food systems. Unlike modern agricultural practices that often rely on codified scientific training, Indigenous farming practices are transmitted orally and experientially across generations. Elders and experienced farmers act as the primary repositories of empirical knowledge—including information about soil composition, water management, pest control, and climatic cues. This intergenerational transfer of knowledge is integral to maintaining the adaptability and sustainability of Indigenous food systems. It ensures that traditional practices continue to evolve in response to environmental changes, preserving not only the techniques of cultivation but also the cultural narratives that inform them.

1.6 Sociobiological Framework for Food System Transformation

The dynamics underlying food systems and their transformation can be compellingly understood through a sociobiological lens that integrates ecological, evolutionary, cultural, and economic perspectives. Sociobiology investigates how evolutionary behavioral pressures and reciprocal social interactions shape food acquisition, processing, and distribution strategies. For instance, studies on food webs provide evidence that ecological constraints, such as stability requirements, can impose limitations on phenotypic diversity, thereby influencing the collective behavior of populations in acquiring and sharing food resources (Barbour, 2021; McCann, 2011). This view supports the notion that the configuration of food networks has an evolutionary basis that extends beyond classical mutation and selection mechanisms, aligning with a sociobiological approach to food systems.

Dynamic interactions in supply chain networks underscore how food systems evolve due to coevolutionary processes among interdependent agents. The reciprocal interactions and adaptive responses among organizations in food supply chains reflect a transformation that mirrors biological coevolution (Ng et al., 2003). In parallel, niche construction theory highlights how organisms actively modify their environments and offers an evolutionary explanation for transformative changes in food systems. Food processing has been shown to increase the net energetic yield of food through cooking (Carmody & Wrangham, 2009), prompting substantial cultural adaptations and coevolutionary feedback loops, as argued within the framework of gene–culture coevolution (Laland et al., 2001; Wollstonecroft, 2011). These theoretical insights suggest that material and cultural innovations with deep evolutionary roots can drive food system transformation.

Moreover, a transformative approach to food systems requires incorporating principles of social justice, resilience, and Indigenous knowledge. Contemporary challenges, such as structural inequities and environmental degradation, necessitate models prioritizing food sovereignty. Empirical work in Indigenous contexts illustrates that community-based strategies, such as intergenerational home gardening, promote resilience and revitalize traditional food practices (Budowle et al., 2019). The Indigenous food sovereignty movement emphasizes decolonial and sustainable self-determination by challenging globalized, industrial agricultural models and re-centering culturally appropriate food practices (Coté, 2016; Hoover, 2017). Incorporating these perspectives into food system transformation policies addresses ecological and evolutionary constraints and sociocultural dimensions central to equitable food distribution and community well-being.

Sociobiology is a field that emerged from evolutionary biology and seeks to explain social behavior through the lens of natural selection. Its intellectual roots can be traced to Charles Darwin's formulation of natural selection, which provided the foundational framework for understanding how traits that promote survival and reproduction—including social behaviors—are selected over generations (Ruse, 2014). Early proponents applied core concepts such as kin selection and reciprocal altruism as mechanisms explaining cooperation and competition among organisms, framing these behaviors as adaptive strategies shaped by evolutionary pressures (Ruse, 2008). This perspective has since evolved from its initial focus on nonhuman species to encompass the full spectrum of human social behavior, thereby integrating biological, cultural, and ecological determinants of action (Ruse, 2008).

The sociobiological approach has been extended to studying food systems transformation in recent years. This novel application examines the evolutionary imperatives that once structured social interactions now influence how food is produced, shared, and consumed. Food system dynamics—ranging from the cooperative foraging strategies observed in animal societies to the complex network interactions seen in human communities—are increasingly recognized as governed by the same adaptive principles that underlie other forms of social behavior (Sear, 2015). Researchers can explore how long-standing evolutionary mechanisms inform contemporary challenges in food sustainability, equity, and resilience by integrating evolutionary theory with food acquisition and distribution analysis. Such an interdisciplinary perspective facilitates understanding food systems as economic or technological constructs and dynamic, adaptive networks influenced by evolved behavioral strategies (Sear, 2015). This expanded view offers promising insights for developing transformative policies to reshape food systems in ecologically sustainable and socially just ways.

The sociobiology of food also draws on the behavioral ecology observed in nonhuman primates to illustrate the evolution of food sharing and competition. Studies show that reciprocal altruism in food sharing among species, such as brown capuchin monkeys, enhances social bonds and resource distribution (Waal, 2000). Such findings, when extrapolated to human behavior, suggest that adaptive mechanisms favor individuals and groups that develop cooperative strategies for food procurement and distribution, ultimately contributing to the stability and transformation of food systems. Evolutionary anthropology provides additional insights into how reproductive, production, and distribution-related behaviors are interwoven with food practices, thereby influencing overall human well-being (Gibson & Lawson, 2015). Furthermore, evolved taste preferences—an adaptive legacy that once helped humans identify essential nutrients—now contribute to public health challenges when food environments favor energy-dense, nutrient-poor options (Breslin, 2013; Lucock et al., 2013). A sociobiological framework for food system transformation necessitates understanding how ecological constraints, coevolutionary dynamics, and cultural niche construction converge to shape food webs and human food behaviors. By integrating evidence from environmental studies (Barbour, 2021; McCann, 2011), coevolutionary models (Ng et al., 2003), and cultural inquiries into food sovereignty (Budowle et al., 2019; Coté, 2016; Hoover, 2017) alongside insights from evolutionary medicine and behavioral ecology (Breslin, 2013; Carmody & Wrangham, 2009; Gibson & Lawson, 2015; Lucock et al., 2013; Waal, 2000), it is possible to envision a multifaceted transformation that stabilizes ecological networks and enhances social equity, cultural resilience, and public health. This interdisciplinary synthesis provides a robust platform for policymakers and researchers to advance sustainable and adaptive food systems in the Anthropocene.

A model sociobiological framework for transforming food systems should be designed to integrate ecological, biological, and cultural dimensions. This framework ought to be suitable for academic applications and policy discussions regarding sustainable and culturally rooted food transitions by including the following components (Table 1.1):

Table 1.1: Sociobiological Framework for Transforming Food Systems

Component	Key Elements	Transformation Goals
Ecological Anchoring	Ecological anchoring focuses on conserving local biodiversity, applying agroecological practices rooted in Indigenous knowledge, and planning food production that responds to local soil, water, and climate conditions.	The goal is to restore ecological balance, ensure long-term sustainability of food sources, and enhance resilience in the face of environmental change.
Biocultural Feedback Loops	This component highlights the interactions between diet, the human microbiome, and health, as well as how food taboos, rituals, and seasonal eating patterns reflect adaptive strategies for survival and well-being.	The aim is to promote community health, preserve cultural traditions that support ecological adaptation, and strengthen intergenerational knowledge transfer.
Knowledge Transmission	Knowledge is passed down through oral traditions, food-related rituals, and apprenticeship models that preserve culinary skills and ecological intelligence within communities.	This ensures that valuable food knowledge is maintained across generations, contributing to cultural identity, resilience, and sustainability.
Social Organization	Food systems are often organized around collective labor practices, clearly defined food roles based on age or gender, and shared governance of natural and food resources.	Strengthening these systems builds social cohesion, empowers local decision-making, and enhances equitable access to resources.

Resilience and Reproduction	Food systems contribute to the reproduction of both ecological and human health through practices that enhance immune resilience, promote environmental regeneration, and support sustainable lifestyles.	The objective is to create food systems that are capable of withstanding shocks while promoting long-term ecological and social stability.
Policy and Planning	Policy frameworks can support food sovereignty by recognizing community ownership of food systems, protecting Indigenous rights, and ensuring ethical governance over land and food resources.	The goal is to align food policies with the needs of local communities, cultural preservation, and ecological sustainability.
Education and Communication	Educational efforts include formal curriculum development, public storytelling, community-based learning, and participatory food mapping to document and revitalize traditional knowledge.	These efforts raise awareness, foster food literacy, and engage youth and stakeholders in sustainability practices.
Innovation and Adaptation	Food system transformation benefits from blending traditional and scientific knowledge, designing foods with microbiome health in mind, and supporting community-based experimentation and innovation.	This supports flexible, adaptive solutions that respect cultural traditions while addressing modern health and environmental challenges.

1. **Foundational Principles:** It should address
   - o   Coevolutionary Logic by recognizing that human food systems have coevolved with specific ecological contexts, species, and microbiomes.
   - o   Embeddedness by understanding that food practices are embedded in biological, social, and symbolic systems.
   - o   Adaptive Feedback to support evolution within food systems through feedback between biological needs and cultural responses (e.g., climate, disease, labor).

2. **Core Components:** This should focus on
   - o   Ecological Anchoring: Local biodiversity conservation (plants, animals, and microbes), agroecological practices rooted in Indigenous knowledge, and water-, soil-, and climate-sensitive food planning.
   - o   Biocultural Feedback Loops: Diet–microbiome–health interactions, food taboos, rituals, and seasonal calendars as adaptive tools and sensory, nutritional, and medicinal food properties encoded in culture.
   - o   Knowledge Transmission: Intergenerational food literacy (oral traditions, rituals, and apprenticeship) and formal and informal learning systems (e.g., seed sharing, storytelling) and culinary heritage as ecological intelligence.
   - o   Social Organization: Collective labor (e.g., harvesting, communal cooking); gendered roles in food production and preservation and, hence, sociobiological division of labor; and community governance over food access, rights, and resources
   - o   Resilience and Reproduction: Food systems designed to support ecological and social reproduction, and diets that reinforce immune resilience, environmental repair, and interdependence and resilience indicators: crop diversity, food sovereignty, and health equity.

3. **Application Pathways**
   - o   Policy and Planning: Support for Indigenous and community-led food sovereignty initiatives; ethical bioprospecting and protection of traditional knowledge and integration of sociobiological insights into food policy and land-use planning.
   - o   Education and Communication: Curriculum development rooted in food ecology and culture, revitalization of ancestral practices through public media and education, and participatory food mapping and ethnobotanical documentation.

     o   Innovation and Adaptation: Hybrid systems blending traditional and scientific knowledge, microbiome-aware food system design (fermented foods, gut-health-based diets), and community labs for coproducing sustainable food technologies.

4. **Transformation Goals:** The transformation goals of Indigenous food systems span multiple interconnected dimensions. In terms of health and nutrition, the primary sociobiological goal is to reconnect diets with evolutionary needs and support a diverse, resilient microbiome. Culturally, these goals center on preserving food-based identity, dignity, and autonomy, recognizing the deep symbolic and communal value of traditional diets. From an environmental perspective, the focus is on restoring ecological cycles and fostering regenerative systems that align with traditional ecological knowledge. Economically, the aim is to promote local food economies and reduce dependence on extractive models, ensuring food justice and sovereignty for Indigenous communities.

5. **Some Guiding Questions**
   o   How has this community's food system coevolved with its environment?
   o   What biocultural food practices enhance ecological and health resilience?
   o   Which cultural expressions (e.g., taboos, festivals, culinary practices) encode adaptive strategies?
   o   How can these be protected, revitalized, and scaled ethically?

1.6.1 Case Studies of the Sociobiology of Food

1.6.1.1 Millet in West Africa

Millet serves as an exemplary case of an Indigenous food crop that is resilient, adaptable, and deeply intertwined with local cultural practices. In West Africa, millet is more than a dietary staple; it is an indicator species for regional ecological health. Archaeobotanical evidence suggests that millet has been continuously cultivated for thousands of years, showing sophisticated agricultural practices that have enabled communities to effectively manage water scarcity and soil degradation (Manning et al., 2011). Moreover, traditional marketplaces dynamically adjust millet pricing in response to seasonal variations and vegetative productivity, underscoring its role as a barometer of environmental change (Brown et al., 2006). The diversity within millet cultivars and localized breeding strategies ensures a robust genetic pool, which is vital for both present-day consumption and future crop resilience (Figure 1.2).

Figure 1.2: Mature Millet in the Field

Source: Das et al. (2019)

1.6.1.2 Quinoa in the Andes

Quinoa (*Chenopodium quinoa*) offers a vivid illustration of an Indigenous food system deeply rooted in its cultural and ecological context. Domesticated over 7,000 years ago, quinoa has evolved within the diverse landscapes of the Andean region. Historically, quinoa was much more than just a food source; it was central to the spiritual and ritual life of Andean communities. Colonial repression sought to suppress its cultivation due to its association with Indigenous rituals, yet the resilience of local traditions ensured its survival. Quinoa's ability to thrive in high-altitude and saline environments underscores its agronomic adaptability, while its ritual significance strengthens community bonds and cultural identity (Bazile et al., 2016; Mizuno et al., 2020). As global interest in quinoa increases due to its nutritional benefits, the challenge remains to balance commercial pressures with preserving its diverse traditional cultural contexts (Figure 1.3).

Figure 1.3: Some Ethnovarieties of Quinoa from the Andes

Source: Cocarico et al. (2024)

1.6.1.3 Goats in the Sahel

Goats exemplify a quintessential form of Indigenous livestock that has long been woven into the socioecological fabric of the Sahel region. For centuries, Sahelian communities have relied on goats not only for meat, milk, and hides but also for wealth storage and social capital. Archaeozoological and ethnographic records trace the domestication and management of goats in West Africa to at least 3,000 to 5,000 years ago, reflecting complex pastoral knowledge systems adapted to varying environmental conditions (Marshall & Hildebrand, 2002). Goats' physiological adaptability enables them to survive in harsh arid and semiarid conditions characterized by water scarcity, poor forage availability, and extreme temperatures (Wilson, 1991). Their browsing behavior allows them to utilize various vegetation types, including shrubs, leaves, and hardy grasses often unsuitable for other livestock species. This feeding flexibility contributes to the resilience of pastoral livelihoods and landscape-level ecological balance, as goats play a role in controlling bush encroachment and seed dispersal.

In many Sahelian communities, goats hold significant cultural and spiritual value. They are often used in rites of passage, religious ceremonies, and bride price negotiations, further embedding them into local governance and identity systems. Genetic diversity among Indigenous goat breeds, such as the Sahelian goat and West African Dwarf goat, is maintained through localized breeding practices, ensuring resilience against emerging diseases and climate variability (FAO, 2015). As global climate change increasingly threatens the viability of conventional livestock systems, goats present a model for sustainable, climate-resilient animal husbandry. Their management illustrates the importance of integrating Indigenous knowledge systems into broader discussions of food security and environmental stewardship in the Global South (Figure 1.4).

Figure 1.4: Color Painting of Sahelian Goat Herds Grazing in Semi-arid Landscape

1.7 Sociobiology and Modern Food Challenges

1.7.1 Breakdown of Traditional Food Systems Through Industrialization

The significant transformations in food systems resulting from industrialization have profoundly affected how food is produced, distributed, and consumed globally. Traditionally, food systems were characterized by localized practices emphasizing diversity, seasonal availability, and community participation. However, the advent of industrial agriculture has triggered a dramatic shift toward monoculture systems that prioritize efficiency, scale, and uniformity over ecological balance and nutrient diversity (Machum, 2015). The industrialization of food production is heavily influenced by the Green Revolution, which introduced high-yield crop varieties, synthetic fertilizers, and extensive irrigation practices to boost agricultural productivity. While these innovations initially yielded increased food supplies, they also resulted in negative consequences, such as environmental degradation, loss of biodiversity, and the marginalization of traditional farming practices and knowledge (Gómez & Ricketts, 2013).

Moreover, the global expansion of fast-food chains and the prevalence of processed food options have further undermined traditional food systems. As urbanization increases, many people opt for convenience foods that often lack essential nutrients, while being high in empty calories, trans fats, and refined sugars. This shift results in dietary patterns that are increasingly disconnected from the agricultural landscapes that initially supported them (Baker et al., 2020). The combined effects of globalization and industrialization have created dietary environments that, while rich in calories, often lack the necessary vitamins, minerals, and phytochemicals vital for health. The repercussions of this transformation present significant social and cultural implications. Food serves not only as sustenance but also as a reflection of cultural identity. As traditional food systems decline, there is a risk of losing important cultural practices associated with food preparation, consumption, and sharing. This loss can lead to weakened community ties and a diminished sense of belonging (Merz & Steinberg, 2014). Furthermore, communities facing disruptions in traditional food systems often confront economic difficulties, as they lose access to local food sources and the stability those sources provide.

1.7.2 Disruption of Coevolved Food Relationships and Consequences for Health

The coevolutionary relationships between humans and their food resources have historically been marked by adaptations that suit ecological contexts and human nutritional needs. As industrialization disrupts these relationships, the health consequences become evident. Traditional diets, which evolved synchronously with local ecological systems, were finely attuned to meet the dietary needs of specific populations. These diets typically included a wide variety of fruits, vegetables, grains, and animal products that reflected local resources (Nyholm et al., 2024). In contrast, the industrial food system promotes reliance on a narrower range of foods, primarily derived from large-scale agricultural production. This shift contributes to a growing disconnection between diet and health, correlating with rising rates of obesity, type 2 diabetes, cardiovascular diseases, and other diet-related conditions, particularly in populations transitioning from traditional to industrialized eating patterns (Baker & Friel, 2016). This is evident in many Indigenous communities globally, where traditional food sources are replaced by processed foods, leading to a marked increase in diet-related health issues (Lang et al., 2014).

The implications extend beyond individual health outcomes; they highlight broader environmental and social challenges. The reduction in dietary diversity undermines human health and diminishes ecosystem resilience. Diverse and sustainable agricultural systems promote ecological balance, allowing populations to access nutritious foods even amid fluctuations caused by climate and environmental changes (Lawrence, 2017). With industrial monocultures prevailing in food production, the resulting agricultural practices often heighten vulnerabilities to pest outbreaks, diseases, and environmental stresses, jeopardizing food security for various populations (Lang et al., 2014).

1.7.3 Sociobiological Insights into Food Insecurity and Dietary Diseases

The framework of sociobiology, which emphasizes the evolutionary underpinnings of behavior and social organization, illuminates contemporary challenges surrounding food insecurity and dietary diseases. The concepts of kin selection and reciprocal altruism provide contextual frameworks for understanding how dietary practices have evolved and how they are influenced by socioeconomic changes in modern contexts (Vittersø et al., 2019). Food insecurity arises not only from a lack of economic access to food but also from systemic inequalities that disrupt traditional food relationships and lead to unhealthy eating practices. Sociobiological perspectives can clarify how community-based solutions that promote local food systems can enhance food security, strengthen community resilience, and improve population health outcomes. For example, initiatives promoting urban gardens, community-supported agriculture, and localized food distribution systems highlight the importance of restoring connections to traditional foods, supporting both nutritional adequacy and community solidarity (Ogwu, 2019). Dietary diseases, which result from the consumption of nutrient-poor processed foods and inadequate access to traditional and nutrient-dense foods, underscore the need for a multifaceted approach to public health. Sociobiological research can help identify behavioral and cultural patterns that affect food choices, guiding policies and public health interventions that address these foundational issues.

This holistic understanding encourages a reconceptualization of food systems that respects the interconnectedness of sociocultural practices, agricultural biodiversity, and health outcomes.

1.8 Methodological Approaches to Studying the Sociobiology of Food

Ethnographic methods have been fundamental in uncovering the cultural dimensions of food practices. These methods, encompassing participant observation, in-depth interviews, and sensory ethnography, provide a rich, localized understanding of how communities produce, distribute, and consume food (Tumilowicz et al., 2015; Wilson, 2023). For example, Wilson's recent work demonstrates that ethnographic inquiry can reveal the intricate relationship between traditional foodways and contemporary sustainable practices (Wilson, 2023). In parallel, ecological methods contribute objective data regarding environmental factors such as resource availability, seasonal variability, and biodiversity. Studies integrate ecological sampling techniques and spatial analysis to contextualize food systems within local ecosystems, thereby explaining how ecological constraints drive adaptive cultural practices. Further reinforcement of this ethnographic–ecological integration is found in the work by Charnley and Durham (2010), who argue that an environmental anthropology perspective must remain grounded in cultural realities rather than solely relying on ecological data. Similarly, Ingram (2011) emphasizes that food systems research should cross scales—from micro-level community practices to broader environmental processes—thus advocating for an integrated approach that marries ecological data with ethnographic insights.

Nutritional anthropology is essential for understanding dietary practices within a biocultural framework. Ulijaszek's studies (Ulijaszek, 2018, 2024) have highlighted that nutritional choices are not random; instead, they are shaped by historical processes, ecological pressures, and cultural transmission. This field combines quantitative dietary assessments with qualitative cultural analyses, offering a comprehensive view of how food practices affect health and nutrition at individual, population, and societal levels. At the same time, evolutionary biology provides a basis for understanding the adaptive significance of dietary behaviors. Research in this area utilizes evolutionary theory to explain the development of human digestive physiology, brain development, and immune function—an idea further explored by McDade (2005) in his discussion on the ecologies of human immune function, which emphasizes how physiological adaptations interact with culturally driven dietary practices. Cohen et al. (2017) expand this conversation into migration studies, demonstrating how biocultural factors influence nutritional transitions among migrant populations in Cameroonian and French contexts. This integrative work underscores the need for anthropological inquiries into nutrition to consider both evolutionary legacies and contemporary cultural adaptations.

1.9 Interdisciplinary Research Challenges and Opportunities in Food Sociobiology

The sociobiology of food presents a classic case of a "wicked problem" due to the inherent complexity of integrating multiple disciplinary perspectives. Interdisciplinary research in this field merges methods from anthropology, ecology, nutrition, and evolutionary biology. However, as noted by Dunn (2022) and further elaborated by Horton et al. (2017), this integration is fraught with challenges such as reconciling divergent epistemological frameworks and methodological scales. Horton and colleagues advocate for an integrated system-wide approach that bridges environmental and social dimensions through collaboration among researchers from diverse backgrounds.

Recent studies have also emphasized the potential of interdisciplinary academic programs and innovative pedagogical frameworks. For instance, Ebel et al. (2020) and Hartle et al. (2017) demonstrate that codesigning curricula and fostering interdisciplinary learning environments can enhance systems thinking and prepare students to tackle complex food system challenges. Moreover, Thoolen et al. (2023) argue that rapid-response research in the food supply chain benefits from the involvement of scholars in behavioral science, food safety, and supply chain management, thereby enriching the methodological portfolio available to researchers.

Additional contributions from Doherty et al. (2019) and Pope et al. (2021) further highlight opportunities for incorporating resilience theory and systems literacy into food systems research. These contributions indicate that adopting nexus approaches—which address the interconnections between food, energy, water, and other resources—can help overcome disciplinary silos and lead to a more comprehensive understanding of food system sustainability. Collectively, the literature suggests that studying the sociobiology of food necessitates a blend of qualitative and quantitative methods that bridge cultural and biological realms. Ethnographic and ecological methods generate localized, context-specific insights; nutritional anthropology situates food practices within a historical and adaptive framework; and evolutionary biology reveals the deep-seated physiological processes underlying dietary behavior. Interdisciplinary collaboration remains both a challenge and an opportunity, demanding innovative approaches that reconcile different scales of analysis and theoretical frameworks.

Future research will benefit from ongoing methodological pluralism, the integration of digital data collection methods, and the development of shared conceptual frameworks that effectively connect cultural, ecological, and biological domains. These efforts promise to tackle global challenges related to food security and environmental sustainability, ultimately informing policy decisions and interventions at various levels.

1.10 Conclusion

This chapter argues that Indigenous food systems are best understood as coevolved sociobiological constructs arising from the dynamic interplay of ecological, biological, and cultural forces. Drawing from evolutionary biology, anthropology, and nutritional science, food can be observed to range from dietary preferences and culinary techniques to symbolic rituals and taboos—are deeply rooted in adaptive responses to environmental and social conditions. By emphasizing the sociobiological foundation of food systems, a conceptual framework integrating ecological sustainability with cultural resilience and human health can be developed. This chapter suggests that traditional food systems, often marginalized by globalized industrial agriculture, offer critical insights for addressing contemporary food insecurity, ecological degradation, and public health crises. Practices such as fermentation, seasonal eating, and communal food governance are historically significant and biologically and environmentally attuned. They exemplify coevolutionary feedback loops that have sustained communities across generations. Furthermore, the sociobiological model for food system transformation presented here outlines principles and applications that promote equity, biodiversity, and adaptive capacity. This model highlights the importance of Indigenous knowledge, intergenerational learning, and biocultural diversity in shaping sustainable futures. As climate change and nutritional transitions increasingly challenge global food systems, re-centering sociobiological thinking in food studies offers a pathway to more integrative, just, and resilient solutions.

Chapter Reflection

1. How does the sociobiological framework help us understand the reciprocal relationship between human cultural practices and the environment, especially in the context of Indigenous food systems?
2. In what ways can the principles of coevolution between humans and their food sources inform sustainable agricultural practices in today's rapidly changing ecological conditions?
3. What are the potential benefits and challenges of integrating Indigenous knowledge and sociobiological insights into modern food policy and sustainability efforts?
4. How do food taboos and rituals function as adaptive strategies in Indigenous communities, and how do they ensure ecological and social stability?
5. How can contemporary societies learn from Indigenous food systems in addressing global issues such as food insecurity, biodiversity loss, and climate change?
6. Think about a food you regularly consume and write down the following:
   1. The ecological or cultural origin of this food.
   2. How it has adapted to its environment or evolved in human culture (e.g., seasonal harvests, preservation methods, or dietary preferences).
   3. Any related rituals or cultural practices they engage in with this food (e.g., special cooking techniques, holiday consumption, or taboos).

REFERENCES

Allen, J. (2012). "Theory of food" as a neurocognitive adaptation. *American Journal of Human Biology*, *24*(2), 123–129. https://doi.org/10.1002/ajhb.22209

Anyanwu, O., Naumova, E., Chomitz, V., Zhang, F., Chui, K., Kartasurya, M., & Folta, S. (2022). The socio-ecological context of the nutrition transition in Indonesia: A qualitative investigation of perspectives from multi-disciplinary stakeholders. *Nutrients*, *15*(1), 25. https://doi.org/10.3390/nu15010025

Baker, P., & Friel, S. (2016). Food systems transformations, ultra-processed food markets and the nutrition transition in Asia. *Globalization and Health*, *12*(1). https://doi.org/10.1186/s12992-016-0223-3

Baker, P., Machado, P., Santos, T., Sievert, K., Backholer, K., Hadjikakou, M., Russell, C., Huse, O., Bell, C., Scrinis, G., Worsley, A., Friel, S., & Lawrence, M. M. (2020). Ultra-processed foods and the nutrition transition: Global, regional and national trends, food systems transformations and political economy drivers. *Obesity Reviews*, *21*(12), e13126. https://doi.org/10.1111/obr.13126

Barbour, M. (2021). Ecological character displacement destabilizes food webs. *The American Naturalist*, *197*(1), 18–28. https://doi.org/10.1086/711875

Bazile, D., Jacobsen, S., & Verniau, A. (2016). The global expansion of quinoa: Trends and limits. *Frontiers in Plant Science*, *7*. https://doi.org/10.3389/fpls.2016.00622

Beery, T., StahlOlafsson, A., Gentin, S., Maurer, M., Stålhammar, S., Albert, C., Bieling, C., Buijs, A., Fagerholm, N., Garcia-Martin, M., Plieninger, T., & Raymond, C. (2023). Disconnection from nature: Expanding our understanding of human–nature relations. *People and Nature*, *5*, 470–488. https://doi.org/10.1002/pan3.10451

Bodinaku, I., Shaffer, J., Connors, A., Steenwyk, J. L., Kastman, E., Rokas, A., Robert, A., & Wolfe, B. (2019). Rapid phenotypic and metabolomic domestication of wild Penicillium molds on cheese. https://doi.org/10.1101/647172

Breslin, P. (2013). An evolutionary perspective on food and human taste. *Current Biology*, *23*(9), R409–R418. https://doi.org/10.1016/j.cub.2013.04.010

Brown, M., Pinzón, J., & Prince, S. (2006). The effect of vegetation productivity on millet prices in the informal markets of Mali, Burkina Faso and Niger. *Climatic Change*, *78*(1), 181–202. https://doi.org/10.1007/s10584-006-9096-4

Budowle, R., Arthur, M., & Porter, C. (2019). Growing intergenerational resilience for Indigenous food sovereignty through home gardening. *Journal of Agriculture Food Systems and Community Development*, *9*, 1–21. https://doi.org/10.5304/jafscd.2019.09b.006

Çakmakçı, R., Salık, M. A., & Çakmakçı, S. (2023). Assessment and principles of environmentally sustainable food and agriculture systems. *Agriculture*, *13*(5), 1073. https://doi.org/10.3390/agriculture13051073

Caporael, L., & Brewer, M. (1991). Reviving evolutionary psychology: Biology meets society. *Journal of Social Issues*, *47*(3), 187–195. https://doi.org/10.1111/j.1540-4560.1991.tb01830.x

Carmody, R., & Wrangham, R. (2009). The energetic significance of cooking. *Journal of Human Evolution*, *57*(4), 379–391. https://doi.org/10.1016/j.jhevol.2009.02.011

Carrus, G., Pirchio, S., & Mastandrea, S. (2018). Social-cultural processes and urban affordances for healthy and sustainable food consumption. *Frontiers in Psychology*, *9*. https://doi.org/10.3389/fpsyg.2018.02407

Charnley, S., & Durham, W. (2010). Anthropology and environmental policy: What counts?. *American Anthropologist*, *112*(3), 397–415. https://doi.org/10.1111/j.1548-1433.2010.01248.x

Cocarico, S., Rivera, D., Beck, S., & Obón, C. (2024). *Qarasiña* culinary tradition: Conserving quinoa (*Chenopodium quinoa*) as an intangible cultural heritage in *Jach'a Puni* (Andean community), Bolivia. *Heritage*, *7*(10), 5390–5412. https://doi.org/10.3390/heritage7100254

Cohen, E., Amougou, N., Ponty, A., Loinger-Beck, J., Nkuintchua, T., Monteillet, N., Bernard, J. Y., Saïd-Mohamed, R., Holdsworth, M., & Pasquet, P. (2017). Nutrition transition and biocultural determinants of obesity among Cameroonian migrants in urban Cameroon and France. *International Journal of Environmental Research and Public Health*, *14*(7), 696. https://doi.org/10.3390/ijerph14070696

Coté, C. (2016). "Indigenizing" food sovereignty. revitalizing Indigenous food practices and ecological knowledges in Canada and the United States. *Humanities*, *5*(3), 57. https://doi.org/10.3390/h5030057

Cronk, L. (1988). Human history as natural history. *Critical Review*, *2*(1), 103–110. https://doi.org/10.1080/08913818908459517

Das, S., Khound, R., Santra, M., & Santra, D. K. (2019). Beyond bird feed: Proso millet for human health and environment. *Agriculture*, *9*(3), 64. https://doi.org/10.3390/agriculture9030064

Doherty, B., Ensor, J., Heron, T., & Prado, P. (2019). Food systems resilience: Towards an interdisciplinary research agenda. *Emerald Open Research*, *1*, 4. https://doi.org/10.12688/emeraldopenres.12850.1

Dunn, R. (2022). The future of food. https://doi.org/10.52750/418989

Ebel, R., Ahmed, S., Valley, W., Jordan, N., Grossman, J., Shanks, C., Stein, M., & Dring, C. (2020). Co-design of adaptable learning outcomes for sustainable food systems undergraduate education. *Frontiers in Sustainable Food Systems*, *4*. https://doi.org/10.3389/fsufs.2020.568743

Fanzo, J., Bellows, A. L., Spiker, M. L., Thorne-Lyman, A. L., & Bloem, M. W. (2021). The importance of food systems and the environment for nutrition. *The American Journal of Clinical Nutrition*, *113*(1), 7–16. https://doi.org/10.1093/ajcn/nqaa313

Food and Agriculture Organization of the United Nations. (2015). The second report on the state of the world's animal genetic resources for food and agriculture. In B. D. Scherf & D. Pilling (Eds.), *FAO commission on genetic resources for food and agriculture assessments* (pp. 1–41). https://openknowledge.fao.org/server/api/core/bitstreams/cb3cd3d4-1c97-49e9-b9e6-c6344ad121c4/content

Foster, K. (2009). A defense of sociobiology. *Cold Spring Harbor Symposia on Quantitative Biology*, *74*, 403–418. https://doi.org/10.1101/sqb.2009.74.041

Georgala, A. (2012). The nutritional value of two fermented milk/cereal foods named "Greek trahanas" and "Turkish tarhana": A review. *Journal of Nutritional Disorders and Therapy*, *3*(1). https://doi.org/10.4172/2161-0509.s11-002

Gibbons, J., & Rinker, D. (2015). The genomics of microbial domestication in the fermented food environment. *Current Opinion in Genetics and Development*, *35*, 1–8. https://doi.org/10.1016/j.gde.2015.07.003

Gibson, M., & Lawson, D. (2015). Applying evolutionary anthropology. *Evolutionary Anthropology Issues News and Reviews*, *24*(1), 3–14. https://doi.org/10.1002/evan.21432

Gómez, M., & Ricketts, K. (2013). Food value chain transformations in developing countries: Selected hypotheses on nutritional implications. *Food Policy*, *42*, 139–150. https://doi.org/10.1016/j.foodpol.2013.06.010

Gonzalez, M. B., Aronson, B. D., Kellar, S., Walls, M. L., & Greenfield, B. L. (2017). Language as a facilitator of cultural connection. *ab-Original: Journal of Indigenous Studies and First Nations' and First Peoples' Culture*, *1*(2), 176–194. https://doi.org/10.5325/aboriginal.1.2.0176

Grimm, E. R., & Steinle, N. I. (2011). Genetics of eating behavior: Established and emerging concepts. *Nutrition Reviews, 69*(1), 52–60. https://doi.org/10.1111/j.1753-4887.2010.00361.x

Harpending, H., Rogers, A., & Draper, P. (1987). Human sociobiology. *American Journal of Physical Anthropology, 30*(S8), 127–150. https://doi.org/10.1002/ajpa.1330300509

Hartle, J., Cole, S., Trepman, P., Chrisinger, B., & Gardner, C. (2017). Interdisciplinary food-related academic programs: A 2015 snapshot of the United States landscape. *Journal of Agriculture Food Systems and Community Development, 7*(4), 1–15. https://doi.org/10.5304/jafscd.2017.074.006

Hertzler, A. A., & Owen, C. (1976). Sociologic study of food habits—A review. II. Differentiation, accessibility and solidarity. *Journal of the American Dietetic Association, 69*(4), 381–384.

Hoover, E. (2017). "You can't say you're sovereign if you can't feed yourself": Defining and enacting food sovereignty in American Indian community gardening. *American Indian Culture and Research Journal, 41*(3), 31–70. https://doi.org/10.17953/aicrj.41.3.hoover

Horton, P., Banwart, S., Brockington, D., Brown, G., Bruce, R., Cameron, D., Holdsworth, M., Lenny Koh, S. C., Ton, J., & Jackson, P. (2017). An agenda for integrated system-wide interdisciplinary agri-food research. *Food Security, 9*(2), 195–210. https://doi.org/10.1007/s12571-017-0648-4

Ingram, J. (2011). A food systems approach to researching food security and its interactions with global environmental change. *Food Security, 3*(4), 417–431. https://doi.org/10.1007/s12571-011-0149-9

Jones, A. (2017). Critical review of the emerging research evidence on agricultural biodiversity, diet diversity, and nutritional status in low- and middle-income countries. *Nutrition Reviews, 75*(10), 769–782. https://doi.org/10.1093/nutrit/nux040

Kitcher, P. (1987). Précis of vaulting ambition: Sociobiology and the quest for human nature. *Behavioral and Brain Sciences, 10*(1), 61–71. https://doi.org/10.1017/s0140525x00056284

Laland, K., Odling-Smee, J., & Feldman, M. (2001). Cultural niche construction and human evolution. *Journal of Evolutionary Biology, 14*(1), 22–33. https://doi.org/10.1046/j.1420-9101.2001.00262.x

Lang, J., Eisen, J., & Zivkovic, A. (2014). The microbes we eat: Abundance and taxonomy of microbes consumed in a day's worth of meals for three diet types. *PeerJ*, *2*, e659. https://doi.org/10.7717/peerj.659

Lawrence, G. (2017). Re-evaluating food systems and food security: A global perspective. *Journal of Sociology*, *53*(4), 774–796. https://doi.org/10.1177/1440783317743678

Leavitt, G. (2007). The incest taboo? *Anthropological Theory*, *7*(4), 393–419. https://doi.org/10.1177/1463499607083427

Lerner, R., & Eye, A. (1992). Sociobiology and human development: Arguments and evidence. *Human Development*, *35*(1), 12–33. https://doi.org/10.1159/000277110

Lewontin, R. (1979). Sociobiology as an adaptationist program. *Behavioral Science*, *24*(1), 5–14. https://doi.org/10.1002/bs.3830240103

Lewontin R. C. (1980). Sociobiology: Another biological determinism. *International Journal of Health Services: Planning, Administration, Evaluation*, *10*(3), 347–363. https://doi.org/10.2190/7826-DPXC-KA90-3MPR

Lieberman, L., Reynolds, L., & Friedrich, D. (1992). The fitness of human sociobiology: The future utility of four concepts in four subdisciplines. *Biodemography and Social Biology*, *39*(1–2), 158–169. https://doi.org/10.1080/19485565.1992.9988812

Liesen, L. (1995). Feminism and the politics of reproductive strategies. *Politics and the Life Sciences*, *14*(2), 145–162. https://doi.org/10.1017/s0730938400018955

Lucock, M., Martin, C., Yates, Z., & Veysey, M. (2013). Diet and our genetic legacy in the recent anthropocene. *Journal of Evidence-Based Complementary and Alternative Medicine*, *19*(1), 68–83. https://doi.org/10.1177/2156587213503345

Luo, B., Li, F., Ahmed, S., & Long, C. (2019). Diversity and use of medicinal plants for soup making in traditional diets of the Hakka in west Fujian, China. *Journal of Ethnobiology and Ethnomedicine*, *15*(1). https://doi.org/10.1186/s13002-019-0335-y

Machum, S. (2015). Shifting practices and shifting discourses: Policy and small-scale agriculture in Canada. *Cahiers Agricultures*, *24*(4), 232–239. https://doi.org/10.1684/agr.2015.0756

Manning, K., Pelling, R., Higham, T., Schwenniger, J., & Fuller, D. (2011). 4500-year-old domesticated pearl millet (*Pennisetum glaucum*) from the Tilemsi Valley, Mali: New insights into an alternative cereal domestication pathway. *Journal of Archaeological Science, 38*(2), 312–322. https://doi.org/10.1016/j. jas.2010.09.007

Mariat, D., Firmesse, O., Levenez, F., Guimarães, V., Sokol, H., Doré, J., Corthier, G., & Furet, J. (2009). The firmicutes/bacteroidetes ratio of the human microbiota changes with age. *BMC Microbiology, 9*(1), 123. https://doi. org/10.1186/1471-2180-9-123

Maroyi, A. (2018). Contribution of Schinziophyton rautanenii to sustainable diets, livelihood needs, and environmental sustainability in Southern Africa. *Sustainability, 10*(3), 581. https://doi.org/10.3390/su10030581

Marshall, F., & Hildebrand, E. (2002). Cattle before crops: The beginnings of food production in Africa. *Journal of World Prehistory, 16*, 99–143. https://doi. org/10.1023/A:1019954903395

McCann, K. (2011). Food webs (mpb-50). https://doi.org/10.23943/ princeton/9780691134178.001.0001

McDade, T. (2005). The ecologies of human immune function. *Annual Review of Anthropology, 34*(1), 495–521. https://doi.org/10.1146/annurev. anthro.34.081804.120348

Merz, C., & Steinberg, M. (2014). Applying a political economy of health standpoint to traditional food acquisition practices and the inequitable prevalence of obesity and diabetes amongst First Nations peoples in British Columbia. *Environmental Health Review, 57*(3), 65–70. https://doi.org/10.5864/d2014-028

Mesoudi, A., & Danielson, P. (2008). Ethics, evolution and culture. *Theory in Biosciences, 127*(3), 229–240. https://doi.org/10.1007/s12064-008-0027-y

Mizuno, N., Toyoshima, M., Fujita, M., Fukuda, S., Kobayashi, Y., Ueno, M., Tanaka, K., Tanaka, T., Nishihara, E., Mizukoshi, H., Yasui, Y., & Fujita, Y. (2020). The genotype-dependent phenotypic landscape of quinoa in salt tolerance and key growth traits. *DNA Research, 27*(4). https://doi.org/10.1093/dnares/dsaa022

Ng, D., Sonka, S., & Westgren, R. (2003). Co-evolutionary processes in supply chain networks. *Journal on Chain and Network Science, 3*(1), 45–58. https:// doi.org/10.3920/jcns2003.x029

Nyholm, J., Walch, A., & Redmond, L. (2024). Traditional food security and food sovereignty in the coastal region of south-central Alaska. *International Journal of Circumpolar Health, 83*(1). https://doi.org/10.1080/22423982.2024.2359161

Obongodot, N. U., & Ogwu, M. C. (2023). Plant food for human health: Case study of Indigenous vegetables in Akwa Ibom State, Nigeria. In S. C. Izah, M. C. Ogwu, & M. Akram (Eds.), *Herbal medicine phytochemistry*. Reference Series in Phytochemistry (pp. 1–38). Springer. https://doi.org/10.1007/978-3-031-21973-3_2-1

Ogwu, M. C. (2019). Towards sustainable development in Africa: The challenge of urbanization and climate change adaptation. In P. B. Cobbinah & M. Addaney (Eds.). *The geography of climate change adaptation in Urban Africa* (pp. 29–55). Springer Nature. http://doi.org/10.1007/978-3-030-04873-0_2

Ogwu, M. C. (2023). Local food crops in Africa: Sustainable utilization, threats, and traditional storage strategies. In S. C. Izah & M. C. Ogwu (Eds.), *Sustainable utilization and conservation of Africa's biological resources and environment*. Sustainable Development and Biodiversity (Vol. 888, pp. 353–374). Springer. https://doi.org/10.1007/978-981-19-6974-4_13

Ogwu, M. C., Osawaru, M. E., & Obahiagbon, G. E. (2017). Ethnobotanical survey of medicinal plants used for traditional reproductive care by Usen people of Edo State, Nigeria. *Malaya Journal of Biosciences, 4*(1), 17–29.

Ogwu, M. C., Thompson, O. P., Kosoe, E. A., Mumuni, E., Akolgo-Azupogo, H., Obahiagbon, E. G., Izah, S. C., & Imarhiagbe, O. (2024). Factors driving the acceptance of genetically modified food crops in Ghana. *Food Safety and Health, 2*(1), 158–168. https://doi.org/10.1002/fsh3.12031

Pope, H., de Frece, A., Wells, R., Borrelli, R., Ajates, R., Arnall, A., Blake, L. J., Dadios, N., Hasnain, S., Ingram, J., Reed, K., Sykes, R., Whatford, L., White, R., Collier, R., & Häsler, B. (2021). Developing a functional food systems literacy for interdisciplinary dynamic learning networks. *Frontiers in Sustainable Food Systems, 5*. https://doi.org/10.3389/fsufs.2021.747627

Rakoczy, H., & Schmidt, M. (2012). The early ontogeny of social norms. *Child Development Perspectives, 7*(1), 17–21. https://doi.org/10.1111/cdep.12010

Remans, R., Flynn, D., DeClerck, F., Diru, W., Fanzo, J., Gaynor, K., Lambrecht, I., Mudiope, J., Mutuo, P.K., Nkhoma, P., Siriri, D., Sullivan, C., & Palm, C. A. (2011). Assessing nutritional diversity of cropping systems in African villages. *Plos One, 6*(6), e21235. https://doi.org/10.1371/journal.pone.0021235

Ruse, M. (2008). The great struggles of life. In C. Crawford, & D. Krebs (Eds.), *Foundations of evolutionary psychology* (pp. 31–43). Taylor & Francis Group/ Lawrence Erlbaum Associates.

Ruse, M. (2014). *Sociobiology – A philosophical analysis*. Wiley. https://doi. org/10.1002/9780470015902.a0003450.pub2

Salgotra, R. K., & Chauhan, B. S. (2023). Genetic diversity, conservation, and utilization of plant genetic resources. *Genes, 14*(1), 174. https://doi.org/10.3390/ genes14010174

Sear, R. (2015). Sociobiology. In P. Whelehan & A. Bolin (Eds.), *The international encyclopedia of human sexuality* (pp. 1115–1354). Wiley. https://doi. org/10.1002/9781118896877.wbiehs491

Soulier, M. (2021). Exploring meat processing in the past: Insights from the Nunamiut people. *Plos One, 16*(1), e0245213. https://doi.org/10.1371/journal.pone.0245213

Thoolen, P., Holmes, P., Jansen, W., Vos, B., & Boer, A. (2023). Interdisciplinary challenges associated with rapid response in the food supply chain. *Supply Chain Management an International Journal, 29*(3), 444–459. https://doi.org/10.1108/ scm-01-2023-0040

Tumilowicz, A., Neufeld, L. M., & Pelto, G. H. (2015). Using ethnography in implementation research to improve nutrition interventions in populations. *Maternal & Child Nutrition, 11*(Suppl. 3), 55–72. https://doi.org/10.1111/mcn.12246

Ulijaszek, S. (2002). Human eating behaviour in an evolutionary ecological context. *Proceedings of the Nutrition Society, 61*(4), 517–526. https://doi. org/10.1079/pns2002180

Ulijaszek, S. (2018). Nutritional anthropology. In P. Whelehan & A. Bolin (Eds.), *The international encyclopedia of human sexuality* (pp. 1–10). Wiley. https://doi.org/10.1002/9781118924396.wbiea1510

Ulijaszek, S. (2024). Nutritional anthropology in the world. *Journal of Physiological Anthropology, 43*(1). https://doi.org/10.1186/s40101-023-00345-0

Vittersø, G., Torjusen, H., Laitala, K., Tocco, B., Biasini, B., Csillag, P., E Labarre, M. D., Lecoeur, J.-L., Maj, A., Majewski, E., Malak-Rawlikowska, A., Menozzi, D., Török, A., & Wavresky, P. (2019). Short food supply chains and their contributions to sustainability: Participants' views and perceptions from 12 European cases. *Sustainability, 11*(17), 4800. https://doi.org/10.3390/su11174800

Waal, F. (2000). Attitudinal reciprocity in food sharing among brown capuchin monkeys. *Animal Behaviour, 60*(2), 253–261. https://doi.org/10.1006/anbe.2000.1471

Wells, J. C. K., & Stock, J. T. (2020). Life history transitions at the origins of agriculture: A model for understanding how niche construction impacts human growth, demography and health. *Frontiers in Endocrinology, 11*, 325. https://doi.org/10.3389/fendo.2020.00325

Wilson, E. O. (1975). *Sociobiology: The new synthesis*. Harvard University Press.

Wilson, M. (2023). The value of ethnographic research for sustainable diet interventions: Connecting old and new foodways in Trinidad. *Sustainability, 15*(6), 5383. https://doi.org/10.3390/su15065383

Wilson, R. T. (1991). *Small Ruminant production and the Small Ruminant genetic resource in tropical Africa* (FAO Animal Production and Health paper 88). FAO.

Wollstonecroft, M. (2011). Investigating the role of food processing in human evolution: A niche construction approach. *Archaeological and Anthropological Sciences, 3*(1), 141–150. https://doi.org/10.1007/s12520-011-0062-3

CHAPTER 2

Indigenous Knowledge Systems and Food Heritage

Abstract

This chapter explores the dynamic relationship between Indigenous Knowledge Systems (IKS) and food heritage, examining how traditional ecological knowledge sustains food practices, biodiversity, and cultural identity across generations. IKS represents holistic, place-based understandings developed through centuries of interaction between Indigenous peoples and their environments. These knowledge systems govern food-related practices—from seed saving, food preparation, and preservation techniques to rituals, seasonal calendars, and land stewardship—forming resilient food systems deeply rooted in spiritual and cultural values. The chapter foregrounds food heritage as a sociobiological legacy reflecting ecological adaptation and cultural continuity. It critically examines the impacts of colonization, globalization, and industrial agriculture on the erosion of Indigenous food practices while emphasizing the contemporary resurgence of food movements led by Indigenous communities. Through literature-based analysis and illustrative case studies—such as taro cultivation in Hawai'i, wild rice governance among the Ojibwe, and enset farming in Ethiopia—this chapter highlights the adaptive strategies, social mechanisms, and intergenerational transmission that underpin sustainable food systems. The chapter also identifies the role of community memory, language, elders, and gendered knowledge in preserving food heritage and advocates for legal, educational, and technological frameworks to support the revitalization of IKS. Ultimately, this chapter calls for re-centering Indigenous perspectives in global food systems discourse as essential to ecological resilience, cultural integrity, and food sovereignty in the face of mounting planetary crises.

Keywords: Indigenous Knowledge Systems, food heritage, sustainability, cultural identity, food sovereignty, biodiversity, resilience

2.1 Introduction

Indigenous Knowledge Systems (IKS) encompass the unique, localized knowledge that Indigenous communities have developed over generations regarding their environment, resources, and practices. These systems integrate cultural, spiritual, and ethical dimensions with practical knowledge, serving as a foundation for the community's identity and a repository of sustainability and resilience solutions. IKS enables communities to respond effectively to ecological changes and social challenges by leveraging their traditional knowledge of biodiversity, land management, and resource-use practices (Rahman et al., 2021; Santoro et al., 2020). In the context of sociobiology, particularly concerning food and food systems, IKS plays a critical role in understanding the relationship between culture, environment, and human behavior. Traditional agricultural practices, food preparation techniques, and dietary choices are manifestations of these systems that showcase intricate relationships among biodiversity, ecology, and cultural heritage. For instance, integrating traditional knowledge with modern scientific practices catalyzes advancements in sustainable agriculture and food security (Rahman et al., 2021). IKS fosters an understanding of the local ecosystem in ways that respect and utilize local biodiversity while promoting genetic diversity in crops, which is vital for resilience against climate change (Santoro et al., 2020).

Food heritage encompasses the traditional, historical, and cultural significance of specific food items and culinary practices. It embodies a community's collective memory and identity as expressed through their food choices and preparation methods. Examples include Italy's regional pasta dishes, which carry centuries-old techniques and local ingredients, or Japan's sushi tradition, which is not only culinary but also deeply embedded in social and ceremonial practices (Guo & Hsu, 2023; Sanip & Mustapha, 2020). Food heritage provides a sense of belonging and continuity, enabling communities to connect with their history while reinforcing their cultural identity (Suremain, 2019). Understanding IKS is paramount for transforming food systems into more sustainable frameworks. Integrating Indigenous knowledge with contemporary scientific approaches can preserve food heritage and enhance food security. For instance, agroecological practices informed by Indigenous knowledge have improved soil health, increased biodiversity, and enhanced productivity without overexploiting resources (Rahman et al., 2021; Santoro et al., 2020). Furthermore, embracing IKS allows for a broader understanding of the social and ecological dimensions of food systems, facilitating cooperation and knowledge sharing among diverse stakeholders—from local farmers to global policymakers. This holistic approach emphasizes the interconnectedness of cultural, social, and environmental factors in promoting sustainable development (Rahman et al., 2021).

The exploration of IKS and food heritages has gained increasing attention, highlighting their essential roles in preserving cultural identities and fostering sustainable livelihoods. IKS, grounded in centuries of lived experience and connection with the land, include practices that significantly contribute to food production, cultural identity, and community well-being. Understanding these systems necessitates acknowledging their legal, social, and ecological dimensions, especially as they encounter threats from globalization and climate change.

Legal frameworks surrounding the protection of Indigenous traditional knowledge have evolved, highlighted by international instruments that emphasize Indigenous rights to cultural heritage. Panjaitan and Roisah (2021) illustrate that various international laws increasingly acknowledge Indigenous peoples' fundamental rights to maintain their traditional knowledge, arguing that such measures are crucial for cultural survival and self-determination. These legal protections validate Indigenous expertise and significantly empower Indigenous communities as they navigate contemporary challenges while maintaining cultural relevance. Food heritage serves as both a means of artistic expression and a vehicle for economic development within Indigenous communities. Giampiccoli and Kalis emphasize the importance of local Indigenous foods in tourism and community development, asserting that such culinary traditions contribute to economic resilience while preserving cultural identity (Giampiccoli & Kalis, 2012).

The impact of Indigenous food practices on health and nutrition is substantial, as noted by Luppens and Power, who emphasize that traditional foods are nutritionally superior and culturally significant, serving as vital components in combating chronic health issues within Indigenous populations (Luppens & Power, 2018). Moreover, Pawera et al. (2020) demonstrate that the increased use of wild food plants within Indigenous communities is linked with enhanced food security, illustrating that traditional ecological knowledge fosters biodiversity and provides adaptive strategies amid changing environmental conditions. Additionally, the role of food heritage in community identity is noteworthy. Ishak et al. highlight how food traditions, such as those of Kedah's Malay heritage, embody distinct cultural narratives that contribute to a sense of belonging among community members (Ishak et al., 2021). Such cultural practices risk erosion due to modern pressures, underscoring the necessity for documentation and preservation efforts to ensure they persist. Moreover, the influence of contemporary challenges such as globalization and the COVID-19 pandemic on Indigenous food practices is significant. The pandemic has increased awareness of food safety and nutrition, prompting individuals to prioritize local and traditional food sources and reinforcing food sovereignty and cultural identity (Janßen et al., 2021; Rodriguez & Pedroso, 2024). The ongoing relevance of conventional culinary knowledge in contemporary settings calls for active engagement and integration into broader societal and educational frameworks. The intertwining of Indigenous knowledge and tourism presents both opportunities and challenges. Fletcher et al. (2016) discuss how Indigenous involvement in tourism development can enhance community benefits while fostering greater respect for cultural heritage. However, the commercialization of Indigenous foods raises the risk of cultural misappropriation, necessitating careful management and collaborative frameworks that empower Indigenous communities (Price et al., 2021).

This chapter explores how IKS shape and sustain food heritage across generations, highlighting their essential contributions to cultural identity, ecological stewardship, and food security. It examines how traditional environmental knowledge, spiritual beliefs, farming practices, and communal food rituals intertwine to create resilient and adaptive food systems. By investigating the transmission of food-related knowledge through oral traditions, rituals, and apprenticeship, the chapter emphasizes the role of IKS in preserving biodiversity, promoting sustainable land use, and ensuring the intergenerational continuity of culinary practices. This chapter also places Indigenous food heritage within broader conversations on cultural resilience, sovereignty, and the urgent need to protect and revitalize local knowledge in the face of globalization and climate change.

2.2 Foundations of Indigenous Knowledge Systems

Indigenous Knowledge Systems (IKS) represent holistic, context-specific bodies of knowledge developed by Indigenous communities over centuries through their interactions with the natural environment. Rooted in lived experiences, observation, and oral traditions, IKS include various disciplines, such as agriculture, medicine, ecology, and cosmology. Unlike Western scientific paradigms, IKS emphasize relationality—between humans, nature, and the spiritual world—viewing knowledge as dynamic, adaptive, and deeply embedded in cultural practices. These systems prioritize sustainability, reciprocity, and respect for natural cycles, often passed down through generations via storytelling, ceremonies, and practical engagement with the land. The foundations of IKS are both epistemological and ethical, guiding how communities perceive their responsibilities toward ecosystems and one another. As such, they provide vital frameworks for managing food systems, preserving biodiversity, and nurturing cultural identity in a rapidly changing world.

IKS are intricate frameworks that reflect the cultural, historical, and ecological contexts of Indigenous communities. Key elements of these systems include oral traditions, experiential learning, and community memory. Oral traditions serve as vital conduits for knowledge transmission in Indigenous cultures, facilitating the sharing of stories that encapsulate spiritual beliefs, historical experiences, and ethical lessons. As highlighted by Richmond et al., the reclamation of cultural values and ways of knowing is crucial, and learning through oral narratives plays a pivotal role in revitalizing Indigenous food knowledge and practices that have been disrupted historically by colonization and assimilation policies (Richmond et al., 2020). Experiential learning—learning through direct experience—is also a foundational aspect of IKS. This approach is often embodied in land-based learning activities such as hunting, fishing, and food preparation, where participants actively engage with their environments to bolster their knowledge and skills (Bowra et al., 2020). Such experiential practices are emphasized by Bowra et al., who articulate that land-based learning not only connects individuals with their cultural heritage but also fosters a sense of belonging to a larger ecological and community framework (Bowra et al., 2020). Community memory, an essential component of IKS, incorporates collective experiences and knowledge passed down through generations, emphasizing continuity and identity within Indigenous cultures. It underscores the importance of maintaining these narratives as they embody historical truths and reinforce community resilience and cohesion.

Another critical aspect of IKS is the role of elders, rituals, and apprenticeship in the transmission of knowledge. Elders hold significant authority and respect within Indigenous societies as custodians of traditional knowledge. They embody the wisdom accumulated over their lifetimes and are essential conduits for teaching younger generations. Fast et al. (2021) support the importance of elders by noting that knowledge-sharing through these generational relationships is vital for cultural preservation and communal integrity. Rituals also play a crucial role in this knowledge transmission; they frame both the learning processes and the cultural significance of knowledge, intertwining experiential learning with community and spiritual dimensions (Burnette et al., 2018). Apprenticeship, involving hands-on practice under the guidance of skilled practitioners or elders, serves as a key educational method that equips learners with specific skills and instills a deeper understanding of the cultural and ethical contexts underlying those skills (Datta, 2024).

The interconnectedness of land, cosmology, and food knowledge forms another foundational pillar of IKS. The land is not merely a physical entity but is inextricably linked to spirituality, identity, and ecological stewardship. Indigenous peoples often view their relationship with the land as a sacred duty to care for and respect the ecosystems they inhabit. As articulated in the works of Miltenburg et al. (2023), fostering relationships with land is central to Indigenous food sovereignty, which requires a return to cultural values and sustainable practices rooted in Indigenous ways of knowing. Food knowledge within Indigenous cultures is therefore deeply tied to cosmological beliefs that guide land management, cultivation, and harvesting practices. Gaudet and Chilton (2018) emphasize that engaging with traditional food systems nurtures community health and identity, bridging the past and present in a culturally meaningful way.

Furthermore, sociopolitical and historical contexts significantly shape the relationship between Indigenous cultures and their lands. Colonial practices have disrupted traditional food systems and intergenerational knowledge transfer, leading to disconnection from cultural practices essential for community well-being and identity (Malli et al., 2023). Environmental repossession is vital in this context, as Indigenous communities strive to reclaim traditional lands and the accompanying knowledge systems that support their cultural practices and health. Integrating contemporary Indigenous knowledge with environmental practices signifies a dynamic approach to sustainability that respects traditional ecological knowledge while adapting to current environmental challenges (Datta et al., 2024) (Table 2.1).

Table 2.1: Indigenous Knowledge Systems (IKS) and Their Food System Value

Foundation	Description	Significance	Food Value of the IKS
Holism	Knowledge integrates spiritual, ecological, social, and cultural dimensions.	Promotes a balanced worldview and interconnected understanding of food, health, and the environment.	Guides food practices that are ecologically sound, socially inclusive, and spiritually meaningful.

Foundation	Description	Significance	Food Value of the IKS
Experiential Learning	Knowledge is gained through observation, participation, and lived experience.	Ensures knowledge is contextually relevant and practically applicable.	Sustains traditional agricultural practices like seed saving, harvesting, and preservation techniques.
Oral Transmission	Knowledge is passed down through stories, songs, rituals, and mentorship.	Preserves cultural identity and continuity across generations.	Maintains culinary traditions, recipes, and the symbolic use of specific foods.
Relationality	Emphasizes relationships among humans, nonhuman beings, and natural elements.	Guides ethical practices in food gathering, land use, and community life.	Encourages respectful harvesting, seasonal eating, and biodiversity conservation in food production.
Adaptability and Resilience	IKS evolves in response to environmental and social changes.	Enables communities to maintain food security and sovereignty amid changing conditions.	Supports crop diversification, local seed varieties, and flexible food systems during climate or social shifts.
Place-based Wisdom	Knowledge is specific to local ecosystems, climates, and cultural landscapes.	Supports sustainable practices tailored to regional biodiversity and agricultural cycles.	Promotes native plants, wild edibles, and local food sources adapted to the environment.
Collective Knowledge	Knowledge is co-created and shared among community members.	Encourages collaboration and inclusivity in food production and resource stewardship.	Facilitates community farming, food sharing, and knowledge exchange about food preparation and storage.

Foundation	Description	Significance	Food Value of the IKS
Spiritual Significance	Recognizes the sacredness of food, land, and natural resources.	Reinforces respect, gratitude, and conservation in all food-related practices.	Embeds moral codes and rituals in food use, fostering sustainable and mindful consumption.

2.3 Food Heritage as Cultural Capital

Food heritage represents a complex interplay of culinary practices, historical narratives, and cultural values that shape a community's identity. It specifically encompasses multiple dimensions, including culinary techniques (the methods employed in food preparation), ingredients (the specific components used in dishes), and the sociocultural meanings attached to both food and culinary practices (Almansouri et al., 2021; Matta, 2016). Culinary techniques form the foundation of food heritage, often reflecting local climatic conditions, available resources, and cultural preferences. For instance, methods such as smoking, fermenting, or pickling serve practical purposes in food preservation and contribute to the flavor profiles that characterize regional cuisines. These techniques can be inherited and adapted over generations, emphasizing the dynamic nature of cultural practices (Raymundo, 2024).

Ingredients, the second dimension, include cultivated crops and Indigenous wild foods that hold cultural significance. They can be categorized as staple, ceremonial, and ethnic foods, each representing unique aspects of a community's heritage. For instance, quinoa holds exceptional value in Andean cultures, serving as a staple food and a symbol of cultural pride and resilience against globalization (Ishak et al., 2021). The sociocultural meanings associated with food significantly enhance food heritage. These meanings often intertwine with concepts of identity, spirituality, and social cohesion, demonstrating how food transcends mere sustenance to embody values and beliefs crucial for community continuity. Thus, food heritage contributes to the construction of collective memories and serves as a vehicle for intergenerational cultural transmission (Ramli et al., 2020).

Food plays a central role in shaping cultural identity, offering individuals a sense of community and belonging. The preparation and consumption of traditional dishes serve as communal rituals that reinforce social ties and evoke shared history. For example, in many Indigenous Canadian communities, traditional foods such as salmon, deer, or pemmican are not merely items of sustenance; they are imbued with rich cultural significance, bearing witness to centuries of heritage and traditional ecological knowledge (Gallegos et al., 2023). Continuity through food heritage is equally vital, providing a sense of stability amid the rapid changes induced by globalization and modernization. Passing recipes through generations ensures that traditional food knowledge is preserved, allowing cultures to maintain their unique characteristics and combat the loss of identity in a rapidly homogenizing world (Alejandro et al., 2022). For instance, the revival of artisan cheese-making in various countries illustrates how communities can reinvigorate traditional practices and adapt them within contemporary contexts, thereby fostering resilience (Jalis et al., 2014). The resilience inherent in food heritage also enables communities to adapt to challenges such as climate change, social upheaval, or economic shifts. By leveraging traditional knowledge alongside modern farming practices, communities have successfully developed strategies that promote food sovereignty and security. For example, permaculture practices rooted in Indigenous knowledge are increasingly recognized for enhancing biodiversity and sustainability in agricultural systems.

2.3.1 Case Examples: Seed Saving, Sacred Foods, Cooking Methods, and Feasting Practices

2.3.1.1 Seed Saving

One poignant example of food heritage preservation is seed saving, which often embodies a community's collective identity and ecological knowledge. In many cultures, such as among the Hopi people in the Southwestern United States, saving seeds from Indigenous crops like blue corn is not merely agricultural; it serves as a cultural link to ancestry and spiritual beliefs. In these communities, seeds are often regarded as relatives, and preserving them signifies maintaining identity and autonomy in the face of modern agricultural pressures. This cultural emphasis on seed saving has significant implications for biodiversity, as it aids in conserving traditional varieties that may otherwise become extinct. Figure 2.1 highlights the key distinctions between orthodox and recalcitrant seeds in size, water content, desiccation tolerance, metabolic activity, and storage potential. These differences are crucial for developing effective seed conservation strategies, especially in biodiversity preservation and agricultural resilience. Orthodox seeds, such as cereals and legumes, are suited for long-term storage and dormancy. In contrast, recalcitrant seeds like jackfruit and durian require more careful handling due to their sensitivity to drying and short viability.

Figure 2.1: Classification of Seeds for Saving Based on their Physiological and Storage Differences

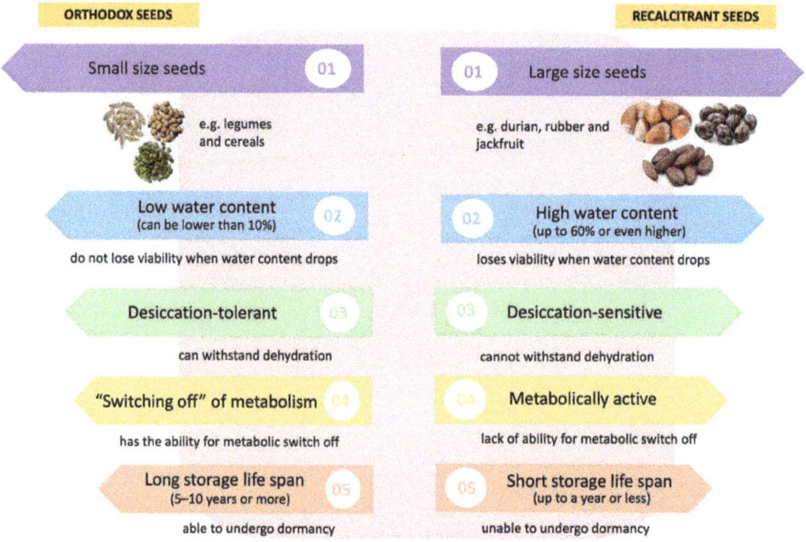

Source: Lah et al. (2023)

2.3.1.2 Sacred Foods

Sacred foods are vital in many cultures, often linked to religious or spiritual significance. These foods are typically central to rituals and ceremonies, reinforcing identity and community bonds. For example, in many Indigenous cultures, foods like maize, acorns, or fish carry sacred meanings, woven into the cultural fabric through mythologies and ceremonial practices. In these contexts, sacred foods often symbolize sustenance and connections to ancestors and spiritual narratives, highlighting their importance beyond mere nutrition. The ritualistic use of these foods in ceremonies cultivates community cohesion and reinforces cultural values across generations. Table 2.2 highlights a selection of sacred foods from various regions, illustrating their symbolic roles and enduring significance in maintaining cultural identity and spiritual well-being.

Table 2.2: Example of Some Sacred Food with High Sociobiological Value

Sacred Food	Cultural Group/Region	Spiritual or Cultural Significance	Traditional Use
Manoomin (Wild Rice)	Ojibwe (Great Lakes, North America)	Considered a sacred gift from the Creator, it is central to migration stories and ceremonies.	Used in feasts, rituals, and as a staple food.
Blue Corn	Hopi (Southwestern United States)	Symbolizes life, spirituality, and sustenance; tied to ceremonial cycles.	Ground into flour for traditional bread and piki.
Taro (Kalo)	Kānaka Maoli (Hawai'i)	Believed to be the ancestor of the Hawaiian people (Hāloa); embodies family and earth.	Grown in lo'i; used to make poi.
Quinoa	Andean peoples (Peru, Bolivia)	Regarded as the "Mother Grain," it was used in Incan rituals and agricultural festivals.	Cooked as porridge, grain dishes, and ceremonial meals.
Yam	Igbo and other West African groups	Central to annual Yam Festivals, and it represents prosperity, ancestry, and renewal.	Offered in rituals, and consumed in community celebrations.
Milk and Blood	Maasai (East Africa)	Sacred sustenance: integral to identity, rituals, and cattle reverence.	Consumed in rites of passage and daily nourishment.
Salmon	Pacific Northwest Tribes (United States/Canada)	Seen as a sacred gift from the river spirits; it represents renewal and reciprocity.	Featured in First Salmon Ceremonies.

Sacred Food	Cultural Group/Region	Spiritual or Cultural Significance	Traditional Use
Dates	Middle Eastern and North African cultures	Mentioned in holy texts (e.g., Quran); symbol of nourishment and blessings.	Consumed during Ramadan and other religious occasions.
Sorghum	Zulu (Southern Africa)	Used in brewing traditional beer for ancestral offerings and communal rites.	Central in rituals, especially for honoring elders
Teff	Ethiopian Orthodox Christians	Used in religious fasting periods and festive dishes like Injera.	Prepared for holidays and spiritual feasts.

2.3.1.3 Cooking Methods

Cooking methods as part of food heritage reveal much about social norms, technology, and the environment. In various Mediterranean cultures, for example, slow cooking reflects a communal approach to food preparation that encourages togetherness and shared experiences. Techniques like braising or stewing allow flavors to meld and deepen over time, embodying the principles of patience, care, and community. Moreover, these methods often incorporate local ingredients that reflect geographical and climatic contexts, reinforcing the connection between food preparation and regional identity. Table 2.3 presents various cooking methods deeply embedded in food heritage worldwide, highlighting their enduring roles in identity, sustainability, and cultural continuity.

Table 2.3: Traditional Cooking Methods that Are Integral Components of Food Heritage across Various Cultures

Cooking Method	Description	Cultural Context	Significance in Food Heritage
Pit Roasting (Earth Oven)	Slow-cooking food in a pit lined with hot stones and covered with leaves or soil	Polynesian (Hāngī), Native American, and Andean	Communal practice, ceremonial use, and ancestral cooking method

Cooking Method	Description	Cultural Context	Significance in Food Heritage
Fermentation	Microbial breakdown of food to enhance preservation and flavor	African (Ogi), Korean (Kimchi), and Nordic (Surströmming)	Adds complexity, enhances shelf life, and reflects ecological wisdom
Stone Grinding	Using stone tools to grind grains or seeds into flour or paste	Indigenous Americans, Africans, and Australian Aboriginals	Preserves ancient techniques, crucial for staple preparation
Smoking	Exposing food to smoke for preservation and flavor	Scandinavian, First Nations, and Southeast Asian	Traditional preservation, sacred meat, and fish preparation
Boiling in Clay Pots	Cooking using unglazed clay for even heat distribution	Indian, African, and Andean	Retains nutrients, links to pottery traditions
Open Fire Grilling	Cooking food directly over flame or embers	Maasai, Amazonian, and Mediterranean	A simple, primal method connects people to natural elements
Steaming with Leaves	Using banana, taro, or corn leaves to steam foods	Southeast Asian, Latin American, and Pacific Islands	Preserves moisture, adds subtle flavors, and is environmentally friendly
Sun Drying	Preserving food by drying it in sunlight	Middle Eastern, African, and South Asian	Sustainable preservation, part of seasonal food cycles

Cooking Method	Description	Cultural Context	Significance in Food Heritage
Salt Curing	Preserving meats or fish with salt	Nordic, Mediterranean, and West African	The traditional preservation method is essential for trade and survival
Ash Cooking	Roasting tubers or roots in hot ashes	Australian Aboriginal, and Pacific Islander	The nomadic cooking style emphasizes resourcefulness

2.3.1.4 Feasting Practices

Feasting practices, often centered around agricultural cycles or seasonal celebrations, serve as an extraordinary repository of cultural knowledge and traditions. For example, many agrarian communities host harvest festivals, during which surplus food is collected and shared in communal meals. Such gatherings not only celebrate the land's bounty but also strengthen social bonds within the community, exemplifying the role of food as social glue. In various cultures, specific dishes served during feasts are rich with symbols and stories that foster a sense of identity and belonging, illustrating how food can capture and communicate cultural narratives. Table 2.4 presents a selection of feasting practices connected to food, showcasing their diverse meanings and roles in sustaining cultural heritage.

Table 2.4: Sacred Feasts and Culinary Rituals Across Cultures

Region/ Community	Name of Feast/Event	Foods Featured	Occasion/ Purpose	Cultural Significance
Yoruba (Nigeria)	Yam Festival	New yams, palm oil, kola nuts	Harvest celebration	Honors earth deities, celebrates abundance

Region/ Community	Name of Feast/Event	Foods Featured	Occasion/ Purpose	Cultural Significance
Haudenosaunee (Iroquois)	Green Corn Festival	Fresh corn, corn soup, berries	Late summer, the first corn harvest	Ceremonial gratitude, renewal of community ties
Maya (Central America)	Hanal Pixán	Tamales, atole, pan de muerto	Day of the Dead observance	Honors the spirits of ancestors
Māori (New Zealand)	Hāngi	Meats and vegetables cooked in an earth oven	Community gatherings, celebrations	Reinforces kinship and hospitality
Native Hawaiians	Lūʻau	Taro, pork, fish, coconut-based dishes	Birthdays, weddings, royal ceremonies	Symbol of aloha, communal generosity
Andean Communities	Pachamama Raymi	Potatoes, maize, guinea pig	Festival to honor Mother Earth (Pachamama)	Reciprocal offering to nature and spiritual guardians
Ethiopian Orthodox Church	Meskel	Injera, doro wat, honey wine	Finding of the True Cross	Religious unity and national identity
Ainu (Japan)	Iyomante (Bear Sending)	Bear meat, millet, sake	Spiritual ceremony to send a bear spirit to the gods	Communion with nature spirits, cultural resilience

2.4 Coevolution of Social Knowledge and Food Ecology

The coevolution of social knowledge and food ecology represents a significant aspect of traditional agricultural systems, emphasizing various elements such as adaptive strategies, environmental cues, and ecological intelligence. These elements collaborate to promote sustainable farming practices that bolster food security and community resilience.

2.4.1 Adaptive Strategies Embedded in Traditional Knowledge

Traditional agricultural methods, such as rotational farming and intercropping, are critical adaptive strategies rooted in the cultural fabric of Indigenous communities. These practices have shown significant advantages for environmental sustainability and food security. For instance, studies in South Africa illustrate how traditional practices like crop rotation and intercropping enhance soil fertility, improve crop yields, and reduce pest infestations (Kom et al., 2024). Similarly, in Nepal, traditional cropping systems greatly bolster household food security by maintaining a diverse array of crops, thereby mitigating risks associated with ecological shocks and market fluctuations (Liu et al., 2023; Gauchan et al., 2020). Integrating modern ecological practices with traditional knowledge can promote improved resource management and agricultural productivity (Lan & Kien, 2021). The role of integrated farming systems also emphasizes this idea, as these systems combine crop and livestock production for optimal resource use, further enhancing resilience to climate variability (Dasgupta et al., 2015; Kumar et al., 2018).

2.4.2 Environmental Cues and Seasonal Calendars

Using environmental cues and seasonal calendars in agriculture represents a crucial intersection of traditional ecological knowledge and sustainable practices. Farmers in various regions have developed calendars based on local climatic conditions, which guide their agricultural activities and empower them to adapt to increasingly erratic climate patterns (Prober et al., 2011). For instance, Aboriginal seasonal knowledge in Australia has been recognized as invaluable for effective environmental management, providing insights that align agricultural practices with ecological rhythms (Prober et al., 2011). The accuracy of these calendars directly correlates with improved agricultural outputs and can inform interventions to mitigate the adverse effects of climate variability (Jiri et al., 2016). A detailed study in Indonesia demonstrates how dynamic crop calendars, which account for finer spatial variations, can significantly enhance agricultural planning and resource allocation (Irawan & Komori, 2024). This approach ensures timely planting and harvesting while promoting sustainability in food production practices by aligning them with ecological cycles.

2.4.3 Ecological Intelligence in Indigenous Food Preservation and Storage Methods

Ecological intelligence, as demonstrated in Indigenous food preservation and storage techniques, plays a vital role in maintaining food security. Methods such as fermentation and traditional drying have been documented across various cultures, contributing to food safety and nutritional preservation (Kom et al., 2024). This knowledge enables communities to maximize their yield while minimizing waste, effectively addressing food scarcity during periods of low production (Ba et al., 2018; Mondal et al., 2023). A review of organic agriculture practices shows that these systems enhance soil health and crop yield, which is critical for long-term sustainability (Cidón et al., 2021). The role of integrated farming systems, which synergistically combine crops, livestock, and aquaculture, aligns with this intelligence, demonstrating efficient resource use and resilience against environmental pressures (Dasgupta et al., 2015; Kumar et al., 2018). Furthermore, enhancing post-harvest practices through traditional wisdom can significantly reduce losses, thereby improving food security (Hussen, 2021; Lutz & Schachinger, 2013).

2.5 Mechanisms for Intergenerational Knowledge Transmission of Social Food Practices and Knowledge Systems

Understanding how food practices and knowledge systems are passed down through generations is essential for grasping social and cultural dynamics. This process involves several interconnected mechanisms, such as storytelling, observational learning, and active participation, that transmit culinary skills and embed deeper cultural meanings. Below, I discuss some factors that contribute to the continuity of food traditions and practices in modern societies.

2.5.1 Storytelling

Storytelling is a rich way through which food traditions are passed down from one generation to the next. It serves as an educational tool within families, helping children grasp the cultural significance and heritage linked to specific dishes or preparation methods. The narratives surrounding food convey practical skills while embedding moral lessons, familial values, and cultural beliefs. For instance, in many Indigenous cultures, stories shared during communal meals can impart ethical teachings about the relationship with nature and food sustainability practices, ensuring that ecological wisdom is passed down through generations (Liu, 2017). Research also emphasizes that these narratives, particularly those reinforced during family gatherings, contribute to a sense of identity and belonging among younger generations, integrating them into their community's food culture (Trofholz et al., 2018).

2.5.2 Observation and Participation

Observation and active participation in food preparation are vital learning methods that enhance the transmission of food-related knowledge. Research indicates that children who cook with their families are more likely to develop positive attitudes toward healthy eating and cooking skills. Studies, such as those by Trofholz et al. (2018), underscore the importance of defining specific family meal practices and routines that young participants can observe and participate in (Organ et al., 2015). Immersing children in the cooking process helps them acquire skills and allows them to connect emotionally with their cultural heritage, laying the groundwork for their future culinary practices.

2.5.3 The Role of Multisensory Learning

Alongside storytelling and participation, the multisensory experience linked to food—including taste, smell, and sight—plays a vital role in memory formation. Engaging in food-related activities stimulates all the senses and helps reinforce the connection between food and its cultural significance. For instance, the act of preparing traditional dishes during holiday celebrations engages family senses and creates lasting memories that resonate throughout one's life.

2.5.4 Role of Festivals

Festivals serve as essential platforms for intergenerational exchanges of food practices, functioning both as celebratory events and spaces for cultural education. Traditional foods are prepared and enjoyed during these festivals, creating opportunities for community members to connect with their food heritage. Research by Organ et al. suggests that children's experiences at food festivals can profoundly influence their understanding of the cultural significance associated with certain foods, fostering pride in their culinary heritage (Sharif et al., 2018). Festivals often feature unique arrangements of communal meals, strengthening social bonds and shared identities. For instance, during harvest festivals in various cultures, communities gather to share the bounty of their labor, reinforcing values of gratitude, communal effort, and traditional agriculture.

2.5.5 Rites of Passage

Rites of passage, including ceremonies such as weddings, births, and funerals, often involve specific food practices with cultural significance. These events provide opportunities for learning about the meanings behind the foods consumed on such occasions. For instance, in many Mediterranean cultures, meals served during weddings symbolize fertility and new beginnings, imparting cultural symbolism to traditional dishes shared during these pivotal life events. Scholars have noted that such rites reinforce traditional food practices and strengthen community ties and identity among participants (Obamwonyi & Onyekuru, 2024).

2.5.6 Gendered Food Roles

Gender roles play a significant role in distributing and transmitting knowledge surrounding food practices. Research indicates that women are often the primary transmitters of culinary knowledge within families, shaping the character of food cultures through their daily practices. Obamwonyi and Onyekuru (2024) elaborate on this dynamic by highlighting specific examples from traditional African festivals, where women's involvement in food preparation translates into safeguarding culinary heritage. In many cultures, the expectation that women will be primary caregivers and food preparers influences how food-related knowledge is passed between generations, with daughters frequently learning cooking skills directly from their mothers or grandmothers. The gendered nature of food preparation roles also reflects broader social structures and cultural norms that extend beyond culinary practices, shaping family dynamics and community identities (Joshi, 2022).

2.5.7 Language as a Repository of Ecological Memory

Language functions as a living repository of ecological memories connected to food practices. The terminology associated with food preparation, ingredients, and cultural customs often encapsulates historical interactions between communities and their environments. For instance, preserving specific names for Indigenous plants or traditional dishes indicates cultural identity and reflects an understanding of the ecological systems in which they exist. Hafiz et al. (2021) note the importance of documenting traditional food knowledge as a means of preserving ecological knowledge and promoting sustainable practices. The decline of Indigenous languages often correlates with the erosion of traditional food knowledge, emphasizing the need for language preservation alongside culinary practices. Language not only serves an educational purpose but also fosters connections between generations, facilitating the transmission of food-related practices. Conversations around the dining table, filled with narratives, terminologies, and traditional wisdom, enable the transfer of knowledge and cultural identity. For example, the dialects of Indigenous Australian communities often contain specific terms related to food gathering and preparation that carry significant ecological insights. The loss of such linguistic nuances can lead to a dilution of traditional knowledge, prompting scholars like Hafiz et al. (2021) to argue for the necessity of conserving and revitalizing these linguistic threads to preserve cultural and ecological wisdom.

2.6 Social Disruption and Erosion of Food Heritage

The relationship between social disruption and the erosion of food heritage is increasingly critical in our globalized world, where factors like colonialism, globalization, and industrial agriculture significantly influence local food practices (Ogwu, 2023; Ogwu et al., 2016). Colonial legacies often involved the displacement of Indigenous populations and their food systems, resulting in a loss of traditional agricultural knowledge and a disconnection from local food heritage. This disruption is exacerbated by globalization, which facilitates the homogenization of food practices through the spread of industrialized agriculture, promoting the adoption of nonlocal commodities and dietary patterns that frequently undermine Indigenous food systems and cultural heritage (Klein, 2018; Ramli et al., 2017).

The impacts of industrial agriculture are profound, as this paradigm tends to prioritize efficiency and profit over biodiversity and cultural preservation. Sidali et al. assert that the effects of industrial agriculture include a degradation of local food systems and a loss of specific agricultural practices, which contradicts local identities associated with traditional foods (Osawaru & Ogwu, 2020; Sidali et al., 2013). As agricultural practices become universally standardized, we observe an increasing trend toward dietary shifts, with individuals migrating from diverse, locally rooted diets toward more standardized global food options (Omar et al., 2015; Osawaru & Ogwu, 2014). These dynamics raise concerns regarding cultural loss, as traditional knowledge surrounding food preparation and consumption diminishes, leading to a phenomenon where the essence of local cuisines can be appropriated and repackaged in ways that often exclude their originating cultures (Omar et al., 2015; Sidali et al., 2013).

The specific mechanisms driving this disruption include land dispossession, which has historically deprived Indigenous communities of their ancestral lands and associated food resources. This disconnection not only threatens the survival of traditional crops and practices but also displaces cultural heritage that is intricately tied to these resources (Crane, 2014; Johnston & Marwood, 2017). For example, Indigenous cheese production in China, as referenced by Klein, illustrates how the heritagization of local foods can clash with state-led agricultural modernization projects prioritizing mass production over local cultural expressions (Klein, 2018; Sanip & Mustapha, 2020). The erosion of food heritage vividly manifests through the loss of unique ingredients and the decline of traditional practices, as seen with the decrease of artisanal cheese production in many regions across the globe (Klein, 2018; Sanip & Mustapha, 2020; Sidali et al., 2013). Moreover, evidence suggests that the commodification of culinary practices through tourism can act both as a protective mechanism and as a potential contributor to the erosion of food heritage. In the context of Malaysia, research shows that while there is a growing interest in heritage foods as a marketing tool for tourism, the sustainability of these heritage practices is often compromised by tourist demands that prioritize novelty over authenticity (Omar et al., 2015; Saad et al., 2021). Tourist engagement with local cuisines can lead to a reimagining of traditional dishes that strips them of their cultural significance, focusing instead on their appeal as consumable commodities (Omar et al., 2015).

2.7 Revitalization and Resurgence of Food Movements Through Social Practices

The revitalization and resurgence of food movements driven by community initiatives, particularly in Indigenous contexts, are closely linked to the principles of food sovereignty and practices that emphasize local and sustainable food systems. This extensive analysis will cover community-led food sovereignty and seed sovereignty initiatives; the roles of Indigenous food champions, scholars, and elders; as well as the integration of educational programs, language revitalization, and digital archiving within these movements.

2.7.1 Community-Led Food Sovereignty and Seed Sovereignty Initiatives

Community-led food sovereignty initiatives are crucial to Indigenous cultures, highlighting the right of communities to produce their own food according to their traditions and values. These initiatives often emerge as responses to globalized agricultural practices that have weakened local food systems. Food sovereignty goes beyond mere access to food; it focuses on the right to define food systems, including production, distribution, and consumption, as articulated by La Via Campesina, a global peasants' movement (Timler et al., 2019). This framework enables communities to reclaim their identities and engage with their unique ecological landscapes while resisting external pressures from industrial and monocultural agricultural systems. Alongside food sovereignty is the notion of seed sovereignty, which pertains to the rights of communities to save, use, and exchange seeds. Initiatives in this area emphasize the preservation of heirloom and Indigenous seed varieties that carry cultural significance and biodiversity value (Ray et al., 2019). Planting traditional varieties not only supports ecological sustainability by enhancing genetic diversity but also reinforces the cultural identity of Indigenous peoples. For instance, the Seed Sovereignty Initiative in Canada represents a collaborative effort among several Indigenous communities to restore traditional seeds and integrate them into local agricultural practices (Prieto et al., 2023). These community-led initiatives often feature collaborative seed banks, where local farmers can exchange and cultivate Indigenous seeds, thus ensuring the continued existence of these varieties (Robin et al., 2020). Furthermore, the participatory nature of such programs promotes community engagement and collective decision-making, which is essential for nurturing social cohesion and cultural revival (Islam et al., 2016).

2.7.2 Role of Indigenous Food Champions, Scholars, and Elders

Indigenous food champions and scholars play a vital role in the resurgence of food movements. Elders, as custodians of traditional ecological knowledge, provide invaluable insights into sustainable practices that have supported their communities for generations. This knowledge encompasses understanding seasonal cycles, cultivation techniques, and the ecological relationships among various species (Swiderska et al., 2022). In many Indigenous cultures, food preparation and consumption rituals are deeply intertwined with spiritual beliefs, highlighting the cultural significance of food beyond mere nutrition. Elders often act as educators within their communities, facilitating workshops and hands-on experiences designed to reconnect younger generations with traditional food systems (Longvah et al., 2017). By weaving teachings about the cultural implications of food alongside practical knowledge about farming and foraging, these leaders are renewing interest in Indigenous food practices, resulting in a revival of traditional diets that prioritize native flora and fauna (John, 2024).

2.7.3 Intergenerational Collaboration

The collaboration between Indigenous elders and youth is essential for ensuring Indigenous practices are preserved and adapted to contemporary contexts. Projects such as "Elders as Teachers," which engage youth in experiential learning about traditional foods, allow for the direct transmission of knowledge and wisdom (Pinedo et al., 2021). Participatory Action Research (PAR) methodologies promote these intergenerational exchanges, enabling younger community members to actively contribute to revitalizing Indigenous foods while nurturing cultural pride (Budiono & Noviani, 2024).

2.7.4 Educational Programs

Educational programs focused on Indigenous food systems are essential for revitalizing local practices. These programs integrate traditional knowledge with contemporary scientific approaches to agriculture, nutrition, and ecological sustainability. For instance, bilingual education initiatives that incorporate Indigenous languages support language revival and emphasize the articulation of culinary practices tied to cultural identity (Brenzinger & Heinrich, 2013). A case study involving the Kānaka Maoli people of Hawai'i illustrates how educational programs that teach traditional fishing and farming techniques empower communities to reclaim their food systems (Ghosh-Jerath et al., 2016).

2.7.5 Language Revitalization

Language is intrinsically linked to food knowledge and cultural identity; therefore, language revitalization efforts are essential to the resurgence of the food movement. As noted by numerous scholars, traditional ecological knowledge is often embedded in linguistic expressions, making language preservation crucial to maintaining food systems. Several Indigenous communities have developed dual-language immersion programs to enhance this connection. These programs use food-related vocabulary and culturally significant narratives as tools for language acquisition, strengthening both linguistic and cultural ties.

2.7.6 Digital Archiving and Technological Integration

Digital archiving acts as a modern tool for preserving and sharing Indigenous food knowledge, facilitating the documentation of traditional practices and recipes. Websites, social media platforms, and mobile applications have emerged as vital spaces for disseminating information about traditional foods to wider audiences. For instance, initiatives like the Indigenous Food Systems Network utilize digital platforms to unite Indigenous communities, provide education, and promote food sovereignty through shared resources, community stories, and maps of Indigenous food practices. Additionally, creating digital repositories that archive oral histories, traditional recipes, and agricultural practices ensures that this knowledge is safeguarded for future generations. These efforts not only document Indigenous food systems but also offer a framework for Indigenous communities to advocate for their rights against encroaching industrial practices. By leveraging technology, communities can connect with a global audience, increasing awareness of their challenges and devising and implementing strategies to sustain their food systems.

The revitalization and resurgence of food movements through social practices highlight the importance of community-led initiatives that align with the principles of food sovereignty and seed sovereignty. The roles of Indigenous food champions, scholars, and elders as educators and knowledge keepers are vital in bridging the gap between traditional and contemporary practices. Furthermore, educational programs and digital archiving efforts serve as powerful tools for language revitalization and the continuity of Indigenous knowledge. Together, these elements create a holistic approach that fosters stronger, more resilient communities capable of navigating and asserting their rights to food systems that reflect their cultural heritage and ecological wisdom. The future of food movements within Indigenous contexts depends on continuing and expanding these practices, providing pathways to ensure food justice, sovereignty, and cultural integrity in an increasingly globalized world.

2.8 Sociobiological Significance of Indigenous Food Knowledge

The sociobiological significance of Indigenous food knowledge is vital in anthropology, ecology, and nutrition. It encompasses survival strategies, social cohesion, ecological adaptation, and health outcomes. Traditional food practices convey essential cultural narratives and environmental stewardship, which are crucial for sustaining both individual and collective well-being. Recognizing food heritage as a sociobiological legacy provides a pathway to resilience, enabling communities to confront contemporary challenges while honoring their historical relationships with the land and food systems. The insights gleaned from this literature emphasize the urgent need to integrate Indigenous food knowledge into modern policies and practices, fostering sustainable and equitable food systems that respect the ancestral wisdom of Indigenous communities. Food heritage serves as a sociobiological legacy that contributes to contemporary resilience in the face of environmental and social challenges.

2.8.1 Indigenous Food Practices as Survival Strategies

Indigenous food systems represent sophisticated survival strategies refined over generations in response to local ecological conditions. Traditional environmental knowledge (TEK) encompasses an intricate understanding of local ecosystems, biodiversity, and resource management practices. Research continues to emphasize the importance of TEK in understanding sustainable practices that ensure food security within Indigenous communities (Berkes, 1993; Oloko & Shukla, 2018). For example, Aboriginal Australians' knowledge of fire management techniques contributes to hunting success and biodiversity conservation, demonstrating an intricate relationship between cultural practices and survival (Bharali et al., 2023). The adaptation and resilience of Indigenous food systems to a changing environment are evident in the case studies explored by Cuni-Sanchez et al.

(2019), which focus on Andean agricultural practices that blend traditional techniques with modern ecological knowledge (Soare et al., 2023). Cultivating native crops such as quinoa and potatoes illustrates the Indigenous understanding of agricultural biodiversity and its role in nutrition security, especially in the context of climate change. The ability to cultivate these crops with minimal external inputs highlights the remarkable sustainable practices embedded in Indigenous food systems.

2.8.2 Social Cohesion Through Food Practices

Food is a powerful medium for promoting social cohesion within communities. The communal nature of food practices often acts as a catalyst for interpersonal relationships, symbiotic exchanges, and cultural identity. Indigenous peoples frequently engage in communal activities such as harvesting, cooking, and feasting, which reinforce social bonds, shared values, and collective identity. For instance, communal feasts in the Maori culture of New Zealand function to strengthen kinship ties and convey cultural heritage through food (Turner et al., 2016). Moreover, the intergenerational transmission of food knowledge plays a crucial role in retaining cultural identity and fostering community cohesion. Studies indicate that traditional food practices are often taught through participation, facilitating mentorship and connection between generations (Chopera et al., 2022; Loth et al., 2019). This knowledge transfer ensures the survival of dietary practices, cultural narratives, and ethical frameworks that govern community life.

2.8.3 Ecological Adaptation and Sustainability

The ecological adaptation embedded within Indigenous food practices goes beyond mere survival; it signifies a storied relationship with the environment that highlights sustainability and stewardship. Indigenous agricultural techniques such as agroecology, polyculture, and permaculture have enhanced local ecosystems and biodiversity. For instance, the study of Liu et al. (2020) on West African Indigenous farming systems illustrates how traditional knowledge of crop rotation and companion planting promotes soil health and boosts agricultural resilience. Furthermore, the concept of food sovereignty provides a critical perspective through which we can comprehend Indigenous food systems as a means of asserting autonomy over food resources. According to La Via Campesina (2007), food sovereignty emphasizes the right of people to define their food systems, including culturally appropriate production methods and the sustainability of local resources. This cultural and ecological adaptation underscores the dual significance of Indigenous food practices in defending against external agricultural pressures and fostering sustainable livelihoods (DeNunzio et al., 2023).

2.8.4 Biological and Cultural Fitness

The interconnectedness of biological and cultural fitness through Indigenous food knowledge highlights the essential role of food practices in human health and societal well-being. Indigenous diets, often rich in diverse, locally sourced, and nutrient-dense foods, contribute to both physical health and cultural vitality. For instance, the traditional Mediterranean diet, which includes fresh vegetables, fruits, legumes, and fish, illustrates the inherent link between healthy food consumption and cultural identity. This diet exemplifies how cultural practices related to food production, preparation, and consumption can enhance health outcomes while affirming cultural identity and heritage. Qualitative studies conducted by Sarkar et al. (2019) argue that cultural fitness derived from Indigenous food practices can serve as a buffer against health disparities. The relational nature of Indigenous food systems addresses nutritional needs while incorporating spiritual aspects and rites that instill value and purpose in community life. This underscores the concept of holistic health, where well-being is viewed through the lens of interconnectedness among physical, social, and ecological dimensions.

2.8.5 Food Heritage as a Sociobiological Legacy for Contemporary Resilience

Recognizing food heritage as a sociobiological legacy is crucial for fostering resilience in contemporary settings. Food heritage serves as a repository of Indigenous knowledge and practices that have historically offered layers of adaptation and survival. These practices empower both individual and communal resilience, especially in the face of global pressures such as climate change, urbanization, and shifts toward processed diets (Ogwu, 2019; Osawaru et al., 2014). The renewed interest in local food systems is evident in Indigenous communities worldwide, where traditional practices are being revitalized to address food insecurity and promote independence from industrialized agricultural systems. Research by Litt et al. (2023) illustrates the beneficial impacts of community-driven food initiatives, such as urban gardening projects in Native American communities, which seek to restore food sovereignty while enhancing cultural and social resilience.

Moreover, several anthropological frameworks highlight that reengaging with Indigenous food practices can enhance adaptive capacities. An approach detailed by Vijayan et al. (2022) emphasizes participatory methodologies to integrate traditional ecological knowledge with contemporary scientific practices, demonstrating the synergistic potential of blending Indigenous wisdom with modern resilience frameworks. This confluence of knowledge systems embodies the sociobiological legacy that prepares communities to navigate contemporary challenges while retaining their cultural identity.

2.9 Some Case Studies of the Relationship Between Indigenous Knowledge and Food Heritage

2.9.1 Taro in Hawaiian Knowledge Systems

Taro (Colocasia esculenta), or kalo in the Hawaiian language, holds profound historical, spiritual, and ecological importance within Native Hawaiian knowledge systems (Figure 2.2). More than a staple crop, taro is viewed as a sacred ancestor in the Hawaiian creation story, believed to be the elder sibling of humankind—born of the union between sky father (Wākea) and earth mother (Papa) (Kameʻeleihiwa, 1992). This genealogical relationship situates taro not simply as food but as kin, making its cultivation and consumption acts of cultural continuity and reverence. The traditional method of taro cultivation in loʻi—flooded terrace systems—testifies to Indigenous ecological engineering. These systems manage water flow from mountain to sea in a way that supports soil fertility, sediment capture, and aquaculture, particularly for native fish species. The intimate understanding of hydrology, landscape, and seasonal cycles embedded in these practices demonstrates the holistic integration of agricultural production with ecosystem stewardship (Lincoln & Vitousek, 2016). Moreover, taro cultivation continues to serve as a site of resistance and resilience, revitalized by Native Hawaiian communities asserting food sovereignty and reclaiming land rights in the face of tourism-driven and settler land use. Efforts to reestablish taro as a dietary and cultural staple closely link to health and well-being among Native Hawaiians. Research by Kaholokula et al. (2018) shows how culturally grounded interventions—such as those that engage in traditional taro cultivation—have improved physical and mental health outcomes. These findings affirm the centrality of food heritage in addressing contemporary issues of chronic disease, cultural alienation, and environmental degradation.

Figure 2.2: Almost Cormless Invasive Hybrid Taro Removed from a Wetland Taro Patch, Hawai'i (*left*), and Runners from Patented Taro Hybrid, Pa'lehua (*right*)

Source: Kagawa-Viviani et al. (2018).

2.9.2 Wild Rice (Manoomin) and Ojibwe Food Governance

Wild rice (*Zizania palustris*), known as manoomin in Anishinaabemowin, serves as both a dietary cornerstone for the Ojibwe and other Anishinaabe peoples of the Great Lakes region and a sacred plant essential to cultural identity, governance, and resistance (Figure 2.3). According to Ojibwe migration stories, their ancestors were instructed to move westward until they found the place "where food grows on the water"—a reference to manoomin (GLIFWC, 2020). The plant symbolizes sustenance and survival, infused with spiritual significance that highlights its role in ceremonies, seasonal rituals, and intergenerational knowledge transfer. The governance of wild rice is based on customary ecological knowledge, which encompasses intricate techniques for harvesting, parching, and reseeding. These practices align with the rhythms of the lake and the lives of nonhuman kin, promoting both ecological sustainability and community autonomy. Ojibwe food governance embodies an ethic of stewardship that prioritizes reciprocity, guardianship, and responsibility toward water and plant relatives (David et al., 2019). It also serves as a political stance against industrial threats such as mining and damming, which compromise the integrity of manoomin ecosystems. Scholars like Whyte (2016) argue that the protection and restoration of manoomin symbolize Indigenous food sovereignty movements that challenge settler-colonial food systems and environmental degradation. The resurgence of wild rice harvesting as a cultural, ecological, and political practice represents a powerful act of resurgence, re-centering Indigenous epistemologies and treaty rights in regional policy discussions.

Figure 2.3: Wild Rice—Manoomin

2.9.3 Enset Cultivation in Ethiopia

Enset (*Ensete ventricosum*), sometimes referred to as the "tree against hunger," is a crucial food crop in Southern Ethiopia, supporting over 20 million people. Though it does not produce edible fruit like its relative, the banana, its pseudostem and corm are processed into high-energy foods such as kocho (a starchy bread) and bulla (a flour-like substance) (Figure 2.4). Enset's resilience to drought, pests, and diseases makes it an anchor of food security in Ethiopia's highland regions, where rainfall variability and soil degradation threaten other staples (Brandt et al., 1997). The cultivation and processing of enset are embedded in rich IKS, which encompass ecological awareness, soil management, varietal selection, and fermentation techniques. Women play a particularly important role in enset food systems. They are the primary processors of the plant and are regarded as the custodians of knowledge related to enset's numerous food, medicinal, and ceremonial uses (Borrell et al., 2020). This gendered knowledge transmission emphasizes the social dimensions of food heritage and reinforces the importance of inclusivity in discussions about sustainability and resilience. Importantly, enset is nutritionally significant and culturally and symbolically resonant. Its cultivation is often linked to social status and household identity, with specific landraces passed down through family lineages. Its perennial nature allows it to be harvested flexibly, providing a crucial buffer during food shortages and famines (Garedew et al., 2009). Enset-based agroecological systems illustrate how Indigenous innovation can support long-term resilience, particularly in the context of climate change and population growth.

Figure 2.4: Two Landraces of Enset (False Banana) in Ethiopia

(A) (B)

Source: Adapted from Yemataw et al. (2017).

2.10 Conclusion

The current state of IKS and food heritages illustrates a dynamic landscape where cultural, ecological, and legal dimensions intersect. Recognizing and supporting these systems foster cultural preservation and promote sustainability and resilience within Indigenous communities. Moving forward, interdisciplinary collaborations, robust legal protections, and community-driven initiatives will be essential to ensure the vitality and recognition of Indigenous knowledge and food heritages. The foundations of IKS are reinforced by oral traditions, experiential learning, and community memory, all intricately woven through the roles of elders, rituals, and apprenticeships. Furthermore, the lived experiences of Indigenous peoples are interconnected with the land, which, along with their cosmological understandings, shapes their food knowledge and cultural practices. Understanding these foundations requires recognizing the ongoing impacts of colonization and a commitment to revitalizing Indigenous knowledge through respectful engagement with Indigenous communities and their teachings. Food heritage is an essential cultural capital, embodying the intricate relationships among culinary techniques, ingredients, and meanings that define communities. It reinforces cultural identity, continuity, and resilience, enabling communities to navigate challenges while maintaining their unique narratives and traditions. The examination of case studies—from seed saving to sacred foods, traditional cooking methods, and feasting practices— highlights the importance of food heritage in sustaining cultural diversity and fostering community ties in an increasingly globalized world. The coevolution of social knowledge and food ecology through traditional agricultural practices illustrates a nuanced understanding of ecological dynamics. The interplay of adaptive strategies, dependence on environmental cues, and food preservation techniques underscores the capacity of IKS to address contemporary agricultural challenges.

As societies confront increasing pressures from climate change, harnessing traditional practices becomes crucial for ensuring sustainable food systems and enhancing resilience in farming communities. The intergenerational transmission of social food practices and knowledge systems is a multidimensional process involving various knowledge-sharing mechanisms, such as storytelling, participation, and observation, illuminated by the celebratory context of festivals and rites of passage. The essential role of language and its related terminologies enriches this transmission, acting as a vessel for ecological memory and cultural identity. Understanding these interconnected processes is vital for preserving culinary heritage in an era marked by rapid globalization and cultural transformation. As future generations navigate their food landscapes, these traditional practices and knowledge systems remain essential in fostering identity, community cohesion, and a sustainable relationship with food. Social disruptions stemming from colonialism, industrial agriculture, and globalization instigate profound changes in food heritage. The intertwining of cultural loss due to land dispossession, dietary shifts, and the standardization of food practices culminates in an ongoing threat to traditional foods. For instance, Malaysian heritage foods confront challenges in balancing tourist engagement with authenticity, reflecting a broader trend toward the commodification of culinary heritage. Thus, it is imperative for future strategies to emphasize the preservation of local food systems and cultural practices amid these global pressures.

Chapter Reflection

1. How do Indigenous Knowledge Systems (IKS) contribute to ecological sustainability and resilience in food systems, particularly in environmental challenges such as climate change?
2. What are some specific examples of Indigenous food heritage practices (such as seed saving, sacred foods, or cooking methods) that highlight the relationship between food, cultural identity, and ecological stewardship?
3. How does the concept of "relationality" in IKS challenge Western scientific approaches to food and agricultural practices? What are the implications of this view for sustainable development?
4. In what ways have colonization and globalization disrupted traditional food practices and cultural identities among Indigenous communities? How are these disruptions being addressed through contemporary food movements?
5. Discuss the role of language in preserving Indigenous food heritage. How does the loss of Indigenous languages affect the transmission of food knowledge across generations?
6. How do gender roles influence food-related knowledge and practice transmission in Indigenous communities? What impact does this have on the preservation of food heritage?
7. What is the sociobiological significance of food heritage? How do traditional food practices contribute to social cohesion and the well-being of Indigenous communities?
8. How do Indigenous food sovereignty and seed sovereignty initiatives help Indigenous communities reclaim control over their food systems and cultural practices?

9. How can modern educational programs and digital platforms contribute to revitalizing Indigenous food knowledge and heritage? What role do community-driven efforts play in these processes?
10. Reflecting on the case studies in the chapter (e.g., taro cultivation in Hawai'i, wild rice governance among the Ojibwe, and enset farming in Ethiopia), how do these practices exemplify the adaptive strategies embedded in IKS?

REFERENCES

Alejandro, A., Bartolo, K., Basallo, Q., Esguerra, M., Mallari, M., Santos, S., & Topacio, R. (2022). Perception of Maloleños on cultural food heritage: An input to gastronomic tourism. *Globus an International Journal of Management and It*, *13*(2), 53–70. https://doi.org/10.46360/globus.mgt.120221009

Almansouri, M., Verkerk, R., Fogliano, V., & Luning, P. (2021). Exploration of heritage food concept. *Trends in Food Science and Technology*, *111*, 790–797. https://doi.org/10.1016/j.tifs.2021.01.013

Ba, Q., Lu, D., Kuo, W., & Lai, P. (2018). Traditional farming and sustainable development of an Indigenous community in the mountain area—A case study of Wutai village in Taiwan. *Sustainability*, *10*(10), 3370. https://doi.org/10.3390/su10103370

Berkes, F. (1993). Traditional ecological knowledge in perspective. In J. T. Inglis (Ed.), *Traditional ecological knowledge: Concepts and cases* (pp. 1–9). Canadian Museum of Nature/International Development Research Centre.

Bharali, A., Deori, C., & Das, J. (2023). A study on the use of fermented rice beverage (sujen) by the deoris of Brahmaputra valley: Socio-cultural values and health related issues. *PARIPEX Indian Journal of Research*, *12*(11), 3–6. https://doi.org/10.36106/paripex/3500267

Borrell, J. S., Biswas, M. K., Goodwin, M., Blomme, G., Schwarzacher, T., Heslop-Harrison, J. S., & Wilkin, P. (2020). Enset in Ethiopia: A poorly characterized but resilient starch staple. *Annals of Botany*, *125*(6), 749–766. https://doi.org/10.1093/aob/mcz051

Bowra, A., Mashford-Pringle, A., & Poland, B. (2020). Indigenous learning on turtle island: A review of the literature on land-based learning. *Canadian Geographer/Le Géographe Canadien, 65*(2), 132–140. https://doi.org/10.1111/cag.12659

Brandt, S. A., Spring, A., Hiebsch, C., McCabe, J. T., Tabogie, E., Diro, M., Wolde-Michael, G., Yntiso, G., Tesfaye, S., & Shigeta, M. (1997). *The "Tree Against Hunger": Enset-based agricultural systems in Ethiopia.* American Association for the Advancement of Science.

Brenzinger, M., & Heinrich, P. (2013). The return of Hawaiian: Language networks of the revival movement. *Current Issues in Language Planning, 14*(2), 300–316. https://doi.org/10.1080/14664208.2013.812943

Budiono, S., & Noviani, E. (2024). Retta language revitalization learning materials in Alor regency. *Bahasa Dan Seni Jurnal Bahasa Sastra Seni Dan Pengajarannya, 51*(2). https://doi.org/10.17977/um015v51i22023p312

Burnette, C., Clark, C., & Rodning, C. (2018). "Living off the land": How subsistence promotes well-being and resilience among Indigenous peoples of the southeastern united states. *Social Service Review, 92*(3), 369–400. https://doi.org/10.1086/699287

Chopera, P., Zimunya, P. R., Mugariri, F. M., & Matsungo, T. M. (2022). Facilitators and barriers to the consumption of traditional foods among adults in Zimbabwe. *Journal of Ethnic Foods, 9*(1), 5. https://doi.org/10.1186/s42779-022-00121-y

Cidón, C., Figueiró, P., & Schreiber, D. (2021). Benefits of organic agriculture under the perspective of the bioeconomy: A systematic review. *Sustainability, 13*(12), 6852. https://doi.org/10.3390/su13126852

Crane, T. (2014). Bringing science and technology studies into agricultural anthropology: Technology development as cultural encounter between farmers and researchers. *Culture Agriculture Food and Environment, 36*(1), 45–55. https://doi.org/10.1111/cuag.12028

Cuni-Sanchez, A., Omeny, P., Pfeifer, M., Olaka, L., Mamo, M. B., Marchant, R., & Burgess, N. D. (2019). Climate change and pastoralists: Perceptions and adaptation in montane Kenya. *Climate and Development*, 11(6), 513–524. https://doi.org/10.1080/17565529.2018.1454880

Dasgupta, P., Goswami, R., Ali, M., Chakraborty, S., & Saha, S. (2015). Multifunctional role of integrated farming system in developing countries. *International Journal of Bio-Resource and Stress Management*, *6*(3), 424. https://doi.org/10.5958/0976-4038.2015.00057.3

Datta, R. (2024). Relationality in Indigenous climate change education research: A learning journey from Indigenous communities in Bangladesh. *Australian Journal of Environmental Education*, *40*(2), 128–142. https://doi.org/10.1017/aee.2024.13

Datta, R., Singha, R., & Hurlbert, M. (2024). Indigenous land-based perspectives on environmental sustainability: Learning from the Khasis Indigenous community in Bangladesh. *Sustainability*, *16*(9), 3678. https://doi.org/10.3390/su16093678

David, P., Anderson, N., & Huberty, B. (2019). *Manoomin: The story of wild rice in Minnesota*. University of Minnesota Press.

DeNunzio, M., Serrano, E., Kraak, V., Chase, M., & Misyak, S. (2023). A feasibility study of the community health worker model for garden-based food systems programming. *Journal of Agriculture Food Systems and Community Development*, *13*(1), 1–19. https://doi.org/10.5304/jafscd.2023.131.005

Fast, E., Lefebvre, M., Reid, C., Deer, W., Swiftwolfe, D., Clark, M., Boldo, V., Mackie, J., & Mackie, R. (2021). Restoring our roots: Land-based community by and for Indigenous youth. *International Journal of Indigenous Health*, *16*(2). https://doi.org/10.32799/ijih.v16i2.33932

Fletcher, C., Pforr, C., & Brueckner, M. (2016). Factors influencing Indigenous engagement in tourism development: An international perspective. *Journal of Sustainable Tourism*, *24*(8–9), 1100–1120. https://doi.org/10.1080/09669582.2016.1173045

Gallegos, R., Aráuz, M., Saeteros-Hernández, A., Chávez, R., & Moyano, M. (2023). The Indigenous bioculture of the Pungalá Parish of Ecuador: An approach to their culinary and medicinal heritage. *Journal of Ethic Foods*. https://doi.org/10.21203/rs.3.rs-3161621/v1

Garedew, E., Sandewall, M., & Söderberg, U. (2009). Land-use and land-cover dynamics in the central rift valley region of Ethiopia. *Environmental Management*, *44*(4), 683–694. https://doi.org/10.1007/s00267-009-9355-z

Gauchan, D., Joshi, B., Sthapit, S., & Jarvis, D. (2020). Traditional crops for household food security and factors associated with on-farm diversity in the mountains of Nepal. *Journal of Agriculture and Environment*, *21*, 31–43. https://doi.org/10.3126/aej.v21i0.38440

Gaudet, J., & Chilton, C. (2018). Milo Pimatisiwin project. *International Journal of Indigenous Health*, *13*(1), 20–40. https://doi.org/10.32799/ijih.v13i1.30264

Ghosh-Jerath, S., Singh, A., Magsumbol, M., Kamboj, P., & Goldberg, G. (2016). Exploring the potential of Indigenous foods to address hidden hunger: Nutritive value of Indigenous foods of Santhal tribal community of Jharkhand, India. *Journal of Hunger and Environmental Nutrition*, *11*(4), 548–568. https://doi.org/10.1080/19320248.2016.1157545

Giampiccoli, A., & Kalis, J. (2012). Tourism, food, and culture: Community-based tourism, local food, and community development in Pondoland. *Culture Agriculture Food and Environment*, *34*(2), 101–123. https://doi.org/10.1111/j.2153-9561.2012.01071.x

GLIFWC (Great Lakes Indian Fish and Wildlife Commission). (2020). Ganawenindiwag Manoomin Anishinaabeg. https://glifwc.org/stewardship/ganawenindiwag-manoomin-anishinaabeg

Guo, Y., & Hsu, F. (2023). Branding creative cities of gastronomy: The role of brand experience and the influence of tourists' self-congruity and self-expansion. *British Food Journal*, *125*(8), 2803–2824. https://doi.org/10.1108/bfj-05-2022-0434

Hafiz, W. Z. S., Gan, M. Y., Mohamad Rahijan, A. W., & Cai, W. (2021). Chinese food culture and festival: Role and symbolic meaning among Hokkien millennials. Proceedings of the Global Tourism Conference 2021, 85–88. https://gala.gre.ac.uk/id/eprint/35141/7/35141_CAI_Chinese_food_culture_and_festival.pdf

Hussen, A. (2021). Important of applying intercropping for sustainable crop production: A review. *International Journal of Research in Agronomy*, *4*(2), 37–40. https://doi.org/10.33545/2618060x.2021.v4.i2a.81

Irawan, A., & Komori, D. (2024). Beyond fixed dates and coarse resolution: Developing a dynamic dry season crop calendar for paddy in Indonesia from 2001 to 2021. *Agronomy*, *14*(3), 564. https://doi.org/10.3390/agronomy14030564

Ishak, N., Ismail, A., Saad, M., & Ramli, A. (2021). Researching Kedah's Malay heritage food tradition and eating culture. *International Journal of Academic Research in Business and Social Sciences*, *11*(16), 140–154. https://doi.org/10.6007/ijarbss/v11-i16/11224

Islam, D., Zurba, M., Rogalski, A., & Berkes, F. (2016). Engaging Indigenous youth to revitalize Cree culture through participatory education. *Diaspora Indigenous and Minority Education, 11*(3), 124–138. https://doi.org/10.1080/155 95692.2016.1216833

Jalis, M., Che, D., & Markwell, K. (2014). Utilising local cuisine to market Malaysia as a tourist destination. *Procedia—Social and Behavioral Sciences, 144*, 102–110. https://doi.org/10.1016/j.sbspro.2014.07.278

Janßen, M., Chang, B., Hristov, H., Pravst, I., Profeta, A., & Millard, J. (2021). Changes in food consumption during the covid-19 pandemic: Analysis of consumer survey data from the first lockdown period in Denmark, Germany, and Slovenia. *Frontiers in Nutrition, 8*. https://doi.org/10.3389/fnut.2021.635859

Jiri, O., Mafongoya, P., Mubaya, C., & Mafongoya, O. (2016). Seasonal climate prediction and adaptation using Indigenous knowledge systems in agriculture systems in southern Africa: A review. *Journal of Agricultural Science, 8*(5), 156. https://doi.org/10.5539/jas.v8n5p156

John, I. (2024). Indigenous or exotic crop diversity? which crops ensure household food security: Facts from Tanzania panel. *Sustainability, 16*(9), 3833. https://doi.org/10.3390/su16093833

Johnston, R., & Marwood, K. (2017). Action heritage: Research, communities, social justice. *International Journal of Heritage Studies, 23*(9), 816–831. https://doi.org/10.1080/13527258.2017.1339111

Joshi, N. (2022). Ethnobotanical study of traditional food in Newar community of Kathmandu valley, central Nepal. *Journal of Plant Resources, 20*(2), 200–210. https://doi.org/10.3126/bdpr.v20i2.57039

Kagawa-Viviani, A., Levin, P., Johnston, E., Ooka, J., Baker, J., Kantar, M., & Lincoln, N. K. (2018). I Ke Ēwe ʻĀina o Ke Kupuna: Hawaiian ancestral crops in perspective. *Sustainability, 10*(12), 4607. https://doi.org/10.3390/su10124607

Kaholokula, J. K., Ing, C. T., Look, M. A., Delafield, R., & Sinclair, K. A. (2018). Culturally responsive approaches to health promotion for Native Hawaiians and Pacific Islanders. *Annals of Human Biology, 45*(3), 249–263. https://doi.org/10.1080/03014460.2018.1465593

Kameʻeleihiwa, L. (1992). *Native land and foreign desires: Pehea Lā E Pono Ai?* Bishop Museum Press.

Klein, J. (2018). Heritagizing local cheese in china: Opportunities, challenges, and inequalities. *Food and Foodways*, *26*(1), 63–83. https://doi.org/10.1080/0 7409710.2017.1420354

Kom, Z., Nicolau, M., & Nenwiini, S. (2024). The use of Indigenous knowledge systems practices to enhance food security in Vhembe District, South Africa. *Agricultural Research*, *13*(3), 599–612. https://doi.org/10.1007/ s40003-024-00716-8

Koohafkan, P. (2018). Globally important agricultural heritage systems (GIAHS): A legacy for food and nutrition security. In B. Burlingame & S. Dernini (Eds.), *Sustainable diets: Linking nutrition and food systems* (pp. 204-214). CABI Digital Library. https://doi.org/10.1079/9781786392848.0204

Kumar, S., Bhatt, B., Dey, A., Khandelwal, S., Kumar, U., Idris, M., Mishra, J. S., & Kumar, S. (2018). Integrated farming system in India: Current status, scope and future prospects in changing agricultural scenario. *The Indian Journal of Agricultural Sciences*, *88*(11), 1661–1675. https://doi.org/10.56093/ijas. v88i11.84880

La Via Campesina. (2007). Food sovereignty: A right for all. https://viacampesina. org/en/wp-content/uploads/sites/2/2008/03/Food-sovereignty-for-Africa-2007.pdf

Lah, N. H. C., El Enshasy, H. A., Mediani, A., Azizan, K. A., Aizat, W. M., Tan, J. K., Afzan, A., Noor, N. M., & Rohani, E. R. (2023). An insight into the behaviour of recalcitrant seeds by understanding their molecular changes upon desiccation and low temperature. *Agronomy*, *13*(8), 2099. https://doi.org/10.3390/ agronomy13082099

Lan, N., & Kien, N. (2021). Back to nature-based agriculture: Green livelihoods are taking root in the Mekong River Delta. *Journal of People Plants and Environment*, *24*(6), 551–561. https://doi.org/10.11628/ ksppe.2021.24.6.551

Lin, Y., Tomi, P., Huang, H., Lin, C., & Chen, Y. (2020). Situating Indigenous resilience: Climate change and Tayal's "millet ark" action in Taiwan. *Sustainability*, *12*(24), 10676. https://doi.org/10.3390/su122410676

Lincoln, N. K., & Vitousek, P. M. (2016). Indigenous Polynesian agricultural systems in Hawai'i: Intensification and ecological sustainability. *Ecology and*

Society, *21*(1), 11. http://www.ulumaupuanui.org/uploads/1/8/2/1/18219029/ indigenous_polynesian_agriculture_in_hawai%CA%BBi.pdf

Litt, J. S., Alaimo, K., Harrall, K. K., Hamman, R. F., Hébert, J. R., Hurley, T. G., Leiferman, J. A., Li, K., Villalobos, A., Coringrato, E., Courtney, J. B., Payton, M., & Glueck, D. H. (2023). Effects of a community gardening intervention on diet, physical activity, and anthropometry outcomes in the USA (CAPS): An observer-blind, randomised controlled trial. *Lancet: Planetary Health, 7*(1), e23–e32. https://doi.org/10.1016/S2542-5196(22)00303-5

Liu, C. (2017). Family-based food practices and their intergenerational geographies in contemporary Guangzhou, China. *Transactions of the Institute of British Geographers*, *42*(4), 572–583. https://doi.org/10.1111/tran.12178

Liu, Y., Yang, Y., Zhang, C., Xiao, C., & Song, X. (2023). Does Nepal have the agriculture to feed its population with a sustainable diet? Evidence from the perspective of human-land relationship. *Foods* (Basel, Switzerland), *12*(5), 1076. https://doi.org/10.3390/foods12051076

Longvah, T., Khutsoh, B., Meshram, I., Krishna, S., Kodali, V., Roy, P., & Kuhnlein, H. (2017). Mother and child nutrition among the Chakhesang tribe in the state of Nagaland, North-East India. *Maternal and Child Nutrition*, *13*(S3). https://doi.org/10.1111/mcn.12558

Loth, K. A., Uy, M. J. A., Winkler, M. R., Neumark-Sztainer, D., Fisher, J. O., & Berge, J. M. (2019). The intergenerational transmission of family meal practices: a mixed-methods study of parents of young children. *Public Health Nutrition, 22*(7), 1269–1280. https://doi.org/10.1017/S1368980018003920

Luppens, L., & Power, E. (2018). "Aboriginal isn't just about what was before, it's what's happening now": Perspectives of Indigenous peoples on the foods in their contemporary diets. *Canadian Food Studies/La Revue Canadienne Des Études Sur L Alimentation*, *5*(2), 142–161. https://doi.org/10.15353/cfs-rcea.v5i2.219

Lutz, J., & Schachinger, J. (2013). Do local food networks foster socio-ecological transitions towards food sovereignty? learning from real place experiences. *Sustainability*, *5*(11), 4778–4796. https://doi.org/10.3390/su5114778

Malli, A., Monteith, H., Hiscock, C., Smith, E., Fairman, K., Galloway, T., & Mashford-Pringle, A. (2023). Impacts of colonization on Indigenous food systems in Canada and the United States: A scoping review. *BMC Public Health*, *23*(1). https://doi.org/10.1186/s12889-023-16997-7

Matta, R. (2016). Food incursions into global heritage: Peruvian cuisine's slippery road to UNESCO. *Social Anthropology*, *24*(3), 338–352. https://doi.org/10.1111/1469-8676.12300

Miltenburg, E., Neufeld, H., Perchak, S., & Skene, D. (2023). "Where Creator Has My Feet, There I Will Be Responsible": Place-making in urban environments through Indigenous food sovereignty initiatives. *International Journal of Environmental Research and Public Health*, *20*(11), 5970. https://doi.org/10.3390/ijerph20115970

Mondal, K., Paul, A., & Mondal, P. (2023). Zero budget natural farming: An agricultural revolution, prospects and problems. *International Journal of Plant and Soil Science*, *35*(20), 228–235. https://doi.org/10.9734/ijpss/2023/v35i203802

Obamwonyi, A., & Onyekuru, J. (2024). The role of traditional African festivals in the sustenance of the ecosystem: Ikenge festival in Utagba-Uno, Southern Nigeria as a paradigm. *Rupkatha Journal on Interdisciplinary Studies in Humanities*, *16*(1). https://doi.org/10.21659/rupkatha.v16n1.14

Ogwu, M. C. (2019). Towards sustainable development in Africa: The challenge of urbanization and climate change adaptation. In P. B. Cobbinah & M. Addaney (Eds.), *The geography of climate change adaptation in Urban Africa* (pp. 29–55). Springer Nature. http://doi.org/10.1007/978-3-030-04873-0_2

Ogwu, M. C. (2023). Local food crops in Africa: Sustainable utilization, threats, and traditional storage strategies. In S. C. Izah & M. C. Ogwu (Eds.), *Sustainable utilization and conservation of Africa's biological resources and environment.* Sustainable Development and Biodiversity (Vol. 888, pp. 353–374). Springer. https://doi.org/10.1007/978-981-19-6974-4_13

Ogwu, M. C., Osawaru, M. E., & Atsenokhai, E. I. (2016). Chemical and microbial evaluation of some uncommon Indigenous fruits and nuts. *Borneo Science*, *37*(1), 54–71.

Oloko, M., & Shukla, S. (2018). "Nya anghuwa che" (our food gives us life): Exploring Indigenous perspectives on traditional food gathering and foraging in an Irigwe community from Nigeria. *Ab-Original*, *2*(1), 1–22. https://doi.org/10.5325/aboriginal.2.1.0001

Omar, S., Karim, S., Bakar, A., & Omar, S. (2015). Safeguarding Malaysian heritage food (MHF): the impact of Malaysian food culture and tourists' food culture involvement on intentional loyalty. *Procedia—Social and Behavioral Sciences*, *172*, 611–618. https://doi.org/10.1016/j.sbspro.2015.01.410

Organ, K., Koenig-Lewis, N., Palmer, A., & Probert, J. (2015). Festivals as agents for behaviour change: A study of food festival engagement and subsequent food choices. *Tourism Management*, *48*, 84–99. https://doi.org/10.1016/j.tourman.2014.10.021

Osawaru, M. E., & Ogwu, M. C. (2014). Ethnobotany and germplasm collection of two genera of cocoyam (*Colocasia* [Schott] and *Xanthosoma* [Schott], Araceae) in Edo State Nigeria. *Science, Technology and Arts Research Journal*, *3*(3), 23–28. http://doi.org/10.4314/star.v3i3.4

Osawaru, M. E., & Ogwu, M. C. (2020). Survey of plant and plant products in local markets within Benin City and environs. In W. Leal Filho, N. Ogugu, D. Ayal, L. Adelake, & I. da Silva (Eds.), *African handbook of climate change adaptation* (pp. 1–24). Filho, Springer Nature. http://doi.org/10.1007/978-3-030-42091-8_159-1

Osawaru, M. E., Ogwu, M. C., & Aigbefue, D. (2014). Survey of ornamental gardens in five local government areas of Southern Edo State Nigeria. *The Bioscientist*, *2*(1), 87–102.

Panjaitan, B., & Roisah, K. (2021). *Protection of Indigenous legal rights towards traditional knowledge used by foreign parties according to international law perspective*. EAI. https://doi.org/10.4108/eai.17-7-2019.2303378

Pawera, L., Khomsan, A., Zuhud, E., Hunter, D., Ickowitz, A., & Polesný, Z. (2020). Wild food plants and trends in their use: From knowledge and perceptions to drivers of change in West Sumatra, Indonesia. *Foods*, *9*(9), 1240. https://doi.org/10.3390/foods9091240

Pinedo, S., Escobar, L., & Neufeld, H. (2021). *The white/wiphala paper on Indigenous peoples' food systems*. FAO. https://doi.org/10.4060/cb4932en

Price, L., García, G., & Narchi, N. (2021). Foods of oppression. *Frontiers in Sustainable Food Systems*, *5*. https://doi.org/10.3389/fsufs.2021.646907

Prieto, M., Sallans, A., Ostertag, S., Wesche, S., Kenny, T., & Skinner, K. (2023). Food programs in Indigenous communities within Northern Canada: A scoping review. *Canadian Geographer/Le Géographe Canadien*, *68*(2), 276–292. https://doi.org/10.1111/cag.12872

Prober, S., O'Connor, M., & Walsh, F. (2011). Australian aboriginal peoples' seasonal knowledge: A potential basis for shared understanding in environmental management. *Ecology and Society*, *16*(2). https://doi.org/10.5751/es-04023-160212

Rahman, D., Moussouri, T., & Alexopoulos, G. (2021). The social ecology of food: Where agroecology and heritage meet. *Sustainability*, *13*(24), 13981. https://doi.org/10.3390/su132413981

Ramli, A., Sapawi, D., Noor, H., & Zahari, M. (2020). Unearthing awareness of food heritage based on age. *Journal of Asian Behavioural Studies*, *5*(17), 41–53. https://doi.org/10.21834/jabs.v5i17.375

Ramli, A., Zahari, M., Halim, N., & Aris, M. (2017). Knowledge on the Malaysian food heritage. *Asian Journal of Quality of Life*, *2*(5), 31–42. https://doi.org/10.21834/ajqol.v2i5.9

Ray, L., Burnett, K., Cameron, A., Joseph, S., LeBlanc, J., Parker, B., Recollet, A., & Sergerie, C. (2019). Examining Indigenous food sovereignty as a conceptual framework for health in two urban communities in Northern Ontario, Canada. *Global Health Promotion*, *26*(3_suppl), 54–63. https://doi.org/10.1177/1757975919831639

Raymundo, A. (2024). Archiving food heritage towards championing food security: A case study of lokalpedia. *Wimaya*, *5*(1), 28–39. https://doi.org/10.33005/wimaya.v5i01.142

Richmond, C., Steckley, M., Neufeld, H., Kerr, R., Wilson, K., & Dokis, B. (2020). First nations food environments: Exploring the role of place, income, and social connection. *Current Developments in Nutrition*, *4*(8), nzaa108. https://doi.org/10.1093/cdn/nzaa108

Robin, T., Dennis, M., & Hart, M. (2020). Feeding Indigenous people in Canada. *International Social Work*, *65*(4), 652–662. https://doi.org/10.1177/0020872820916218

Rodriguez, M., & Pedroso, J. (2024). Exploring culinary heritage: Insights from local gatekeepers in the province of antique Philippines. https://doi.org/10.21203/rs.3.rs-4023763/v1

Saad, M., Kamarizzaman, N., Ishak, N., & Pratt, T. (2021). Pahang's heritage food: Do consumption values play a role in the development of behavioral intentions? *Journal of Asian Behavioural Studies*, *6*(20), 45–58. https://doi.org/10.21834/jabs.v6i20.400

Sanip, M., & Mustapha, R. (2020). The role of gastronomic tourism education in sustaining Malaysian heritage food, pp. 14–24.

Santoro, A., Venturi, M., Bertani, R., & Agnoletti, M. (2020). A review of the role of forests and agroforestry systems in the FAO globally important agricultural heritage systems (GIAHS) programme. *Forests, 11*(8), 860. https://doi.org/10.3390/f11080860

Sarkar, D., Walker-Swaney, J., & Shetty, K. (2019). Food diversity and Indigenous food systems to combat diet-linked chronic diseases. *Current Developments in Nutrition, 4*(Suppl. 1), 3–11. https://doi.org/10.1093/cdn/nzz099

Sharif, M., Rahman, A., Zahari, M., & Abdullah, K. (2018). Malay traditional food knowledge transfer. *Asian Journal of Quality of Life, 3*(10), 79–88. https://doi.org/10.21834/ajqol.v3i10.103

Sidali, K., Kastenholz, E., & Bianchi, R. (2013). Food tourism, niche markets and products in rural tourism: Combining the intimacy model and the experience economy as a rural development strategy. *Journal of Sustainable Tourism, 23*(8–9), 1179–1197. https://doi.org/10.1080/09669582.2013.836210

Soare, I., Privitera, D., Lupu, C., & Ganuşceac, A. (2023). Enhancing rural integration into European agriculture: Rediscovering sustainable agri-food in Romania Dealu Mare region, Romania. *Central European Journal of Geography and Sustainable Development, 5*(2), 46–61. https://doi.org/10.47246/cejgsd.2023.5.2.3

Suremain, C. (2019). From multi-sited ethnography to food heritage: What theoretical and methodological challenges for anthropology?. *Revista Del Cesla International Latin American Studies Review*, (24), 7–32. https://doi.org/10.36551/2081-1160.2019.24.7-32

Swiderska, K., Argumedo, A., Wekesa, C., Ndalilo, L., Song, Y., Rastogi, A., & Ryan, P. (2022). Indigenous peoples' food systems and biocultural heritage: Addressing Indigenous priorities using decolonial and interdisciplinary research approaches. *Sustainability, 14*(18), 11311. https://doi.org/10.3390/su141811311

Timler, K., Varcoe, C., & Brown, H. (2019). Growing beyond nutrition. *International Journal of Indigenous Health, 14*(2), 95–114. https://doi.org/10.32799/ijih.v14i2.31938

Trofholz, A., Thao, M., Donley, M., Smith, M., Isaac, H., & Berge, J. (2018). Family meals then and now: A qualitative investigation of intergenerational transmission of family meal practices in a racially/ethnically diverse and immigrant population. *Appetite*, *121*, 163–172. https://doi.org/10.1016/j.appet.2017.11.084

Turner, K., Davidson-Hunt, I., Desmarais, A., & Hudson, I. (2016). Creole hens and Ranga-Ranga: Campesino foodways and biocultural resource-based development in the Central Valley of Tarija, Bolivia. *Agriculture*, *6*(3), 41. https://doi.org/10.3390/agriculture6030041

Whyte, K. (2016). Indigenous food systems, environmental justice, and settler-industrial states. In M. Rawlinson & C. Ward (Eds.), *The Routledge handbook of food ethics* (pp. 354–365). Routledge.

Yemataw, Z., Chala, A., Ambachew, D., Studholme, D. J., Grant, M. R., & Tesfaye, K. (2017). Morphological variation and inter-relationships of quantitative traits in enset (*Ensete ventricosum* (welw.) Cheesman) Germplasm from South and South-Western Ethiopia. *Plants*, *6*(4), 56. https://doi.org/10.3390/plants6040056

CHAPTER 3

Evolutionary Perspectives of Indigenous Diets and Food Systems

Abstract

This chapter explores the evolutionary foundations of Indigenous diets and food systems, emphasizing their adaptive, ecological, and cultural dimensions. Drawing on interdisciplinary literature, it traces how early human subsistence strategies—rooted in foraging, hunting, and gathering—evolved into complex agroecological practices deeply interwoven with local environments and spiritual beliefs. Indigenous food systems represent a dynamic interplay between human biology and culture, shaped by natural selection, sociocultural transmission, and environmental pressures over millennia. These systems are often nutritionally superior, emphasizing biodiversity, seasonality, and sustainability. However, colonization and the global spread of ultra-processed foods have disrupted traditional dietary patterns, leading to profound health disparities, including rising rates of obesity, diabetes, and cardiovascular diseases in many Indigenous populations. Through an evolutionary lens, this chapter examines key dietary adaptations such as lactase persistence, starch digestion, and microbiome diversity to understand the mismatch between ancestral diets and modern food environments. It also highlights how food taboos, ritualized consumption, and gendered knowledge transmission preserve dietary norms and reinforce identity. Notably, the chapter discusses contemporary efforts toward food sovereignty, revitalization of Indigenous foodways, and integration of traditional and scientific knowledge systems in climate-resilient agriculture. Ultimately, it argues for the urgent need to reclaim Indigenous dietary practices as a pathway to health equity, cultural continuity, and ecological sustainability.

Keywords: Indigenous food systems, evolutionary nutrition, dietary adaptation, traditional ecological knowledge, food sovereignty, cultural identity, health disparities

3.1 Introduction

The essential role of Indigenous diets in maintaining health and culture cannot be overstated, as these diets were finely tuned through generations of interactions with local ecosystems. Understanding the evolution of these food systems is crucial for addressing contemporary health disparities observed in Indigenous populations, particularly regarding the increasing reliance on modern, processed food sources. This review synthesizes existing literature on Indigenous diets from an evolutionary perspective while examining the historical, cultural, and environmental factors that influence these dietary patterns.

Indigenous food systems demonstrate a profound understanding of local flora and fauna, which has developed over centuries of practice and environmental adaptation. These traditional diets include foods that typically provide essential nutrients (Table 3.1). According to McCartan et al. (2020), the significance of traditional foods for Indigenous communities is considerable, as these foods have historically played a vital role in energy and nutrient intake. The nutritional quality of traditional foods often contrasts sharply with that of modern, store-bought foods, which are mainly characterized by high sugar and unhealthy fats, linked to an increase in chronic health issues such as obesity, diabetes, and cardiovascular diseases in Indigenous populations (McCartan et al., 2020).

Traditional diets often include wild game, fish, fruits, and vegetables native to the region. For instance, Fyfe et al.'s (2018) research highlights the nutritional benefits of specific Indigenous foods like the Australian Green Plum, which possesses high nutritional value and holds significant cultural importance. Therefore, integrating seasonal, local foods into diets has been crucial for promoting health among Indigenous populations, as these foods contribute to the dietary diversity often lacking in modern diets. The cultural significance of traditional foods extends beyond their nutritional value. Auger (2016) asserts that traditional diets are integral to Indigenous identities and community well-being, directly linking health outcomes to cultural continuity. The preparation and sharing of traditional foods are rich with sociocultural practices, fostering community bonds and reinforcing identities. The decline in the consumption of traditional foods due to colonization has negatively affected community health and cohesion.

Brimblecombe et al. (2013) illustrate that economic barriers and limited access to traditional foods have resulted in alarming health statistics among Aboriginal communities in Australia, where many individuals now face higher rates of preventable chronic diseases. This grave situation highlights the need to reclaim and

Table 3.1: Indigenous Food Systems and Traditional Diets

Indigenous Group/Region	Ecological Context	Primary Food Staples	Methods of Procurement	Nutritional and Cultural Value	Adaptive Features/ Resilience Mechanisms
Hopi (Southwestern United States)	Arid desert environment	Blue corn, beans, and squash	Dryland farming, seed saving	High in protein and fiber, spiritual symbolism, a staple in ceremonial life	Drought-resistant crop varieties, and ecological forecasting
Inuit (Arctic regions)	Arctic tundra and marine	Seal, caribou, fish, and whale blubber	Hunting, fishing, preservation (freezing, fermenting)	High-fat diets for energy, rich in Omega-3, and deep cultural identity	Seasonal mobility and traditional ecological knowledge (TEK) of animal behavior
Maasai (East Africa)	Semiarid savannah	Milk, blood, and meat	Pastoralism (livestock herding)	Nutritionally adapted to low-carb diets, and culturally tied to cattle	Seasonal migration, communal grazing rights
Ojibwe (Great Lakes region)	Boreal forest and freshwater ecosystems	Wild rice (manoomin), berries, and fish	Canoe-based harvesting, foraging, and fishing	High in fiber and protein, sacred food, and central to ceremonies	Rotational harvesting, traditional water stewardship
Andean Quechua (Peru, Bolivia)	High-altitude mountainous terrain	Quinoa, potatoes, and corn	Terrace farming, seed exchanges, barter systems	Diverse micronutrients, ceremonial foods, and crop diversity linked to cosmology	Andean agro-biodiversity, climate-resilient landraces

Indigenous Group/Region	Ecological Context	Primary Food Staples	Methods of Procurement	Nutritional and Cultural Value	Adaptive Features/ Resilience Mechanisms
Aboriginal Australians (various nations)	Varied: desert, savannah, and coastal	Bush tucker (yams, kangaroo, witchetty grub, and nuts)	Foraging, hunting, fire-stick farming	Locally adapted nutrition, and linked to Dreamtime stories	Fire ecology, seasonal mobility, multilingual ecological classification
Zulu (Southern Africa)	Subtropical savannah	Sorghum, maize, beans, and wild greens	Subsistence farming, wild harvesting	High in carbohydrates and micronutrients; has a role in rituals	Intercropping, drought-resistant seed varieties
Igbo (Southeastern Nigeria)	Tropical rainforest	Yam, cassava, plantain, and palm oil	Swidden agriculture, gendered farming roles	Staples are energy-rich, and yams are tied to social status and festivals	Mixed cropping systems, agroforestry integration
Ainu (Japan)	Temperate forest and riverine areas	Salmon, deer, millet, and wild roots	Seasonal hunting, fishing, and gathering	Proteins and essential fats, and rituals around food	Ritual ecology, rotational resource use

Indigenous Group/Region	Ecological Context	Primary Food Staples	Methods of Procurement	Nutritional and Cultural Value	Adaptive Features/ Resilience Mechanisms
Kānaka Maoli (Hawai'i)	Volcanic islands and marine-coastal	Taro (kalo), breadfruit, and fish	Loʻi (irrigated terrace farming), fishpond aquaculture	Staple starch, taro considered an ancestor, food as medicine	Biocultural restoration, water management infrastructure
Tibetan Plateau Communities	Cold alpine region	Barley (tsampa), yak meat, and butter tea	Animal husbandry, terrace farming	Calorically dense and central to Buddhist customs	Adaptations to low oxygen, rotational grazing
Maya (Central America)	Tropical forest	Maize, beans, chili, and squash	Milpa (shifting agriculture), agroforestry	Balanced amino acid profile (maize–bean combo) and sacred calendar foods	Soil fertility cycling, polyculture systems
Batak (Philippines)	Mountain–forest ecosystem	Root crops, bananas, and forest game	Slash-and-burn cultivation, hunting, and gathering	Food is closely tied to spiritual beliefs and ecology	Knowledge of forest cycles, flexible farming
Zulu (South Africa)	Subtropical grasslands	Sorghum, pumpkins, and wild greens	Mixed farming, cattle herding	Key for ceremonial and social gatherings	Indigenous drought forecasting, communal labor systems

revitalize traditional food systems to improve dietary quality and reinvigorate cultural practices and identities. Food sovereignty is vital when discussing Indigenous diets, as it encompasses the rights of Indigenous communities to maintain control over their food systems. Indigenous groups frequently encounter structural inequalities that restrict their access to traditional foods, raising concerns about autonomy and self-determination. The literature indicates that food sovereignty is linked to cultural identity and collective health outcomes; therefore, empowering Indigenous populations to manage and engage with their food resources is essential for ensuring food security and health (Harding & Oetzel, 2019; Livingstone et al., 2023). Adopting an evolutionary perspective on diets provides a framework for understanding and addressing health issues. Basile et al. (2018) emphasize the benefits of recognizing the evolutionary context of human diets, particularly regarding how modern dietary patterns diverge from those consumed historically during human evolution. The mismatch hypothesis suggests that contemporary diets rich in processed foods are harmful because they do not align with the genetic predispositions shaped by ancestral eating habits. To Indigenous diets, this evolutionary perspective underscores the importance of traditional culinary practices sustained over generations. Incorporating foods that align with human nutritional needs can be beneficial in combating the epidemic of chronic diseases faced by many Indigenous populations today.

Evolutionary nutrition is an interdisciplinary framework that examines the relationship between human evolution, dietary patterns, and health outcomes, emphasizing the mismatch between ancestral diets and modern eating behaviors. This perspective is grounded in the understanding that humans have evolved in specific environments, leading to adaptations that favor certain food types prevalent in ancestral diets. The relevance of this paradigm to Indigenous diets is pronounced, clarifying how traditional eating practices have developed over centuries of natural selection and cultural evolution, contributing to the health and resilience of Indigenous communities (Akinola et al., 2020; Gluckman et al., 2011). Indigenous dietary patterns often reflect the ecological and geographical contexts in which these communities have lived. Many traditional Indigenous foods are nutrient-dense, offering beneficial components usually missing in contemporary diets that rely on processed foods, as highlighted by Akinola et al. (2020), who emphasize the rich nutritional profiles of Indigenous food crops compared to nonnative alternatives. These foods, recognized for their contributions to health and sustainability, underscore the evolutionary importance of maintaining traditional food systems to preserve health and nutritional security in Indigenous contexts. Natural selection has played a pivotal role in shaping human dietary habits. As early humans adapted to diverse environments, dietary choices were influenced by the availability of local resources and the nutritional needs that arose in different contexts. The work of Carrera-Bastos et al. (2011) supports this notion, explaining that optimal gene expression, which impacts health spans, can often be linked to dietary practices aligned with our evolutionary history. Natural selection favored whole foods, including fruits, vegetables, nuts, seeds, and lean proteins, which have been integral to human dietary evolution.

Environmental pressures, such as climate variability and resource availability, have driven adaptations in dietary practices among Indigenous populations. For instance, reliance on seasonal and locally sourced foods reflects an understanding of nutrition and a strong awareness of the need for sustainability. Govender et al. (2016) further illustrate this by highlighting the significance of Indigenous crops in diversifying diets and addressing nutritional gaps among economically disadvantaged populations. Thus, integrating ecological wisdom into food practices has been crucial for fostering food security, resilience, and community health. Sociocultural evolution has also been pivotal in reinforcing dietary traditions and transmitting food knowledge within Indigenous communities (Ogwu et al., 2025). Riediger et al. (2021) discuss that food choices and dietary patterns among Indigenous children and youth have been influenced by familial and community contexts and sociopolitical structures that impact food security and health access. The persistence of traditional food customs, even amid external pressures and socioeconomic changes, emphasizes the importance of cultural heritage and identity in sustaining robust dietary systems.

Moreover, the loss of traditional food sources has been shown to contribute to adverse health outcomes in Indigenous populations, such as increased obesity and diabetes rates, highlighting the disruptions caused by modernization and colonial legacies. McCartan et al. (2020) emphasize that Indigenous communities, despite their deep historical reliance on traditional diets, face challenges as their energy intake increasingly shifts toward processed foods due to the effects of colonization and the loss of sovereignty over food systems. This shift illustrates the critical need for policies and interventions that facilitate access to traditional foods and incorporate Indigenous knowledge systems to address health disparities (Akudugu & Ogwu, 2024; Kosoe et al., 2023). As such, evolutionary nutrition provides a significant lens to understand the importance of restoring Indigenous food systems that honor ecological wisdom and cultural practices. It emphasizes the fundamental connection between health, diet, and identity among Indigenous peoples, revealing pathways toward enhanced well-being through a reconnection to ancestral dietary patterns (Abdul et al., 2023; Coté, 2016). Effective public health nutrition interventions for Indigenous populations must be culturally sensitive and aligned with traditional practices. Research shows that culturally adapted interventions have shown promise in improving dietary behaviors among Indigenous groups (Harding & Oetzel, 2019; Livingstone et al., 2023). These interventions promote the incorporation of traditional foods, community knowledge, and cultural beliefs into health programs. Such a focus addresses dietary deficiencies while fostering community engagement and revitalizing Indigenous cultural practices, promoting well-being through the revival of traditional culinary arts (Oladeji et al., 2024).

This chapter explores the historical and adaptive evolution of Indigenous diets and food systems through the lens of ecological, cultural, and biological coevolution. It investigates how Indigenous communities have developed sustainable food practices finely attuned to their environments, spiritual worldviews, and health needs over millennia. By tracing the journey from foraging and early cultivation to contemporary adaptations and revivals, the chapter highlights the resilience and ingenuity inherent in Indigenous foodways. It also examines the complex interplay between traditional ecological knowledge, nutritional strategies, and cultural identity, emphasizing the importance of food as both sustenance and symbolism. The contribution of this chapter lies in its interdisciplinary approach— bridging anthropology, evolutionary biology, nutrition science, and Indigenous studies to offer a holistic understanding of food heritage. It underscores how Indigenous food systems serve as repositories of cultural memory and models for sustainable diets and climate-resilient agriculture. The chapter critically engages with historical patterns and modern transformations and contributes to ongoing dialogues on food sovereignty, biodiversity conservation, and health equity in Indigenous and global contexts.

3.2 Human Evolution and Dietary Adaptation

3.2.1 Early Hominins and the Foraging Niche

The dietary practices of early hominins have increasingly been recognized as fundamental components in the trajectory of human evolution. The foraging niche, characterized predominantly by hunter-gatherer subsistence models, illustrates the adaptive strategies employed by early human populations in diverse environments. Research has extensively documented how dietary diversity was crucial for early human survival, innovation, and cognitive development. Hunter-gatherer societies depended on an intricate understanding of their ecological surroundings while exhibiting remarkable flexibility in their diet and resource exploitation. The hunter-gatherer subsistence model encompasses a range of strategies that extend from foraging for fruits, nuts, and tubers to hunting animals, indicating a versatile dietary approach. These subsistence strategies allowed early hominins, such as those from the genera Homo and Australopithecus, to exploit various ecological niches effectively. Research by Sponheimer et al. (2013) has shown evidence of dietary resourcefulness in early hominins through stable isotopic analysis, revealing varied consumption patterns consistent with changes in environment and available resources. The diversity of diet contributed to the nutritional intake necessary for metabolic processes underpinning core evolutionary changes, including increased brain size and complexity (Sterelny, 2007).

Furthermore, the social organization among early hominins created an enabling framework for cooperative foraging. Multiple studies, such as that of Kaplan et al. (2000), emphasize that social structures and relationships built around foraging could have significantly influenced population dynamics and the transmission of knowledge across generations, thereby facilitating cultural evolution rooted in shared dietary practices. This interaction between sociality and diet under-scores the significance of cooperative behavior and strategy in the success of early hominins. As hunter-gatherers, populations depended on food availability and diverse foraging techniques designed to maximize caloric intake and ensure food security amid environmental changes. Research highlights that variability in seasonal food sources compelled early humans to adapt their hunting and gathering tactics, resulting in adaptive strategies that were both ecological and social in nature (Szilágyi et al., 2023). This foraging adaptability is crucial for understanding the cultural complexities that emerged as hominins established regional settlements and social groupings, further reinforcing the necessity of collaboration for survival.

3.2.2 Evolution of Digestive Traits and Metabolism

The evolutionary trajectory of *Homo sapiens* has been intricately linked to significant adaptations in digestive traits and metabolic efficiency. Notable adaptations, such as lactose tolerance and the duplication of amylase genes, are directly correlated with dietary shifts that arose alongside changes in subsis-tence strategies, particularly following the advent of agriculture (Migliano & Vinicius, 2021). The gene responsible for lactase persistence in adult humans is a model example showcasing the significant impact of dietary practices on genetic evolution. Populations that historically consumed dairy products, primarily in pastoralist societies, demonstrated a genetic adaptation that permitted the breakdown of lactose into adulthood, thus ensuring a continued source of nutrition from milk (Carlson & Kingston, 2007). Furthermore, gene duplications for amylase, the enzyme responsible for starch digestion, indicate a biological response to incorporating higher starch diets stemming from agriculture (Berger et al., 2015). As early humans began cultivating cereal grains, those with multiple copies of the amylase gene likely had a nutritional advantage in digesting these new dietary staples. This evolutionary trait reflects the dynamic relationship between food production practices and human physiological adaptability. Such selective pressure underscores the critical role of diet in shaping metabolic pathways and steering human evolutionary history toward greater resilience and adaptability to varied diets.

In addition to genetic adaptations, the role of the gut microbiome has become increasingly recognized as an essential factor in human evolutionary processes. Coevolution between the human gut microbiome and dietary selections has significant implications for health and digestion (Proffitt et al., 2023). With each dietary shift, the composition and functionality of gut microbiota adapt, optimizing nutrient absorption and metabolic processes essential for supporting the human brain's energy demands. A study by David et al. indicates that human-associated microbiomes differ notably in composition based on dietary habits, further demonstrating that the evolution of digestible foods coevolved with our microbial partners (Cornélio et al., 2016). This intimate relationship between diet and the gut microbiome holistically reveals fundamental insights into how diet has aided in the evolutionary path of Homo sapiens, enabling adaptability not just through genetic changes but also via microbial partnerships promoting health and efficiency in nutrient uptake.

3.2.3 Dietary Patterns and Migration

The interplay between migration, climatic shifts, and dietary patterns has been a fundamental theme in the evolution of human diets. Migration events, spurred by environmental changes and resource availability, forced hominins to adapt their subsistence strategies to cope with new ecological challenges. According to anthropological studies, climate change influenced food availability, compelling early human groups to modify their dietary practices based on local resource accessibility (Lee-Thorp & Sponheimer, 2006). Isotopic analysis of archeological remains provides persuasive evidence of how dietary adaptations mirrored significant climatic events. For instance, research has uncovered fluctuations in ecological conditions due to glacial and interglacial periods that profoundly impacted the types and availability of food sources for early humans (Domínguez-Rodrigo et al., 2021). As demonstrated by studies on hominin populations in diverse biogeographic regions, such as the Arctic and tropical environments, the adaptation of dietary practices showcases the versatility of early humans to exploit diverse foodways based on environmental exigencies (Rodríguez-Hidalgo et al., 2015). The stark contrast in diet observed between Arctic and tropical populations illustrates distinct adaptations that arose through migration and ecological variation. Arctic populations developed diets rich in high-fat seals and fish to cope with resource scarcity during long winters, while tropical populations often utilized a variety of fruits, nuts, and visible plant resources readily available in their environment (Ogwu, 2024; Wauchope et al., 2016). These contrasting dietary practices underscore the broader theme of biogeographical influences in shaping human diets, as regions dictated foodways that were both nutritional and essential for survival.

3.3 Transition from Foraging to Farming in Indigenous Societies

3.3.1 Independent Centers of Plant and Animal Domestication

The transition from foraging to farming represents one of the most significant shifts in human history, characterized by the domestication of various plants and animals across different regions. Notable independent centers of plant domestication include the cultivation of maize in Mesoamerica, potatoes in the Andean region, and yams in West Africa. Mesoamerican societies particularly developed diverse strains of maize that became a staple of their diet, reflecting a long history of cultivation and selective breeding (Gepts et al., 2012; Ogwu, 2020; Smith, 2005). Similarly, the Andean potato has been cultivated for thousands of years, leading to a multitude of varieties adapted to various climatic conditions (Gepts et al., 2012). In West Africa, yam cultivation has been crucial for food security and has also acquired strong cultural significance, illustrating the diverse agricultural practices tailored to regional environments (Degla & Sourokou, 2020; Tamiru et al., 2007). These independent centers highlight the unique pathways through which different societies transitioned from foraging lifestyles to complex agricultural societies.

3.3.2 Coevolution of Crops and Cultural Practices

The domestication of plants did not occur in isolation; rather, it was intertwined with evolving cultural practices that shaped agricultural methods. For instance, the selective breeding of maize in Mesoamerica was driven by the demand for nutritional value, taste, and storability, allowing communities to secure food resources over extended periods (Fuller et al., 2010; Milla et al., 2015). This coevolution extended to determining the cultural meanings ascribed to these staple crops, with yam holding a sacred status in West African societies, embodying economic and sociocultural values (Degla & Sourokou, 2020). As these crops were bred for desirable traits, they influenced social interactions, agricultural techniques, and even religious practices, illustrating the complex relationships between human societies and their cultivated resources. Thus, the domestication process can be seen as a biological and cultural evolution, where agricultural advancements were essential for developing societal structures and cultural identity.

3.3.3 Agroecological Knowledge and Sustainability

The shift from foraging to agriculture led to significant advancements in agroecological knowledge, which is essential for sustainable farming practices. Traditional methods such as soil fertility management, polyculture, fallowing, and water management emerged as key strategies for maintaining soil health and maximizing crop yields (Larson et al., 2014; Zhang et al., 2021). These practices can be seen in cultivating diverse crops, enabling Indigenous societies to sustain ecological balance while supporting food production systems (Fuller et al., 2010). For instance, using yam in polycultural systems has shown a strong resilience against pests and diseases, bolstered by farmers' intrinsic knowledge of their ecological environments (Tamiru et al., 2007). Additionally, agroecological knowledge includes a comprehensive understanding of the interactions among crops, soil, climate, and cultural practices, resulting in sustainable agricultural systems that have endured through generations.

3.4 Nutritional Ecology of Indigenous Diets

3.4.1 Macronutrient Balance and Micronutrient Density

Indigenous diets are often characterized by a balanced intake of complex carbohydrates, healthy fats, and proteins from diverse sources, contrasting with the processed, high-glycemic-index diets in modern societies. Traditional Indigenous diets typically feature nutrient-dense foods that are low in calories and high in fiber, contributing to improved health outcomes. For instance, consuming traditional foods has been associated with increased dietary diversity, which is crucial for achieving adequate micronutrient intake and enhancing overall health (Kasimba et al., 2017; Sidiq et al., 2022). Recent studies highlight that traditional foods can offer significant health benefits and improve food security by providing essential nutrients that are often lacking in modern diets dominated by processed foods (Ghosh-Jerath et al., 2020; Sidiq et al., 2022). Moreover, modern diets often lead to deficiencies in essential vitamins and minerals due to reliance on high-caloric, low-nutrient foods. Indigenous foods, such as tubers, greens, and animal-based proteins, supply essential macronutrients and a rich array of micronutrients that promote health and prevent chronic diseases (Phillips et al., 2014). This emphasizes the urgent need for reintroducing and promoting Indigenous dietary practices to combat the rising incidences of obesity and noncommunicable diseases among Indigenous populations worldwide (Brimblecombe et al., 2013; Sidiq et al., 2022).

3.4.2 Role of Wild Foods and Seasonal Diets

Wild foods play an indispensable role in the nutrition of Indigenous communities, as they are often more nutrient-dense than domesticated varieties. The gathering of wild foods, such as greens, tubers, berries, game meat, and insects, significantly contributes to the dietary intake and food sovereignty of these communities (Banna & Bersamin, 2018; Kagie et al., 2019). Seasonal variation in dietary choices is a hallmark of Indigenous food systems, which align with local ecosystems and the availability of natural resources. This practice ensures a varied diet rich in essential nutrients while supporting sustainable harvesting practices that promote biodiversity and ecological balance (Ghosh-Jerath et al., 2021; Schembri et al., 2016). Research indicates that the consumption of wild foods by Indigenous peoples correlates with better nutritional outcomes and food security (Banna & Bersamin, 2018; Sidiq et al., 2022). For instance, communities that prioritize seasonal foraging often experience lower rates of dietary deficiencies, showcasing their reliance on diverse food sources resilient to climate variations and agricultural pressures (Kagie et al., 2019; Kapoor et al., 2024).

3.4.3 Fermentation, Preservation, and Cooking Methods

Fermentation, preservation, and traditional cooking methods are vital practices in Indigenous food systems that enhance the nutritional quality of foods. These methods increase the bioavailability of nutrients while reducing the presence of antinutrients that can hinder nutrient absorption. For instance, fermentation techniques break down complex compounds, improving the digestibility and nutrient composition of Indigenous foods like tubers and vegetables (Irakoze, 2024; Owade et al., 2019). Additionally, these methods can prolong the shelf life of foods, ensuring a steady supply during lean seasons (Kasimba et al., 2017; Owade et al., 2019). Research supports that traditional cooking and preservation methods enhance nutrient intake while perpetuating cultural practices and community health. The successful integration of fermentation and preservation into Indigenous food systems exemplifies their role in promoting food security and public health (Ellena & Nongkynrih, 2017; Irakoze, 2024; Kagie et al., 2019). As Indigenous communities navigate modern dietary challenges, these traditional methods provide a resilient framework for maintaining nutritional health amid globalization and dietary shifts (Lee & Lewis, 2018; Schembri et al., 2016).

3.5 Cultural Evolution of Dietary Norms and Taboos

3.5.1 Food Taboos and Sociobiological Functions

Food taboos serve various sociobiological functions, including influences on reproductive health, ecological conservation, and social cohesion within communities. For instance, certain dietary restrictions can enhance reproductive health by avoiding foods that are considered harmful during pregnancy (Teixidor-Toneu et al., 2021). Additionally, food taboos often act as ecological conservation mechanisms; by prohibiting the consumption of specific fauna or flora, communities may ensure the sustainable use of local resources, preserving biodiversity for future generations (Middleton et al., 2024). These norms not only promote individual well-being but also strengthen group cohesion, as shared taboos help forge a collective identity, reinforcing social bonds among members (Ibrahim & Howarth, 2016).

3.5.2 Ritualized Consumption and Food Symbolism

The consumption of food within ritual contexts plays a significant role in social and cultural practices, translating into rich symbolic meanings during lifecycle ceremonies such as births, weddings, and funerals. Rituals involving food, such as communal feasts or fasting practices, serve as tools for social cohesion and identity reinforcement (Moscato & Ozanne, 2019). For example, the preparation and sharing of traditional foods during feasts can symbolize community unity and cultural heritage (Shokeran et al., 2023). Furthermore, the act of eating itself often transcends basic nourishment, intertwining with sacredness and community values, as seen in various cultures' lifecycle events (THM, 2023). Ritualized behaviors surrounding food consumption can elevate the importance of certain food items, thereby enhancing their perceived value and significance within the cultural narrative (Nurbaya, 2023).

3.5.3 Gendered Food Roles in Knowledge Transmission

The transmission of dietary knowledge and practices often occurs along gender lines, with women frequently taking on the role of nutritional gatekeepers and knowledge custodians. This gendered division of labor is evident in various contexts where women are primarily responsible for food procurement and preparation, resulting in greater nutritional knowledge (Middleton et al., 2024). Research indicates that women are more likely to pass down medicinal knowledge related to food and health to their female offspring, thereby reinforcing traditional practices and ensuring the continuity of herbal and dietary wisdom across generations (Montanari & Teixidor-Toneu, 2021; Ogwu et al., 2023; Torres-Avilez et al., 2019). In contrast, men in many societies often participate in decision-making roles concerning resources, which influences the availability of food and nutrition without necessarily possessing the same depth of knowledge about its preparation and implications for health (Achiro et al., 2023).

3.6 Microbiome and Coevolutionary Diets

The gut microbiota plays a crucial role in health and disease, particularly regarding the comparative diversity observed between Indigenous and industrialized populations. Indigenous populations generally exhibit a richer and more diverse gut microbiota compared to their industrialized counterparts, which often contain less microbial diversity due to lifestyle changes, dietary habits, and antibiotic exposure. For instance, findings from studies indicate that traditional populations harbor unique taxa such as *Prevotella* and *Treponema*, which are associated with healthier metabolic patterns, while industrialized populations show higher levels of *Bacteroides* and *Bifidobacterium* linked to obesity and metabolic diseases (Cuesta-Zuluaga et al., 2018). Moreover, these variations contribute to understanding metabolic pathways that relate gut microbiota composition to health outcomes such as obesity and chronic diseases, emphasizing the importance of dietary patterns in microbiome configuration (Gautam et al., 2024). The health implications of gut microbiota diversity are significant. A diverse gut microbiota is associated with enhanced immune function and resilience against chronic diseases such as obesity, diabetes, and cardiovascular disorders (Walsh et al., 2023). Research indicates that microbial diversity contributes to the modulation of inflammation and metabolic pathways, suggesting that higher diversity can protect against dysbiosis, which is often observed in industrialized nations where diets are heavily processed (Cuesta-Zuluaga et al., 2018). Therefore, understanding these differences is crucial for developing dietary interventions aimed at restoring a healthy microbiota composition, particularly by integrating aspects of traditional diets into modern nutrition.

In terms of fermented foods and their role in promoting microbial health, items like yogurt and traditional fermented beverages are central. The fermentation process enriches foods with diverse microbial communities, which can enhance gut health through various routes. For instance, the metabolites produced during fermentation—such as organic acids, vitamins, and bacteriocins—exhibit probiotic properties that foster microbial balance in the gut (Chileshe et al., 2020; Nagarajan et al., 2022; Obahiagbon & Ogwu, 2024; Ogwu et al., 2024). Studies suggest that fermented foods improve digestion and contribute to the prevention of various chronic diseases by influencing gut microbiota composition and functionality ("Fermented and Microbial Foods," 2015). The presence of beneficial metabolites in these foods plays a pivotal role in their health-promoting potential; for instance, lactic acid bacteria present in yogurt can alleviate gastrointestinal discomfort and enhance the body's immune responses (Ashaolu, 2019).

Additionally, a meta-analysis has shown that traditional fermented foods, regarded as repositories of microbial diversity, have the potential to facilitate beneficial changes in the gut microbiome (Xu et al., 2022). The variation in microbial communities across different fermented foods indicates substrate-specific effects on health, underscoring the importance of dietary patterns linked to cultural practices (Walsh et al., 2023). Thus, consuming fermented foods can help address the microbiota diversity gaps often seen in industrialized diets, promoting overall gut health and well-being. Consequently, the relationship between gut microbiota diversity in Indigenous populations compared to industrialized individuals, along with the functional attributes of fermented foods, emphasizes significant implications for health and disease prevention. Establishing dietary patterns that emphasize fermented foods could enhance our understanding of microbial health and alleviate the consequences associated with reduced micro-biome diversity in modern diets.

3.7 Colonization, Dietary Disruption, and Evolutionary Mismatch

The introduction of processed foods and monoculture crops has significantly disrupted traditional dietary patterns among Indigenous populations, leading to various health issues, including nutritional deficiencies and metabolic disorders. The shift toward ultra-processed food consumption has contributed to a decline in local food systems. Monteiro et al. (2013) illustrate that ultra-processed prod-ucts now dominate global food systems, characterized by higher energy density, unhealthy fats, added sugars, and salt compared to their minimally processed counterparts. This widespread transition is particularly alarming as dietary pat-terns have increasingly been linked to the prevalence of chronic diseases such as obesity and cardiovascular diseases among Indigenous populations in North America and Australia. Epidemiological studies have highlighted the adverse health outcomes associated with these dietary changes. For instance, rates of cardiovascular disease and diabetes have surged among Native American and Aboriginal populations, often correlating with the introduction of Western diets high in ultra-processed foods (Browne et al., 2021; Sherriff et al., 2019). The prevalence of obesity is particularly concerning within these communities, especially among children, as reported by Thurber et al., which underscores the importance of addressing dietary behaviors from an early age to mitigate these trends (Thurber et al., 2017). Sayers et al. indicate that low birth weights and subsequent overweight issues may pose a risk for chronic diseases in Ab-original cohorts, emphasizing the compounded effects of dietary transitions on long-term health outcomes (Sayers et al., 2013). Cultural displacement due to colonization has also led to significant loss of food sovereignty and traditional ecological knowledge, which are critical for maintaining Indigenous health. The Aboriginal Rethink Sugary Drink media campaign highlights that the arrival of colonial systems disrupted traditional food practices and restructured the socio-economic dynamics that had supported Indigenous food systems for millennia (Browne et al., 2021). The erosion of traditional knowledge directly affects food procurement and dietary options, subsequently diminishing the overall health of these communities (Browne et al., 2021; Ng et al., 2011). Moreover, significant food insecurity, often accompanied by obesity in Aboriginal settings, illustrates the complexity of these interrelations and underscores the influence of socioeconomic factors on dietary habits and health (Ng et al., 2011; Sherriff et al., 2019).

3.8 Resurgence of Indigenous Diets and Food Sovereignty

The resurgence of Indigenous diets and the focus on food sovereignty represent significant movements to restore cultural heritage and enhance resilience in the face of climate and health challenges. These movements incorporate traditional practices, community empowerment, and integration of innovative Indigenous and scientific knowledge systems.

3.8.1 Return-to-Tradition Movements

The return-to-tradition movements, which encompass local food revivals, slow food initiatives, seed banks, and culinary tourism, play a crucial role in reestablishing Indigenous diets. These movements aim to revive traditional agricultural practices and promote greater reliance on local biodiversity and culturally relevant food options (Frank & Durden, 2017). For instance, in rural areas, outreach programs that highlight the advantages of traditional diets—characterized by a lower dependence on meat and processed foods—can positively impact public health by reducing the risks of diseases such as Type II diabetes (Frank & Durden, 2017). Additionally, culinary tourism and the creation of seed banks not only support local economies but also reinforce cultural identities linked to food, enabling communities to reclaim their culinary heritage (Coté, 2016).

3.8.2 Integration of Indigenous and Scientific Knowledge

The intersection of Indigenous knowledge and scientific research has gained traction, particularly in the nutritional study of various traditional crops like teff, fonio, and quinoa. These crops, often classified as "orphan crops," hold potential for climate resilience and food security due to their adaptability to harsh environmental conditions (Figueroa-Helland et al., 2018; Mabhaudhi et al., 2019). Research indicates that the inclusion of Indigenous agricultural practices and crop varieties in scientific studies can enhance biodiversity and strengthen food systems against the backdrop of climate change (Ma & Rahut, 2024; Noort et al.,

2022). As Indigenous communities assert their rights and deepen connections with their ancestral lands, they employ traditional ecological knowledge alongside modern scientific methods to build more resilient food systems (Joseph et al., 2022; Schneider, 2022).

3.8.3 Role in Resilience to Climate and Health Crises

Indigenous diets and practices significantly contribute to the resilience of communities facing climate and health crises. Traditional food systems inherently promote sustainability, as they often utilize local resources and optimize agricultural practices suited to the local environment (Ogwu, 2023; Sayre et al., 2017). The adaptive potential of these systems enhances not only food security but also community health, offering natural solutions to mitigate the impacts of climate change (Favas et al., 2024; Naik et al., 2024). Additionally, the emphasis on agroecological methods, including crop diversification and sustainable land management practices, strengthens the resilience of food systems, empowering communities to better navigate environmental uncertainties (Gifawesen et al., 2020; Ma & Rahut, 2024). Moreover, traditional food practices serve as a means of social cohesion and cultural identity, reinforcing communities against the homogenizing forces of globalization while fostering a sense of belonging and purpose within health and ecological contexts (Coté, 2016; Schneider, 2022). The collective management of traditional landscapes, as seen in various Indigenous-led projects, illustrates how food sovereignty can mobilize communities toward effective adaptation and resilience strategies in the face of multiple crises (Farooq et al., 2019; Sayre et al., 2017).

3.9 Conclusion

Exploring Indigenous diets and food systems from an evolutionary perspective highlights the critical intersections between nutrition, culture, and health. Recognizing the historical context and ongoing challenges Indigenous communities face is essential for promoting health equity and restoring traditional food practices. Future research should prioritize collaboration with Indigenous communities to revitalize traditional food systems, ensuring that interventions are culturally relevant and effectively address the unique dietary needs of these populations. In conclusion, human evolution is deeply rooted in a complex interplay between dietary adaptation and environmental pressures. The adaptability inherent in hunter-gatherer subsistence models illustrates the vital role of social organization and cooperation, while genetic adaptations emphasize the long-standing evolutionary implications of dietary changes. Moreover, migration and climate-induced shifts pushed early humans into diverse ecosystems, creating a dynamic relationship with food sources that significantly impacted their evolution. Together, these themes reflect the intricate and multifaceted narrative of human dietary adaptation, fundamental to our species' success and longevity.

The transition from foraging to farming is a complex process characterized by the independent domestication of plants and animals, the coevolution of crops with cultural practices, and the development of sustainable agroecological knowledge. Understanding these dynamics offers insight into the formation of complex Indigenous societies, illustrating that agriculture was not merely a technological advancement but also a significant cultural evolution that shaped human history over millennia. The resurgence of Indigenous diets and the movement toward food sovereignty are deeply connected to cultural revitalization and the pursuit of sustainable and resilient food systems. These dynamics not only celebrate traditional practices but also provide practical solutions to contemporary challenges posed by climate change and public health. The dietary disruptions caused by colonization and the influx of ultra-processed foods have triggered an epidemiological transition marked by rising rates of metabolic disorders among Indigenous populations. This interplay of cultural displacement and nutritional change presents serious risks to health and well-being, underscoring the urgent need for culturally sensitive interventions to restore food sovereignty and traditional dietary practices.

Chapter Reflection

1. How does your current diet align—or clash—with the diets humans evolved to eat?
2. In what ways do traditional foods act as carriers of cultural identity and intergenerational memory?
3. What do adaptations like lactase persistence or amylase gene duplication reveal about the relationship between food and evolution?
4. How has colonization reshaped Indigenous foodways, and what health consequences have followed?
5. What can the rich microbiomes of Indigenous communities teach us about modern dietary health?

6.   How do rituals and taboos around food reflect deeper ecological or spiritual knowledge?
7.   How do Indigenous farming practices offer solutions to today's climate and food crises?
8.   How is food knowledge passed down—and what's lost when this chain breaks?
9.   What does the "evolutionary mismatch" teach us about chronic diseases in Indigenous communities?
10. How can reviving Indigenous foodways be a path toward sovereignty, sustainability, and healing?

REFERENCES

Abdul, M., Ingabire, A., Lam, C., Bennett, B., Menzel, K., MacKenzie-Shalders, K., & van Herwerden, L. (2023). Indigenous food sovereignty assessment—A systematic literature review. *Nutrition and Dietetics*, *81*(1), 12–27. https://doi. org/10.1111/1747-0080.12813

Achiro, E., Okidi, L., Echodu, R., Alarakol, S., Nassanga, P., & Ongeng, D. (2023). Status of food safety knowledge, attitude, and practices of caregivers of children in Northern Uganda. *Food Science and Nutrition*, *11*(9), 5472–5491. https://doi.org/10.1002/fsn3.3504

Akinola, R., Pereira, L., Mabhaudhi, T., Bruin, F., & Rusch, L. (2020). A review of Indigenous food crops in Africa and the implications for more sustainable and healthy food systems. *Sustainability*, *12*(8), 3493. https://doi.org/10.3390/su12083493

Akudugu, M. A., & Ogwu, M. C. (2024). Sustainable development policies and interventions: A bibliometric analysis of the contributions of the academic community. *Journal of Cleaner Production*, *434*, 139919. http://doi.org/10.1016/j. jclepro.2023.139919

Ashaolu, T. (2019). A review on selection of fermentative microorganisms for functional foods and beverages: The production and future perspectives. *International Journal of Food Science and Technology*, *54*(8), 2511–2519. https:// doi.org/10.1111/ijfs.14181

Auger, M. (2016). Cultural continuity as a determinant of Indigenous peoples' health: A metasynthesis of qualitative research in Canada and the United States. *International Indigenous Policy Journal*, *7*(4). https://doi.org/10.18584/iipj.2016.7.4.3

Banna, J., & Bersamin, A. (2018). Community involvement in design, implementation and evaluation of nutrition interventions to reduce chronic diseases in Indigenous populations in the U.S.: A systematic review. *International Journal for Equity in Health, 17*(1), 116. https://doi.org/10.1186/s12939-018-0829-6

Basile, A., Schwartz, D., Rigdon, J., & Stapell, H. (2018). Status of evolutionary medicine within the field of nutrition and dietetics. *Evolution Medicine and Public Health, 2018*(1), 201–210. https://doi.org/10.1093/emph/eoy022

Berger, L. R., Hawks, J., de Ruiter, D. J., Churchill, S. E., Schmid, P., Delezene, L. K., Kivell, T. L., Garvin, H. M., Williams, S. A., DeSilva, J. M., Skinner, M. M., Musiba, C. M., Cameron, N., Holliday, T. W., Harcourt-Smith, W., Ackermann, R. R., Bastir, M., Bogin, B., Bolter, D., & Zipfel, B. (2015). Homo naledi, a new species of the genus homo from the Dinaledi Chamber, South Africa. *Elife, 4*. https://doi.org/10.7554/elife.09560

Brimblecombe, J., Ferguson, M., Liberato, S., & O'Dea, K. (2013). Characteristics of the community-level diet of aboriginal people in remote northern Australia. *The Medical Journal of Australia, 198*(7), 380–384. https://doi.org/10.5694/mja12.11407

Browne, J., MacDonald, C., Egan, M., Carville, K., Delbridge, R., & Backholer, K. (2021). Relevance of the Aboriginal rethink sugary drink media campaign to Aboriginal and non-Aboriginal audiences in regional Victoria. *Australian and New Zealand Journal of Public Health, 45*(3), 263–269. https://doi.org/10.1111/1753-6405.13086

Carlson, B., & Kingston, J. (2007). Docosahexaenoic acid biosynthesis and dietary contingency: Encephalization without aquatic constraint. *American Journal of Human Biology, 19*(4), 585–588. https://doi.org/10.1002/ajhb.20683

Carrera-Bastos, P. (2011). The Western diet and lifestyle and diseases of civilization. *Research Reports in Clinical Cardiology, 15*, 15–35. https://doi.org/10.2147/rrcc.s16919

Chileshe, J., Heuvel, J., Handema, R., Zwaan, B., Talsma, E., & Schoustra, S. (2020). Nutritional composition and microbial communities of two non-alcoholic traditional fermented beverages from Zambia: A study of Mabisi and Munkoyo. *Nutrients, 12*(6), 1628. https://doi.org/10.3390/nu12061628

Cornélio, A., Bittencourt-Navarrete, R., Brum, R., Queiroz, C., & Costa, M. (2016). Human brain expansion during evolution is independent of fire control and cooking. *Frontiers in Neuroscience, 10*. https://doi.org/10.3389/fnins.2016.00167

Coté, C. (2016). "Indigenizing" food sovereignty revitalizing Indigenous food practices and ecological knowledges in Canada and the United States. *Humanities*, *5*(3), 57. https://doi.org/10.3390/h5030057

Cuesta-Zuluaga, J., Corrales-Agudelo, V., Velásquez-Mejía, E., Carmona, J., Abad, J., & Escobar, J. (2018). Gut microbiota is associated with obesity and cardiometabolic disease in a population in the midst of westernization. *Scientific Reports*, *8*(1). https://doi.org/10.1038/s41598-018-29687-x

Degla, P., & Sourokou, N. (2020). Food, socio-cultural and economic importance of yam in the north-east of Benin. *Journal of Agriculture and Environmental Sciences*, *9*(2). https://doi.org/10.15640/jaes.v9n2a12

Domínguez-Rodrigo, M., Courtenay, L., Cobo-Sánchez, L., Baquedano, E., & Mabulla, A. (2021). A case of hominin scavenging 1.84 million years ago from Olduvai Gorge (Tanzania). *Annals of the New York Academy of Sciences*, *1510*(1), 121–131. https://doi.org/10.1111/nyas.14727

Ellena, R., & Nongkynrih, K. (2017). Changing gender roles and relations in food provisioning among matrilineal Khasi and patrilineal Chakhesang Indigenous rural people of North-East India. *Maternal and Child Nutrition*, *13*(S3), e12560. https://doi.org/10.1111/mcn.12560

Farooq, M., Rehman, A., & Pisante, M. (2019). Sustainable agriculture and food security. In M. Farooq & M. Pisante (Eds.), *Innovations in sustainable agriculture* (pp. 3–24). Springer. https://doi.org/10.1007/978-3-030-23169-9_1

Favas, C., Cresta, C., Whelan, E., Smith, K., Manger, M., Chandrasenage, D., Singhkumarwong, A., Kawasaki, J., Moreno, S., & Goudet, S. (2024). Exploring food system resilience to the global polycrisis in six Asian countries. *Frontiers in Nutrition*, *11*. https://doi.org/10.3389/fnut.2024.1347186

Figueroa-Helland, L., Thomas, C., & Aguilera, A. (2018). Decolonizing food systems: Food sovereignty, Indigenous revitalization, and agroecology as counter-hegemonic movements. *Perspectives on Global Development and Technology*, *17*(1–2), 173–201. https://doi.org/10.1163/15691497-12341473

Frank, S., & Durden, T. (2017). Two approaches, one problem: Cultural constructions of type ii diabetes in an Indigenous community in Yucatán, Mexico. *Social Science and Medicine*, *172*, 64–71. https://doi.org/10.1016/j.socscimed.2016.11.024

Fuller, D., Allaby, R., & Stevens, C. (2010). Domestication as innovation: The entanglement of techniques, technology and chance in the domestication of cereal crops. *World Archaeology*, *42*(1), 13–28. https://doi.org/10.1080/00438240903429680

Fyfe, L., Netzel, M. E., Tinggi, U., Biehl, E. M., & Sultanbawa, Y. (2018). Buchanania obovata: An Australian Indigenous food for diet diversification. *Nutrition and Dietetics*, *75*(5), 527–532. https://doi.org/10.1111/1747-0080.12437

Gautam, A., Poopalarajah, R., Ahmad, A., Rana, B., Denekew, T., Anh, N., Utenova, L., Kunwar, Y. S., Bhandari, N. N., & Jha, A. (2024). Ecological factors that drive microbial communities in culturally diverse fermented foods. https://doi.org/10.1101/2024.08.20.608727

Gepts, P., Bettinger, R., Brush, S., Damania, A., Famula, T., McGuire, P., & Qualset, C. (2012). Introduction: The domestication of plants and animals: Ten unanswered questions. In P. Gepts, T. R. Famula, R. L. Bettinger, S. B. Brush, A. B. Damania, P. E. McGuire, & C. O. Qualset (Eds.), *Biodiversity in agriculture* (pp. 1–8). Cambridge University Press. https://doi.org/10.1017/cbo9781139019514.002

Ghosh-Jerath, S., Kapoor, R., Barman, S., Singh, G., Singh, A., Downs, S., & Fanzo, J. (2021). Traditional food environment and factors affecting Indigenous food consumption in Munda tribal community of Jharkhand, India. *Frontiers in Nutrition*, *7*. https://doi.org/10.3389/fnut.2020.600470

Ghosh-Jerath, S., Kapoor, R., Singh, A., Downs, S., Barman, S., & Fanzo, J. (2020). Leveraging traditional ecological knowledge and access to nutrient-rich Indigenous foods to help achieve SDG 2: An analysis of the Indigenous foods of Sauria Paharias, a vulnerable tribal community in Jharkhand, India. *Frontiers in Nutrition*, *7*. https://doi.org/10.3389/fnut.2020.00061

Gifawesen, S., Tola, F., & Duguma, M. (2020). Review on role of home garden agroforestry practices to improve livelihood of small scale farmers and climate change adaptation and mitigation. *Journal of Plant Sciences*, *8*(5), 134. https://doi.org/10.11648/j.jps.20200805.15

Gluckman, P., Low, F., Buklijaš, T., Hanson, M., & Beedle, A. (2011). How evolutionary principles improve the understanding of human health and disease. *Evolutionary Applications*, *4*(2), 249–263. https://doi.org/10.1111/j.1752-4571.2010.00164.x

Govender, L., Pillay, K., Siwela, M., Modi, A., & Mabhaudhi, T. (2016). Food and nutrition insecurity in selected rural communities of KwaZulu-Natal, South Africa—linking human nutrition and agriculture. *International Journal of Environmental Research and Public Health*, *14*(1), 17. https://doi.org/10.3390/ijerph14010017

Harding, T., & Oetzel, J. (2019). Implementation effectiveness of health interventions for Indigenous communities: A systematic review. *Implementation Science*, *14*(1). https://doi.org/10.1186/s13012-019-0920-4

Ibrahim, Y., & Howarth, A. (2016). Contamination, deception and "othering": The media framing of the horsemeat scandal. *Social Identities*, *23*(2), 212–231. https://doi.org/10.1080/13504630.2016.1207512

Irakoze, M. (2024). The role of lactic fermentation in ensuring the safety and extending the shelf life of African Indigenous vegetables and its economic potential. *Applied Research*, *4*(1). https://doi.org/10.1002/appl.202400131

Joseph, L., Cuerrier, A., & Mathews, D. (2022). Shifting narratives, recognizing resilience: New anti-oppressive and decolonial approaches to ethnobotanical research with Indigenous communities in Canada. *Botany*, *100*(2), 65–81. https://doi.org/10.1139/cjb-2021-0111

Kagie, R., Lin, S., Hussain, M., & Thompson, S. (2019). A pragmatic review to assist planning and practice in delivering nutrition education to Indigenous youth. *Nutrients*, *11*(3), 510. https://doi.org/10.3390/nu11030510

Kaplan, H. S., Hill, K., Lancaster, J., & Hurtado, A. M. (2000). A theory of human life history evolution: Diet, intelligence, and longevity. *Evolutionary Anthropology*, *9*(4), 156–185. https://doi.org/10.1002/1520-6505(2000)9:4<156::AID-EVAN5>3.0.CO;2-7

Kapoor, R., Sabharwal, M., & Ghosh-Jerath, S. (2024). Co-existence of potentially sustainable Indigenous food systems and poor nutritional status in ho Indigenous community, India: An exploratory study. *Environmental Research Letters*, *19*(6), 064033. https://doi.org/10.1088/1748-9326/ad4b44

Kasimba, S., Motswagole, B., Covic, N., & Claasen, N. (2017). Household access to traditional and Indigenous foods positively associated with food security and dietary diversity in Botswana. *Public Health Nutrition*, *21*(6), 1200–1208. https://doi.org/10.1017/s136898001700369x

Kosoe, E. A., Achana, G. T. W., & Ogwu, M. C. (2023). Regulations and policies for herbal medicine and practitioners. In S. C. Izah, M. C. Ogwu, & M. Akram (Eds.), *Herbal medicine phytochemistry* (pp. 1–23). Reference Series in Phytochemistry. Springer. https://doi.org/10.1007/978-3-031-21973-3_33-1

Larson, G., Piperno, D., Allaby, R., Purugganan, M., Andersson, L., Arroyo-Kalin, M., Barton, L., Vigueira, C. C., Denham, T., Dobney, K., Doust, A. N., Gepts, P., Gilbert, M. T. P., Gremillion, K. J., Lucas, L., Lukens, L., Marshall, F. B., Olsen, K. M., Pires, J. C., & Fuller, D. (2014). Current perspectives and the future of domestication studies. *Proceedings of the National Academy of Sciences, 111*(17), 6139–6146. https://doi.org/10.1073/pnas.1323964111

Lee, A., & Lewis, M. (2018). Testing the price of healthy and current diets in remote aboriginal communities to improve food security: Development of the Aboriginal and Torres Strait Islander healthy diets ASAP (Australian Standardised Affordability and Pricing) methods. *International Journal of Environmental Research and Public Health, 15*(12), 2912. https://doi.org/10.3390/ijerph15122912

Lee-Thorp, J., & Sponheimer, M. (2006). Contributions of biogeochemistry to understanding hominin dietary ecology. *American Journal of Physical Anthropology, 131*(S43), 131–148. https://doi.org/10.1002/ajpa.20519

Livingstone, K., Love, P., Mathers, J., Kirkpatrick, S., & Olstad, D. (2023). Cultural adaptations and tailoring of public health nutrition interventions in Indigenous peoples and ethnic minority groups: Opportunities for personalised and precision nutrition. *Proceedings of the Nutrition Society, 82*(4), 478–486. https://doi.org/10.1017/s002966512300304x

Ma, W., & Rahut, D. (2024). Climate-smart agriculture: Adoption, impacts, and implications for sustainable development. *Mitigation and Adaptation Strategies for Global Change, 29*(5). https://doi.org/10.1007/s11027-024-10139-z

Mabhaudhi, T., Chimonyo, V., Hlahla, S., Massawe, F., Mayes, S., Nhamo, L., & Modi, A. T. (2019). Prospects of orphan crops in climate change. *Planta, 250*(3), 695–708. https://doi.org/10.1007/s00425-019-03129-y

McCartan, J., Burgel, E., McArthur, I., Testa, S., Thurn, E., Funston, S., Kho, A., McMahon, E., & Brimblecombe, J. (2020). Traditional food energy intake among Indigenous populations in select high-income settler-colonized countries: A systematic literature review. *Current Developments in Nutrition, 4*(11), nzaa163. https://doi.org/10.1093/cdn/nzaa163

Middleton, L., Astuti, P., Brown, B., Brimblecombe, J., & Stacey, N. (2024). "We don't need to worry because we will find food tomorrow": Local knowledge and drivers of mangroves as a food system through a gendered lens in West Kalimantan, Indonesia. *Sustainability*, *16*(8), 3229. https://doi.org/10.3390/su16083229

Migliano, A., & Vinicius, L. (2021). The origins of human cumulative culture: From the foraging niche to collective intelligence. *Philosophical Transactions of the Royal Society B Biological Sciences*, *377*(1843). https://doi.org/10.1098/rstb.2020.0317

Milla, R., Osborne, C., Turcotte, M., & Violle, C. (2015). Plant domestication through an ecological lens. *Trends in Ecology and Evolution*, *30*(8), 463–469. https://doi.org/10.1016/j.tree.2015.06.006

Montanari, B., & Teixidor-Toneu, I. (2021). Mountain isolation and the retention of traditional knowledge in the High Atlas of Morocco. *The Journal of North African Studies*, *27*(5), 977–997. https://doi.org/10.1080/13629387.2021.1901690

Monteiro, C., Moubarac, J., Cannon, G., Ng, S., & Popkin, B. (2013). Ultra-processed products are becoming dominant in the global food system. *Obesity Reviews*, *14*(S2), 21–28. https://doi.org/10.1111/obr.12107

Moscato, E., & Ozanne, J. (2019). Rebellious eating: Older women misbehaving through indulgence. *Qualitative Market Research an International Journal*, *22*(4), 582–594. https://doi.org/10.1108/qmr-07-2018-0082

Nagarajan, M., Rajasekaran, B., & Venkatachalam, K. (2022). Microbial metabolites in fermented food products and their potential benefits. *International Food Research Journal*, *29*(3), 466–486. https://doi.org/10.47836/ifrj.29.3.01

Naik, A., Jogi, M., & Shreenivas, B. (2024). Assessing the impact of climate change on global crop yields and farming practices. *Archives of Current Research International*, *24*(5), 696–712. https://doi.org/10.9734/acri/2024/v24i5743

Ng, C., Corey, P., & Young, T. (2011). Socio-economic patterns of obesity among Aboriginal and non-Aboriginal Canadians. *Canadian Journal of Public Health*, *102*(4), 264–268. https://doi.org/10.1007/bf03404046

Noort, M., Renzetti, S., Linderhof, V., Rand, G., Marx-Pienaar, N., de Kock, H., Magano, N., & Taylor, J. (2022). Towards sustainable shifts to healthy diets and food security in sub-Saharan Africa with climate-resilient crops in bread-type products: A food system analysis. *Foods*, *11*(2), 135. https://doi.org/10.3390/foods11020135

Nurbaya, N. (2023). Expressions language of saro eating ritual ternate ethnic. *Jurnal Sains Sosio Humaniora*, *7*(2), 225–235. https://doi.org/10.22437/jssh. v7i2.21293

Obahiagbon, E. G., & Ogwu, M. C. (2024). Organic food preservatives: The shift towards natural alternatives and sustainability in the Global South's markets. In M. C. Ogwu, S. C. Izah, & N. R. Ntuli (Eds.), *Food safety and quality in the Global South* (pp. 299–329). Springer. https://doi.org/10.1007/978-981-97-2428-4_10

Ogwu, M. C. (2020). Value of *Amaranthus* [L.] species in Nigeria. In V. Wais-undara (Ed.), *Nutritional value of Amaranth* (pp. 1–21). IntechOpen. http://doi.org/10.5772/intechopen.86990

Ogwu, M.C. (2023). Local food crops in Africa: Sustainable utilization, threats, and traditional storage strategies. In S. C. Izah, & M. C. Ogwu (Eds.), *Sustainable utilization and conservation of Africa's biological resources and environment.* Sustainable development and biodiversity (Vol. 888, pp. 353–374). Springer. https://doi.org/10.1007/978-981-19-6974-4_13

Ogwu, M. C. (2024). Plants as monitors and managers of pollution. In A. L. Srivastav, A. S. Grewal, M. Markandeya, & T. D. Pham (Eds.), *Advances in pollution research: Role of green chemistry in ecosystem restoration to achieve environmental sustainability* (pp. 51–60). Elsevier. https://doi.org/10.1016/B978-0-443-15291-7.00022-5

Ogwu, M. C., Odozi, I. P., Ahonsi, O. C., Uleanya, K. O., & Odozi, E. B. (2024). Organic acid production from cassava. In M. C. Ogwu, S. C. Izah, A. A. Cunha Alves, & S. C. Babu (Eds.), *Plant biology, sustainability and climate change, sustainable cassava* (pp. 395–418). Academic Press. https://doi.org/10.1016/B978-0-443-21747-0.00009-6

Ogwu, M. C., Ojo, A. O., & Osawaru, M. E. (2025). Quantitative ethnobotany of Afenmai people of Southern Nigeria: An assessment of their crop utilization, and preservation methods. *Genetic Resources and Crop Evolution*, *72*, 5807–5829. https://doi.org/10.1007/s10722-024-02302-x

Ogwu, M. C., Osawaru, M. E., Amodu, E., & Osamo, F. (2023). Comparative morphology, anatomy and chemotaxonomy of two *Cissus* Linn. species. *Brazilian Journal of Botany*, *46*, 397–412. https://doi.org/10.1007/s40415-023-00881-0

Oladeji, O. A., Karigidi, K. O., & Ogwu, M. C. (2024). Indices for monitoring and measuring the physicochemical properties of safe and quality food. In M. C.

Ogwu, S. C. Izah, & N. R. Ntuli (Eds.), *Food safety and quality in the Global South* (pp. 123–150). Springer. https://doi.org/10.1007/978-981-97-2428-4_5

Owade, J., Abong, G., Okoth, M., & Mwang'ombe, A. (2019). A review of the contribution of cowpea leaves to food and nutrition security in East Africa. *Food Science and Nutrition*, *8*(1), 36–47. https://doi.org/10.1002/fsn3.1337

Phillips, K., Pehrsson, P., Agnew, W., Scheett, A., Follett, J., Lukaski, H. C., & Patterson, K. (2014). Nutrient composition of selected traditional United States northern plains native American plant foods. *Journal of Food Composition and Analysis*, *34*(2), 136–152. https://doi.org/10.1016/j.jfca.2014.02.010

Proffitt, T., Reeves, J., Braun, D., Malaivijitnond, S., & Luncz, L. (2023). Wild macaques challenge the origin of intentional tool production. *Science Advances*, *9*(10). https://doi.org/10.1126/sciadv.ade8159

Riediger, N., LaPlante, J., Mudryj, A., & Clair, L. (2021). Diet quality among Indigenous and non-Indigenous children and youth in Canada in 2004 and 2015: A repeated cross-sectional design. *Public Health Nutrition*, *25*(1), 123–132. https://doi.org/10.1017/s1368980021002561

Rodríguez-Hidalgo, A., Saladié, P., Ollé, A., & Carbonell, E. (2015). Hominin subsistence and site function of td10.1 bone bed level at Gran Dolina Site (Atapuerca) during the late Acheulean. *Journal of Quaternary Science*, *30*(7), 679–701. https://doi.org/10.1002/jqs.2815

Sayers, S., Mott, S., Mann, K., Pearce, M., & Singh, G. (2013). Birthweight and fasting glucose and insulin levels: Results from the aboriginal birth cohort study. *The Medical Journal of Australia*, *199*(2), 112–116. https://doi.org/10.5694/mja13.10200

Sayre, J. W., Toklu, H. Z., Ye, F., Mazza, J., & Yale, S. (2017). Case reports, case series—From clinical practice to evidence-based medicine in graduate medical education. *Cureus, 9*(8), e1546. https://doi.org/10.7759/cureus.1546

Schembri, L., Curran, J., Collins, L., Pelinovskaia, M., Bell, H., Richardson, C., & Palermo, C. (2016). The effect of nutrition education on nutrition-related health outcomes of aboriginal and Torres Strait Islander People: A systematic review. *Australian and New Zealand Journal of Public Health*, *40*, S42–S47. https://doi.org/10.1111/1753-6405.12392

Schneider, L. (2022). Decolonizing conservation? Indigenous resurgence and buffalo restoration in the American West. *Environment and Planning E Nature and Space*, *6*(2), 801–821. https://doi.org/10.1177/25148486221119158

Sherriff, S., Baur, L., Lambert, M., Dickson, M., Eades, S., & Muthayya, S. (2019). Aboriginal childhood overweight and obesity: The need for aboriginal designed and led initiatives. *Public Health Research and Practice, 29*(4). https://doi.org/10.17061/phrp2941925

Shokeran, D., Hanis, H., Assim, M., & Shokeran, A. (2023). A case study of the exploration of implicit symbolic meaning and practices of the native people of Sematan areas, Lundu, Sarawak. *International Journal of Academic Research in Business and Social Sciences, 13*(15). https://doi.org/10.6007/ijarbss/v13-i15/18681

Sidiq, F., Coles, D., Hubbard, C., Clark, B., & Frewer, L. (2022). The role of traditional diets in promoting food security for Indigenous peoples in low- and middle-income countries: A systematic review. *IOP Conference Series Earth and Environmental Science, 978*(1), 012001. https://doi.org/10.1088/1755-1315/978/1/012001

Smith, B. (2005). Reassessing Coxcatlan cave and the early history of domesticated plants in Mesoamerica. *Proceedings of the National Academy of Sciences, 102*(27), 9438–9445. https://doi.org/10.1073/pnas.0502847102

Sponheimer, M., Alemseged, Z., Cerling, T. E., Grine, F. E., Kimbel, W. H., Leakey, M. G., Lee-Thorp, J. A., Manthi, F. K., Reed, K. E., Wood, B. A., & Wynn, J. G. (2013). Isotopic evidence of early hominin diets. *Proceedings of the National Academy of Sciences of the United States of America, 110*(26), 10513–10518. https://doi.org/10.1073/pnas.1222579110

Sterelny, K. (2007). Social intelligence, human intelligence and niche construction. *Philosophical Transactions of the Royal Society B Biological Sciences, 362*(1480), 719–730. https://doi.org/10.1098/rstb.2006.2006

Szilágyi, A., Kovács, V., Czárán, T., & Szathmáry, E. (2023). Evolutionary ecology of language origins through confrontational scavenging. *Philosophical Transactions of the Royal Society B Biological Sciences, 378*(1872). https://doi.org/10.1098/rstb.2021.0411

Tamiru, M., Becker, H., & Maass, B. (2007). Diversity, distribution and management of yam landraces (*Dioscorea* spp.) in Southern Ethiopia. *Genetic Resources and Crop Evolution, 55*(1), 115–131. https://doi.org/10.1007/s10722-007-9219-4

Teixidor-Toneu, I., Elgadi, S., Zine, H., Manzanilla, V., Ouhammou, A., & D'Ambrosio, U. (2021). Medicines in the kitchen: Gender roles shape ethnobotanical knowledge in Marrakshi households. *Foods, 10*(10), 2332. https://doi.org/10.3390/foods10102332

THM, P. (2023). Symbols in Jrai people's life-cycle rituals across Vietnam's central highlands. *Proceedings of the 9th International Conference on Arts and Humanities*, *9*(1), 68–85. https://doi.org/10.17501/23572744.2022.9105

Thurber, K., Dobbins, T., Neeman, T., Banwell, C., & Banks, E. (2017). Body mass index trajectories of Indigenous Australian children and relation to screen time, diet, and demographic factors. *Obesity*, *25*(4), 747–756. https://doi.org/10.1002/oby.21783

Torres-Avilez, W., Nascimento, A., Santoro, F., Medeiros, P., & Albuquerque, U. (2019). Gender and its role in the resilience of local medical systems of the Fulni-ô people in NE Brazil: Effects on structure and functionality. *Evidence-Based Complementary and Alternative Medicine*, *2019*(1), 1–15. https://doi.org/10.1155/2019/8313790

Walsh, A., Leech, J., Huttenhower, C., Delhomme-Nguyen, H., Crispie, F., Chervaux, C., & Cotter, P. (2023). Integrated molecular approaches for fermented food microbiome research. *Fems Microbiology Reviews*, *47*(2). https://doi.org/10.1093/femsre/fuad001

Wauchope, H., Shaw, J., Varpe, Ø., Lappo, E., Boertmann, D., Lanctot, R., & Fuller, R. (2016). Rapid climate-driven loss of breeding habitat for arctic migratory birds. *Global Change Biology*, *23*(3), 1085–1094. https://doi.org/10.1111/gcb.13404

Xu, M., Su, S., Zhang, Z., Jiang, S., Zhang, J., Xu, Y., & Hu, X. (2022). Two sides of the same coin: Meta-analysis uncovered the potential benefits and risks of traditional fermented foods at a large geographical scale. *Frontiers in Microbiology*, *13*. https://doi.org/10.3389/fmicb.2022.1045096

Zhang, Y., Zhang, Y., Hu, S., Zhou, X., Liu, L., Liu, J., Zhao, K., & Li, X. (2021). Pastoralism and millet cultivation during the bronze age in the temperate steppe region of Northern China. *Frontiers in Earth Science*, *9*. https://doi.org/10.3389/feart.2021.748327

CHAPTER 4

Sustainability of Indigenous Food Systems

Abstract

This chapter explores the sustainability of Indigenous food systems, emphasizing their complex integration of cultural identity, ecological knowledge, and community resilience. Indigenous food systems extend far beyond basic sustenance, serving as vital expressions of cultural heritage, social cohesion, and environmental stewardship. Rooted in Traditional Ecological Knowledge and practices such as polyculture, agroforestry, crop rotation, and wild harvesting, these have long enabled Indigenous communities to preserve biodiversity, adapt to changing climatic conditions, and promote soil health and water conservation. Central to these systems is the principle of food sovereignty, which empowers Indigenous peoples to maintain control over their food sources and protect their cultural autonomy amid colonial disruption, globalization, and industrial agriculture. This chapter highlights the cultural, social, and gender dimensions of Indigenous food systems, emphasizing the roles of oral traditions, intergenerational learning, and language in preserving food knowledge and environmental ethics. Case studies using fonio and millets in West Africa, teff and enset in Ethiopia, and maize in Mesoamerica illustrate the resilience and adaptability of these systems under environmental and sociopolitical pressures. The chapter also examines how Indigenous communities mobilize food sovereignty movements and advocate for policies that recognize their rights to land, resources, and traditional knowledge. Indigenous food systems offer valuable models for addressing global challenges related to food security, biodiversity conservation, and climate resilience, calling for their integration into contemporary sustainability frameworks while honoring Indigenous self-determination.

Keywords: food sovereignty, ecological sustainability, cultural identity, intergenerational knowledge, agroecology, climate resilience

4.1 Introduction

Indigenous food systems are an intricate tapestry woven through the sociocultural, ecological, and historical contexts of Indigenous communities. They extend beyond mere sustenance, embodying practices that reflect the identity, governance, and deep connections these communities have with their ancestral lands. These systems sustain not only the physical well-being of Indigenous populations but also serve as vital conduits for cultural expression and collective identity, flood control methodologies, resilience strategies, and community governance (Delormier et al., 2017; Jernigan et al., 2021; Whyte, 2018). The interdependence of culture and food practices manifests in ways that preserve Indigenous languages, stories, and traditions through gathering and consuming food (Domingo et al., 2021; Swiderska et al., 2022).

Cultural identity and ecological knowledge are critically intertwined in Indigenous food systems. This interdependence reinforces a deep connection to the land and the broader ecological health of their environments (Kuhnlein, 2014; Robin & Hart, 2025). Such knowledge underscores the cultural significance of traditional diets, which include food sovereignty as a core aspect of community governance (Brant et al., 2023; Delormier et al., 2017). Moreover, Indigenous food practices inform dietary choices that reflect the ecological diversity of their respective territories, emphasizing local species and traditional preparation methods that have evolved over generations (Kenny et al., 2018; Whyte, 2018). The essence of food systems in Indigenous contexts embodies the labor and love invested in preserving these relationships—an aspect often marginalized in mainstream discussions of food sovereignty in dominant food systems and structures (Robin, 2019). Indigenous food systems adopt a holistic perspective that interlinks various domains of knowledge, including botany, ecology, and anthropology, thus offering comprehensive insights into how communities interact with plants, animals, and ecosystems (Kapoor et al., 2024; Levkoe et al., 2019). This concept recognizes the synergistic relationships Indigenous peoples maintain with their environments, where food production extends to hunting, fishing, and foraging activities that honor TEK (Jernigan et al., 2021; Robin, 2019). Furthermore, these systems are adaptive; they have evolved through centuries of engagement with the land and its ecosystems, continually negotiating the challenges posed by environmental changes, colonial practices, and modern socioeconomic pressures (Delormier et al., 2017; Malli et al., 2023).

A focus on sustainability emerges from this understanding of Indigenous food systems, where ecological health and community well-being are seen as inseparable (Domingo et al., 2021; Molinos et al., 2022). Implementing regenerative practices aligned with Indigenous values—such as biodiversity conservation, seasonal cycles, and reciprocity—lays the foundation for resilient food production systems that can withstand external pressures (Akinola et al., 2020; Michnik et al., 2021). These sustainable approaches resonate with the priorities articulated by Indigenous leaders, emphasizing the necessity of integrating Indigenous wisdom and knowledge within broader sustainability agendas to ensure long-term ecological and community health (Heaney et al., 2024; Hutchinson et al., 2023). Furthermore, when addressing contemporary food insecurity, it is vital to recognize the potential of Indigenous food systems to heal and restore not just physical health but also mental and spiritual well-being—the holistic nature of health in many Indigenous cultures (Domingo et al., 2021; Swiderska et al., 2022). As recent studies reveal, the reclamation of land and traditional foods closely ties to improved health outcomes, reflecting a broader definition of food sovereignty that asserts the right to healthy, traditional, and regenerative food practices (Brant et al., 2023; Malli et al., 2023). These connections present compelling arguments for policies that honor and promote Indigenous food systems as pathways toward health equity and sustainability (Oladeji et al., 2024).

Food systems encompass far more than nutrition for many Indigenous communities—they serve as a framework for communal living, knowledge exchange, and spiritual fulfillment (Browne et al., 2020; Domingo et al., 2021; Ogwu & Kosoe, 2024). The shared experiences surrounding food shape social structures, community dynamics, and cultural practices, strengthening the resilience of these societies (Delormier et al., 2017; Heaney et al., 2024). Moreover, integrating Indigenous methodologies into contemporary food research acts as a reclamation of agency, providing avenues for Indigenous voices to influence and direct their food sovereignty narratives and practices effectively (Domingo et al., 2021; Lopes et al., 2024). Indigenous food systems thus challenge conventional agriculture and food security paradigms, proposing a more inclusive and holistic approach that aligns with local ecologies and cultural integrity (Michnik et al., 2021; Whyte, 2018). By redefining the goals and metrics of food production, communities can create sustainable models that meet the needs of people and the health of the ecosystems they depend on (Ghosh-Jerath et al., 2019; Malli et al., 2023). This conceptual shift ultimately recognizes the multilayered significance of food as a cultural and ecological cornerstone, which seems at odds with the expansive industrialized food systems that dominate global discourses (Kuhnlein, 2014; Robin, 2019; Weiler et al., 2014).

This chapter aims to explore the complex relationship between Indigenous communities and their food sources, highlighting the cultural, ecological, and sustainable significance of these systems. It seeks to define the stakeholders of Indigenous food systems and their activities, emphasizing their role not only in food security but also in maintaining cultural identity, ecological knowledge, and community well-being. The chapter explores the resilience of these systems, illustrating how Indigenous peoples have developed adaptive strategies to sustain their food sources amid climate change and environmental degradation. Additionally, it delves into the concept of food sovereignty, examining how Indigenous communities assert control over their food systems and protect their cultural heritage. The chapter discusses the challenges these systems face due to globalization and industrial agriculture while also looking at opportunities for revitalization through grassroots movements, policy change, and intercultural collaboration. By integrating ecological and cultural perspectives, this chapter contributes to the book by expanding the sociobiological framework of food systems and offering real-world case studies that illustrate the relevance of Indigenous food practices in modern sustainability efforts. Ultimately, the chapter aims to bridge the gap between Indigenous knowledge and contemporary global food systems, positioning these practices as valuable models for ecological stewardship and sustainable food security.

4.2 Indigenous Food Systems: Stakeholders and Activities

Indigenous food systems are foundational expressions of cultural identity and ecological wisdom intricately woven into the fabric of Indigenous communities. These systems embody a complex interplay of traditional agricultural practices, food-related ceremonies, and ecological stewardship that reflect deep-seated relationships with the land and its ecosystems. The stakeholders involved in preserving, advocating for, and revitalizing these food systems include Indigenous communities, governmental bodies, nongovernmental organizations (NGOs), academic institutions, and various private sector entities, each contributing unique expertise and resources to enhance the sustainability and resilience of Indigenous food practices.

4.2.1 Indigenous Communities

Indigenous communities are the primary stakeholders and custodians of their traditional food systems. Their role as stewards of Indigenous knowledge, cultural practices, and agricultural techniques is essential in preserving the integrity of these food systems amid globalization and environmental challenges. Activities undertaken by these communities include food production, knowledge transmission, cultural practices, and active advocacy for food sovereignty. Indigenous farmers implement sustainable agriculture techniques to maintain biodiversity and ensure ecological health. Practices such as crop rotation, agroforestry, and polyculture illustrate how Indigenous farming enhances local ecosystems while providing nutritious food to the community (Kuhnlein & Chotiboriboon, 2022). These approaches preserve soil health, which is vital for long-term agricultural productivity and exemplify Indigenous resilience to climate change. The cultural significance of food is also critical within these communities, where food plays a central role in many rituals and ceremonies, reinforcing social ties and cultural identity (Figueroa-Helland et al., 2018). Furthermore, the intergenerational transmission of knowledge is crucial for sustaining Indigenous food systems. Elders within these communities pass down their profound ecological and culinary insights through oral traditions and storytelling. Such practices ensure that the connection between communities and their environments endures across generations (Drawson et al., 2017; Vijayan et al., 2022). Additionally, food sovereignty movements, led by Indigenous communities, focus on reclaiming their rights to control food systems, resisting the pressures of industrial agriculture, and advocating for policies that mitigate the impacts of climate change on traditional lands (Akudugu & Ogwu, 2024; Figueroa-Helland et al., 2018).

4.2.2 Government Agencies

Government agencies at different levels play a significant role in supporting and regulating Indigenous food systems. These agencies can develop favorable policies that safeguard Indigenous land rights and promote food sovereignty. Recognizing intellectual property rights related to Indigenous agricultural practices is crucial for protecting the unique foods and knowledge developed over centuries (Afful-Arthur et al., 2021). Furthermore, environmental regulations imposed by national agencies protect Indigenous ecosystems from the threats posed by industrial agriculture and deforestation, which could otherwise devastate Indigenous food systems. Supporting community-based conservation initiatives that prioritize Indigenous land stewardship ensures sustainable practices prevail (Figueroa-Helland et al., 2018). Resource allocation also plays a key role, as governments provide funds and technical assistance to strengthen Indigenous agricultural development, enhance food security, and encourage the preservation of traditional knowledge surrounding Indigenous crops and practices (Arizona et al., 2019).

4.2.3 Nongovernmental Organizations (NGOs)

NGOs represent another important stakeholder group in the ecosystem surrounding Indigenous food systems. They often advocate for Indigenous rights, providing legal support to stabilize land claims and combat biopiracy while ensuring community engagement in preserving their traditional knowledge (Figueroa-Helland et al., 2018). Various NGOs participate in capacity-building activities, conducting training workshops that equip Indigenous producers with sustainable management practices to enhance the resilience of their food systems against climate challenges (Arizona et al., 2019). Moreover, NGOs' documentation and research efforts are crucial for recording Indigenous agricultural practices and understanding their ecological benefits. By collaborating closely with Indigenous communities, NGOs serve as conduits for sharing insights regarding traditional knowledge with broader audiences, ameliorating the implications of knowledge loss resulting from globalization (Drawson et al., 2017; Vijayan et al., 2022). In times of crisis, whether due to food insecurity or natural disasters, NGOs often step in with emergency relief efforts, helping communities rebuild their food systems and regain their footing (Kuhnlein & Chotiboriboon, 2022).

4.2.4 Academic Institutions and Researchers

The involvement of academic institutions and researchers in studying and revitalizing Indigenous food systems is essential. Scholars in fields such as anthropology, ethnobotany, and agricultural science conduct research that documents traditional farming techniques, cultural practices, and the nutritional aspects of Indigenous diets (Kuhnlein & Chotiboriboon, 2022; Vijayan et al., 2022). Collaborative projects between universities and Indigenous communities promote the revitalization of traditional practices, acknowledging the significant insights that Indigenous wisdom offers to contemporary agricultural challenges (Figueroa-Helland et al., 2018). Education about Indigenous food systems in academic courses can help cultivate a broader understanding of the value of TEK, ultimately enhancing student awareness and advocacy for Indigenous food sovereignty (Chatwood et al., 2015). Furthermore, the commitment of academic institutions to support Indigenous students through mentorship and culturally relevant programs contributes to the preservation of knowledge systems and community engagement (Trudgett et al., 2016).

4.2.5 Private Sector and Industry Stakeholders

While the private sector's interests in Indigenous food systems can sometimes conflict with the needs of Indigenous communities, some companies and initiatives align with the values of sustainability and ethical sourcing (Kuhnlein & Chotiboriboon, 2022). Industries such as fair trade, which respect Indigenous land rights and promote local crops, can provide much-needed economic support to these communities, facilitating sustainable livelihoods while preserving traditional practices (Figueroa-Helland et al., 2018). Additionally, the culinary sector plays a pivotal role in promoting Indigenous foods and practices, increasing demand for traditional ingredients and dishes. This not only supports economic opportunities for Indigenous communities but also fosters pride in cultural identities connected to food heritage (Kuhnlein & Chotiboriboon, 2022). Corporate social responsibility initiatives can serve as platforms to support food sovereignty movements, contributing financially to seed preservation and sustainable agriculture projects that empower Indigenous communities (Figueroa-Helland et al., 2018).

4.2.6 International Organizations

The impact of international organizations, including the United Nations and the Food and Agriculture Organization (FAO), should not be overlooked when considering Indigenous food systems. Global policy frameworks created by these entities can advocate for the protection of Indigenous food sovereignty and traditional agricultural practices on an international scale, ensuring that Indigenous rights are respected amid global agricultural pressures (Kuhnlein & Chotiboriboon, 2022). Moreover, these organizations contribute to the sustainable development goals that closely align with Indigenous food systems, recognizing their importance in achieving global objectives related to hunger, health, and environmental sustainability. By collaborating with governments and NGOs, international organizations can integrate Indigenous knowledge and practices into broader sustainability agendas (Figueroa-Helland et al., 2018). Financial assistance and technical support from these organizations help Indigenous communities enhance their food systems, adapt to climate change challenges, and implement sustainable practices that ensure the longevity of Indigenous food systems (Figueroa-Helland et al., 2018).

Figure 4.1 illustrates the interconnected roles and activities of these key stake-holders. Indigenous communities are at the core, driving their food systems' production, transmission, and cultural significance. Government agencies and NGOs provide policy, legal, and technical support to protect and promote food sovereignty. Academic institutions contribute through research and education on TEK and food systems. The private sector promotes sustainable economic prac-tices, while international organizations advocate for global policy frameworks that align with the rights and needs of Indigenous peoples. Through these collective efforts, Indigenous food systems continue to evolve and adapt to modern chal-lenges, ensuring cultural heritage survival and ecological practices' sustainability.

Figure 4.1: Indigenous Food Systems Stakeholders and Activities

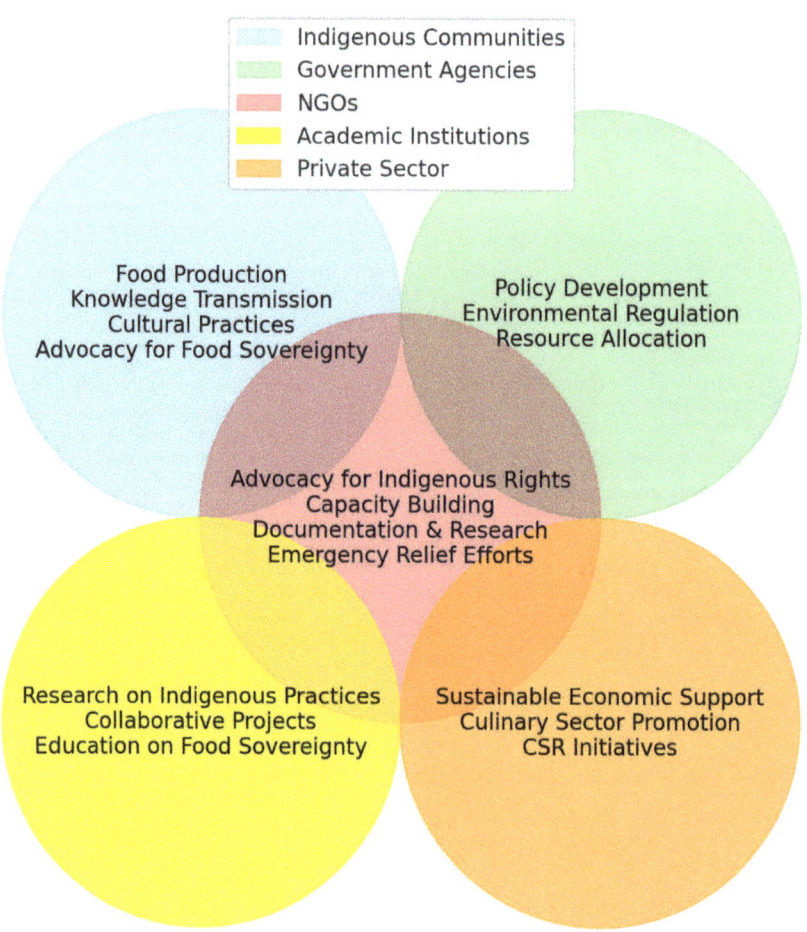

4.3 Roles of Ecological Knowledge and Sustainable Practices in Indigenous Food Systems

Indigenous food systems are deeply rooted in TEK that has evolved through close environmental interaction over generations. These systems are vital for food and nutritional security and embody sustainability principles that promote ecosystem balance and cultural continuity. Table 4.1 highlights the interconnected roles of ecological knowledge and sustainable practices that underpin the resilience and effectiveness of Indigenous food systems. The relationship between environmental knowledge, sustainable practices, and Indigenous food systems is a crucial area of interdisciplinary research, emphasizing the resilience and adaptability of Indigenous communities in the face of climate change and environmental degradation. Central to this synthesis is the concept of agroecology, which encompasses sustainable agricultural practices deeply rooted in TEK. Indigenous food systems often integrate diverse practices such as polyculture, crop rotation, and the cultivation of native varieties, which intrinsically promote biodiversity and contribute to soil health. These practices starkly contrast with industrial agriculture, which typically relies on monoculture and synthetic inputs, potentially leading to soil depletion and reduced biodiversity (Swiderska et al., 2022). Many Indigenous communities utilize polyculture systems that allow various crops to be grown simultaneously. This enhances resilience against pests and diseases and improves soil fertility through complementary planting strategies (Aich et al., 2022; Swiderska et al., 2022). Additionally, traditional crop varieties, often referred to as landraces, are more resilient to environmental stressors, making them essential for sustaining food sovereignty in the face of climate change (Carrasco-Torrontegui et al., 2021; Swiderska et al., 2022). Indigenous peoples have cultivated these varieties over millennia, adapting them to their local climatic conditions and ensuring they meet nutritional and cultural needs (Carrasco-Torrontegui et al., 2021). Integrating traditional knowledge into agroecological practices exemplifies how communities maintain their food systems while supporting ecosystem health and diversity (Whyte et al., 2015).

Moreover, Indigenous food systems demonstrate remarkable adaptability to local climatic conditions, often reflecting a profound awareness of environmental changes. Traditional agricultural practices frequently incorporate drought-resistant crops and specific water management techniques tailored to their local ecosystems (Deaconu et al., 2021; Tweheyo et al., 2024). For instance, Indigenous farmers have historically developed sophisticated systems for rainwater harvesting

Table 4.1: Roles of Ecological Knowledge and Sustainable Practices in Indigenous Food Systems

Role	Ecological Knowledge	Sustainable Practices
Biodiversity Conservation	Understanding native plant and animal species, seasonal cycles, and habitat interactions	Intercropping, crop rotation, and wild harvesting without overexploitation
Resource Management	Knowledge of soil fertility, water flow, and forest dynamics	Agroforestry, controlled burns, and rainwater harvesting
Climate Resilience	Awareness of weather patterns, drought-resistant crops, and climate adaptation strategies	Use of traditional drought-resistant seed varieties and timing of planting/harvesting
Nutritional Security	Identification of nutrient-rich Indigenous foods and their preparation	Diversified diets and seasonal food planning
Cultural Transmission	Oral traditions and storytelling about food-related rituals and land stewardship	Community feasts, seed exchanges, and ceremonial planting/harvesting
Soil Health Maintenance	Understanding of natural fertilization through composting, animal manure, etc.	Use of organic fertilizers, cover cropping, and minimal tillage
Pest and Disease Control	Recognizing pest cycles and plant–medicinal relationships	Botanical pesticides, companion planting, and biological control
Seed Sovereignty and Conservation	Preservation of heirloom and wild seed varieties through generational knowledge	Community seed banks and traditional seed exchange systems
Water Stewardship	Knowledge of watershed ecosystems and moisture retention techniques	Use of mulch, swales, and local irrigation customs
Landscape Stewardship	Deep understanding of terrain, microclimates, and ecological indicators	Zoning for foraging, cultivation, and sacred groves

and soil moisture conservation, which are critical for maintaining agricultural productivity during drought (Deaconu et al., 2021; Johnson et al., 2021). These strategies underscore the importance of integrating Indigenous knowledge into contemporary climate adaptation discussions, as they provide valuable insights into sustainable land management that enhances resilience (Sithole, 2019).

Various Indigenous groups use ecological calendars to synchronize agricultural activities with seasonal changes, which aids in planning planting and harvesting in response to climate variations (Ruelle et al., 2022). These calendars are essential for monitoring environmental conditions and identifying optimal times for cultivating crops, thereby enhancing food security and sovereignty within the community (Kimani et al., 2014; Ruelle et al., 2022). Such time-honored practices, refined through centuries of observation and adaptation, illustrate how Indigenous knowledge not only preserves cultural heritage but also promotes environmental sustainability (Ruelle et al., 2022). Numerous examples of ecological stewardship exist within Indigenous practices that actively contribute to food production and conservation. Controlled burns like those of the Lumbee Indians in the Southeastern United States exemplify traditional land management techniques that sustain food-producing landscapes by rejuvenating soil nutrients and encouraging the growth of specific plant species vital for food (Cloete & Idsardi, 2013; Robin, 2019). Furthermore, many Indigenous communities have developed sustainable fishing and hunting methods that align with natural cycles, ensuring wildlife populations remain healthy and abundant while safeguarding their food sources and cultural traditions (Hatfield et al., 2018). These practices demonstrate a deep understanding of ecological interdependencies and highlight the critical role of stewardship in sustaining Indigenous food systems.

Increasingly, there is recognition of women's integral role in sustainable practices and the transmission of ecological knowledge (Johnson et al., 2021; Stein et al., 2018). Women often act as custodians of traditional knowledge, leading community efforts toward sustainable food systems and asserting their authority over local resources (Stein et al., 2018). Through initiatives centered on food sovereignty, women foster communal leadership, promote equitable access to productive resources, and enhance community resilience to economic and environmental challenges (Robin, 2019). These dynamics emphasize the interconnectedness of ecological sustainability, gender equity, and community empowerment within Indigenous food systems, reinforcing the need to acknowledge and amplify the voices of all community members to enhance Indigenous food sovereignty (Williams & Hardison, 2013). The interplay between ecological knowledge and sustainable practices within Indigenous food systems offers vital lessons for contemporary agricultural practices and climate adaptation strategies. By integrating TEK with current scientific methods, there is potential to improve food security and restore ecological balance across diverse landscapes (Aich et al., 2022; Swiderska et al., 2022). As Indigenous peoples continue to lead the way in sustainable practices, their holistic and adaptive approaches provide a roadmap for achieving resilience in the face of climate change while advancing ecological integrity and cultural continuity.

4.4 Cultural Significance of Indigenous Food Systems

Indigenous food systems are deeply rooted in cultural frameworks beyond mere sustenance. They serve as crucial expressions of social identity, cosmology, and environmental stewardship. Table 4.2 highlights the core cultural roles that food plays within Indigenous communities across various geographies. Investigating Indigenous food systems reveals a profound interconnection between food, cultural identity, and community well-being among Indigenous peoples. Food practices extend beyond simple nourishment; they embody significant meanings that anchor cultural identity and facilitate continuity across generations. In many Indigenous cultures, food functions as a tangible link to ancestry, with traditional food practices acting as vessels that transport knowledge, rituals, and collective memories from one generation to the next. For instance, research indicates that engagement with traditional foods in the Syilx Okanagan community in British Columbia enhances cultural connectedness and overall well-being among its members, reinforcing the concept that Indigenous food sovereignty encompasses more than just access to food; it entails the right to self-governance and the expression of culture surrounding food practices (Blanchet et al., 2021; Chakraborty et al., 2021).

Food as a bearer of cultural identity lies at the heart of Indigenous food systems. This significance can be seen through the preparation, sharing of meals, and consumption practices. Traditional foods provide nutritional benefits and strengthen community bonds through shared practices and beliefs (Jennings et al., 2020; Oster et al., 2014). The Māori of New Zealand exemplify this by promoting a broader definition of food security that includes communal sharing and traditional food acquisition practices, contrasting with Western definitions that overlook the nuances of Indigenous communal traditions. This highlights the importance of food in Indigenous societies, which goes beyond mere nutritional needs (Chanza & Musakwa, 2022; McKerchar et al., 2014).

Table 4.2: Key Cultural Roles of Indigenous Food Systems

Cultural Roles	Function and Meaning	Illustrative Examples
Cultural Identity and Belonging	Foods are tied to tribal, clan, and community identity, often defining "who we are" as a people.	The consumption of teff in Ethiopia is strongly linked to cultural identity, pride, and heritage.
Spiritual and Ceremonial Role	Many foods are integral to rituals, offerings, and seasonal ceremonies that honor ancestors, spirits, and the Earth.	Maize in Mesoamerican rituals or yam festivals among the Igbo people of Nigeria.
Oral Tradition and Knowledge Transmission	Food-related knowledge is passed down through generations via stories, songs, proverbs, and rituals.	Indigenous recipes and farming songs among the Aymara and Quechua communities.
Social Cohesion	Shared food practices strengthen kinship ties, intergenerational relationships, and social obligations.	Communal harvesting and food sharing in Sámi reindeer herding or West African village feasts.
Gender Roles and Empowerment	Traditional food production often highlights women's roles in planting, harvesting, preparing, and preserving, contributing to social status.	Women seed keepers in Indigenous North American tribes, and matrilineal farming among the Khasi.
Resilience and Adaptation	Traditional food systems promote adaptive responses to environmental and cultural change while preserving core values.	Use of wild edible plants during food shortages, and crop rotation in Indigenous Sahelian farming.
Cultural Continuity	Food practices provide continuity across generations, maintaining cultural cohesion in the face of globalization and colonization.	Foraging and preparing native plants like bush tomatoes by Aboriginal Australians.
Ecological Ethics	Indigenous foodways often reflect a sacred responsibility to care for the land, water, and all living beings.	Seasonal hunting taboos or sustainable harvesting of non-timber forest products.

Rituals and ceremonies entwined with food further highlight its importance within Indigenous cultures. Communal gatherings such as feasts and harvest celebrations often revolve around specific foods, enhancing spiritual connections and social cohesion. For example, the Gaddis of Himachal Pradesh illustrate how cultural rituals are linked to their food systems, which remain connected to their histories and social traditions (Johnson-Jennings et al., 2020; Malhotra et al., 2021). These events serve as vital markers of identity and continuity, facilitating intergenerational knowledge transfer crucial for cultural preservation. Furthermore, the cyclical nature of food-related rituals often reflects agricultural cycles, reinforcing a symbiotic relationship between communities and their environments that fosters resilience and sustainability (Chanza & Musakwa, 2022; Osawaru & Ogwu, 2020). Additionally, certain foods embody rich symbolic meanings within Indigenous societies. For instance, maize in Mesoamerican cultures is a staple food and a central symbol of life, fertility, and ancestry (Burnette et al., 2018; Jernigan et al., 2023). Similarly, cassava in Central Africa and teff in Ethiopia are crucial to cultural identity and community cohesion, encapsulating numerous historical and spiritual narratives. These culturally significant foods act as repositories of collective memories and heritage, highlighting their essential role in preserving cultural identity (Briggs & Moyo, 2012; Gonçalves et al., 2021).

The influence of climate change on Indigenous food systems highlights an urgent need for revitalization practices rooted in traditional knowledge. Adaptation strategies within Indigenous food systems tackle challenges posed by climate change while improving food security. Indigenous communities have demonstrated resilience by conserving biodiversity and leveraging TEK (Briggs & Moyo, 2012; Chanza & Musakwa, 2022). The interaction between climate adaptability and traditional practices shows that Indigenous knowledge is essential for addressing modern environmental challenges. These connections with the land emphasize the importance of food sovereignty in fostering ecological sustainability and cultural continuity (Lemke & Delormier, 2017; Oster et al., 2014).

Food sovereignty emerges as a crucial factor in the ongoing efforts of Indigenous peoples to reclaim their food systems and enhance their health and well-being. It emphasizes their rights to define and engage with food practices founded on relational values and respect for natural systems (Blanchet et al., 2021; Jernigan et al., 2023). This framework challenges dominant societal norms surrounding food that have marginalized Indigenous practices, highlighting the importance of culturally specific strategies centered on community needs and aspirations (Maudrie et al., 2023; Robin et al., 2020). Recognizing Indigenous perspectives on food systems illuminates the losses experienced due to colonial disruptions and paves the way toward healing and revitalization (Deen et al., 2024; Johnson-Jennings et al., 2020).

The role of women in Indigenous food systems is particularly significant, as they often serve as custodians of food knowledge and family continuity. Research indicates that female tribal members are crucial in preserving culinary traditions and facilitating communal gatherings that strengthen family ties (Burnette et al., 2020; Robin et al., 2020). This centrality reflects a broader theme in which food serves not only as sustenance but also as a means of transmitting cultural values and fostering social connections. Therefore, initiatives to enhance food sovereignty must acknowledge women's roles and ensure women's active participation in revitalizing Indigenous food systems (Kolahdooz et al., 2024; Law-Ay et al., 2022). Exploring the relationship between food, health, and cultural continuity among Indigenous populations reveals that well-being is fundamentally connected to the reclamation of traditional food practices. Increased access to traditional foods correlates with improved health outcomes and cultural revitalization. Studies indicate that communities engaging with their traditional food systems experience physical health benefits and the revitalization of cultural identity tied to historical practices (Burnette et al., 2018; McKerchar et al., 2014; Oster et al., 2014). This underscores a critical intersection among health, identity, and cultural practices, advocating for policies that promote Indigenous food sovereignty and equitable access to traditional food systems as foundational for holistic health. The cultural significance of Indigenous food systems is intricate and deeply woven into the identities of these communities. Food practices serve as sources of nourishment and vital connections to heritage, identity, and community well-being. Revitalizing these food systems through culturally anchored practices offers insights into resilience amid environmental and sociopolitical challenges while nurturing intergenerational continuity and cultural integrity. Reclaiming food as a cultural cornerstone—rooted in the principles of sovereignty and community—remains essential for Indigenous peoples as they navigate their evolving identities within their unique cultural landscapes.

4.5 Indigenous Knowledge Systems in Local Food Practices

Indigenous Knowledge Systems (IKS) in local food practices represent a complex mosaic of traditions, beliefs, and experiential knowledge that serve as essential frameworks for ensuring the sustainability of food systems within

Indigenous communities. IKS, particularly those connected to food practices, are vital to the survival, identity, and resilience of Indigenous communities (Ogwu, 2023). These knowledge systems address food production and the deeper relationships between people, culture, and the environment. In Indigenous societies, food systems are intricately woven into social structures, ecological management, and spirituality.

4.5.1 Transmission of Knowledge

The transmission of knowledge regarding local food practices, including methods of planting, harvesting, and food preparation, is mainly facilitated through oral traditions, storytelling, and direct engagement under the guidance of elders (Swiderska et al., 2022). This process of passing down Indigenous food systems knowledge is complex and multifaceted. Research indicates that the primary methods for sharing this knowledge are oral traditions and storytelling. Elders and community leaders are crucial in this role, serving as the main custodians of traditional food knowledge. Oral narratives and stories provide practical guidance on growing, harvesting, and preparing food while imbuing these practices with cultural significance. Studies highlight that storytelling often acts as a pedagogical tool that links food practices to broader cosmologies, ecological systems, and historical contexts (Berkes, 2008). For example, certain planting or harvesting methods often come with stories that underscore their spiritual importance, reinforcing the reciprocal relationship between the community and nature (Kimmerer, 2013). Furthermore, hands-on experience is vital in transmitting Indigenous food knowledge. Elders frequently involve younger community members directly in agricultural activities, food harvesting, and preparation. These practical learning experiences allow younger generations to connect with their heritage in a tactile, embodied manner. In many Indigenous agricultural systems, food cultivation is seen as a collective responsibility, and community involvement in planting, maintaining, and harvesting crops ensures the continuity of this knowledge (Gonzalez, 2022).

4.5.2 Intergenerational Learning

Intergenerational learning is central to many Indigenous food systems. Food knowledge transmission occurs through formal instruction and daily immersion in food practices. Mentorship is key to this learning process, where older community members guide younger generations in food-related activities. This relationship between mentors and learners is often rooted in long-term, personal bonds and community participation (Barton et al., 2014). Through hands-on experience in food gathering and preparation, younger generations gain a deep understanding of food systems, including planting techniques, seasonal cycles, and ecological balance. Research also indicates that food systems learning occurs within the context of community life. Elders frequently teach younger generations about the rhythms of the environment, such as identifying seasonal changes that influence food harvesting. This fosters a strong connection with the natural world, as food practices are inherently linked to the cycles of the land, water, and animals. Consequently, young learners internalize practical food knowledge and ecological wisdom, recognizing the interconnectedness between humans and nature (du Bray et al., 2023). This relationship is vital for cultivating environmental stewardship and instilling a profound respect for the environment, which is essential for maintaining biodiversity and sustainable food systems (Nabhan, 2009).

This intergenerational exchange reflects time-honored practices in which elders play a pivotal role in mentoring younger generations, fostering a deep connection to the land and food sources that are critical for community resilience (Obongodot & Ogwu, 2023). This hands-on experience not only transfers practical knowledge but also instills a profound respect for ecological stewardship in younger members of the community, which is an integral part of Indigenous culture (Young et al., 2024).

Intergenerational learning within Indigenous food systems is significantly shaped by community participation, enabling youth to immerse themselves in food gathering and preparation practices. This experiential learning approach fosters a sense of belonging and responsibility essential for the perpetuation of cultural identity (Domingo et al., 2021; Young et al., 2024). Younger generations engaging in food-related activities are often taught the importance of sustainable harvesting and recognizing the interconnectedness of ecological systems, a relationship reinforced by TEK (Batal & Decelles, 2019). Thus, the food systems become a conduit for community bonding, helping to bridge the gap between individuals and their cultural heritage while also addressing challenges posed by modern food environments (Domingo et al., 2021).

4.5.3 Linguistic Connections

Linguistic connections between Indigenous languages and food systems are profound and multifaceted. Many Indigenous languages encapsulate rich terminologies related to local flora, fauna, and food preparation techniques, serving as repositories of ecological knowledge (Sidiq et al., 2022; Trott & Mulrennan, 2024). Language plays a crucial role in transmitting Indigenous food knowledge, as many Indigenous languages contain specialized vocabulary related to food production, plant species, and ecological practices. Linguistic studies have shown that words associated with food describe the biological aspects of plants and animals and encode cultural practices and relationships with the land (Thomas & Kirby, 2018). For example, in many Indigenous languages, food terms often reflect the seasonal cycles and ecological niches of different foods, providing essential insights into local environmental management systems. The loss of Indigenous languages is seen as a significant threat to the erosion of food knowledge. As Indigenous languages become endangered or extinct, the nuanced understandings of local ecosystems embedded within them fade as well. Research indicates that when Indigenous languages are lost, the specialized ecological and agricultural knowledge tied to those languages becomes harder to transmit (Maffi, 2005). For instance, in some Indigenous communities in the Americas, the names of plants and animals are linked to their properties and uses, and losing these terms diminishes the community's ability to manage and conserve these species effectively. Furthermore, language revitalization efforts often intertwine with the revival of Indigenous food systems. Efforts to document and teach Indigenous languages have been shown to contribute to the preservation and revitalization of food practices. Community-led language programs have aimed to reintroduce traditional food-related vocabulary and practices, thus ensuring that younger generations can access a fuller understanding of their ecological and cultural heritage (Hermes & Goebel, 2013).

IKS surrounding food practices are sustained through complex, intergenerational processes of learning and transmission. Elders are primary knowledge holders, passing down practices and ecological wisdom through oral traditions, storytelling, and hands-on experiences. Younger generations learn these systems through immersive experiences, mentorship, and community engagement, which helps them develop a deep understanding of the land and its cycles. Additionally, language plays a crucial role in preserving food knowledge, as many Indigenous languages encapsulate critical ecological and food-related information. As languages face the threat of extinction, so too does the intricate knowledge associated with Indigenous food systems. Protecting and revitalizing these systems require safeguarding linguistic heritage and fostering environments where knowledge can be passed down through direct, experiential learning. This integrated approach ensures that Indigenous food systems remain vibrant and adaptive, offering valuable insights for sustainable food practices in the face of modern challenges.

4.6 Resilience and Adaptation of Indigenous Food Systems to Modern Food Challenges

In examining the resilience and adaptation of Indigenous food systems to contemporary challenges, it is essential to acknowledge the multifaceted threats facing Indigenous communities, often exacerbated by historical and ongoing colonial practices. These external pressures include land displacement, loss of biodiversity, climate change, and the encroachment of industrial agriculture, each contributing to a profound transformation of traditional food systems. The historical context reveals how colonial policies and capitalist agricultural practices have marginalized Indigenous methods of food production, often labeled as inferior or outdated by dominant societies. Lys discusses how colonial preferences for exotic vegetables instilled a sense of inferiority regarding African Indigenous crops, complicating their acceptance and utilization within contemporary food systems (Lys, 2024). Malli et al. (2023) emphasize that colonial impacts have led to reduced harvest quality and diminishing yields of traditional food sources due to environmental degradation and pollution, highlighting a troubling trend of declining food security among Indigenous populations.

Climate change exacerbates these vulnerabilities by altering traditional ecosystems and impacting the availability of vital food resources. Indigenous food systems are especially sensitive to climate changes due to their close ties to local environmental conditions. Rampersad et al. (2023) note that many Indigenous communities inhabit regions experiencing rapid environmental changes, threatening their subsistence. A study by Tandoh et al. (2023) points out that Indigenous plants adapt well to marginal conditions, yet these resources are increasingly threatened by the pressures of modern agricultural expansions and changing climates. The loss of biodiversity, particularly the disappearance of traditional crops, compromises the nutritional and cultural integrity that these food systems provide (Argumedo et al., 2021). As modern agricultural systems introduce mainstream crops, the knowledge related to Indigenous crop diversity faces extinction, underscoring the need for protective measures and revitalization efforts (Tandoh et al., 2023).

Resistance to these challenges is evident through revitalization efforts led by Indigenous communities. Movements advocating for food sovereignty serve as mechanisms for reclaiming control over food systems, allowing communities to integrate traditional knowledge with modern science. Chanza and Musakwa (2022) underscore the significance of conventional food management practices that contribute to food security and play roles in cultural preservation. This intersectionality between tradition and modernity manifests in efforts to promote Indigenous food systems as sustainable alternatives amid changing dynamics. Food sovereignty movements encompass the broader rights of Indigenous peoples to maintain and protect their traditional food systems, as emphasized by Swiderska et al. (2022), who discuss Indigenous peoples' biocultural heritage.

As Indigenous communities engage in revitalization efforts, governmental policies that favor industrial systems present persistent challenges. Wagner illustrates the necessity for food governance agreements to recognize and accommodate Indigenous needs alongside conventional water management practices, arguing that a disconnect exists between settler policies and the rights of Indigenous peoples to food sovereignty (Wagner, 2023). This disconnect is echoed in the sentiments expressed by Robin and Hart (2025), emphasizing that the imposition of colonial food systems has historically deprived Indigenous peoples of their integral relationship with land and food, which are critical for cultural and community well-being. By confronting these colonial legacies, Indigenous communities are actively working toward reintegrating TEK with contemporary practices, thereby sustaining their food systems against modern challenges.

The ongoing dialogue surrounding food sovereignty highlights a path toward resilience, where Indigenous communities strive to adapt and assert their rights in dietary practices against a backdrop of colonial exploitation. Dennis and Robin (2020) point out that the Indigenous paradigm for health extends beyond basic nutrition, encompassing cultural significance, environmental sustainability, and community relationships. Food systems are viewed through a lens of medicinality, emphasizing holistic approaches that nurture both land and spirit. Akinola et al. (2020) assert that recognizing Indigenous crops and their nutritional merits could lead to sustainable food systems that encourage biodiversity and local adaptation.

In examining these aspects, it becomes evident that revitalizing Indigenous food systems is as much about cultural preservation as ecological integrity. The resurgence of traditional practices intermingled with modern perspectives fosters an environment where resilience can thrive. The importance of storytelling and oral tradition in transferring knowledge relevant to local ecosystems and food sources is invaluable, as described by Timler and Sandy regarding the persistence of diverse foodways in the face of colonization (Timler & Sandy, 2020). Herein lies a critical explanation for the adaptability of Indigenous food systems: Their components of resilience are deeply rooted in cultural continuity and communal participation.

Moreover, the intersection of social justice issues and food security cannot be overlooked in discussions about Indigenous food systems. Sherriff et al. (2022) note a growing body of literature addressing food insecurity among Aboriginal populations, linking it to systemic inequities resulting from historical injustices. Tackling these inequities calls for multidimensional solutions that recognize the complexities of food systems while advocating for decolonization practices and the return of land to Indigenous stewardship (Robin et al., 2021). By doing so, communities can foster both food sovereignty and ecological sustainability simultaneously—a dual objective essential for long-term resilience. Indigenous food systems have inherent adaptive capacities, as demonstrated by traditional practices that have endured environmental fluctuations over generations. These systems are deeply connected with ecological knowledge and reflect a sophisticated understanding of local biodiversity. Studies affirm that Indigenous agricultural practices, characterized by crop diversity and low external inputs, are generally more resilient to climate changes (Argumedo et al., 2021). The resurgence of interest in Indigenous crops highlights the role of these systems in alleviating global food and nutrition crises amid climate challenges (Tak et al., 2024). Through this lens, the broader potential of Indigenous systems to inform sustainable agricultural practices becomes evident, offering a counter-narrative to dominant industrial models that frequently disregard ecological considerations in favor of productivity. A crucial focus for future research and policy must involve integrating Indigenous perspectives within scientific discussions about food systems, especially as they face the pressures of a rapidly changing world. Michnik et al. (2021) outline that participatory action research frameworks foster the inclusion of Indigenous voices, promoting cultural and ecological revitalization. Such collaborative approaches are critical in addressing food security while honoring the traditions and knowledge systems of Indigenous peoples. There is a clear need to form alliances that advocate for Indigenous rights and sovereignty in food while confronting the systemic barriers imposed by colonial frameworks. The resilience and adaptation of Indigenous food systems amid modern challenges reflect both the determination of Indigenous communities and the necessity for broader societal acknowledgment of their rights. While facing external threats such as land displacement and climate change, Indigenous peoples are reclaiming their identities and knowledge through food sovereignty initiatives and revitalization efforts. Integrating traditional practices with contemporary insights protects biodiversity and revitalizes community health and cultural integrity.

4.7 Indigenous Food Sovereignty and Self-Determination

Food sovereignty refers to the right of people to healthy and culturally appropriate food produced through ecologically sound and sustainable methods, controlled by local communities, and in harmony with their social, economic, and cultural contexts (Via Campesina, 1996). This concept challenges the dominance of globalized industrial food systems, advocating for community-led control over food production, distribution, and consumption (Obahiagbon & Ogwu, 2023). Food sovereignty emphasizes the need for ecologically sustainable and culturally appropriate practices that not only ensure food security but also restore power to local communities, especially Indigenous peoples, over their own food systems. Indigenous food systems are foundational to food sovereignty, as they are built on generations of ecological knowledge and cultural practices that prioritize sustainability, community well-being, and self-determination. These systems represent not just agricultural or food production practices; they are also integrally tied to cultural heritage, spiritual values, and environmental stewardship. Indigenous food sovereignty seeks to reclaim control over food systems disrupted or marginalized by colonialism, industrialization, and the global food economy (Gonzalez, 2022). By recognizing and revitalizing Indigenous food systems, communities can reclaim their right to access and control food that is healthy, nutritious, and deeply tied to their cultural and ecological contexts.

4.7.1 Indigenous Rights and Policies

The protection of Indigenous food sovereignty is inextricably linked to the recognition of Indigenous rights—particularly land rights, access to natural resources, and the safeguarding of traditional knowledge. Indigenous communities have historically been dispossessed of their land, and their food systems have faced severe disruption due to colonial and capitalist expansion. The imposition of foreign agricultural practices, along with the encroachment of industrial agriculture, has displaced traditional methods and undermined community control over food production. Therefore, the fight for food sovereignty must be viewed as part of a larger struggle for Indigenous rights. One of the most significant international frameworks supporting these rights is the United Nations Declaration on the Rights of Indigenous Peoples (UNDRIP), adopted in 2007. UNDRIP recognizes Indigenous peoples' rights to self-determination, land, culture, and food sovereignty. Article 24 of UNDRIP, for example, highlights the right of Indigenous peoples to maintain and strengthen their distinct spiritual and cultural practices, including those related to food. International initiatives, such as biocultural rights frameworks, advocate for the protection of TEK, which encompasses knowledge of food production, sustainable harvesting, and land management (Prist et al., 2024). These frameworks call for policies that support Indigenous autonomy over their food systems, ensuring that they have the authority to manage and safeguard their agricultural lands and food practices from external threats. Moreover, national policies aimed at land reclamation, biodiversity protection, and agricultural sustainability are also vital to Indigenous food sovereignty (Abajue & Ogwu, 2024). In countries like Bolivia, Ecuador, and Brazil, constitutional reforms have granted Indigenous communities greater control over their land and resources, promoting the resurgence of traditional food systems as part of their broader cultural and environmental protection agendas (Mobetty et al., 2025).

4.7.2 Examples of Indigenous Food Systems and Sovereignty

1. **Fonio and Millets in West Africa**—Fonio (*Digitaria exilis*) and millets (*Pennisetum glaucum* and *Eleusine coracana*) are two essential staple crops in the Sahelian region of West Africa. These crops have historically played a crucial role in Indigenous food systems due to their drought resistance and adaptability to the region's harsh climate. Fonio, for instance, is one of the oldest cultivated grains in West Africa and has been a cornerstone of local diets and cultural practices for centuries (Ballogou et al., 2014). Millets, too, are deeply integrated into local agricultural systems, providing resilience against drought and offering vital nutrition in a region that is prone to food insecurity as a result of climate change. The cultural significance of fonio and millet extends beyond food security. These grains appear in local rituals, feasts, and community gatherings, further solidifying their place in the social fabric of West African communities. Furthermore, gendered labor divisions in the cultivation of these crops reflect the societal roles and knowledge systems passed down through generations. As West Africa confronts the impacts of climate change, these crops are increasingly recognized for their contributions to ecological sustainability, promoting agroecological practices that safeguard soil health and biodiversity (Abrouk et al., 2020).

2. **Teff and Enset in Ethiopia**—Teff (*Eragrostis tef*) and enset (*Ensete ventricosum*) are two vital crops that form the backbone of Ethiopia's food security, nutrition, and culture. Teff is the primary ingredient in injera, a

staple food consumed across the country, while enset, often referred to as "false banana," is crucial for subsistence in Ethiopia's southern highlands. Both crops are essential to Ethiopian identity and hold long-standing cultural and spiritual significance. Teff cultivation is highly adapted to the diverse ecological conditions of the Ethiopian highlands, ranging from lowland areas to high-altitude zones. It is particularly suited for sustainable farming systems as it is drought-resistant, produces multiple harvests, and helps maintain soil fertility. Enset, on the other hand, is known for providing food security during times of drought, making it essential for the survival of rural communities in Ethiopia (Negash & Niehof, 2004). Both crops are integral to Ethiopia's food sovereignty, as they are cultivated and consumed by Indigenous communities that have developed intricate knowledge of their cultivation, harvesting, and storage methods.

3. **Maize in Mesoamerican Civilizations**—Maize (*Zea mays*) is perhaps one of the most iconic foods in the history of Indigenous food sovereignty. Originally domesticated in Mesoamerica, maize has shaped the culture, economy, and spirituality of Indigenous peoples throughout the Americas. For Mesoamerican societies such as the Aztecs and Maya, maize was not only a staple crop but also a sacred food, deeply embedded in their creation myths, rituals, and cosmology (Vavilov, 1992). The cultivation of maize is central to Mesoamerican food sovereignty, as it remains a staple crop for Indigenous communities in countries like Mexico, Guatemala, and Honduras. These communities have developed highly specialized agricultural techniques, such as milpas, which integrate maize with beans, squash, and other crops in a diverse, multi-crop system. This agroecological system is crucial for promoting soil fertility and pest control, maintaining biodiversity, and providing a balanced diet. In addition to its practical role in food security, maize continues to hold profound cultural significance. It symbolizes life, growth, and community, and its cultivation and consumption are central to spiritual practices and communal identity (Alcorn, 2000).

Indigenous food sovereignty is a powerful movement that emphasizes the right of Indigenous peoples to control their food systems, ensuring that food production, distribution, and consumption align with ecological sustainability and cultural identity. By protecting land rights, recognizing traditional knowledge, and promoting self-determination, Indigenous communities can reclaim control over their food systems and preserve their rich cultural heritage. Examples like fonio and millets in West Africa, teff and enset in Ethiopia, and maize in Meso-american civilizations illustrate how Indigenous food systems are essential to food sovereignty and ecological sustainability. These systems not only provide resilience in the face of climate change but also serve as cultural and spiritual anchors for communities, highlighting the profound interconnectedness of food, culture, and identity.

4.8 Conclusion: The Future of Indigenous Food Systems

In exploring Indigenous food systems, it becomes clear that they represent much more than a means of subsistence; they articulate a community's history, identity, and ongoing relationships with their lands. The emphasis on cultural specificity and ecological knowledge highlights the need to understand food systems as crucial areas of inquiry and action in the pursuit of sustainable futures. Recognizing the intricate interplay between cultural practices and ecological dynamics not only illuminates the challenges faced by these communities but also opens pathways for regenerative and equitable food practices that embrace Indigenous world-views and methodologies. Indigenous food systems provide invaluable insights and solutions for addressing some of the most pressing global sustainability challenges. Their deep integration of ecological knowledge, cultural values, and sustainable practices offers a comprehensive framework for tackling issues such as food security, biodiversity loss, and climate resilience. Indigenous food systems serve not only to nourish bodies but also to nurture the environment, foster social cohesion, and preserve cultural identity. The sustainable agricultural practices rooted in these systems—such as crop diversity, soil conservation, and water management—provide a pathway for building more resilient food systems in the face of climate change. As the world grapples with increasing environ-mental degradation, Indigenous food systems can function as models for how we can work in harmony with nature to ensure both the health of the planet and the communities that depend on it.

However, the future of Indigenous food systems is not assured. They face sig-nificant threats from climate change, industrial agriculture, land dispossession, and the erosion of traditional knowledge. To ensure these systems continue to thrive, it is imperative to prioritize their protection and revitalization. This can be achieved through policies that recognize and respect Indigenous land rights, protect biodiversity, and promote food sovereignty. Community-based initiatives, academic research, and collaborations between Indigenous peoples and external stakeholders can play a vital role in preserving these systems, ensuring that future generations can benefit from the ecological and cultural wealth embedded in Indigenous food practices. By supporting the revitalization of Indigenous food systems, we not only protect cultural heritage but also embrace sustainable practices that are critical for the global sustainability agenda. Ultimately, the preservation and strengthening of Indigenous food systems are essential for creating a more sustainable and equitable future for all.

Chapter Reflection

1. How do Indigenous food systems integrate cultural identity and ecological sustainability?
2. How does food sovereignty differ from conventional food security frameworks in the context of Indigenous communities?
3. What role does Traditional Ecological Knowledge (TEK) play in developing resilient and adaptive agricultural practices?
4. How are oral traditions and intergenerational learning central to preserving Indigenous food knowledge?
5. What threats does the erosion of Indigenous languages pose to food system sustainability and ecological knowledge?
6. How have colonial histories and industrial agriculture disrupted Indigenous food sovereignty?
7. What are Indigenous women's unique leadership roles in sustaining food systems?
8. How do specific examples like fonio, teff, enset, and maize reflect the resilience and adaptability of Indigenous food systems?
9. In what ways can international legal frameworks such as UNDRIP support Indigenous food sovereignty?
10. How can Indigenous food systems inform broader global strategies for climate resilience, biodiversity conservation, and sustainable development?
11. What policy and research gaps need to be addressed to strengthen Indigenous food systems in the face of modern challenges?

REFERENCES

Abajue, M. C., & Ogwu, M. C. (2024). Nutritional status of Indigenous and traditional edible insects: Challenges and limitations. In M. C. Ogwu, S. C. Izah, & N. R. Ntuli (Eds.), *Food safety and quality in the Global South* (pp. 711–729). Springer. https://doi.org/10.1007/978-981-97-2428-4_23

Abrouk, M., Ahmed, H. I., Cubry, P., Šimoníková, D., Cauet, S., Pailles, Y., Bettgenhaeuser, J., Gapa, L., Scarcelli, N., Couderc, M., Zekraoui, L., Kathiresan, N., Čížková, J., Hřibová, E., Doležel, J., Arribat, S., Bergès, H., Wieringa, J. J., Gueye, M., … Krattinger, S. G. (2020). Fonio millet genome unlocks African orphan crop diversity for agriculture in a changing climate. *Nature Communications, 11*(1), 4488. https://doi.org/10.1038/s41467-020-18329-4

Afful-Arthur, P., Kwafoa, P., Ampah-Johnston, M., & Mensah, V. (2021). Managing and accessing Indigenous knowledge for national development: The role of academic libraries in Ghana. *Information Development, 38*(4), 535–548. https://doi.org/10.1177/02666669211009916

Akinola, R., Pereira, L., Mabhaudhi, T., Bruin, F., & Rusch, L. (2020). A review of Indigenous food crops in Africa and the implications for more sustainable and healthy food systems. *Sustainability, 12*(8), 3493. https://doi.org/10.3390/su12083493

Akudugu, M. A., & Ogwu, M. C. (2024). Sustainable development policies and interventions: A bibliometric analysis of the contributions of the academic community. *Journal of Cleaner Production, 434*, 139919. http://doi.org/10.1016/j.jclepro.2023.139919

Alcorn, J. B. (2000). Development or decolonization? The case of Indigenous peoples' food systems in Mexico. *The Food Systems in Mexico, 13*(2), 75–88.

Argumedo, A., Song, Y., Khoury, C., Hunter, D., Dempewolf, H., Guarino, L., & de Haan, S. (2021). Biocultural diversity for food system transformation under global environmental change. *Frontiers in Sustainable Food Systems, 5.* https://doi.org/10.3389/fsufs.2021.685299

Arizona, Y., Wicaksono, M., & Vel, J. (2019). The role of indigeneity NGOs in the legal recognition of Adat communities and customary forests in Indonesia.

The Asia Pacific Journal of Anthropology, *20*(5), 487–506. https://doi.org/10 .1080/14442213.2019.1670241

Ballogou, V. Y., Soumanou, M. M., Toukourou, F., & Hounhouigan, J. D. (2014). Indigenous knowledge on landraces and fonio-based food in Benin. *Ecology of Food and Nutrition, 53*(4), 390–409. https://doi.org/10.1080/03670 244.2013.811388

Barton, L. A., Williams, P., & Bowers, J. (2014). Learning and transmission of traditional ecological knowledge in Indigenous communities. *Environmental Education Research, 20*(5), 645–662. https://doi.org/10.1080/13504622.2013. 823719

Batal, M., & Decelles, S. (2019). A scoping review of obesity among Indigenous peoples in Canada. *Journal of Obesity*, *2019*, 1–20. https://doi. org/10.1155/2019/9741090

Berkes, F. (2008). *Sacred ecology: Traditional ecological knowledge and resource management*. Routledge.

Blanchet, R., Batal, M., Johnson-Down, L., Johnson, S., Okanagan Nation Salmon Reintroduction Initiatives & Willows, N. (2021). An Indigenous food sovereignty initiative is positively associated with well-being and cultural connectedness in a survey of Syilx Okanagan adults in British Columbia, Canada. *BMC Public Health*, *21*(1), 1405. https://doi.org/10.1186/ s12889-021-11229-2

Brant, S., Williams, K., Andrews, J., Hammelman, C., & Levkoe, C. (2023). Indigenous food systems and food sovereignty: A collaborative conversation from the American Association of Geographers 2022 Annual Meeting. *Journal of Agriculture, Food Systems, and Community Development, 12*(3), 141–154. https://doi.org/10.5304/jafscd.2023.123.012

Briggs, J., & Moyo, B. (2012). The resilience of Indigenous knowledge in small-scale African agriculture: Key drivers. *Scottish Geographical Journal*, *128*(1), 64–80. https://doi.org/10.1080/14702541.2012.694703

Browne, J., Gilmore, M., Lock, M., & Backholer, K. (2020). First Nations Peoples' participation in the development of population-wide food and nutrition policy in Australia: A political economy and cultural safety analysis. *International Journal of Health Policy and Management*, *10*, 871–885. https://doi. org/10.34172/ijhpm.2020.175

Burnette, C., Clark, C., & Rodning, C. (2018). "Living off the land": How subsistence promotes well-being and resilience among Indigenous peoples of the southeastern united states. *Social Service Review*, *92*(3), 369–400. https://doi.org/10.1086/699287

Burnette, C., Lesesne, R., Temple, C., & Rodning, C. (2020). Family as the conduit to promote Indigenous women and men's enculturation and wellness: "I wish I had learned earlier." *Journal of Evidence-Based Social Work*, *17*(1), 1–23. https://doi.org/10.1080/26408066.2019.1617213

Carrasco-Torrontegui, A., Gallegos-Riofrío, C., Espinoza, F., & Swanson, M. (2021). Climate change, food sovereignty, and ancestral farming technologies in the Andes. *Current Developments in Nutrition*, *5*, 54–60. https://doi.org/10.1093/cdn/nzaa073

Chakraborty, J., Parida, B., & Singh, N. (2021). Future food sustainability can be traced back into local people's socio-cultural roots in Uttarakhand Himalaya, India. *Sustainability*, *13*(13), 7060. https://doi.org/10.3390/su13137060

Chanza, N., & Musakwa, W. (2022). Revitalizing Indigenous ways of maintaining food security in a changing climate: Review of the evidence base from Africa. *International Journal of Climate Change Strategies and Management*, *14*(3), 252–271. https://doi.org/10.1108/ijccsm-06-2021-0065

Chatwood, S., Paulette, F., Baker, R., Eriksen, A., Hansen, K., Eriksen, H., Hiratsuka, V., Lavoie, J., Lou, W., Mauro, I., Orbinski, J., Pabrum, N., Retallack, H., & Brown, A. (2015). Approaching Etuaptmumk—Introducing a consensus-based mixed method for health services research. *International Journal of Circumpolar Health*, *74*(1), 27438. https://doi.org/10.3402/ijch.v74.27438

Cissé, H., Sow, A., & Coulibaly, O. (2017). Climate resilience and agroecological practices in West African food systems: The case of fonio and millet cultivation. *Journal of Agricultural Sustainability*, *32*(1), 35–47. https://doi.org/10.1007/s12573-017-0165-1

Cleveland, D. A., Soleri, D., & Smith, S. E. (2015). Food sovereignty and the Indigenous right to grow. *Environmental Studies Journal*, *43*(2), 51–63. https://doi.org/10.1007/978-1-4939-3582-7_6

Cloete, P., & Idsardi, E. (2013). Consumption of Indigenous and traditional food crops: Perceptions and realities from South Africa. *Agroecology and Sustainable Food Systems*, *37*(8), 902–914. https://doi.org/10.1080/21683565.2013.805179

Deaconu, A., Mercille, G., & Batal, M. (2021). Promoting traditional foods for human and environmental health: Lessons from agroecology and Indigenous communities in Ecuador. *BMC Nutrition, 7*(1). https://doi.org/10.1186/s40795-020-00395-y

Deen, C., Sherriff, S., Shelling, M., Gall, A., Cubillo, B., Te Morenga, L., Brimblecombe, J., & Matthews, V. (2024). Measuring Indigenous food security—A case for Indigenous designed tools. *Health Promotion Journal of Australia, 36*(1), e945. https://doi.org/10.1002/hpja.945

Delormier, T., Horn-Miller, K., McComber, A., & Marquis, K. (2017). Reclaiming food security in the Mohawk community of Kahnawake through Haudenosaunee responsibilities. *Maternal and Child Nutrition, 13*(S3), e12556. https://doi.org/10.1111/mcn.12556

Dennis, M., & Robin, T. (2020). Healthy on our own terms. *Journal of Critical Dietetics, 5*(1), 4–11. https://doi.org/10.32920/cd.v5i1.1333

Domingo, A., Charles, K., Jacobs, M., Brooker, D., & Hanning, R. (2021). Indigenous community perspectives of food security, sustainable food systems and strategies to enhance access to local and traditional healthy food for partnering Williams treaties first nations (Ontario, Canada). *International Journal of Environmental Research and Public Health, 18*(9), 4404. https://doi.org/10.3390/ijerph18094404

Drawson, A., Toombs, E., & Mushquash, C. (2017). Indigenous research methods: A systematic review. *International Indigenous Policy Journal, 8*(2). https://doi.org/10.18584/iipj.2017.8.2.5

du Bray, M., Sanchez, A., & Ogwu, M. (2023). *Time immemorial: Water governance approaches and Indigenous rights.* River Field Studies Network. QUBES Educational Resources. https://doi.org/10.25334/2HPW-H547

Figueroa-Helland, L., Thomas, C., & Aguilera, A. (2018). Decolonizing food systems: Food sovereignty, Indigenous revitalization, and agroecology as counter-hegemonic movements. *Perspectives on Global Development and Technology, 17*(1–2), 173–201. https://doi.org/10.1163/15691497-12341473

Ghosh-Jerath, S., Downs, S., Singh, A., Paramanik, S., Goldberg, G., & Fanzo, J. (2019). Innovative matrix for applying a food systems approach for developing interventions to address nutrient deficiencies in Indigenous communities in India: A study protocol. *BMC Public Health, 19*(1). https://doi.org/10.1186/s12889-019-6963-2

Gonçalves, C., Schlindwein, M., & Martinelli, G. (2021). Agroforestry systems: A systematic review focusing on traditional Indigenous practices, food and nutrition security, economic viability, and the role of women. *Sustainability*, *13*(20), 11397. https://doi.org/10.3390/su132011397

Gonzalez, B. C. (2022). *Understanding resilience in food systems for food security and nutrition in Latin America and the Caribbean* [Unpublished master's project paper]. Cornell University. https://ecommons.cornell.edu/server/api/core/bitstreams/f2204616-a086-4aa7-92ff-91e3e1f1d942/content

Hatfield, S., Marino, E., Whyte, K., Dello, K., & Mote, P. (2018). Indian time: Time, seasonality, and culture in traditional ecological knowledge of climate change. *Ecological Processes*, *7*(1). https://doi.org/10.1186/s13717-018-0136-6

Heaney, D., Padilla-Zakour, O., & Chen, C. (2024). Processing and preservation technologies to enhance Indigenous food sovereignty, nutrition security and health equity in north America. *Frontiers in Nutrition*, *11*. https://doi.org/10.3389/fnut.2024.1395962

Hermes, M., & Goebel, R. (2013). Revitalizing Indigenous food systems through language revitalization. *Journal of Indigenous Studies*, *35*(2), 45–60.

Hutchinson, P., McIlduff, C., Legare, M., Keewatin, M., Hagel, M., Chapados, M., & Acharibasam, J. (2023). Indigenous knowledge mobilization: Reflection on context, content, and relationship. *Alternative an International Journal of Indigenous Peoples*, *19*(4), 902–913. https://doi.org/10.1177/11771801231198082

Jennings, D., Paul, K., Little, M., Olson, D., & Johnson-Jennings, M. (2020). Identifying perspectives about health to orient obesity intervention among urban, transitionally housed Indigenous children. *Qualitative Health Research*, *30*(6), 894–905. https://doi.org/10.1177/1049732319900164

Jernigan, V., Maudrie, T., Nikolaus, C., Benally, T., Johnson, S., Teague, T., Mayes, M., Jacob, T., & Taniguchi, T. (2021). Food sovereignty indicators for Indigenous community capacity building and health. *Frontiers in Sustainable Food Systems*, *5*. https://doi.org/10.3389/fsufs.2021.704750

Jernigan, V., Nguyen, C., Maudrie, T., Demientieff, L., Black, J., Mortenson, R., Wilbur, R. E., Clyma, K. R., Lewis, M., & Lopez, S. (2023). Food sovereignty and health: A conceptual framework to advance research and practice. *Health Promotion Practice*, *24*(6), 1070–1074. https://doi.org/10.1177/15248399231190367

Johnson, D., Parsons, M., & Fisher, K. (2021). Indigenous climate change adaptation: New directions for emerging scholarship. *Environment and Planning E Nature and Space, 5*(3), 1541–1578. https://doi.org/10.1177/25148486211022450

Johnson-Jennings, M., Jennings, D., Paul, K., & Little, M. (2020). Identifying needs and uses of digital Indigenous food knowledge and practices for an Indigenous food wisdom repository. *Alternative an International Journal of Indigenous Peoples, 16*(4), 290–299. https://doi.org/10.1177/1177180120954446

Kapoor, R., Sabharwal, M., & Ghosh-Jerath, S. (2024). Co-existence of potentially sustainable Indigenous food systems and poor nutritional status in ho Indigenous community, India: An exploratory study. *Environmental Research Letters, 19*(6), 064033. https://doi.org/10.1088/1748-9326/ad4b44

Kenny, T., Hu, X., Kuhnlein, H., Wesche, S., & Chan, H. (2018). Dietary sources of energy and nutrients in the contemporary diet of Inuit adults: Results from the 2007–08 Inuit health survey. *Public Health Nutrition, 21*(7), 1319–1331. https://doi.org/10.1017/s1368980017003810

Kimani, E., Ogendi, G., & Makenzi, P. (2014). An evaluation of climate change Indigenous coping and adaptation strategies for sustainable agro-pastoral based livelihoods in Baringo county, Kenya. *IOSR Journal of Environmental Science Toxicology and Food Technology, 8*(8), 38–58. https://doi.org/10.9790/2402-08833858

Kimmerer, R. W. (2013). *Braiding sweetgrass: Indigenous wisdom, scientific knowledge, and the teachings of plants.* Milkweed Editions.

Kolahdooz, F., Jang, S. L., Deck, S., Ilkiw, D., Omoro, G., Rautio, A., Pirkola, S., Møller, H., Ferguson, G., Evengård, B., Mantla-Look, L., DeLancey, D., Corriveau, A., Irlbacher-Fox, S., Wagg, A., Roache, C., Rittenbach, K., Conter, H. J., Falk, R., & Sharma, S. (2024). A scoping review of the current knowledge of the social determinants of health and infectious diseases (specifically covid-19, tuberculosis, and h1n1 influenza) in Canadian arctic Indigenous communities. *International Journal of Environmental Research and Public Health, 22*(1), 1. https://doi.org/10.3390/ijerph22010001

Kuhnlein, H. (2014). Food system sustainability for health and well-being of Indigenous peoples. *Public Health Nutrition, 18*(13), 2415–2424. https://doi.org/10.1017/s1368980014002961

Kuhnlein, H., & Chotiboriboon, S. (2022). Why and how to strengthen Indigenous peoples' food systems with examples from two unique Indigenous communities. *Frontiers in Sustainable Food Systems, 6.* https://doi.org/10.3389/fsufs.2022.808670

Law-Ay, S., Fermil, F., Bernaldez, E., Cantere, B., Lanoy, J., & Santamaria, J. (2022). Binuhat: Inventory and documentation of the Indigenous products of Ata-Manobo at Talaingod, Davao Del Norte. *Davao Research Journal, 13*(1), 76–90. https://doi.org/10.59120/drj.v13i1.6

Lemke, S., & Delormier, T. (2017). Indigenous peoples' food systems, nutrition, and gender: Conceptual and methodological considerations. *Maternal and Child Nutrition, 13*(S3), e12499. https://doi.org/10.1111/mcn.12499

Levkoe, C., Ray, L., & McLaughlin, J. (2019). The Indigenous food circle: Reconciliation and resurgence through food in northwestern Ontario. *Journal of Agriculture Food Systems and Community Development, 9,* 1–14. https://doi.org/10.5304/jafscd.2019.09b.008

Lopes, C., Mihrshahi, S., Hunter, J., Ronto, R., & Cawthorne, R. (2024). Co-designing research for sustainable food systems and diets with aboriginal communities: A study protocol. *International Journal of Environmental Research and Public Health, 21*(3), 298. https://doi.org/10.3390/ijerph21030298

Lys, I. M. (2024). The role of lactic fermentation in ensuring the safety and extending the shelf life of African Indigenous vegetables and its economic potential. *Applied Research, 4*(1). https://doi.org/10.1002/appl.202400131

Maffi, L. (2005). Linguistic, cultural, and biological diversity. *Annual Review of Anthropology, 34,* 599–617. https://doi.org/10.1146/annurev.anthro.34.081804.120437

Malhotra, A., Nandigama, S., & Bhattacharya, K. (2021). Food, fields and forage: A socio-ecological account of cultural transitions among the Gaddis of Himachal Pradesh in India. *Heliyon, 7*(7), e07569. https://doi.org/10.1016/j.heliyon.2021.e07569

Malli, A., Monteith, H., Hiscock, C., Smith, E., Fairman, K., Galloway, T., & Mashford-Pringle, A. (2023). Impacts of colonization on Indigenous food systems in Canada and the United States: A scoping review. *BMC Public Health, 23*(1), 2105. https://doi.org/10.1186/s12889-023-16997-7

Maudrie, T. L., Nguyen, C. J., Wilbur, R. E., Mucioki, M., Clyma, K. R., Ferguson, G. L., & Jernigan, V. (2023). Food security and food sovereignty: The difference between surviving and thriving. *Health Promotion Practice, 24*(6), 1075–1079. https://doi.org/10.1177/15248399231190366

McKerchar, C., Bowers, S., Heta, C., Signal, L., & Matoe, L. (2014). Enhancing Māori food security using traditional Kai. *Global Health Promotion, 22*(3), 15–24. https://doi.org/10.1177/1757975914543573

Michnik, K., Thompson, S., & Beardy, B. (2021). Moving your body, soul, and heart to share and harvest food. *Canadian Food Studies / La Revue Canadienne Des Études Sur L Alimentation, 8*(2). https://doi.org/10.15353/cfs-rcea.v8i2.446

Mobetty, F., Batal, M., Levacher, V., Sebai, I., & Mercille, G. (2025). Exploring Indigenous food sovereignty and food environments characteristics through food interventions in Canada: A scoping review. *International Journal of Circumpolar Health, 84*(1), 2438428. https://doi.org/10.1080/22423982.2024.2438428

Molinos, J. G., Gavrilyeva, T., Joompa, P., Narita, D., Chotiboriboon, S., Parilova, V., Sirisai, S., Okhlopkov, I., Zhang, Z., Yakovleva, N., Kongpunya, P., Gowachirapant, S., Gabyshev, V., & Kriengsinyos, W. (2022). Study protocol: International joint research project "climate change resilience of Indigenous socioecological systems" (RISE). *Plos One, 17*(7), e0271792. https://doi.org/10.1371/journal.pone.0271792

Nabhan, G. P. (2009). *Coming home to eat: The pleasures and politics of local foods*. W.W. Norton and Company.

Negash, A., & Niehof, A. (2004). The significance of enset culture and biodiversity for rural household food and livelihood security in southwestern Ethiopia. *Agriculture and Human Values 21*, 61–71. https://doi.org/10.1023/B:AHUM.0000014023.30611.ad

Obahiagbon, E. G., & Ogwu, M. C. (2023). Consumer perception and demand for sustainable herbal medicine products and market. In S. C. Izah, M. C. Ogwu, & M. Akram (Eds.), *Herbal medicine phytochemistry*. Reference Series in Phytochemistry (pp. 1–34). Springer. https://doi.org/10.1007/978-3-031-21973-3_65-1

Obongodot, N. U., & Ogwu, M. C. (2023). Plant food for human health: Case study of Indigenous vegetables in Akwa Ibom State, Nigeria. In S. C. Izah, M. C. Ogwu, & M. Akram (Eds.), *Herbal medicine phytochemistry*. Reference Series in Phytochemistry (pp. 1–38). Springer. https://doi.org/10.1007/978-3-031-21973-3_2-1

Ogwu, M. C. (2023). Local food crops in Africa: Sustainable utilization, threats, and traditional storage strategies. In S. C. Izah & M. C. Ogwu (Eds.), *Sustainable utilization and conservation of Africa's biological resources and environment.* Sustainable Development and Biodiversity (Vol. 888, pp. 353–374). Springer. https://doi.org/10.1007/978-981-19-6974-4_13

Ogwu, M. C., & Kosoe, E. A. (2024). Place of cultural diversity in sustainable water resource management in Ghana. In S. C. Izah, M. C. Ogwu, A. Loukas, & H. Hamidifar (Eds.), *Water crises and sustainable management in the Global South* (pp. 423–460). Springer. https://doi.org/10.1007/978-981-97-4966-9_14

Oladeji, O. A., Karigidi, K. O., & Ogwu, M. C. (2024). Indices for monitoring and measuring the physicochemical properties of safe and quality food. In M. C. Ogwu, S. C. Izah, & N. R. Ntuli (Eds.), *Food safety and quality in the Global South* (pp. 123–150). Springer. https://doi.org/10.1007/978-981-97-2428-4_5

Osawaru, M. E., & Ogwu, M. C. (2020). Survey of plant and plant products in local markets within Benin City and environs. In L. W. Filho, N. Ogugu, D. Ayal, L. Adelake, & I. da Silva, (Eds.), *African handbook of climate change adaptation* (pp. 1–24). Springer Nature. http://doi.org/10.1007/978-3-030-42091-8_159-1

Oster, R., Grier, A., Lightning, R., Mayan, M., & Toth, E. (2014). Cultural continuity, traditional Indigenous language, and diabetes in Alberta first nations: A mixed methods study. *International Journal for Equity in Health, 13*(1). https://doi.org/10.1186/s12939-014-0092-4

Prist, P. R., Seppelt, R., Hayman, D. T. S., Molua, E. L., Arneth, A., Biber-Freudenberger, L., Bukvareva, E., Chaudhary, S., Dubey, P. K., Földvári, G., Godoy-Faúndez, A., Fischer, J., Howe, C., Hussain, A., Lorilla, R. S., Maire, E., Materu, S. F., Miyake, Y., Türkmen, A., and Vanham, D. (2024). Chapter 2: Status and past trends of interactions in the nexus. In P. A. Harrison, P. D. McElwee, and T. L. van Huysen (Eds.), *Thematic assessment of the interlinkages among biodiversity, water, food and health of the intergovernmental science policy platform on biodiversity and ecosystem services.* IPBES Secretariat. https://doi.org/10.5281/zenodo.13850301

Rampersad, C., Geto, T., Samuel, T., Abebe, M., Gomez, M., Pironon, S., Büchi, L., Haggar, J., Stocks, J., Ryan, P., Buggs, R. J. A., Demissew, S., Wilkin, P., Abebe, W. M., & Borrell, J. (2023). Indigenous crop diversity maintained despite the introduction of major global crops in an African Centre of Agrobiodiversity. *Plants People Planet, 5*(6), 985–996. https://doi.org/10.1002/ppp3.10407

Robin, T. (2019). Our hands at work: Indigenous food sovereignty in western Canada. *Journal of Agriculture Food Systems and Community Development, 9,* 1–15. https://doi.org/10.5304/jafscd.2019.09b.007

Robin, T., Burnett, K., Parker, B., & Skinner, K. (2021). Safe food, dangerous lands? traditional foods and Indigenous peoples in Canada. *Frontiers in Communication, 6*. https://doi.org/10.3389/fcomm.2021.749944

Robin, T., Dennis, M., & Hart, M. (2020). Feeding Indigenous people in Canada. *International Social Work, 65*(4), 652–662. https://doi.org/10.1177/0020872820916218

Robin, T., & Hart, M. (2025). Cree food knowledge and being well. *International Journal of Environmental Research and Public Health, 22*(2), 181. https://doi.org/10.3390/ijerph22020181

Ruelle, M., Skye, A., Collins, E., & Kassam, K. (2022). Ecological calendars, food sovereignty, and climate adaptation in standing rock. *Geohealth, 6*(12). https://doi.org/10.1029/2022gh000621

Sherriff, S., Kalucy, D., Tong, A., Naqvi, N., Nixon, J., Eades, S., Ingram, T., Slater, K., Dickson, M., Lee, A., & Muthayya, S. (2022). Murradambirra dhangaang (make food secure): Aboriginal community and stakeholder perspectives on food insecurity in urban and regional Australia. *BMC Public Health, 22*(1). https://doi.org/10.1186/s12889-022-13202-z

Sidiq, F., Coles, D., Hubbard, C., Clark, B., & Frewer, L. (2022). The role of traditional diets in promoting food security for Indigenous peoples in low- and middle-income countries: A systematic review. *IOP Conference Series Earth and Environmental Science, 978*(1), 012001. https://doi.org/10.1088/1755-1315/978/1/012001

Sithole, A. (2019). Women's use of Indigenous knowledge systems to cope with climate change. *Advances in Social Sciences Research Journal, 6*(6), 111–119. https://doi.org/10.14738/assrj.66.6470

Stein, K., Mirosa, M., & Carter, L. (2018). Māori women leading local sustainable food systems. *Alternative an International Journal of Indigenous Peoples, 14*(2), 147–155. https://doi.org/10.1177/1177180117753168

Swiderska, K., Argumedo, A., Wekesa, C., Ndalilo, L., Song, Y., Rastogi, A., & Ryan, P. (2022). Indigenous peoples' food systems and biocultural heritage: Addressing Indigenous priorities using decolonial and interdisciplinary research approaches. *Sustainability, 14*(18), 11311. https://doi.org/10.3390/su141811311

Tak, M., Hussain, S., Zargar, H., & Blake, L. (2024). Food systems in protracted crises: Examining Indigenous food sovereignty amid de-development in Kashmir. *Disasters, 49*(1), e12666. https://doi.org/10.1111/disa.12666

Tandoh, P. K., Idun, I. A., & Bemanu, B. (2023). Mitigating global food and nutritional insecurity: Role of Indigenous crops. In A. González-Reyna & P. Kaushik (Eds.), *Landraces - Its productive conservation in animals and plants*. IntechOpen. https://doi.org/10.5772/intechopen.109394

Thomas, J., & Kirby, S. (2018). Self domestication and the evolution of language. *Biology & Philosophy, 33*(1), 9. https://doi.org/10.1007/s10539-018-9612-8

Timler, K., & Sandy, D. (2020). Gardening in ashes: The possibilities and limitations of gardening to support Indigenous health and well-being in the context of wildfires and colonialism. *International Journal of Environmental Research and Public Health, 17*(9), 3273. https://doi.org/10.3390/ijerph17093273

Trott, N., & Mulrennan, M. (2024). "Part of Who We Are...": A review of the literature addressing the sociocultural role of traditional foods in food security for Indigenous people in Northern Canada. *Societies, 14*(3), 34. https://doi.org/10.3390/soc14030034

Trudgett, M., Page, S., & Harrison, N. (2016). Brilliant minds: A snapshot of successful Indigenous Australian doctoral students. *The Australian Journal of Indigenous Education, 45*(1), 70–79. https://doi.org/10.1017/jie.2016.8

Tweheyo, R., Bamwesigye, D., & Kiconco, M. (2024). Climate change mitigation in southwestern Uganda using Indigenous knowledge. https://doi.org/10.20944/preprints202401.0422.v1

Vavilov, N. I. (1992). *The origin and geography of cultivated plants*. Cambridge University Press.

Via Campesina. 1996. *Declaration on food sovereignty: A future without hunger*. Presented at the NGO Forum on World Food Security, Rome. https://viacampesina.org/en/what-is-food-sovereignty/#:~:text=1996%20Declaration:%20A%20Future%20without,governance%20and%20policy%2Dmaking%20circles

Vijayan, D., Ludwig, D., Rybak, C., Kaechele, H., Hoffmann, H., Schönfeldt, H. C., Mbwana, H. A., Rivero, C. V., & Löhr, K. (2022). Indigenous knowledge in food system transformations. *Communications Earth and Environment, 3*(1), 213. https://doi.org/10.1038/s43247-022-00543-1

Wagner, J. (2023). Changing the narrative: Settler colonialism, food and the Columbia river treaty. *Anthropologica, 65*(1). https://doi.org/10.18357/anthropologica65120231312

Weiler, A., Hergesheimer, C., Brisbois, B., Wittman, H., Yassi, A., & Spiegel, J. (2014). Food sovereignty, food security and health equity: A meta-narrative mapping exercise. *Health Policy and Planning, 30*(8), 1078–1092. https://doi.org/10.1093/heapol/czu109

Whyte, K. (2018). Food sovereignty, justice, and Indigenous peoples: An Essay on Settler Colonialism and Collective Continuance. In A. Barnhill, M. Budolfson, & T. Doggett (Eds.), *The Oxford handbook of food ethics*. Oxford Academic. https://doi.org/10.1093/oxfordhb/9780199372263.013.34

Whyte, K., Brewer, J., & Johnson, J. (2015). Weaving Indigenous science, protocols and sustainability science. *Sustainability Science, 11*(1), 25–32. https://doi.org/10.1007/s11625-015-0296-6

Williams, T., & Hardison, P. (2013). Culture, law, risk and governance: Contexts of traditional knowledge in climate change adaptation. In J. K. Maldonado, B. Colombi, & R. Pandya (Eds.), *Climate change and Indigenous peoples in the United States* (pp. 23–36). Springer. https://doi.org/10.1007/978-3-319-05266-3_3

Young, L., Shukla, S., & Wilson, T. (2024). Indigenous values and perspectives for strengthening food security and sovereignty: Learning from a community-based case study of Misko-Ziibiing (Bloodvein River First Nation), Manitoba, Canada. *Frontiers in Sustainable Food Systems, 8*. https://doi.org/10.3389/fsufs.2024.1321231

CHAPTER 5

Fonio and Millets in West Africa

Abstract

This chapter offers an in-depth examination of fonio (*Digitaria* spp.) and millets—particularly pearl millet (*Pennisetum glaucum*) and finger millet (*Eleusine coracana*)—as cornerstones of sociobiological resilience in West African food systems. Drawing on interdisciplinary perspectives from agroecology, ethnobotany, genetics, gender studies, and food anthropology, it explores the coevolution of these grains with Sahelian and Sudano-Sahelian communities under conditions of climatic uncertainty, ecological marginality, and cultural continuity. Fonio and millets thrive under minimal inputs, short growth cycles, and poor soils, making them climate-resilient crops well suited to food security and nutritional equity in the face of climate change. Their cultivation is deeply gendered, with women playing critical roles in seed selection, processing, knowledge transmission, and market integration, thereby linking agricultural production to cultural heritage, ceremonial life, and economic empowerment. The chapter also interrogates contemporary threats to millet cultivation, including market pressures, urbanization, dietary transitions, and the commodification of fonio as a global "superfood." It emphasizes the urgent need for integrated policy frameworks that balance market innovation with food sovereignty and the protection of Indigenous knowledge systems. Finally, the chapter highlights recent genomic advances and community-led seed initiatives as key pathways for revitalizing these underutilized grains, positioning fonio and millets not as relics of subsistence agriculture but as dynamic agents of sustainable development, biodiversity conservation, and sociocultural resilience in an era of global climate and food system transformations.

Keywords: fonio, pearl millet, Indigenous crops, West Africa, climate resilience, agroecology, food sovereignty, sociobiology

5.1 Introduction

The ecological and cultural significance of fonio (*Digitaria* spp.) and millets, particularly *Pennisetum glaucum* and *Eleusine coracana*, in West Africa is profound, reflecting their Indigenous origins and adaptations to the region's diverse agroecological conditions. Fonio is considered one of the oldest cultivated crops in West Africa, with evidence of its domestication dating back to 500 BC, and is often referred to as "hungry rice" due to its rapid maturation and ability to be grown in poor soils (Abrouk et al., 2020; Adoukonou-Sagbadja et al., 2007; Tabassum et al., 2024). The importance of fonio and millets in the socioeconomic fabric of the region is underscored by their central role in local diets and agricultural practices, as well as their adaptability to climate variability, which makes them vital for food security, especially in light of climate change implications (Abrouk et al., 2020; Vetriventhan et al., 2020; Yerima & Achigan-Dako, 2021). Historically, the processes of domestication and diffusion of these cereal crops within West Africa demonstrate their resilience in marginal agroecological zones. Fonio has been an integral part of the region's agricultural heritage since its domestication near the Mopti region in Mali. Millets, particularly pearl millet, have also been cultivated across West Africa for centuries and serve as staples that support both human and animal nutrition (Burton et al., 2024; Yerima & Achigan-Dako, 2021). This long trajectory of cultivation suggests that West African agricultural practices have evolved alongside these crops, incorporating local ecological knowledge and adapting to environmental changes (Pujarula et al., 2021; Yerima et al., 2020).

The persistence of fonio and millet cultivation can be attributed to their inherent characteristics as climate-resilient crops. Fonio's drought tolerance and ability to thrive in nutrient-poor soils contribute to its successful cultivation in unpredictable weather patterns—a trait particularly valuable in the semiarid and subhumid regions of West Africa (Abrouk et al., 2020; Tabassum et al., 2024; Vetriventhan et al., 2020). The adaptability of these millets to varying climatic and soil conditions underscores their potential role as crops of choice in developing sustainable agricultural practices amid climate change (Longueville et al., 2016). Therefore, contemporary agronomic research focuses on leveraging the genetic diversity within these crops to enhance their productivity and resilience (Yerima & Achigan-Dako, 2021; Yerima et al., 2020). The recovery and improvement of these underutilized crops are crucial not only for enhancing food security but also for fostering cultural heritage and biodiversity conservation across West Africa (Abrouk et al., 2020; Tabassum et al., 2024; Wang et al., 2021). The sociobiological framing of this discourse facilitates a multifaceted exploration of the roles fonio and millet play beyond mere agricultural products; they are intertwined with identity, tradition, and the sociocultural fabric of the communities that cultivate them. The cultural importance of fonio is exemplified in the creation myths of the Dogon people, establishing a deep-rooted reverence for these crops in regional traditions and ethnic identities (Adoukonou-Sagbadja et al., 2006; Burton et al., 2024). The interplay of ecological knowledge—that is, the Indigenous wisdom surrounding these crops—and modern agricultural practices can foster resilience against climatic challenges, promoting sustainable agricultural development (Gigou et al., 2009; Tabassum et al., 2024; Yerima & Achigan-Dako, 2021).

Furthermore, the historical trajectories of fonio and millets illuminate the current challenges in their production and use, necessitating innovative approaches to their cultivation practices. Economic pressures and globalization have contributed to a decline in the cultivation of traditional crops in favor of more industrialized agriculture, often overlooking the nutritional and cultural values these Indigenous crops represent (Kanlindogbè et al., 2020). As global markets evolve, comprehending the historical allure and contemporary relevance of fonio and millets may inspire renewed interest among farmers, researchers, and policymakers, thereby facilitating the revitalization of these traditional crops (Abrouk et al., 2020; Pujarula et al., 2021; Yerima & Achigan-Dako, 2021). As we progress through this chapter, it will become clear that a renewed focus on fonio and millets is not merely for the sake of agriculture; instead, it is vital for supporting food sovereignty, enhancing community resilience, and preserving the rich cultural heritage of West Africa (Osawaru & Ogwu, 2020). The subsequent sections will explore genetic diversity studies, agronomic practices, and the socioeconomic implications that arise from the cultivation of these important crops, ultimately contributing to a holistic understanding of their role in contemporary West African society.

This chapter examines the sociobiological significance of *Digitaria* spp., *P. glaucum*, and *Eleusine coracana* as emblematic crops of resilience and cultural heritage throughout West Africa. By drawing on interdisciplinary perspectives from agroecology, ethnobotany, nutritional science, and anthropology, the chapter investigates how these grains have coevolved with local ecological systems and sociocultural practices. By analyzing their genetic diversity, adaptive traits, and roles in traditional farming systems, it highlights how fonio and millets support climate resilience, food security, and biodiversity in arid and semiarid regions. Special attention is given to gendered labor, Indigenous knowledge systems, and ritual uses that connect these crops to West African lifeways. The chapter contributes to the broader framework of the book by demonstrating how Indigenous food systems are not only repositories of biodiversity and resilience but are also living expressions of sociobiological adaptation. In doing so, it underscores the importance of preserving underutilized crops amid climate change, shifting food preferences, and global superfood markets while advocating for renewed policy, research, and institutional support for their sustainable revitalization.

5.2 Agroecological Context and Adaptation of Millet in West Africa

The agroecological context of millet in West Africa is significantly influenced by the climatic and soil conditions specific to the Sahelian and Sudano-Sahelian environments. These regions are characterized by a semiarid climate that features hot, dry conditions, with rainfall variability that can result in prolonged periods of drought. A key characteristic of the Sahel is the bimodal distribution of rainfall, which, although traditionally reliable, has recently undergone erratic shifts, significantly impacting agricultural productivity (Pierre et al., 2011). Rainfall in these areas averages between 300 and 900 mm annually, often occurring unpredictably, which necessitates crops that are particularly resilient to drought and poor soil conditions (Abrouk et al., 2020).

In the face of these environmental challenges, fonio (*D. exilis*) and pearl millet (*P. glaucum*) have emerged as vital crops due to their adaptations to harsh conditions. Both crops are closely linked to food security in the region and have been traditionally cultivated by local populations because of their ability to thrive in marginal soils with limited water availability (Abrouk et al., 2020). Fonio, often referred to as "African rice," is notable for its resilience, being a drought-resistant cereal crop that not only withstands adversity in its growing conditions but also offers high nutritional value, containing essential amino acids that are often lacking in the diets of populations in the Sahel (Nwokocha et al., 2024; Zhu, 2020). Similarly, pearl millet is known for its advantages in dry, nutrient-poor soils where other staple crops might fail (Carcedo et al., 2024).

The comparative phenology of fonio and millets reveals several advantageous traits that contribute to their success under challenging agricultural conditions. Both species exhibit short maturation periods, which allows them to capitalize on minimal rainfall during growing periods and reduces the risk of crop failure (Carcedo et al., 2024). This short lifecycle is complemented by their low input needs, requiring less fertilizer and irrigation compared to more water-intensive crops such as maize and rice (Sonkamble et al., 2024). Furthermore, their environmental resilience extends beyond drought tolerance to include adaptability to poor soil fertility, characterized by low organic matter and high acidity, common in many West African regions (Sonkamble et al., 2024). Research comparing the phenologies of these crops indicates that they possess physiological structures well adapted to the ecological conditions of the Sahel. For instance, evidence suggests that millet varieties can efficiently utilize available moisture through deep root systems that access water not available to other shallow-rooted crops (Saleh et al., 2013). This adaptation is critical given the soil constraints present in many regions where these crops are cultivated, largely composed of sandy or clayey soils that diminish moisture retention (Carcedo et al., 2024). Pearl millet produces a cylindrical, spike-like panicle densely packed with small grains, while sorghum forms a loose, branched panicle with rounded seeds (Figure 5.1). Understanding these morphological distinctions is essential for crop identification, breeding, and agronomic research, particularly in regions where these cereals play a crucial role in food security and climate resilience.

Moreover, the nutritional composition of these grains enhances food security narratives; fonio and millet serve as staple foods that support the dietary intake of populations reliant on them, especially where other options are not viable (Nwokocha et al., 2024; Zhu, 2020). As climatic conditions continue to shift unpredictably, the role of fonio and pearl millet as adaptive responses becomes increasingly essential. The ongoing need for agricultural diversification in these environments aims to not only improve food availability but also bolster resilience against climate variability. Cultivating these crops offers strategic advantages in terms of food security and helps foster sustainable agricultural practices that mitigate the impacts of climate change on farming communities (Sonkamble et al., 2024).

In addition to their direct agricultural importance, fonio and millets support broader socioeconomic structures by enabling local communities to maintain their cultural heritage and practice traditional agricultural methods that are sometimes overlooked in favor of more competitive cash crops. This sociocultural aspect of agroecology demonstrates how deeply intertwined local identity and land-use practices are, encapsulated in the rich ethnobotanical uses of these millets beyond sustenance—ranging from rituals to medicinal applications (Dansi et al., 2010;

Ravichandran et al., 2023). Thus, the sustainability of fonio and millets is not merely about climate resilience; it also encapsulates the complexities of farmer preferences, cultural significance, and community practices (Kanlindogbè et al., 2020). Through a combination of genetic diversity, favorable agricultural traits, and cultural significance, the millets of West Africa, especially fonio and pearl millet, represent an essential component of agroecological resilience in light of increasingly unpredictable climate patterns. Research efforts aimed at studying and enhancing these crops could potentially lead to new breeding programs that prioritize climate resilience traits, ensuring that both critical crop diversity and food security are upheld for future generations (Adoukonou-Sagbadja et al., 2007).

Figure 5.1: Ear of Pearl Millet from West Africa

100 mm

Source: Adapted from Pérez-Jordà et al. (2024)

5.3 Genetic and Nutritional Diversity of Millet in West Africa

Millets are a diverse group of small-seeded grasses grown worldwide as cereal crops or grains for food and fodder. They are recognized for their resilience to harsh climates, nutritional richness, and crucial role in food security, particularly in arid and semiarid regions. Millets are broadly categorized into three groups: Major Millets, Minor Millets, and Pseudo Millets (Figure 5.2). This classification is based on their grain size, usage, and botanical characteristics. While major millets like sorghum, finger millet, and pearl millet are more widely cultivated and consumed, minor millets include nutrient-dense varieties such as foxtail, kodo, and little millet. Pseudo millets, including amaranth and buckwheat, are not true cereals but are used similarly due to their grain-like properties and high nutritional value. This diagram illustrates the key representatives within each category. Pearl millet (Pennisetum glaucum) is a vital crop in West Africa, known for its significant genetic, nutritional, and cultural diversity. Its long-standing cultivation across the region is reflected in a wealth of genetic variability that underscores both its agronomical importance and its adaptation to the ecological constraints of the Sahel. Studies indicate substantial varietal diversity across regions and among ethnic groups, highlighting pearl millet's adaptability to diverse conditions, particularly through traits such as early maturity, which provide a drought escape mechanism typical of this semiarid region (Drabo et al., 2018; Pucher et al., 2015).

Figure 5.2: Classification of Millets: Major, Minor, and Pseudo Millets

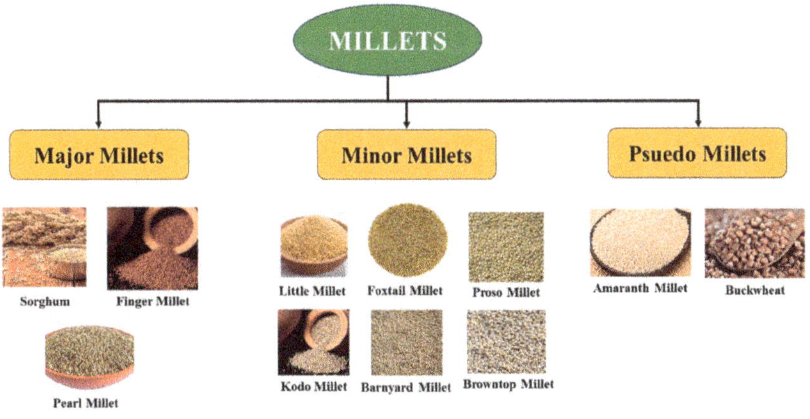

Source: Mohanan et al. (2025)

The genetic diversity of pearl millet in West Africa is particularly pronounced. Research has identified a broad range of agromorphological traits among pearl millet accessions, with significant variation in characteristics such as panicle structure and grain size (Figure 5.1; Pucher et al., 2015; Upadhyaya et al., 2016). The Iniadi landrace is notable for its predominance in countries such as Benin and Burkina Faso, reflecting selection pressures exerted by local farmers over generations. Pearl millet's ability to thrive in poor, drought-prone soils has led to a diverse array of phenotypes that local farmers traditionally select for resilience and yield under challenging environmental conditions (Hu et al., 2015; Rhoné et al., 2020).

Traditional seed systems and in situ conservation practices are crucial for sustaining the genetic diversity of pearl millet. These systems leverage local knowledge and practices to retain traditional varieties and facilitate germplasm sharing among farming communities. Continuous cultivation within family farming systems, often without modern fertilization techniques, reinforces the role of these Indigenous seed systems in maintaining the inherent diversity of local landraces (Abdoul-Karim et al., 2022; Rhoné et al., 2020). This agricultural practice supports a broad genetic base necessary to cope with environmental stresses, thus securing food sovereignty in vulnerable regions (Serba et al., 2020). From a nutritional perspective, pearl millet is highly valued for its composition, which includes essential micronutrients, protein, fiber, and a low glycemic index, making it particularly favorable for health in local diets (Ogwu, 2023; Ogwu et al., 2024). Research indicates that various millet cultivars exhibit high mineral content, especially iron and zinc, which are crucial in regions facing micronutrient deficiencies (Moussa et al., 2022). Studies have shown that pearl millet's nutritional profile can surpass that of commonly consumed cereals such as rice, maize, and wheat, especially regarding fiber content and associated health benefits. For instance, pearl millet has a lower glycemic index compared to these cereals, suggesting that its incorporation into the diet can be beneficial for blood sugar management (Dussert et al., 2015; Moussa et al., 2022).

When comparing the nutritional values of pearl millet to major cereals, the advantages become clear. Studies support the idea that pearl millet contains higher levels of certain essential amino acids that are often deficient in staple grains like maize and rice. Furthermore, the dietary fiber in pearl millet significantly contributes to digestive health, enhancing its role in a balanced diet. As a drought-resistant crop that thrives in less favorable soils, pearl millet is a crucial food source in arid climates, thereby addressing nutritional needs and fostering economic stability in local communities (Abdoul-Karim et al., 2022; Moussa et al., 2022; Yameogo et al., 2022). Exploring pearl millet in West Africa reveals a complex narrative involving genetic diversity, traditional agricultural practices, and exceptional nutritional value. The interaction between these components underscores the need to conserve traditional varieties while integrating modern agricultural practices to ensure food security and health. Considering climate change and food scarcity, recognizing pearl millet's historical and ongoing significance may encourage its expanded cultivation and use in both local foods and global food systems (Dussert et al., 2015; Kushwaha et al., 2014; Singh et al., 2017).

5.4 Indigenous Knowledge and Gendered Practices of Millet in West Africa

The role of women in the agricultural practices associated with millet cultivation in West Africa is profoundly significant, encompassing various stages from planting to marketing. In many rural communities, women are the backbone of millet production. They are primarily responsible for planting, nurturing, harvesting, processing, and marketing millet, which is a vital source of nutrition and livelihood in the region. The active participation of women in these processes not only supports their families but also sustains local economies and preserves traditional agricultural practices (Bombom et al., 2024; Mason et al., 2015). Specifically, studies highlight that women's roles are often interlinked with cultural rituals and social responsibilities, showcasing the integration of gender within agricultural practices (Mason et al., 2015; Rouamba et al., 2021). Table 5.1 presents a synthesis of Indigenous knowledge systems and local practices associated with millet cultivation, processing, and use. It highlights how ecological wisdom, gendered labor roles, spiritual beliefs, and intergenerational knowledge transmission have collectively shaped the resilience and sustainability of millet-based food systems. These traditional practices reflect sophisticated agroecological insights and underscore the sociobiological relevance of millet in contemporary food security and climate adaptation efforts.

Women's hands are the first to connect with the soil when planting millet, and their knowledge of seed selection and planting techniques is often passed down through generations. This Indigenous knowledge is crucial for maintaining the genetic diversity of millet crops, which are increasingly important for food security amid climate change (Bombom et al., 2024; Sonkamble et al., 2024). Furthermore, women utilize specific planting calendars synchronized

Table 5.1: Some Indigenous Knowledge and Practices Associated with Millet Cultivation in West Africa

Domain	Indigenous Knowledge and Practices	Gendered Dimensions
Seed Selection and Breeding	Farmers select seeds based on drought tolerance, taste, storability, and growth cycle. Local landraces are conserved.	Women often manage seed banks and pass on seed selection knowledge through generations.
Planting and Land Preparation	Use of traditional lunar calendars to time planting; intercropping with legumes to improve soil fertility.	Men typically prepare land and perform initial planting in communal labor groups.
Harvesting Techniques	Hand harvesting using sickles; rituals performed to ensure abundance and protection from pests.	Women lead post-harvest rituals and organize communal harvesting events.
Processing and Storage	Use of grinding stones and calabashes; sun-drying and clay granaries used for long-term preservation.	Women dominate all post-harvest processing, from threshing to storage management.
Culinary Use and Nutrition	Used in porridges, fermented drinks, and ceremonial dishes; known for high iron, calcium, and protein content.	Women are primary knowledge holders of recipes and traditional nutritional practices.
Cultural and Ritual Significance	Millet is central to initiation rites, weddings, and harvest festivals; associated with fertility and ancestral spirits.	Both men and women lead rituals, but women often serve as spiritual custodians of grain.
Transmission of Knowledge	Knowledge passed orally through proverbs, songs, and storytelling; practical learning during communal labor.	Elder women are key transmitters of knowledge to younger generations, especially girls.

with local weather patterns, which they have learned from their mothers and grandmothers. These calendars are embedded in various cultural practices, including songs and festivals that celebrate the agricultural cycle, marking important milestones such as planting and harvest times (Chanza & Musakwa, 2022; Egeruoh-Adindu, 2022).

Ethnographic studies reveal that these seasonal festivals are not merely celebratory but in fact serve as important educational gatherings where knowledge is transmitted across generations (Banda et al., 2024; Leipe et al., 2019). In many West African cultures, songs and rituals associated with millet cultivation hold collective memory and convey vital information regarding agricultural techniques, pest management, and soil fertility. This oral tradition acts as a vehicle for the transmission of agricultural knowledge, ensuring that valuable practices are retained and adapted to changing environmental conditions (Mason et al., 2015; Subedi, 2022). For instance, community events may often include demonstrations of traditional practices such as hand-weeding, which is critical for maintaining crop health and enhancing yields in ways that respect and preserve ecological balances (Adeyemo, 2022).

In terms of processing millet, the post-harvest phase is primarily managed by women, who utilize traditional methods for milling and preparing millet for consumption or sale. Their expertise not only enhances household food security but also plays a vital role in the market, where processed millet products are sold. This has significant implications for both community income and women's empowerment, as they gain financial independence and agency in decision-making processes related to household finances and broader community initiatives (Mabhaudhi et al., 2019; Sangappa et al., 2023). The intertwining of women's labor in these areas highlights gendered divisions of labor, which are often reinforced by societal norms and cultural expectations (Chanza & Musakwa, 2022; Zossou et al., 2016).

Furthermore, women often employ innovative marketing strategies to sell millet and its by-products, effectively adapting to market signals and consumer demands. By working together within networks, such as farmers' producer organizations, women can strengthen their bargaining power and negotiate better prices, thereby securing a stronger foothold in the local agricultural economy (Sangappa et al., 2023). Studies argue that by bolstering these networks and addressing logistical and market-related challenges, women's contributions to millet farming could be maximized, not only improving household incomes but also reinforcing the cultural heritage associated with millet cultivation (Jaiswal et al., 2023; Sangappa et al., 2023). The transmission of knowledge regarding millet farming is also evident in environmental practices vital for sustaining production amid a changing climate. Indigenous knowledge systems emphasize sustainable farming techniques, such as crop rotation and intercropping millet with other crops, which are essential for soil health and pest management (Solomon et al., 2016; Vetriventhan et al., 2020). These practices underscore the resilience of local communities who utilize traditional ecological knowledge to cope with significant challenges, including climate variability and market fluctuations (Fiaveh, 2019; Rouamba et al., 2021). Consequently, integrating Indigenous agricultural wisdom with modern practices offers substantial opportunities for enhancing millet production sustainably.

Celebratory events centered around millet, such as harvest festivals, further emphasize the social and cultural dimensions of farming, where communal identities are forged. Women often lead these celebrations, solidifying their role as custodians of agricultural knowledge and traditional practices (Chanza & Musakwa, 2022; Watson & Knight-Manuel, 2017). These festivals serve as crucial platforms for discussing agricultural challenges and innovations, ensuring that women's voices are heard in decision-making circles that influence agricultural policy and practice. The orchestration of these communal gatherings underscores the interconnectedness of individual, cultural, and ecological narratives in the discourse surrounding millet in West Africa. The resilience of millet as a crop is reinforced by women's roles in nurturing and sustaining its cultural significance through various traditional practices that reflect generational knowledge. Such practices are essential for enhancing food security and nutrition within households, as millets are recognized for their high nutrient content and adaptability (Bombom et al., 2024; Mabhaudhi et al., 2019). The culture surrounding millet cultivation not only promotes community cohesion but also addresses the broader socioeconomic challenges faced by West African societies. The gendered practices of millet cultivation in West Africa illustrate a rich tapestry of cultural, social, and economic dynamics that are intricately connected to women's roles in agriculture. Normalizing women's expertise in planting, harvesting, processing, and marketing millet bolsters community resilience and supports food security. Their contributions highlight the importance of integrating Indigenous knowledge systems into broader agricultural policies and practices, creating pathways for sustainable development that honor the traditions and practices that have nurtured these essential crops for generations.

5.5 Cultural Symbolism and Ritual Uses of Millet in West Africa

Millet, particularly pearl millet and fonio, plays a significant role in the cultural and social landscape of West Africa. This importance is prominently displayed during key life events such as naming ceremonies, weddings, and funerals, where millet serves not only as a staple food but also as a symbol of abundance and prosperity. The ritual use of millet in these ceremonies signifies its deep-rooted importance in societal values and traditions, making it more than just a food commodity; it is a cultural substance central to local identities. Research shows that these rituals often intertwine millet-based foods with ancestral reverence and community bonding, with the act of sharing millet dishes reinforcing social ties and cultural identities within the community (Burton et al., 2024; Drabo et al., 2018).

Moreover, the symbolic meanings of fonio and millet are deeply rooted in local belief systems and oral traditions. In many communities, millet is associated with fertility and prosperity, often offered in spiritual rites as a gesture of appreciation to deities or ancestors. Oral traditions recount tales that celebrate the qualities of millet, linking its resilience and growth to the blessings of ancestral spirits. These narratives not only preserve the cultural significance of millet but also emphasize its role in sustainable practices, illustrating how traditional knowledge systems aid in the preservation of agricultural biodiversity and cultural heritage (Drabo et al., 2018; Styring et al., 2019). The culinary arts in West Africa reflect the vital role of millet in shaping regional cuisines such as Tô, Dèguè, and Acha pudding. These dishes serve as more than just sustenance; they encapsulate the cultural essence and shared histories of various ethnic groups. Tô, a staple in several West African households, is made from millet flour and traditionally served with stew, showcasing the versatility and nutritional value of millet grains. Meanwhile, Dèguè and Acha pudding are celebrated for their delightful textures and flavors, inviting communal sharing and reinforcing family bonds during traditional feasts and gatherings (Gigou et al., 2009; Hayes et al., 2020; Moussa et al., 2022).

In the Sahelian region, the cultivation of millet fits within a broader agricultural framework that includes various millet species as integral components of food security and dietary diversity. Farmers strategically plant pearl millet and fonio due to their drought resilience and adaptability to poor soils, making them reliable crops in the face of climate variability. The integration of millet into local farming systems provides not only sustenance and nourishment but also serves as a cultural expression, preserving traditional farming methods amid modern agricultural challenges (Gigou et al., 2009; Manning et al., 2011). Additionally, the rhythmic cycles of planting and harvesting millet are deeply intertwined with rich cultural practices and communal rituals that signify the seasons and agricultural milestones. These agricultural cycles cultivate a collective identity among communities, with communal support structures emerging around millet cultivation, processing, and consumption. Traditional ceremonies and festivals dedicated to millet harvesting honor the value placed on ecological stewardship, the preservation of sociocultural heritage, and the reinforcement of community solidarity through shared labor and celebration (Burton et al., 2024; Manning et al., 2011).

Millet's redundancy as a local staple also emphasizes its potential in addressing food insecurity, given the increasing climatic pressures that challenge agricultural productivity. The promotion of traditional varieties, such as fonio, recognized for their nutritional benefits, aligns with contemporary efforts to strengthen regional food systems amid globalization. This interplay of traditional practices and modern agricultural strategies positions millet not only as a food resource but also as a cultural emblem encompassing a wealth of heritage that West African countries are eager to protect and promote (Abdoul-Karim et al., 2022; Yameogo et al., 2022).

Additionally, as climate conditions continue to evolve, the resilience of millet cultivation and its sociocultural significance is illustrated by ongoing agricultural research and local initiatives. Studies exploring the genetic diversity and adaptive capabilities of pearl millet reveal how traditional landraces possess essential traits for maintaining food security under changing climate scenarios. This scientific understanding, combined with local knowledge systems, emphasizes the importance of integrating cultural practices with scientific advancements to ensure sustainable millet production (Rhoné et al., 2020; Sultan et al., 2019). The multifaceted role of millet, particularly in ritual contexts, reflects profound cultural values, communal identities, and agricultural resilience in West Africa. As societies evolve, the enduring significance of millet remains a defining characteristic of cultural heritage, embodying the rich tapestry of traditions, celebrations, and agricultural practices that bind communities together. The ongoing recognition of millet's role in local cuisines and its adaptive qualities in agricultural systems underscores the commitment to preserving cultural and natural heritage in an increasingly globalized world (Abdoul-Karim et al., 2022; Sultan et al., 2019).

5.6 Effects of Challenges and Changing Landscapes on the Sociobiology of Millet

The sociobiology of millet has increasingly been affected by various challenges and changing landscapes. A decline in millet cultivation, along with the marginalization of fonio, has primarily been observed due to market pressures, inadequate policies, and shifting dietary preferences. These factors have significantly changed agricultural practices and the social dynamics surrounding this traditional crop. Market forces play a crucial role in shaping the agricultural landscape of fonio. The growing preference for major staple crops like rice and wheat has led to the neglect of Indigenous crops such as fonio, despite its recognized nutritional value and resilience to climate change (Burton et al., 2024; Diop et al., 2018). Agricultural policies frequently favor these dominant crops, resulting in insufficient investment and research focused on fonio, which remains underutilized despite its potential benefits as a climate-resilient crop (Michael et al., 2025). The commodification of fonio, particularly as a "superfood" in global markets, reflects a double-edged sword; while it can enhance visibility and generate income, it risks undermining local food sovereignty and traditional farming practices that have sustained rural communities for generations (Diop et al., 2018).

In addition to market influences, dietary preferences have shifted considerably, reflecting broader social and economic transformations. As urban migration increases so too does a change in dietary habits, wherein populations gravitate toward processed foods and away from traditional staples like fonio. This evolution in food consumption is driven by various factors, including convenience, availability, and perceived status associated with modern diets (Burton et al., 2024). Higher-income households, for instance, have been observed to favor diverse diets that often exclude traditional crops, thereby exacerbating the decline in fonio cultivation (Zhang et al., 2024). Furthermore, those with greater dietary knowledge tend to adopt healthier eating patterns, prioritizing nutritional value over tradition, which can further marginalize staple crops (Verdugo et al., 2023). The implications of these dietary shifts extend beyond individual preferences, impacting community health and food sovereignty. The marginalization of fonio not only threatens dietary diversity but also undermines local agricultural practices that have historically supported food security (Diop et al., 2018). As fonio's status as a healthful grain gains attention globally, local farmers may transform their practices to cater to external markets, leading to a detachment from the cultural significance of fonio (Ijeomah et al., 2023). This commodification has implications for local food sovereignty, as communities may lose the autonomy to define their food systems and prioritize nutrition according to local needs rather than external market demands (Verdugo et al., 2023).

Simultaneously, climate change presents substantial threats to millet cultivation, particularly in regions where fonio is traditionally grown. Changes in precipitation patterns and rising temperatures challenge the already marginal environments conducive to fonio cultivation, raising concerns over crop yields and food security (Abrouk et al., 2020). Moreover, urban migration often results in a loss of traditional agricultural knowledge as young people leave rural areas for urban centers, further jeopardizing fonio cultivation and conservation (Burton et al., 2024; Ogwu, 2019). The interplay of climate change with shifting socioeconomic dynamics creates a complex web of challenges for the sociobiology of millet and fonio cultivation. The role of fonio as a potential "superfood" may present a paradox. While its nutritional benefits can be emphasized to boost demand, there is a risk of commodifying the crop at the expense of traditional agricultural practices and community food systems (Diop et al., 2018). As fonio is promoted as a healthful alternative in urban centers and global markets, local farmers might shift their focus from sustaining local food systems to prioritizing production for external markets (Ijeomah et al., 2023). This shift requires a careful balance between recognizing fonio's potential in combating malnutrition and ensuring that local food sovereignty is not compromised (Verdugo et al., 2023). To address these challenges, a multifaceted approach is essential. Integrated conservation efforts that consider the sociocultural roles of fonio, along with supportive agricultural policies, could revitalize interest in millet cultivation. Such strategies must focus on promoting the benefits of fonio for both nutrition and sustainability while empowering local communities to reclaim their food systems (Diop et al., 2025; Ogwu, 2020). Additionally, encouraging research into fonio's genetic diversity can support breeding efforts aimed at enhancing its resilience to climate stresses, ensuring its viability as an agricultural staple (Abrouk et al., 2020).

Moreover, promoting agricultural practices that emphasize agroecology and traditional knowledge can enhance the resilience of fonio cultivation against climate change impacts (Diop et al., 2018). Community-led initiatives aimed at promoting dietary diversity and integrating fonio into modern diets can improve local consumption patterns. By highlighting the crop's cultural significance and nutritional benefits, communities can potentially engage new generations in the cultivation and appreciation of fonio (Diop et al., 2023). Ultimately, the future of fonio and millet cultivation depends on properly addressing these complex interrelations. Climate adaptation strategies, informed agricultural policies, and a commitment to local food sovereignty are essential to navigate the declining trends in fonio cultivation and to mitigate the effects of emerging challenges. As incentives for market engagement grow alongside climate threats, reevaluating traditional crops like fonio presents opportunities for creating sustainable food systems that honor both ancient practices and contemporary health demands. The sociobiology of millet, especially fonio, is being reshaped by various factors, including declining cultivation due to market and policy pressures, urban migration altering dietary preferences, and the potential commodification of fonio as a superfood. Effectively addressing these challenges requires an integrated approach that balances market opportunities with the necessity of maintaining local food sovereignty and sustainability, paving the way for a resilient agricultural future for millet and fonio.

5.7 Revitalization and Innovation of Millet in West Africa

The revitalization of millet cultivation in West Africa is crucial for enhancing food security, improving livelihoods, and ensuring environmental sustainability. The integration of agroecological practices, farmer-led seed initiatives, and institutional policies support this effort (Izah & Ogwu, 2023; Ogwu & Izah, 2023). Promoting millet through agroecological management aligns with a broader ecological focus, emphasizing soil health, biodiversity, and ecosystem resilience. For instance, research highlights that pearl millet is resilient in drought-stricken areas due to its adaptive traits, which benefit from agroecological management practices (Sultan et al., 2013; Yadav et al., 2010). The combination of sustainable agriculture and natural resource management creates an environment where millet can thrive, thereby reinforcing not only crop production but also addressing climate mitigation and adaptation strategies (Diack et al., 2017; Upadhyaya et al., 2016).

Farmer-led seed initiatives represent another vital pillar of millet revitalization. These grassroots programs empower local communities to maintain and enhance their crop varieties, ensuring that genetic diversity is preserved and utilized to develop resilient strains suited to local climatic conditions (Ogwu et al., 2024). Studies have shown that local landraces of pearl millet possess significant genetic diversity, which is crucial for breeding programs that improve drought tolerance and yield stability (Ibrahim et al., 2015; Rhoné et al., 2020). By promoting genetic improvements alongside traditional practices, communities can significantly increase their resilience against climate variability, securing their food systems for the future.

Institutional support from research centers, NGOs, and governmental bodies is essential in this revitalization process. For example, increased collaboration among research institutions can lead to sharing best practices in millet husbandry, breeding, and pest management strategies tailored to specific regional contexts (Trail et al., 2016). Additionally, extension services aimed at millet farmers, providing science-based information on sustainable practices, can significantly enhance crop yields (Jaiswal et al., 2023). Moreover, policies that prioritize millet in regional agricultural development plans can catalyze investments in infrastructure and supply chains, giving farmers access to seeds and markets for their harvested products (Jaiswal et al., 2023; Sultan et al., 2013).

The commercialization of millet and its value-added products is another aspect of revitalization that holds great promise. Innovations in processing and product development have led to a proliferation of millet-based goods such as fonio flour and gluten-free products. This diversification in usage helps recapture consumer interest while promoting nutritional benefits inherent in millet, which is acknowledged for its high protein and dietary fiber content (Bombom et al., 2024; Saleh et al., 2013). The growing trend of gluten-free diets in global markets adds an additional layer of opportunity for millet commercialization, as consumers seek alternatives to conventional grain products (Saleh et al., 2013). Furthermore, the development of local markets for millet products fuels economic growth and enhances food sovereignty within communities. By establishing cooperatives that process and market millet-based foods, farmers can enhance their income prospects while fostering community solidarity (Saleh et al., 2013). Such initiatives also pave the way for cross-regional trade, enriching the local economy and potentially expanding the consumer base through regional partnerships (Drabo et al., 2018; Yameogo et al., 2022).

In light of climate change and its adverse effects on agricultural productivity, efforts to innovate within the millet value chain are more urgent than ever. The genetic improvement of pearl millet through advanced breeding techniques, including genomic selection, shows considerable promise. Research indicates a strong potential for developing varieties that can withstand expected climatic stressors, promoting not just yield increases but also sustainability (Mariac et al., 2010; Yadav et al., 2010). At the same time, community-based seed banks contribute to conserving local varieties, thereby securing food sources amid climatic uncertainties (Dussert et al., 2015). This multifaceted approach to revitalizing millet includes integrating scientific research, traditional knowledge, market dynamics, and community engagement. Addressing challenges faced by millet cultivators—including limited access to resources, environmental fluctuations, and market competition—requires a holistic framework that leverages local insights while incorporating scientific advancements to build resilient and adaptive agricultural systems (Jaiswal et al., 2023; Stich et al., 2010).

As we look ahead, the importance of regional policies cannot be overstated. These policies should aim to highlight the role of millet within the broader context of food security and nutrition strategies. Ensuring that millet receives the same level of attention as other staple crops in governmental agricultural policies will support efforts for its revival (Jaiswal et al., 2023). Moreover, international collaborations may further enhance these efforts by sharing knowledge and resources that could boost millet production and consumption on a global scale (Sultan et al., 2013). Revitalizing millet in West Africa is a comprehensive endeavor that requires the convergence of agroecological practices, community-led initiatives, institutional support, and innovative commercialization strategies. The future of millet cultivation offers resilience against climate change, improved food security, and economic empowerment for local communities. By continuing to emphasize the unique attributes of millet, stakeholders can ensure its crucial role in future sustainable agricultural paradigms.

5.8 Sociobiological Reflections on Millet

The interplay between agricultural practices and sociocultural dynamics surrounding millet, particularly fonio, is a significant aspect of food sovereignty, sustainability, and resilience in West Africa. Understanding the coevolution of farmers and these crops illustrates a sustainable framework developed over centuries of adaptation and cultivation strategies that arise from an intricate connection between Indigenous knowledge and biodiversity. This symbiotic relationship has fostered resilient farming systems that provide reliable food sources even in marginal conditions, thereby reinforcing the concept of food sovereignty—a principle that empowers local communities to reclaim control over their food systems and resources (Burton et al., 2024; Gigou et al., 2009).

Millet serves as an emblematic crop that not only sustains local diets but also embodies ecological adaptability to the challenging environmental conditions of West Africa. The resilience of fonio farming systems is tied to their ability to thrive in low-nutrient and poorly managed soils, where other cereals may struggle to grow (Champion et al., 2021). This inherent adaptability is underscored by fonio's rapid growth cycle and minimal agricultural inputs, enabling communities to rely on this crop for food security when climate variability threatens more conventional crops (Animasaun & Lawrence, 2023). Numerous studies have shown that fonio is compatible with these environmental challenges, owing in part to its drought resistance and its ability to flourish under low-rainfall scenarios (Saleh et al., 2013; Sonkamble et al., 2024). Moreover, fonio and other minor millets contribute significantly to local diet diversity, providing not only basic carbohydrates but also essential micronutrients and health benefits (Ogwu et al., 2024). Their high nutritional value, characterized by beneficial amino acids like methionine and lysine, along with important minerals such as iron and calcium, enhances dietary diversity and addresses pressing malnutrition issues—especially in many West African communities (Bhat et al., 2018; Satyavathi et al., 2021). As global interest shifts toward sustainable and nutritious food systems, the resurgence of fonio reflects an increasing recognition of its potential, resulting in policy initiatives that promote millets as integral to food security strategies (Ceasar & Maharajan, 2022; Nwokocha et al., 2024).

Further elaboration on fonio's genetic and agronomic attributes reveals another layer of this millet's importance. Studies have highlighted its unique genome, which provides the genetic diversity needed to bolster agricultural resilience as climate change affects traditional farming practices (Doggalli et al., 2024; Mohod et al., 2023). Understanding these genetic characteristics opens pathways for breeding programs aimed at enhancing yield, nutritional content, and climate readiness (Abrouk et al., 2020). The sociocultural dimensions of fonio cultivation also deserve attention, as they reflect the traditional practices and cultural signifiers attached to this crop. Fonio is often celebrated in local ceremonies and viewed as a staple within various culinary traditions, connecting ancestral wisdom with modern nutritional needs (Burton et al., 2024; Choudhary et al., 2023). As communities adapt to ongoing climatic and economic pressures, the choices surrounding fonio cultivation embody a rejection of industrial agricultural models that prioritize high-yield monocultures in favor of ecologically sustainable and culturally resonant agricultural practices (Burton et al., 2025).

In addition, social movements advocating for the revival of millets center around recognizing their ecosystem services, such as soil health and local biodiversity conservation. Historically pushed aside in agricultural development due to the dominant focus on staple cereals like rice and maize, millets—including fonio—are now recognized as key players in addressing both food insecurity and environmental degradation (Adoukonou-Sagbadja et al., 2010; Animasaun & Lawrence, 2023). Agricultural diversification strategies that emphasize millets not only enhance local resilience but also promote community engagement, which is essential for sustainable agricultural practices (Ceasar & Maharajan,

2022; Choudhary et al., 2023). The implications reach far beyond immediate food security; fonio's contribution to climate adaptation is highlighted by its lower dependence on synthetic fertilizers and its ability to thrive in less-than-ideal conditions. Promoting cashew crops helps to mitigate climate impacts, which is crucial as communities face the realities of climate-induced agricultural decline (Saleh et al., 2013). Thus, incorporating fonio and other minor millets into modern agricultural systems aligns with international goals for sustainable development and nutritional security (Adoukonou-Sagbadja et al., 2010; Nwokocha et al., 2024).

Public policies and market interventions play a crucial role in enhancing fonio's contribution to food sovereignty and community resilience. Efforts to create favorable market conditions for fonio, such as improving supply chains and facilitating direct sales from farmers to consumers, can empower local producers and stabilize their livelihoods amid fluctuating market prices (Goron & Raizada, 2015; Tiwari et al., 2023). Education and outreach programs focusing on the nutritional benefits and sustainability of fonio can further increase consumer awareness and demand, facilitating its reintegration into mainstream diets and preferences (Ceasar & Maharajan, 2022; Nwokocha et al., 2024). Advocacy surrounding fonio and similar crops can help reshape perceptions, promoting their utilization as both nourishment and heritage. Such initiatives can position millets as critical components of national food policies that aim to address the emerging challenges of food insecurity, environmental degradation, and climate adaptation, particularly in vulnerable regions within West Africa where traditional cereals struggle to maintain productivity (Choudhary et al., 2023; Singh et al., 2023).

5.9 Conclusion

Fonio and millet are not just crops—they are keystones in the sociobiological heritage of West African communities. Their ongoing cultivation reflects a sophisticated system of ecological adaptation, nutritional ingenuity, and cultural resilience that predates colonial agricultural models and remains vital in an era of climate uncertainty. This chapter has shown that fonio and millet's drought tolerance, rapid maturation, and rich micronutrient profiles uniquely position them as strategic resources for ensuring food security and environmental sustainability in the Sahel and beyond. The gendered dimensions of their production, ceremonial and symbolic roles, and integration into local economies underscore their broader sociocultural and ecological significance. Yet, these grains remain marginalized in agricultural policy and research agendas. Re-centering fonio and millet in national food strategies, academic inquiry, and market development is crucial for fostering food sovereignty rooted in Indigenous knowledge systems. Moving forward, interdisciplinary and participatory approaches that integrate genomic research, agroecological principles, and value chain innovations will be essential. By reframing fonio and millet as agents of resilience rather than relics of subsistence, this chapter advocates for a paradigm shift—one that honors the wisdom of traditional farming systems while leveraging scientific advances to reposition these crops as cornerstones of sustainable development in West Africa.

Chapter Reflection

1. How have fonio and millets coevolved with the challenging Sahelian and Sudano-Sahelian environments?
2. What roles do Indigenous knowledge systems play in sustaining millet cultivation and ensuring food security?
3. Why are gendered labor roles, particularly women's contributions, so central to millet farming systems?
4. How do rituals and cultural practices surrounding fonio and millet reinforce community identity and resilience?
5. In what ways are fonio and millets superior nutritionally compared to global staples like rice, maize, and wheat?
6. How are market globalization and the commodification of fonio as a "superfood" both empowering and potentially threatening to local farmers?
7. How is urban migration contributing to the erosion of millet cultivation knowledge and practices?
8. What role can agroecological management and farmer-led seed systems play in revitalizing millet production?
9. How can genetic diversity in fonio and pearl millet be harnessed for future climate-resilient agriculture?
10. What are the key policy and institutional gaps that must be addressed to elevate millets within national and regional food security frameworks?
11. How does the sociobiological framing of fonio and millet challenge dominant industrial agricultural models in Africa?
12. How might global dietary shifts and health movements create new opportunities for promoting fonio and millet cultivation in both local and international markets?

REFERENCES

Abdoul-Karim, T., Atta, S., Soulé, M., & Bakasso, Y. (2022). Combined effect of fertilizer micro-dosing and intercropped millet/cowpea effect on agronomic and economic advantages in prone Sahel area, Niger. *Discover Sustainability*, *3*(1), 31. https://doi.org/10.1007/s43621-022-00099-2

Abrouk, M., Ahmed, H., Cubry, P., Šimoníková, D., Cauet, S., Bettgenhaeuser, J., Gapa, L., Pailles, Y., Scarcelli, N., Couderc, M., Zekraoui, L., Kathiresan, N., Čížková, J., Hřibová, E., Doležel, J., Arribat, S., Bergès, H., Wieringa, J. J., Gueye, M., … Krattinger, S. (2020). Fonio millet genome unlocks African orphan crop diversity for agriculture in a changing climate. https://doi.org/10.1101/2020.04.11.037671

Adeyemo, O. (2022). Editorial: A call to strengthen eco-innovation using Indigenous resources and waste products. *Proceedings of the Nigerian Academy of Science, 15*(1). https://doi.org/10.57046/rmoc6397

Adoukonou-Sagbadja, H., Dansi, A., Vodouhè, R., & Akpagana, K. (2006). Indigenous knowledge and traditional conservation of fonio millet (*Digitaria exilis, Digitaria iburua*) in Togo. *Biodiversity and Conservation, 15*(8), 2379–2395. https://doi.org/10.1007/s10531-004-2938-3

Adoukonou-Sagbadja, H., Wagner, C., Dansi, A., Ahlemeyer, J., Daïnou, O., Akpagana, K., Ordon, F., & Friedt, W. (2007). Genetic diversity and population differentiation of traditional fonio millet (*Digitaria* spp.) landraces from different agro-ecological zones of West Africa. TAG. *Theoretical and Applied Genetics. Theoretische und angewandte Genetik, 115*(7), 917–931. https://doi.org/10.1007/s00122-007-0618-x

Animasaun, D., & Lawrence, J. (2023). Phenotypic characterisation and comparative non-targeted gc-ms-based metabolomic profiling of two contrasting seedling fonio millet (*Digitaria exilis*) accessions: An insight to drought tolerance in small millets. https://doi.org/10.21203/rs.3.rs-3369465/v1

Banda, L. O. L., Banda, C. V., Banda, J. T., & Singini, T. (2024). Preserving cultural heritage: A community-centric approach to safeguarding the Khulubvi Traditional Temple Malawi. *Heliyon, 10*(18), e37610. https://doi.org/10.1016/j.heliyon.2024.e37610

Bhat, S., Nandini, C., & Tippeswamy, V. (2018). Significance of small millets in nutrition and health-a review. *Asian Journal of Dairy and Food Research, 37*(1), 35–40. https://doi.org/10.18805/ajdfr.dr-1306

Bombom, A., Kaweesi, T., Walugembe, F., Bhebhe, S., & Maphosa, M. (2024). Millets: Traditional "poor man's" crop or future smart nutri-cereals? In L. Yadav & Upasana (Eds.), *Millets - Rediscover ancient grains*. IntechOpen. https://doi.org/10.5772/intechopen.110534

Burton, G., Ceci, P., MacKinnon, L., Masters, L., Ryan, P., Turnbull, C., Ulian, T., & Vorontsova, M. (2025). Phylogenetics, evolution and biogeography of four digitaria food crop lineages across West Africa, India, and Europe. https://doi.org/10.1101/2025.03.27.645812

Burton, G., Gori, B., Camara, S., Ceci, P., Condé, N., Couch, C., Magassouba, S., Vorontsova, M. S., Ulian, T., & Ryan, P. (2024). Landrace diversity and heritage of the Indigenous millet crop fonio (*Digitaria exilis*): Socio-cultural and climatic drivers of change in the Fouta Djallon region of Guinea. *Plants People Planet*, *7*(3), 704–718. https://doi.org/10.1002/ppp3.10490

Carcedo, A., Maddonni, G., Ramalingam, A. P., Parray, S. A., Tugoo, M. Z., Pereira, T. A., Perumal, R., Vara Prasad, P. V., & Ciampitti, I. (2024). Pearl millet phenology assessment: An integration of field, a review, and in Silico approach. *Crop Science*, *64*(6), 3028–3042. https://doi.org/10.1002/csc2.21352

Ceasar, S., & Maharajan, T. (2022). The role of millets in attaining united nation's sustainable developmental goals. *Plants People Planet*, *4(*4), 345–349. https://doi.org/10.1002/ppp3.10254

Champion, L., Fuller, D., Ozainne, S., Huysecom, É., & Mayor, A. (2021). Agricultural diversification in West Africa: An archaeobotanical study of the site of Sadia (Dogon Country, Mali). *Archaeological and Anthropological Sciences*, *13*(4), 60. https://doi.org/10.1007/s12520-021-01293-5

Chanza, N., & Musakwa, W. (2022). Revitalizing Indigenous ways of maintaining food security in a changing climate: Review of the evidence base from Africa. *International Journal of Climate Change Strategies and Management*, *14*(3), 252–271. https://doi.org/10.1108/ijccsm-06-2021-0065

Choudhary, S., Singh, Y. K., Moond, V., Bahadur Singh, D., Kant, A., Tejasree, P., & Pandey, S. K. (2023). Unearthing the nutritional and agricultural value through scientific innovation in the natural gene pool of millets. *International Journal of Environment and Climate Change*, *13*(12), 1188–1201. https://doi.org/10.9734/ijecc/2023/v13i123783

Dansi, A., Adoukonou-Sagbadja, H., & Vodouhè, R. (2010). Diversity, conservation and related wild species of fonio millet (*Digitaria* spp.) in the northwest of Benin. *Genetic Resources and Crop Evolution, 57*(6), 827–839. https://doi.org/10.1007/s10722-009-9522-3

Diack, O., Kane, N., Berthouly-Salazar, C., Guèye, M., Diop, B., Fofana, A., Sy, O., Tall, H., Zekraoui, L., Piquet, M., Couderc, M., Vigouroux, Y., Diouf, D., & Barnaud, A. (2017). New genetic insights into pearl millet diversity as revealed by characterization of early- and late-flowering landraces from Senegal. *Frontiers in Plant Science, 8*. https://doi.org/10.3389/fpls.2017.00818

Diop, B. M., Guèye, M. C., Agbangba, C. E., Cissé, N., Deu, M., Diack, O., Fofana, A., Kane, N. A., Ndir, K. N., Ndoye, I., Ngom, A., Leclerc, C., Piquet, M., Vigouroux, Y., Zekraoui, L., Billot, C., & Barnaud, A. (2018). Fonio (*Digitaria exilis* (kippist) stapf): A socially embedded cereal for food and nutrition security in Senegal. *Ethnobiology Letters, 9*(2), 150–165. https://doi.org/10.14237/ebl.9.2.2018.1072

Diop, B. M., Guèye, M. C., Leclerc, C., Deu, M., Zekraoui, L., Calatayud, C., Rivallan, R., Kaly, J. R., Cissé, M., Piquet, M., Diack, O., Ngom, A., Berger, A., Ndoye, I., Ndir, K., Vigouroux, Y., Kane, N. A., Barnaud, A., & Billot, C. (2025). Ethnolinguistic and genetic diversity of fonio (*Digitaria exilis*) in Senegal. *Plants, People, Planet, 7*(3), 666–678. https://doi.org/10.1002/ppp3.10428

Doggalli, G., Devi, O., Hazarika, S., Tiwari, A., Angmo, D., Laishram, B., Singh, S. S., & Singh, O. B. (2024). Minor millets: An underutilized grains to ensure food security. *International Journal of Research in Agronomy, 7*(5), 546–550. https://doi.org/10.33545/2618060x.2024.v7.i5g.735

Drabo, I., Zangre, R., Danquah, E., Ofori, K., Witcombe, J., & Hash, C. (2018). Identifying farmers' preferences and constraints to pearl millet production in the Sahel and North-Sudan zones of Burkina Faso. *Experimental Agriculture, 55*(5), 765–775. https://doi.org/10.1017/s0014479718000352

Dussert, Y., Snirc, A., & Robert, T. (2015). Inference of domestication history and differentiation between early- and late-flowering varieties in pearl millet. *Molecular Ecology, 24*(7), 1387–1402. https://doi.org/10.1111/mec.13119

Egeruoh-Adindu, I. (2022). Leveraging Indigenous knowledge for effective environmental governance in West Africa. *Beijing Law Review, 13*(4), 931–947. https://doi.org/10.4236/blr.2022.134060

Fiaveh, D. (2019). Masculinity, male sexual virility, and use of aphrodisiacs in Ghana. *The Journal of Men S Studies, 28*(2), 165–182. https://doi.org/10.1177/1060826519887510

Gigou, J., Stilmant, D., Diallo, T., Cissé, N., Sanogo, M., Vaksmann, M., & Dupuis, B. (2009). Fonio millet (*Digitaria exilis*) response to n, p and k fertilizers under varying climatic conditions in West Africa. *Experimental Agriculture*, *45*(4), 401–415. https://doi.org/10.1017/s0014479709990421

Goron, T., & Raizada, M. (2015). Genetic diversity and genomic resources available for the small millet crops to accelerate a new green revolution. *Frontiers in Plant Science*, *6*. https://doi.org/10.3389/fpls.2015.00157

Hayes, A. M. R., Gozzi, F., Diatta, A., Gorissen, T., Swackhamer, C., Bellmann, S., & Hamaker, B. R. (2020). Some pearl millet-based foods promote satiety or reduce glycaemic response in a crossover trial. *British Journal of Nutrition*, *126*(8), 1168–1178. https://doi.org/10.1017/s0007114520005036

Hu, Z., Mbacké, B., Perumal, R., Guèye, M., Sy, O., Bouchet, S., Vara Prasad, V., & Morris, G. P. (2015). Population genomics of pearl millet (*Pennisetum glaucum* (L.) R. Br.): Comparative analysis of global accessions and Senegalese landraces. *BMC Genomics*, *16*(1), 1048. https://doi.org/10.1186/s12864-015-2255-0

Ibrahim, A., Abaidoo, R., Fatondji, D., & Opoku, A. (2015). Determinants of fertilizer microdosing-induced yield increment of pearl millet on an acid sandy soil. *Experimental Agriculture*, *52*(4), 562–578. https://doi.org/10.1017/s0014479715000241

Ijeomah, O., Ndukwu, C., & Azubuike, C. (2023). Safety evaluation of fonio (digitaria exilis)/ricebean (vigna umbellata) based complementary food. *European Journal of Nutrition and Food Safety*, *15*(4), 16–22. https://doi.org/10.9734/ejnfs/2023/v15i41303

Izah, S. C., & Ogwu, M. C. (2023). *Sustainable utilization and conservation of Africa's biological resources and environment*. Sustainable Development and Biodiversity Series (1st ed., Vol. 32, 691p.) (Ramawat K.G.). Springer Nature. https://doi.org/10.1007/978-981-19-6974-4

Jaiswal, N., Meghalatha, K., Nishad, T., Khan, A., & Jatav, D. (2023). Revitalizing agriculture extension services for millets: A comprehensive review of strategies, challenges, and innovations. *International Journal of Plant and Soil Science*, *35*(23), 307–315. https://doi.org/10.9734/ijpss/2023/v35i234245

Kanlindogbè, C., Sêkloka, E., & Kwon-Ndung, E. (2020). Genetic resources and varietal environment of grown fonio millets in West Africa: Challenges and perspectives. *Plant Breeding and Biotechnology*, *8*(2), 77–88. https://doi.org/10.9787/pbb.2020.8.2.77

Kushwaha, H., Woliy, K., Singh, V., Kumar, A., & Yadav, D. (2014). Assessment of genetic diversity among cereals and millets based on PCR amplification using Dof (DNA binding with one finger) transcription factor gene-specific primers. *Österreichische Botanische Zeitschrift*, *301*(2), 833–840. https://doi.org/10.1007/s00606-014-1095-8

Leipe, C., Long, T., Sergusheva, E. A., Wagner, M., & Tarasov, P. E. (2019). Discontinuous spread of millet agriculture in eastern Asia and prehistoric population dynamics. *Science Advances*, *5*(9), eaax6225. https://doi.org/10.1126/sciadv.aax6225

Longueville, F., Hountondji, Y., Kindo, I., Gemenne, F., & Ozer, P. (2016). Long-term analysis of rainfall and temperature data in Burkina Faso (1950–2013). *International Journal of Climatology*, *36*(13), 4393–4405. https://doi.org/10.1002/joc.4640

Mabhaudhi, T., Chimonyo, V., Hlahla, S., Massawe, F., Mayes, S., Nhamo, L., & Modi, A. T. (2019). Prospects of orphan crops in climate change. *Planta*, *250*(3), 695–708. https://doi.org/10.1007/s00425-019-03129-y

Manning, K., Pelling, R., Higham, T., Schwenniger, J., & Fuller, D. (2011). 4500-year old domesticated pearl millet (*Pennisetum glaucum*) from the Tilemsi Valley, Mali: New insights into an alternative cereal domestication pathway. *Journal of Archaeological Science*, *38*(2), 312–322. https://doi.org/10.1016/j.jas.2010.09.007

Mariac, C., Jehin, L., Saïdou, A., Thuillet, A., Couderc, M., Sire, P., Jugdé, H., Adam, H., Bezançon, G., Pham, J.-L., & Vigouroux, Y. (2010). Genetic basis of pearl millet adaptation along an environmental gradient investigated by a combination of genome scan and association mapping. *Molecular Ecology*, *20*(1), 80–91. https://doi.org/10.1111/j.1365-294x.2010.04893.x

Mason, S., Maman, N., & Pale, S. (2015). Pearl millet production practices in semi-arid West Africa: A review. *Experimental Agriculture*, *51*(4), 501–521. https://doi.org/10.1017/s0014479714000441

Michael, T., Ozersky, P., Hartwick, N., Colt, K., Allsing, N., Marmerto, A., Kitony, J., Carroll, E., Van Buren, R., de Gracia Coquerel, M., Gano, B., Eck, N., Adams, J., Beyene, G., Taylor, N., Villmer, J., Kane, N., Guèye, M., Diop, B., … Shakoor,

N. (2025). Trait development and CRISPR-enabled improvement in the African orphan crop fonio (*Digitaria exilis*). https://doi.org/10.21203/rs.3.rs-5743774/v1

Mohanan, M. M., Vijayakumar, A., Bang-Berthelsen, C. H., Mudnakudu-Nagaraju, K. K., & Shetty, R. (2025). Millets: Journey from an ancient crop to sustainable and healthy food. *Foods, 14*(10), 1733. https://doi.org/10.3390/foods14101733

Mohod, N., Ashoka, P., Borah, A., Goswami, P., Koshariya, A., Sahoo, S., & Prabhavathi, N. (2023). The international year of millet 2023: A global initiative for sustainable food security and nutrition. *International Journal of Plant and Soil Science, 35*(19), 1204–1211. https://doi.org/10.9734/ijpss/2023/v35i193659

Moussa, M., Ponrajan, A., Campanella, O., Okos, M., Martínez, M., & Hamaker, B. (2022). Novel pearl millet couscous process for West African markets using a low-cost single-screw extruder. *International Journal of Food Science and Technology, 57*(7), 4594–4601. https://doi.org/10.1111/ijfs.15797

Nwokocha, K., Oboh, G., Ademosun, A., Adefegha, S., & Akindahunsi, A. (2024). Sensory qualities, nutritional properties and glycaemic indices of biscuits produced from processed fonio millet flour. *Tropical Journal of Phytochemistry and Pharmaceutical and Sciences, 3*(1). https://doi.org/10.26538/tjpps/v3i1.2

Ogwu, M. C. (2019). Towards sustainable development in Africa: The challenge of urbanization and climate change adaptation. In P. B. Cobbinah & M. Addaney (Eds.), *The geography of climate change adaptation in Urban Africa* (pp. 29–55). Springer Nature. http://doi.org/10.1007/978-3-030-04873-0_2

Ogwu, M. C. (2020). Value of *Amaranthus* [L.] species in Nigeria. In V. Waisundara (Ed.), *Nutritional value of Amaranth* (pp. 1–21). IntechOpen. http://dx.doi.org/10.5772/intechopen.86990

Ogwu, M. C. (2023). Local food crops in Africa: Sustainable utilization, threats, and traditional storage strategies. In S. C. Izah & M. C. Ogwu (Eds.), *Sustainable utilization and conservation of Africa's biological resources and environment.* Sustainable Development and Biodiversity (Vol. 888, pp. 353–374). Springer. https://doi.org/10.1007/978-981-19-6974-4_13

Ogwu, M. C., & Izah, S. C. (2023). *One health implications of agrochemicals and their sustainable alternatives.* Sustainable Development and Biodiversity Series (1st ed., Vol. 34, 826p.) (Ramawat K.G.). Springer. https://doi.org/10.1007/978-981-99-3439-3

Ogwu, M. C., Izah, S. C., Alves, A. A. C., & Babu, S. (2024). *Sustainable Cassava: Strategies from production through waste management*. Elsevier.

Ogwu, M. C., Izah, S. C., & Ntuli, N. R. (2024). *Food safety and quality in the Global South*. Springer Nature. https://doi.org/10.1007/978-981-97-2428-4

Osawaru, M. E., & Ogwu, M. C. (2020). Survey of plant and plant products in local markets within Benin City and environs. In L. W. Filho, N. Ogugu, D. Ayal, L. Adelake, & I. da Silva (Eds.), *African handbook of climate change adaptation* (pp. 1–24). Springer Nature. http://doi.org/10.1007/978-3-030-42091-8_159-1

Pérez-Jordà, G., Peña-Chocarro, L., Sabato, D., Peralta Gómez, A., Ribera, A., García Borja, P., Negre, J., & Martín Civantos, J. M. (2024). The path of African Millets (*Pennisetum glaucum* and *Sorghum bicolor*) to Iberia. *Agronomy*, *14*(10), 2375. https://doi.org/10.3390/agronomy14102375

Pierre, C., Bergametti, G., Marticorena, B., Mougin, É., Lebel, T., & Ali, A. (2011). Pluriannual comparisons of satellite-based rainfall products over the Sahelian belt for seasonal vegetation modeling. *Journal of Geophysical Research Atmospheres*, *116*(D18). https://doi.org/10.1029/2011jd016115

Pucher, A., Sy, O., Angarawai, I., Gondah, J., Zangre, R., Ouédraogo, M., Sanogo, M. D., Boureima, S., Tom Hash, C., & Haussmann, B. I. G. (2015). Agro-morphological characterization of West and Central African pearl millet accessions. *Crop Science*, *55*(2), 737–748. https://doi.org/10.2135/cropsci2014.06.0450

Pujarula, V., Pusuluri, M., Bollam, S., Das, R., Ratnala, R., Adapala, G., Thuraga, V., Rathore, A., Srivastava, R. K., & Gupta, R. (2021). Genetic variation for nitrogen use efficiency traits in global diversity panel and parents of mapping populations in pearl millet. *Frontiers in Plant Science*, *12*. https://doi.org/10.3389/fpls.2021.625915

Ravichandran, A., Nallusamy, S., Mannu, J., Bharathi, N., Senthil, N., Venu-gopal, A., & Sowdhamini, R. (2023). Deciphering millet diversity: Proteomic clusters and phylogenetic insights. *International Journal of Plant and Soil Science*, *35*(20), 125–133. https://doi.org/10.9734/ijpss/2023/v35i203792

Rhoné, B., Defrance, D., Berthouly-Salazar, C., Mariac, C., Cubry, P., Couderc, M., Dequincey, A., Assoumanne, A., Kane, N. A., Sultan, B., Barnaud, A., & Vigouroux, Y. (2020). Pearl millet genomic vulnerability to climate change in West Africa highlights the need for regional collaboration. *Nature Communications*, *11*(1). https://doi.org/10.1038/s41467-020-19066-4

Rouamba, A., Shimelis, H., Drabo, I., Laing, M., Gangashetty, P., Mathew, I., Mrema, E., & Shayanowako, A. (2021). Constraints to pearl millet (*Pennisetum glaucum*) production and farmers' approaches to striga hermonthica management in Burkina Faso. *Sustainability, 13*(15), 8460. https://doi.org/10.3390/su13158460

Saleh, A., Zhang, Q., Chen, J., & Shen, Q. (2013). Millet grains: Nutritional quality, processing, and potential health benefits. *Comprehensive Reviews in Food Science and Food Safety, 12*(3), 281–295. https://doi.org/10.1111/1541-4337.12012

Sangappa, Abbuseat, Kailashnath, D. Rafi E. Charishma, K. Ramakiran and Ravi, S.C. (2023). An investigation into consumer preferences regarding Millet and Millet-based value-added products. *Biological Forum—An International Journal, 15*(10): 1346–1350. https://www.researchtrend.net/bfij/pdf/An-Investigation-into-Consumer-Preferences-Regarding-Millet-and-Millet-Based-Value-Added-Products-Sangappa-247.pdf

Satyavathi, C., Ambawat, S., Khandelwal, V., & Srivastava, R. (2021). Pearl millet: A climate-resilient nutricereal for mitigating hidden hunger and provide nutritional security. *Frontiers in Plant Science, 12*. https://doi.org/10.3389/fpls.2021.659938

Serba, D., Sy, O., Sanogo, M., Issaka, A., Ouédraogo, M., Ango, I., Drabo, I., & Kanfany, G. (2020). Performance of dual-purpose pearl millet genotypes in West Africa: Importance of morphology and phenology. *African Crop Science Journal, 28*(4), 481–498. https://doi.org/10.4314/acsj.v28i4.1

Singh, P., Boote, K., Kadiyala, M., Nedumaran, S., Gupta, S., Srinivas, K., & Bantilan, M. (2017). An assessment of yield gains under climate change due to genetic modification of pearl millet. *The Science of the Total Environment, 601–602*, 1226–1237. https://doi.org/10.1016/j.scitotenv.2017.06.002

Singh, S., Yadav, R., Tripathi, A., Kumar, M., Kumar, M., Yadav, S., Kumar, D., Kumar, S., & Yadav, R. (2023). Current status and promotional strategies of millets: A review. *International Journal of Environment and Climate Change, 13*(9), 3088–3095. https://doi.org/10.9734/ijecc/2023/v13i92551

Solomon, D., Lehmann, J., Fraser, J., Leach, M., Amanor, K., Frausin, V., Kristiansen, S. M., Millimouno, D., & Fairhead, J. (2016). Indigenous African soil enrichment as a climate-smart sustainable agriculture alternative. *Frontiers in Ecology and the Environment, 14*(2), 71–76. https://doi.org/10.1002/fee.1226

Sonkamble, S., Kumbhar, A., Pawar, K., & Nalawade, S. (2024). Sustainable cultivation of millets. *International Journal of Multidisciplinary Research and Analysis*, *7*(5). https://doi.org/10.47191/ijmra/v7-i05-20

Stich, B., Haussmann, B. I. G., Pasam, R., Bhosale, S., Hash, C. T., Melchinger, A. E., & Parzies, H. K. (2010). Patterns of molecular and phenotypic diversity in pearl millet [pennisetum glaucum (l.) r. br.] from West and Central Africa and their relation to geographical and environmental parameters. *BMC Plant Biology*, *10*(1), 216. https://doi.org/10.1186/1471-2229-10-216

Styring, A., Diop, A., Bogaard, A., Champion, L., Fuller, D., Gestrich, N., Macdonald, K. C., & Neumann, K. (2019). Nitrogen isotope values of pennisetum glaucum (pearl millet) grains: Towards a reconstruction of past cultivation conditions in the Sahel, West Africa. *Vegetation History and Archaeobotany*, *28*(6), 663–678. https://doi.org/10.1007/s00334-019-00722-9

Subedi, B. (2022). Whose knowledge counts? a reflection on the field narratives of Indigenous health knowledge and practices. *Dhaulagiri Journal of Sociology and Anthropology*, *16*, 59–69. https://doi.org/10.3126/dsaj.v16i01.50947

Sultan, B., Defrance, D., & Iizumi, T. (2019). Evidence of crop production losses in West Africa due to historical global warming in two crop models. *Scientific Reports*, *9*(1), 12834. https://doi.org/10.1038/s41598-019-49167-0

Sultan, B., Roudier, P., Quirion, P., Alhassane, A., Muller, B., Dingkuhn, M., Ciais, P., Guimberteau, M., Traore, S., & Baron, C. (2013). Assessing climate change impacts on sorghum and millet yields in the Sudanian and Sahelian savannas of West Africa. *Environmental Research Letters*, *8*(1), 014040. https://doi.org/10.1088/1748-9326/8/1/014040

Tabassum, N., Ahmed, H. I., Parween, S., Sheikh, A. H., Saad, M. M., Krattinger, S. G., & Hirt, H. (2024). Host genotype, soil composition, and geo-climatic factors shape the fonio seed microbiome. *Microbiome*, *12*(1), 11. https://doi.org/10.1186/s40168-023-01725-5

Tiwari, H., Singh, P., Naresh, R., M., M., Monika, S., Islam, A., Kumar, S., Veer Singh, K., Pandey, A. K., & Shukla, A. (2023). Millets based integrated farming system for food and nutritional security, constraints and agro-diversification strategies to fight global hidden hunger: A review. *International Journal of Plant and Soil Science*, *35*(19), 630–643. https://doi.org/10.9734/ijpss/2023/v35i193593

Trail, P., Abaye, O., Thomason, W., Thompson, T., Guèye, F., Diédhiou, I., Diatta, M. B., & Faye, A. (2016). Evaluating intercropping (living cover) and mulching (desiccated cover) practices for increasing millet yields in Senegal. *Agronomy Journal, 108*(4), 1742–1752. https://doi.org/10.2134/agronj2015.0422

Upadhyaya, H., Reddy, K., Ahmed, M., Kumar, V., Gumma, M., & Ramachandran, S. (2016). Geographical distribution of traits and diversity in the world collection of pearl millet [*Pennisetum glaucum* (L.) R. Br., synonym: *Cenchrus americanus* (L.) morrone] landraces conserved at the ICRISAT genebank. *Genetic Resources and Crop Evolution, 64*(6), 1365–1381. https://doi.org/10.1007/s10722-016-0442-8

Verdugo, G., Cuadrado, G., & Ortega, Y. (2023). Family farming as a contribution to food sovereignty, case Guarainag Parish. *Agriculture, 13*(9), 1827. https://doi.org/10.3390/agriculture13091827

Vetriventhan, M., Azevedo, V., Upadhyaya, H., Raguchander, T., Kane-Potaka, J., Anitha, S., Ceasar, S. A., Muthamilarasan, M., Venkatesh Bhat, B., Hariprasanna, K., Bellundagi, A., Cheruku, D., Backiyalakshmi, C., Santra, D., Vanniarajan, C., & Tonapi, V. A. (2020). Genetic and genomic resources, and breeding for accelerating improvement of small millets: Current status and future interventions. *The Nucleus, 63*(3), 217–239. https://doi.org/10.1007/s13237-020-00322-3

Wang, X., Chen, S., Ma, X., Yssel, A. E. J., Chaluvadi, S. R., Johnson, M. S., Gangashetty, P., Hamidou, F., Sanogo, M. D., Zwaenepoel, A., Wallace, J., de Peer, Y., Bennetzen, J. L., & Van Deynze, A. (2021). Genome sequence and genetic diversity analysis of an under-domesticated orphan crop, white fonio (*Digitaria exilis*). *Gigascience, 10*(3). https://doi.org/10.1093/gigascience/giab013

Watson, V., & Knight-Manuel, M. (2017). Challenging popularized narratives of immigrant youth from West Africa: Examining social processes of navigating identities and engaging civically. *Review of Research in Education, 41*(1), 279–310. https://doi.org/10.3102/0091732x16689047

Yadav, R., Sehgal, D., & Vadez, V. (2010). Using genetic mapping and genomics approaches in understanding and improving drought tolerance in pearl millet. *Journal of Experimental Botany, 62*(2), 397–408. https://doi.org/10.1093/jxb/erq265

Yameogo, P., Zigani, S., Jiao, X., Zhang, H., & Zhang, J. (2022). Improving fertilization methods and cropping systems for sustainable production of pearl millet (*Pennisetum glaucum*) in West Africa: A review. *Frontiers of Agricultural Science and Engineering, 9*(4), 588. https://doi.org/10.15302/j-fase-2021422

Yerima, A., & Achigan-Dako, E. (2021). A review of the orphan small grain cereals improvement with a comprehensive plan for genomics-assisted breeding of fonio millet in West Africa. *Plant Breeding*, *140*(4), 561–574. https://doi.org/10.1111/pbr.12930

Yerima, A. R. I. B., Achigan-Dako, E. G., Aissata, M., Sêkloka, E., Billot, C., Adje, C. O. A., Barnaud, A., & Bakasso, Y. (2020). Agromorphological characterization revealed three phenotypic groups in a region-wide germplasm of fonio (*Digitaria exilis* (Kippist) Stapf) from West Africa. *Agronomy*, *10*(11), 1653. https://doi.org/10.3390/agronomy10111653

Zhang, L., Ye, L., Qian, L., & Zuo, X. (2024). The impact of dietary preference on household food waste: Evidence from China. *Frontiers in Nutrition*, *11*. https://doi.org/10.3389/fnut.2024.1415734

Zhu, F. (2020). Fonio grains: Physicochemical properties, nutritional potential, and food applications. *Comprehensive Reviews in Food Science and Food Safety*, *19*(6), 3365–3389. https://doi.org/10.1111/1541-4337.12608

Zossou, E., Arouna, A., Diagne, A., & Agboh-Noameshie, R. (2016). Gender gap in acquisition and practice of agricultural knowledge: Case study of rice farming in West Africa. *Experimental Agriculture*, *53*(4), 566–577. https://doi.org/10.1017/s0014479716000582

CHAPTER 6

Teff and Enset in Ethiopia

Abstract

Teff (*Eragrostis tef*) and enset (*Ensete ventricosum*) are fundamental components of Ethiopia's agroecological and sociocultural systems, providing critical nutrition, economic stability, and cultural identity. This chapter presents a sociobiological investigation of their genetic diversity, agronomic practices, functional food properties, and cultural significance. Teff, a gluten-free grain rich in iron and essential amino acids, addresses micronutrient deficiencies and supports global markets, while enset—termed the "tree against hunger"—contributes to year-round food security through its starchy pseudostems and fermentation-based preservation techniques. Drawing on interdisciplinary research and Indigenous knowledge, this chapter examines gendered labor divisions, seed selection processes, and intergenerational knowledge transfer that sustain these complex farming systems. Women play pivotal roles in processing, culinary traditions, and the conservation of varietal diversity, underscoring the gendered dimensions of food sovereignty. The chapter also explores how spiritual practices, communal rituals, and ecological stewardship are embedded in the cultivation of teff and enset, reinforcing community resilience. However, these systems face mounting challenges from climate variability, land-use change, market-driven homogenization, and policy gaps that threaten genetic resources and traditional knowledge. This chapter advocate for integrative, transdisciplinary approaches that elevate farmer-led innovation, prioritize agrobiodiversity conservation, and develop policy frameworks sensitive to both ecological sustainability and sociocultural integrity. Teff and enset exemplify how Indigenous food systems can inform global discussions on sustainable agriculture, nutrition, and climate adaptation within the Anthropocene.

Keywords: teff, enset, Indigenous knowledge, agroecology, gendered practices, food sovereignty, fermentation, Ethiopian food systems

6.1 Introduction

Ethiopia is known for its diverse agroecological zones, which contribute to rich ethnobotanical diversity. This diversity showcases the remarkable adaptability of various crops, particularly teff (*Eragrostis tef—E. tef*) and enset (*Ensete ventricosum—E. vent*), to different climatic and geographical conditions throughout the nation. Ethiopia's climatic variety, ranging from the highlands that support coffee and enset cultivation to arid lowlands suited for other crops, creates a unique backdrop for these two staples (Gebrehiwot et al., 2024; Wang & Çakır, 2021). Enset is particularly important in the southern regions, where its cultivation is vital to the livelihoods of local communities, ensuring food security and aiding traditional practices (Olango et al., 2014; Quinlan et al., 2014). In contrast, teff, with its origins tracing back over 4,000 years, plays a crucial role in the diets of many Ethiopians (Abewa et al., 2019; Hrušková et al., 2012). The dual significance of teff and enset encompasses nutritional value, cultural identity, socioeconomic resilience, and traditional agricultural practices that have endured over generations.

The role of teff and enset goes beyond mere sustenance; they embody elements of Ethiopian identity and cultural heritage. Teff is foundational in the production of injera, a staple food that serves both as a primary dietary staple and as an accompaniment to various traditional dishes, representing the heart of Ethiopian cuisine (Gebru et al., 2020; Habte et al., 2022). Enset, often referred to as the "false banana," is crucial not only for its nutritional contributions but also as part of rituals and social gatherings within ethnolinguistic groups like the Sidama and Wolaita (Olango et al., 2014; Quinlan et al., 2014). Both crops are deeply intertwined with the cultural practices and social fabric of Ethiopian society, and their continued cultivation illustrates a profound connection between agriculture and identity in Ethiopia.

Ethiopia's diverse agroecological zones encompass a wide range of climatic conditions, elevations, and soils, thereby supporting various crops and plant species that have shaped the country's agriculture and cuisine. The highland regions are characterized by temperate climates that favor the cultivation of enset, while the midland and lowland areas often support teff production due to its resilience against various environmental stressors (Abewa et al., 2019; Nascimento et al., 2018). Each zone harbors distinct traditional knowledge systems and practices that enhance the cultivation of these crops, such as soil management and inter-cropping strategies that leverage complementary crop traits (Olango et al., 2014;

Quinlan et al., 2014). Ethiopia is home to over 6,000 plant species, of which around 1,200 are utilized for food or medicinal purposes, showcasing the ethnobotanical wealth of the nation (Borrell et al., 2019). This rich botanical diversity is maintained through intricate cultural practices and Indigenous knowledge systems (Olango et al., 2014). In particular, the Wolaita and Sidama communities cultivate multiple varieties of enset, tailored to different climatic conditions and culinary uses, underscoring the adaptability and resilience of these agricultural practices (Olango et al., 2014). Similarly, tef's genetic and phenotypic diversity allows it to be cultivated across varying altitudes and microclimates, making it a staple that fulfills dietary needs throughout Ethiopia (Nascimento et al., 2018).

In addition to enhancing food security, the diverse cultivation practices of teff and enset contribute significantly to the socioeconomic fabric of Ethiopian society. Over 20 million people depend on enset as a staple food source, providing essential calories and nutrients, while teff is increasingly recognized not only for its ethnic significance but also for its export potential to increase the income of smallholder farmers (Fanta & Satheesh, 2019; Gidelew et al., 2022). The interdependence of these crops with the cultural identity of their cultivating communities offers a comprehensive framework for understanding Ethiopia's agroecological landscape. Furthermore, recognizing the intrinsic connection between agroecology and ethnobotany is vital for sustainable practices that support biodiversity while enhancing food security (Borrell et al., 2019).

Teff and enset are essential in the dietary habits of Ethiopians, serving unique roles in culinary practices and enhancing nutritional intake. Teff is particularly revered for its high nutrient density, providing essential amino acids, iron, and calcium, which are especially vital in the context of nutritional deficiencies commonly seen in poorer regions (Gebru et al., 2020; Habte et al., 2022). Incorporating teff into traditional meals, particularly through injera, helps manage various health conditions, including diabetes and high cholesterol, due to its low glycemic index (Habte et al., 2022). Additionally, its gluten-free nature makes it a safe option for individuals with celiac disease, increasing its appeal in global markets amid rising health consciousness. Enset provides numerous benefits not only as a food staple but also as a source of fiber and feed for livestock, supporting the agricultural economy of rural households (Fanta & Satheesh, 2019). The diverse uses of enset extend beyond nutrition; its pseudostems are commonly utilized as building materials, while its leaves serve as natural wrappers for food (Fanta & Satheesh, 2019). Enset is integrated into the cultural and ritual practices of many ethnic groups, emphasizing its significance beyond mere sustenance. The cultural heritage surrounding these crops reflects Ethiopia's unique identity and traditions. Festivals and rituals often focus on cultivating and harvesting teff and enset, fostering community solidarity and reinforcing cultural practices (Olango et al., 2014; Quinlan et al., 2014). The symbolism associated with these staple crops extends to storytelling, music, and art, expressing the deep relationships that Ethiopians maintain with their agrarian landscapes. Such cultural connections underscore the necessity of preserving both teff and enset, particularly in the face of globalization and modernization, which threaten traditional farming practices (Olango et al., 2014).

This chapter explores the significance of teff and enset, delving into their contributions to Ethiopian diets, cultural heritage, and socioeconomic structures. It highlights the adaptive agricultural techniques used in the cultivation of these crops and their importance amid changing environmental and market conditions. Moreover, the chapter evaluates the contemporary challenges and prospects related to the cultivation of teff and enset, including policy implications for enhancing food security and sustaining agricultural practices.

6.2 Botanical and Agronomic Profiles of Teff (*E. tef*) and Enset (*E. vent*)

Teff (*E. tef*) and enset (*E. vent.*) are crucial crops in Ethiopia, recognized for their nutritional contributions and cultural significance. An examination of their botanical and agronomic profiles reveals important insights into their taxonomy, morphology, agroecological requirements, and the genetic diversity that contributes to their resilience against climate change.

6.2.1 Taxonomy and Morphology of Teff and Enset

Teff (*E. tef*) belongs to the Poaceae family and falls under the *Eragrostis* genus, which is known for its fine seeds and remarkable adaptability to diverse environmental conditions. Morphologically, teff is identified by its small and hard seeds, which can vary in color from white to reddish-brown (Borrell et al., 2019; White et al., 2022). The plant typically reaches heights of 0.4 to 1.2 meters and features slender, flowering heads that bend gracefully under their own weight. In contrast, enset (*E. vent*), a member of the Musaceae family, thrives in the highlands of Ethiopia and is recognized for its large, banana-like pseudostems that support broad leaves. Figure 6.1 showcases the remarkable diversity in grain color and morphology among teff accessions collected from various regions of Ethiopia

(Abewa et al., 2019). Each sample represents a specific locality or agroecological zone, such as Denbia, Minjar, Yilmana Densa, and Bahir Dar, reflecting both traditional cultivation practices and localized adaptation. The variation in grain appearance—from creamy white to brownish hues—underscores the genetic richness of teff, Ethiopia's staple cereal crop known for its nutritional value and gluten-free properties. This visual documentation is vital for plant breeders, agronomists, and food scientists aiming to enhance teff productivity, market value, and climate resilience through regionally adapted varieties.

Figure 6.1: Grain Color and Morphological Diversity of Teff (*E. tef*) Accessions Across Ethiopian Agroecological Zones

Source: Abewa et al. (2019)

The enset plant can grow up to 3 meters tall, featuring a notable corm that serves as the primary storage organ for carbohydrates, which allows it to be a staple source of nutrition for millions (Figure 6.2; Borrell et al., 2019; Mengesha et al., 2022; Worojie & Geremew, 2025). Figures 6.3 and 6.4 illustrate the phenotypic diversity in midrib and petiole (Figure 6.2) as well as pseudostem coloration (Figure 6.3) among different enset (*E. vent*) accessions collected from various regions of Ethiopia. These color differences represent important agromorphological traits used in local cultivar classification, and they may reflect underlying

Figure 6.2: Enset Plant

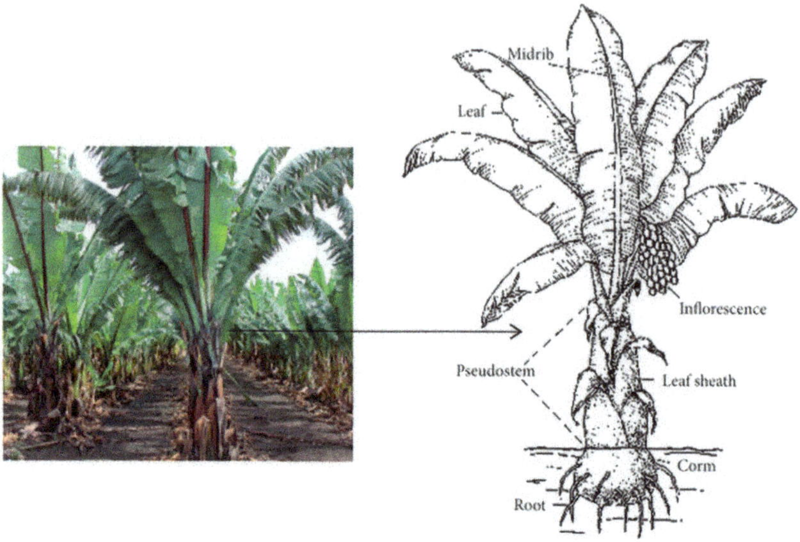

Source: Yemataw et al. (2020)

Figure 6.3: Phenotypic Variation in Midrib and Petiole Among Enset
(*E. vent*) Accessions from Ethiopia Used by Local Farmers
Key: A *Korina*; **B** *Boicho*; **C** *Qebere*; **D** *Separa*; **E** *Qombotra*;
F *Hella*; **G** *Etine*; **H** *Astara*; **I** *Kaseta*

Source: Dilebo et al. (2024)

Figure 6.4: Pseudostem Variation in Color and in End-Use Value: **A-E** Landraces with Sweet *Amicho, Astara, Leqeqa, Qebere, Xessa,* and *Orada,* respectively: **F-H** Landraces with Nonedible and Hard *Amicho, Sisqella, Gishira,* and *Hella* from Left to Right; **I-K** Landraces Considered Best in Fermented Products (*Qocho* and *Bulla*), *Gimbo, Separa,* and *Hiniba* from Left to Right

Source: Dilebo et al. (2024)

genetic diversity or adaptation to specific ecological conditions. Understanding such morphological traits supports breeding efforts, germplasm conservation, and the promotion of farmer-preferred varieties in enset-based farming systems, which are central to food security in Southern Ethiopia. The taxonomic classification of teff and enset crops highlights their deep-rooted significance in the region's agriculture, with teff being a cereal grain native to Ethiopia and enset, often called "false banana," showcasing an essential evolutionary adaptation as a drought-resistant crop. This unique morphological adaptation gives enset a competitive edge in the highland regions where moisture conservation is critical for survival (Borrell et al., 2019; Feleke & Tekalign, 2022; Mengesha et al., 2022; Worojie & Geremew, 2025).

6.2.2 Agroecological Requirements and Cultivation Cycles

Teff flourishes in diverse agroecological zones, thriving particularly well in well-drained soils at altitudes ranging from 1,500 to 2,500 meters above sea level.

Its cultivation cycle spans approximately three to four months, during which it requires full sunlight and moderate rainfall (Chase et al., 2025; Dilebo et al., 2023; Yemataw et al., 2018). Enset, conversely, has slightly different agroecological needs. This crop is predominantly found in moist, fertile highlands and requires significant inputs of organic matter, especially manure, for optimal growth. The enset plant demonstrates a cultivation cycle that can extend several years from planting to harvesting, enabling it to produce food consistently during months when other crops may fail (Nuraga et al., 2022; Olango et al., 2014; Yemataw et al., 2018). The agronomic practices surrounding both crops exhibit remarkable adaptability in response to evolving climatic conditions. Farmers in Ethiopia have engaged in various cropping systems to ensure food security amid unpredictable rainfall patterns. Teff's quick maturation allows for multiple cropping cycles, while enset's longevity ensures a stable food source, affirming its role in traditional farming systems (Borrell et al., 2019; Chase et al., 2025; Nuraga et al., 2022).

6.2.3 Genetic Diversity and Climate Resilience Attributes

Genetic diversity is a defining feature of both teff and enset, significantly contributing to their resilience against climate challenges. Teff exhibits considerable genetic variability, with landraces adapted to distinct ecological niches, enhancing its resistance to diseases and pests (Haile et al., 2023; White et al., 2022). This genetic variability enables selection and breeding programs aimed at improving yield and environmental adaptation, particularly in light of rising global temperatures and erratic precipitation patterns (Feleke & Tekalign, 2022; Mengesha et al., 2022). Regarding enset, recent studies have highlighted the genetic diversity present within its landraces, showcasing traits that enhance drought tolerance and resistance to biotic stressors (Haile et al., 2023; Nuraga et al., 2020; Yemataw et al., 2018). The cultivation of different enset landraces based on local environmental conditions reflects a complex system of farmer selection that prioritizes genetic traits beneficial for survival under climatic stress (Dilebo et al., 2023; Yemataw et al., 2018). Indigenous knowledge plays a vital role in managing this diversity, as farmers utilize traditional methods to select and propagate the most resilient varieties (Gerura et al., 2019; Olango et al., 2014). Current genetic research emphasizes the need for enhanced conservation efforts surrounding these crops, as the agroecological zones where teff and enset thrive may shift with the global climate crisis. Protecting genetic resources will be crucial for maintaining the resilience of these crops and ensuring sustained food security in Ethiopia (Biswas et al., 2020; Getachew et al., 2014; Mengesha et al., 2022). Conservation strategies must include both in situ and ex situ practices to support the genetic base necessary for future breeding and adaptation efforts (Biswas et al., 2020, p. 20; Haile et al., 2023).

6.3 Nutritional and Functional Values

Teff (*E. tef*) and enset (*E. ventricosum*) are crucial food sources with notable nutritional and functional profiles. Both are particularly significant in addressing global nutritional challenges, especially in regions where they serve as staple foods.

6.3.1 Macronutrient and Micronutrient Profiles

Teff has a well-rounded nutritional profile, consisting of high amounts of carbohydrates, proteins, and dietary fiber. Specifically, teff contains significant levels of essential amino acids, notably lysine—often lacking in many grains (Nespeca et al., 2020). Its micronutrient content is also noteworthy; teff boasts higher concentrations of essential minerals compared to many other grains. Studies indicate that teff can contain approximately 150 mg of iron per 100 g, making it an excellent source for combating iron deficiency (Alaunyte et al., 2014; Inglett et al., 2015). Conversely, enset is known for its starchy pseudostems and tubers, which serve as a significant energy source in Ethiopian diets. While enset is lower in protein compared to teff, it provides a substantial amount of carbohydrates and dietary fiber, making it valuable for sustaining energy levels and promoting digestive health (Hassona, 2023; Mezemir, 2015). Enset is also rich in micronutrients such as calcium and potassium, which positively contribute to overall dietary quality (Mezemir, 2015).

6.3.2 Functional Foods and Dietary Roles

Both grains serve vital roles in functional food applications. Teff is inherently gluten-free, making it an appealing alternative for those with celiac disease or gluten intolerance (Barretto et al., 2020; Wojas et al., 2020). The high fiber content of teff supports digestive health and may aid in preventing chronic diseases (Mekonen et al., 2019). Additionally, the unique starch properties of teff make it suitable for various culinary applications, including gluten-free bread and traditional Ethiopian injera (Homem et al., 2022; Nascimento et al., 2018). Enset also plays a crucial role in food security due to its adaptability and resistance to adverse environmental conditions. The fermentation process of enset can enhance its nutritional value and probiotic potential, thereby promoting gut health (Hassona, 2023; Mezemir, 2015). The high dietary fiber content of enset contributes positively to managing blood sugar levels, improving metabolic health for consumers (Hassona, 2023).

6.3.3 Role in Addressing Malnutrition and Food Security

Both teff and enset are increasingly recognized for their potential to enhance food security and address malnutrition. Teff, often dubbed a "supergrain," offers promise in alleviating micronutrient deficiencies, particularly iron, which is prevalent in populations suffering from malnutrition (Alaunyte et al., 2010; Inglett et al., 2015). Incorporating teff into the diets of vulnerable populations can lead to improved health outcomes, especially for those with limited access to diverse diets. Similarly, enset's resilience to harsh growing conditions enables it to be cultivated in areas where other crops cannot thrive, ensuring stable food supplies (Adepoju et al., 2024; Hassona, 2023). As a nutrient-dense food source, enset can directly combat food insecurity by providing reliable nutrition to communities that primarily rely on it for sustenance.

6.4 Indigenous Cultivation and Processing Knowledge of Teff and Enset

Indigenous knowledge regarding the cultivation and processing of teff and enset plays a crucial role in agriculture and food security in Ethiopia. This knowledge encompasses traditional cultivation techniques, soil management, harvesting methods, processing practices, and the fermentation essential for preserving enset-based foods. Teff is traditionally cultivated using methods that have evolved over millennia, particularly in Ethiopia, where it is a staple grain. The cultivation practices vary significantly across regions, primarily influenced by local environmental conditions. Reports indicate that teff can thrive in diverse soil types and climatic conditions, enhancing its adaptability and making it suitable for different agricultural practices throughout Ethiopia, from sea level to elevations of up to 2,800 meters (Lee, 2018; Mottaleb & Rahut, 2018). Farmers employ Indigenous methods such as crop rotation to maintain soil fertility, although some continue with monoculture practices that may lead to soil degradation (Mottaleb & Rahut, 2018). Traditional practices, including selective breeding and knowledge of soil nutrient requirements, have been pivotal in the long-term sustainability of teff agriculture (Merchuk-Ovnat et al., 2020). Indigenous techniques often involve the application of locally available organic matter as fertilizers, aligning with contemporary permaculture principles focused on ecological awareness (Derso, 2020).

Harvesting practices for teff involve traditional methods that minimize loss and preserve the grain's nutritional quality. Harvesting typically occurs when the teff plant is ripe, marked by a change in color and hardening of the grains (Barretto et al., 2020). Indigenous methods often prefer manual harvesting with sickles, ensuring that the grains remain undamaged (Barretto et al., 2020). Post-harvest processing of teff includes activities such as threshing, traditionally performed with wooden sticks or oxen, to separate the grains from the chaff (Barretto et al., 2020). For storage, teff grains are usually kept in tightly sealed containers to guard against pests and moisture, with knowledge passed down through generations regarding the most effective storage techniques (Assefa et al., 2018). Unlike many other cereal grains, teff has a high mineral content, which can be maintained through this traditional post-harvest handling methods, thereby enhancing its status as a nutritional staple (Gebru et al., 2020).

Enset is primarily valued for its starch-rich corm and pseudostem. The role of fermentation in enset processing exemplifies Indigenous knowledge used for food preservation and flavor enhancement. Fermentation transforms the enset pulp into a product known as "kocho," which can be stored for long periods without refrigeration (Haile et al., 2021; Yemataw et al., 2018). This preservation method, deeply embedded in Ethiopian culture, utilizes lactic acid bacteria to prevent spoilage and enhance the flavor profile of enset products (Yemataw et al., 2018). In addition to kocho, fermentation is also essential in the preparation of a beverage called "bulla," made from the same plant, showcasing the versatility and nutritional importance of enset in Ethiopian diets (Yemataw et al., 2018). The fermentation process not only extends the shelf life of enset but also makes it a critical component for food security, especially in regions prone to food shortages due to climatic fluctuations (Haile et al., 2021).

6.5 Gendered Knowledge and Labor Practices

The division of labor in cultivating teff and enset crops reveals clear gender roles that reflect broader societal norms and expectations. In Ethiopia, where both crops are staples, women play a crucial role in agricultural production, although their contributions often go unrecognized. Research shows that women are significantly involved in all aspects of teff production, including sowing, weeding, harvesting, and processing (Tekalign et al., 2020). For example, a study in the Borecha District demonstrated that women not only contribute more labor overall but also take on tasks related to the post-harvest processing of teff—functions often overlooked in discussions about agricultural productivity and economic contributions (Tekalign et al., 2020). Additionally, the agricultural landscape of Ethiopia has increasingly recognized the importance of integrating gender perspectives into policy frameworks (Paudyal et al., 2019). It is noted that, despite their involvement in routine agricultural practices, women often encounter barriers to resources and decision-making opportunities, perpetuating a cycle of marginalization in agricultural roles (Huyer, 2016). While men traditionally dominate the wholesale market dynamics of teff, women frequently occupy the processing and retailing spaces, highlighting a clear division of labor that suggests underlying power structures (Tekalign et al., 2020). The labor burden on women is intensified by their dual roles in reproductive spheres, which men typically do not share, thereby affecting overall productivity and efficiency (Doss & Gottlieb, 2025; Tekalign et al., 2020).

Seed selection and processing are critical components of agricultural practice, particularly for crops like teff and enset. Women are often viewed as key custodians of agronomic knowledge, especially in local seed selection practices. They apply Indigenous knowledge and skills passed down through generations regarding the desirable traits of seeds—traits that enhance crop resilience and yield (Khatri-Chhetri et al., 2019). This knowledge transmission is not only essential for maintaining seed diversity but also plays a pivotal role in food security for families and communities at large (Beuchelt & Badstue, 2013). Regarding processing, women's roles extend to the preparation of food that utilizes both teff and enset. The culinary uses of these crops highlight women's labor contributions, as they bear the brunt of time-consuming tasks such as grinding teff flour or fermenting enset for cooking. Research shows that women's culinary activities are often undervalued in economic assessments of agricultural output, despite their significance in nutrition and health for households (Huyer, 2016).

Furthermore, women's expertise in food preparation has significant implications for nutritional outcomes, which are crucial in rural Ethiopian settings where food security is often precarious. The intergenerational knowledge transfer regarding recipes and the uses of these staple foods shapes community dietary practices and contributes to a collective cultural identity (Atreya & Gartaula, 2022). Knowledge transmission in agricultural contexts typically reflects strong intergenerational links within families. In rural Ethiopian settings, women act as primary educators for both agricultural skills and culinary techniques, ensuring that knowledge about teff and enset is preserved and adjusted over time (Chanana & Aggarwal, 2018). Women's unique experiences in agriculture, including a nuanced understanding of seasonal cycles and local environmental conditions, substantially enrich the knowledge base in agricultural communities.

However, as modern agricultural practices and climate-smart agriculture become increasingly adopted, there is a risk that traditional knowledge systems may be devalued or lost. Disruptions in these systems exacerbate the challenges faced by women who are balancing innovative practices with Indigenous methods (Zerssa et al., 2021). Studies have shown that women's adaptability in managing these shifts is vital for ensuring food security while fostering community resilience against climate variabilities (Khatri-Chhetri et al., 2019; Zerssa et al., 2021). Effective communication about agricultural innovations, ideally tailored to women's understanding, can enhance their ability to actively participate in new farming strategies (Christie et al., 2016). Bridging traditional knowledge with modern agricultural techniques is essential for cultivating a progressive agricultural framework that respects women's contributions while promoting sustainable practices (Giller et al., 2015).

6.6 Cultural and Spiritual Dimensions of Teff and Enset

Enset plays a crucial role in the sociocultural fabric and agricultural practices in Ethiopia, especially in the southern and southwestern regions. This crop serves as a staple food and is a central element in various rituals, festivals, and oral traditions, reflecting its cultural and spiritual dimensions. The cultural significance of enset and teff, another staple crop, is evident through their integral roles in traditional ceremonies, social rituals, and community identity among numerous Ethiopian ethnic groups, particularly among the Wolaita and Gurage communities (Dilebo et al., 2023; Olango et al., 2014). The rituals surrounding enset utilization mark significant life events, from birth to marriage to death. The entire lifecycle of the enset plant is revered, culminating in different food products such as kocho (the fermented pulp), bulla (the flour), and amicho (the corm), all of which have specific roles in various ceremonies (Blomme et al., 2023; Daba & Shigeta, 2016). Festivals celebrating the harvest of these crops often involve communal gatherings, dances, and sharing of traditional dishes, fostering a sense of community and continuity of cultural practices (Borrell et al.,

2020; Kidane et al., 2021). The use of enset within festivals reinforces its status as a symbol of sustenance and prosperity, linking food security with cultural identity (Daba & Shigeta, 2016; Dilebo et al., 2023).

Teff holds significant importance in Ethiopian culture, primarily as the main ingredient in injera, a traditional flatbread that serves as the foundation of Ethiopian cuisine (Borrell et al., 2020). The significance of teff is deeply intertwined with rituals and celebratory feasts. Similar to enset, festivals centered around teff incorporate local crops and livestock, reflecting the community's agricultural practices and socioeconomic conditions (Dilebo et al., 2023; Kidane et al., 2021). The dual cultivation of teff and enset supports households nutritionally and fosters agricultural biodiversity, which is essential to Ethiopian heritage (Borrell et al., 2019; Mengesha et al., 2022). The symbolic meanings associated with enset and teff extend beyond mere nutritional value. In Ethiopian cosmology, food often signifies more than sustenance; it embodies relationships, ancestry, and identity (Mengesha et al., 2022; Quinlan et al., 2014). The diversity of enset landraces cultivated across various regions reflects adaptive agricultural practices connected to local ecosystems and cultural preferences. Local farmers' ecological knowledge guides the selection of specific enset varieties for desired traits, including resilience to climate variations and suitability for particular culinary uses (Gebre & Lemma, 2019; Haile et al., 2021; Yemataw et al., 2016). Such practices highlight the interconnection between culture, agriculture, and spirituality in Ethiopian society.

Local governance of landraces is another critical aspect of enset cultivation. Communal management practices emphasize preserving genetic diversity while promoting ecological sustainability (Borrell et al., 2019, 2020). This participatory governance reflects local values, prioritizing environmental stewardship and cultural preservation. Moreover, the significance of sacred landscapes showcases how agricultural practices influence the spiritual beliefs of communities, intertwining the land's natural characteristics with cultural practices and rites (Mengesha et al., 2022; Yemataw et al., 2016). In community governance, enset cultivation symbolizes resilience. Farmers possess extensive Indigenous knowledge that allows them to effectively manage enset diversity, adapting to changes such as climate variability or socioeconomic shifts (Dilebo et al., 2023; Gebre & Lemma, 2019). This governance ensures the availability of enset varieties, which are crucial for food security and maintain cultural practices and social cohesion within communities (Blomme et al., 2023; Quinlan et al., 2015). Elders play a significant role in transmitting knowledge about enset and teff cultivation, securing future generations' access to these culturally revered crops (Dilebo et al., 2023; Yemataw et al., 2016).

Moreover, enset production systems are intricately linked to soil management practices that enhance soil fertility while addressing modern agricultural challenges (Borrell et al., 2019; Haileslassie et al., 2006). Local approaches to cultivating both teff and enset include crop rotation and intercropping, promoting diversity and ecological stability. This traditional ecological wisdom reflects the deep spiritual connection that communities have with their land, viewing it as a sacred entity that sustains life (Borrell et al., 2019; Dilebo et al., 2023). Additionally, women's roles in enset cultivation are particularly significant. Enset is often referred to as a "women's crop" due to the labor-intensive processes associated with its cultivation and processing, which are typically managed by women in the community (Daba & Shigeta, 2016; Woyesa & Kumar, 2021). This gendered division of labor highlights women's critical contributions to agricultural practices and food security in Ethiopian society. The experiences surrounding enset cultivation and harvesting also serve as rich narratives within oral traditions. Stories recounted from generation to generation about enset's cultivation and cultural importance contribute to communal identity, enabling societies to retain intrinsic knowledge and practices amid modernization (Blomme et al., 2023; Olango et al., 2014). This oral tradition underscores the spiritual reverence held for crops like enset and teff, embedding them within the community's historical narrative.

The significance of nature and its bountiful gifts is woven into the fabric of Ethiopian spiritual beliefs, where land and the crops it produces are often linked to ancestral worship and reverence (Forsido et al., 2013; Mengesha et al., 2022). The interplay of spirituality with agrarian practices reflects a deep recognition of the interdependence between communities and their environment. Rituals of giving thanks during harvest festivals honor the land and strengthen community ties, fostering shared identities rooted in agricultural practices (Borrell et al., 2020; Negassa et al., 2025; Yemataw et al., 2017). Enset and teff embody rich cultural and spiritual dimensions within Ethiopian society; their roles transcend mere sustenance, interlocking with community identity, resilience, and governance. As local farmers confront global challenges such as climate change, the capacity to adapt agricultural practices while safeguarding traditional knowledge becomes increasingly critical. The cultural heritage associated with these crops nourishes both the body and the spirit, reflecting the complex relationships between food, culture, and community in Ethiopia.

6.7 Challenges and Threats to Teff and Enset in Ethiopia

The cultivation of teff and enset in Ethiopia plays a crucial role in the country's agriculture, providing food security for millions. However, these crops are facing significant challenges driven by land-use changes, climate variability, commercialization pressures, and policy neglect, leading to various socioeconomic and environmental implications. Land-use changes and climate variability are notable factors threatening teff and enset cultivation. The Ethiopian Highlands, which serve as the primary region for teff production, are undergoing rapid land-use transitions due to urbanization, forest encroachment, and agricultural expansion (Elias et al., 2022; Murgatroyd et al., 2024). These changes can result in soil degradation and disrupt traditional farming practices that have supported these staple crops for centuries. Research indicates that declines in soil quality due to inappropriate land management threaten agricultural sustainability (Elias et al., 2022). Climate variability exacerbates this threat, causing unpredictable weather patterns and sporadic rainfall, which directly affect crop yields (Murgatroyd et al., 2024). Fluctuations in temperature and precipitation can lead to decreased resilience of the agricultural systems, making it increasingly difficult to produce adequate food for the growing population (Murgatroyd et al., 2024).

The commercialization of teff has presented both opportunities and challenges, contributing to genetic erosion and the risk of losing traditional landraces. As demand for teff increases, particularly for export, farmers face pressure to adopt high-yield varieties and commercial practices that may overlook or displace traditional landraces (Sankaranarayanan et al., 2020; Sirany et al., 2022). This commercialization process can diminish the genetic diversity of teff, rendering crops more vulnerable to disease and pests. Research indicates that many traditional landraces are experiencing a reduced presence as commercial varieties gain dominance, impacting the resilience of the agricultural system (Olango et al., 2014). This trend is particularly worrisome given the essential role that genetic diversity plays in adapting to changing environmental conditions and ensuring food security in the face of climate change (Borrell et al., 2020).

Policy neglect significantly hampers farmers' ability to effectively manage teff and enset production. Notable policy gaps exist regarding agricultural extension services, which are crucial for disseminating knowledge and innovative agricultural practices to smallholder farmers (Elemineh et al., 2020). The absence of structured policy approaches can result in insufficient support for teff and enset farmers in terms of access to improved seeds, pest management practices, and financial resources. Government policies have failed to adequately address the needs for sustainable agricultural practices, particularly amid the increasing pressures of climate change and commercialization (Sankaranarayanan et al., 2020). Moreover, as Ethiopia faces challenges related to food security, it becomes increasingly important for policy frameworks to include strategies that specifically target the promotion and support of Indigenous crops like enset, which has tremendous potential to enhance dietary diversity and resilience (Koch et al., 2021; Sirany et al., 2022). Micro-level studies reveal that farmers' decisions to adopt new technologies and modern farming techniques are often influenced by socioeconomic and environmental factors, many of which are exacerbated by inadequate policy support (Ayele et al., 2020; Fikadu et al., 2020). For instance, farmers have reported that their productivity of teff is significantly affected by both personal circumstances and neighboring socioeconomic factors, emphasizing the importance of collaborative approaches to problem-solving within farming communities (Fikadu et al., 2020). Such socioeconomic dynamics necessitate a reevaluation of agricultural policies to ensure they are inclusive and responsive to local conditions.

Moreover, the Indigenous knowledge that farmers have about enset cultivation is crucial for maintaining its diversity and adapting to changes, yet this knowledge is vulnerable amid modern agricultural pressures (Olango et al., 2014; Yemataw et al., 2016). With rapid urbanization and a shift toward market-oriented production, there is a risk that this knowledge may be lost, hindering efforts to fully exploit the potential of enset as a resilient food security crop (Olango et al., 2014). Consequently, programs aimed at facilitating traditional practices and integrating them into modern agricultural frameworks must be prioritized to achieve the sustainability of both teff and enset. The interrelated challenges posed by land-use changes, climate variability, commercialization, and policy neglect profoundly affect the future of teff and enset in Ethiopia. These threats underline the necessity for comprehensive agricultural policies that bolster resilience, promote genetic diversity, and enhance the Indigenous knowledge systems of farmers. Addressing these multidimensional challenges through collaborative efforts among farmers, policymakers, and stakeholders is critical to securing the viability of essential staple crops and, consequently, the food security of Ethiopia's population.

6.8 Sociobiological Reflection on Teff and Enset in Ethiopia

Both teff and enset are deeply rooted in the sociocultural practices of various communities, reflecting an intricate coevolution of human societies with these crops. Teff, a staple grain used to make injera, serves as a primary food source, while enset, often referred to as "false banana," provides diverse nutritional options, including food types like "Kocho," "Bulla," and "Amicho" (Olango et al., 2014; Yemata, 2020). Agricultural studies indicate that the domestication of these crops dates back thousands of years, emphasizing their integral role in the livelihoods and cultural identity of communities, particularly among ethnic groups such as the Sidama and Wolaita (Oh et al., 2022; Tsegaye & Struik, 2002). These interactions also highlight traditional ecological knowledge and Indigenous practices that significantly shape crop diversity and resilience in the Ethiopian highlands (Olango et al., 2014; Tsegaye & Struik, 2002).

The importance of teff and enset in landscape resilience and sustainable livelihoods stems from their cultivation within mixed-farming systems that integrate crop and livestock production. Enset's role as a perennial staple is critical, especially in mitigating seasonal food shortages and enhancing resilience to climate variability (Azeze et al., 2024; Bassa et al., 2024). Furthermore, the agroecological practices surrounding these crops improve soil fertility and biodiversity, contributing to a sustainable agricultural landscape (Blomme et al., 2023; Tsegaye & Struik, 2002). Numerous studies indicate that incorporating enset and teff into diversified agricultural systems helps ensure food security and nutritional balance for millions of Ethiopians, reinforcing the necessity of these crops for sustaining local economies (Bassa et al., 2024; Blomme et al., 2023).

Systems thinking and agroecological integration provide essential frameworks for understanding the dynamics of cultivating teff and enset. This perspective emphasizes the interdependence of agricultural practices, environmental sustainability, and socioeconomic factors (Sahle et al., 2023). For example, research indicates that smallholders who actively engage in integrated management practices—balancing crop cultivation and livestock rearing—are better equipped to face challenges posed by climate change and market fluctuations (Sahle et al., 2023; Tsegaye & Struik, 2002). Furthermore, maintaining genetic diversity in these crops supports adaptive responses to emerging challenges, highlighting the importance of traditional knowledge in modern agricultural strategies (Haile et al., 2023; Woldeyohannes et al., 2022). Integrating these crops within local agroecosystems illustrates a sustainable model that leverages both environmental and sociocultural dimensions for resilience building (Borrell et al., 2019). Focusing on systems thinking allows for a deeper understanding of agroecological integration strategies that enhance food security and community resilience in the face of rapid socioeconomic and environmental changes.

6.9 Conclusion

Teff and enset exemplify the intricate interplay between biology, culture, and sustainability in Ethiopia's highland food systems. As climate-resilient crops deeply rooted in Indigenous ecological knowledge, their continued cultivation and utilization reveal adaptive strategies that have evolved over millennia. This chapter has shown how these staples support not only nutrition and food security but also social organization, gender-specific labor practices, and agroecological stewardship. The fermentation technologies associated with enset and the nutrient-dense grains of teff reflect profound empirical wisdom and innovation among smallholder farmers, particularly women, who are the custodians of these knowledge systems. However, these traditional systems face increasing pressure from climate uncertainties, genetic erosion, land-use changes, and shifting consumption patterns influenced by globalization. While teff has entered global markets, risks of biopiracy, monocropping, and cultural appropriation are also present. Enset, in contrast, remains marginalized despite its immense potential as a food security buffer crop. Moving forward, research must concentrate on documenting and valuing Indigenous practices, enhancing crop breeding without displacing traditional varieties, and integrating farmer voices into national and global food policies. Policy frameworks that support community-led conservation, agrobiodiversity, and inclusive value chains are urgently needed. By centering teff and enset in sociobiological and interdisciplinary dialogues, we not only honor Ethiopia's food heritage but also advance broader discussions on sustainable and sovereign food systems in the Anthropocene.

REFERENCES

Abewa, A., Adgo, E., Yitaferu, B., Alemayehu, G., Assefa, K., Solomon, J. K. Q., & Payne, W. (2019). Teff grain physical and chemical quality responses to soil physicochemical properties and the environment. *Agronomy*, *9*(6), 283. https://doi.org/10.3390/agronomy9060283

Adepoju, M., Verheecke-Vaessen, C., Pillai, L. R., Phillips, H., & Cervini, C. (2024). Unlocking the potential of Teff for sustainable, Gluten-free diets and unravelling its production challenges to address global food and nutrition security: A review. *Foods, 13*(21), 3394. https://doi.org/10.3390/foods13213394

Alaunyte, I., Stojceska, V., Plunkett, A., & Derbyshire, E. (2014). Dietary iron intervention using a staple food product for improvement of iron status in female runners. *Journal of the International Society of Sports Nutrition, 11*(1), 50. https://doi.org/10.1186/s12970-014-0050-y

Assefa, Y., Emire, S., Villanueva, M., Abebe, W., & Ronda, F. (2018). Influence of milling type on tef injera quality. *Food Chemistry, 266*, 155–160. https://doi.org/10.1016/j.foodchem.2018.05.126

Atreya, K., & Gartaula, H. (2022). Changing gender role declines maize yield, but remittances offset: Findings from migrant households in the central Himalayas, Nepal. *Outlook on Agriculture, 51*(2), 247–259. https://doi.org/10.1177/00307270221097984

Ayele, A., Awel, M., Oljirra, A., Bayessa, M., Aliyi, I., & Faris, A. (2020). Determinants of productivity and efficiency of teff production in Southwest Ethiopia: Parametric approach. https://doi.org/10.21203/rs.3.rs-95125/v1

Azeze, T., Eshetu, M., Yilma, Z., & Berhe, T. (2024). Typification and differentiation of smallholder dairy production systems in smallholder mixed farming in the highlands of Southern Ethiopia. *Plos One, 19*(8), e0307685. https://doi.org/10.1371/journal.pone.0307685

Barretto, R., Buenavista, R., Rivera, J., ShuYu, W., Prasad, P., & Siliveru, K. (2020). Teff (*Eragrostis tef*) processing, utilization, and future opportunities: A review. *International Journal of Food Science and Technology, 56*(7), 3125–3137. https://doi.org/10.1111/ijfs.14872

Bassa, Z., Ketema, M., Kuma, B., & Mehary, A. (2024). Impact of level of enset (*Ensete ventricosum, Musaceae*) production on food and nutrition security: Empirical evidences from Wolaita and Kembata Tambaro zones of Southern Ethiopia. https://doi.org/10.21203/rs.3.rs-3898290/v1

Beuchelt, T., & Badstue, L. (2013). Gender, nutrition- and climate-smart food production: Opportunities and trade-offs. *Food Security, 5*(5), 709–721. https://doi.org/10.1007/s12571-013-0290-8

Biswas, M., Darbar, J., Borrell, J., Bagchi, M., Biswas, D., Nuraga, G., Demissew, S., Wilkin, P., Schwarzacher, T., & Heslop-Harrison, J. (2020). The landscape of microsatellites in the enset (*Ensete ventricosum*) genome and web-based marker resource development. *Scientific Reports*, *10*(1), 15312. https://doi.org/10.1038/s41598-020-71984-x

Blomme, G., Elizabeth, K., Buta, S., Chala, A., Kebede, R., Addis, T., & Yemataw, Z. (2023). Enset production system diversity across the Southern Ethiopian highlands. *Sustainability*, *15*(9), 7066. https://doi.org/10.3390/su15097066

Borrell, J., Biswas, M., Goodwin, M., Blomme, G., Schwarzacher, T., Heslop-Harrison, J., Wendawek, A. M., Berhanu, A., Kallow, S., Janssens, S., Molla, E. L., Davis, A. P., Woldeyes, F., Willis, K., Demissew, S., & Wilkin, P. (2019). Enset in Ethiopia: A poorly characterized but resilient starch staple. *Annals of Botany*, *123*(5), 747–766. https://doi.org/10.1093/aob/mcy214

Borrell, J., Goodwin, M., Blomme, G., Jacobsen, K., Wendawek, A., Gashu, D., Lulekal, E., Asfaw, Z., Demissew, S., & Wilkin, P. (2020). Enset-based agricultural systems in Ethiopia: A systematic review of production trends, agronomy, processing, and the wider food security applications of a neglected banana relative. *Plants People Planet*, *2*(3), 212–228. https://doi.org/10.1002/ppp3.10084

Chanana, N., & Aggarwal, P. (2018). Women in agriculture and climate risks: Hotspots for development. *Climatic Change*, *158*(1), 13–27. https://doi.org/10.1007/s10584-018-2233-z

Chase, R., Borrell, J. S., Rodenburg, J., Roux, N., Wendawek, A., & Büchi, L. (2025). Farmer selection of drought-tolerant enset landraces reduces trait diversity in drier environments. *Plants People Planet*. https://doi.org/10.1002/ppp3.70032; https://nph.onlinelibrary.wiley.com/doi/full/10.1002/ppp3.70032

Christie, M., Parks, M., & Mulvaney, M. (2016). Gender and local soil knowledge: Linking farmers' perceptions with soil fertility in two villages in the Philippines. *Singapore Journal of Tropical Geography*, *37*(1), 6–24. https://doi.org/10.1111/sjtg.12134

Daba, T., & Shigeta, M. (2016). Enset (*Ensete ventricosum*) production in Ethiopia: Its nutritional and socio-cultural values. *Agriculture and Food Sciences Research*, *3*(2), 66–74. https://doi.org/10.20448/journal.512/2016.3.2/512.2.66.74

Derso, C. (2020). Farmers' Indigenous knowledge practice for crop protection: A case study of ant protection on teff (Eragrostis teff) crop in North Wollo, Ethiopia. *Asian Journal of Research in Crop Science*, *5*(1), 14–20. https://doi.org/10.9734/ajrcs/2020/v5i130086

Dilebo, T., Feyissa, T., & Asfaw, Z. (2024). Farmers' local knowledge on classification, utilization, and on-farm management of enset (*Ensete ventricosum* (welw.) cheesman) landraces diversity in Hadiya, Southern Ethiopia. *Genetic Resources and Crop Evolution, 71*, 1575–1603. https://doi.org/10.1007/s10722-023-01714-5

Dilebo, T., Feyissa, T., Asfaw, Z., & Zewdu, A. (2023). On-farm diversity, use pattern, and conservation of enset (*Ensete ventricosum*) genetic resources in Southern Ethiopia. *Journal of Ethnobiology and Ethnomedicine, 19*(1). https://doi.org/10.1186/s13002-022-00569-x

Doss, C., & Gottlieb, C. (2025). Gendered patterns of labor in agriculture. *Agricultural Economics, 56*(3), 431–445. https://doi.org/10.1111/agec.70012

Elemineh, D., Merie, H., & Kassa, M. (2020). Prevalence and associated factors of agricultural technology adoption and teff productivity in Basso Liben District, East Gojjame Zone, North West Ethiopia. https://doi.org/10.1101/2020.10.28.358770

Elias, E., Biratu, G., & Smaling, E. (2022). Vertisols in the Ethiopian highlands: Interaction between land use systems, soil properties, and different types of fertilizer applied to teff and wheat. *Sustainability, 14*(12), 7370. https://doi.org/10.3390/su14127370

Fanta, S., & Satheesh, N. (2019). A review on nutritional profile of the food from enset. *Nutrition and Food Science, 49*(5), 824–843. https://doi.org/10.1108/nfs-11-2018-0306

Feleke, N., & Tekalign, W. (2022). The neglected traditional enset (*Ensete ventricosum*) crop landraces for the sustainable livelihood of the local people in Southern Ethiopia. *International Journal of Food Science, 2022*, 1–10. https://doi.org/10.1155/2022/6026763

Fikadu, A., Heckelei, T., & Woldeyohanes, T. (2020). Technical efficiency of teff farms controlling for neighborhood effects in Ethiopia. https://doi.org/10.21203/rs.3.rs-30863/v1

Forsido, S., Rupasinghe, H., & Astatkie, T. (2013). Antioxidant capacity, total phenolics, and nutritional content in selected Ethiopian staple food ingredients. *International Journal of Food Sciences and Nutrition, 64*(8), 915–920. https://doi.org/10.3109/09637486.2013.806448

Gebre, T., & Lemma, A. (2019). Diversity and distribution of enset landraces in Amaro Special District, Southern Ethiopia. *Omo International Journal of Sciences*, 2(1), 82–104. https://doi.org/10.59122/134d609

Gebrehiwot, N. T., Gezahegn, T. W., Abbay, A. G., Entehabu, T. G., Beyene, A. T., Tesfay, A. H., & Sebhatu, K. T. (2024). Does membership in seed producer cooperatives improve smallholders' teff productivity? A comparative analysis in north Ethiopia. *Annals of Public and Cooperative Economics*, 95(4), 1121–1137. https://doi.org/10.1111/apce.12466

Gebru, Y., Sbhatu, D., & Kim, K. (2020). Nutritional composition and health benefits of teff (*Eragrostis tef* (zucc.) Trotter). *Journal of Food Quality*, 2020, 1–6. https://doi.org/10.1155/2020/9595086

Gerura, F., Meressa, B., Kyallo, M., Tesfaye, A., Olango, T., & Nasser, Y. (2019). Genetic diversity and population structure of enset (*Ensete ventricosum* Welw Cheesman) landraces of the Gurage Zone, Ethiopia. *Genetic Resources and Crop Evolution*, 66(8), 1813–1824. https://doi.org/10.1007/s10722-019-00825-2

Getachew, S., Mekbib, F., Admassu, B., Kelemu, S., Kidane, S., Negisho, K., Djikeng, A., & Nzuki, I. (2014). A look into genetic diversity of enset (*Ensete ventricosum* (Welw.) Cheesman) using transferable microsatellite sequences of banana in Ethiopia. *Journal of Crop Improvement*, 28(2), 159–183. https://doi.org/10.1080/15427528.2013.861889

Gidelew, G., Tefera, T., & Aweke, C. (2022). From staple food to market-oriented crop: Commercialization level of smallholder teff (*Eragrostis teff*) growers in Jamma District, Ethiopia. *Cabi Agriculture and Bioscience*, 3(1). https://doi.org/10.1186/s43170-022-00123-5

Giller, K., Andersson, J., Corbeels, M., Kirkegaard, J., Mortensen, D., Erenstein, O., & Vanlauwe, B. (2015). Beyond conservation agriculture. *Frontiers in Plant Science*, 6. https://doi.org/10.3389/fpls.2015.00870

Habte, M., Beyene, E., Feyisa, T., Admasu, F., Tilahun, A., & Diribsa, G. (2022). Nutritional values of teff (*Eragrostis tef*) in diabetic patients: Narrative review. *Diabetes Metabolic Syndrome and Obesity Targets and Therapy*, 15, 2599–2606. https://doi.org/10.2147/dmso.s366958

Haile, A., Kovi, M., Sagen-Johnsen, S., Tesfaye, B., Hvoslef-Eide, A., & Rognli, O. (2023). Genetic diversity, population structure, and selection signatures in enset (*Ensete ventricosum* (Welw.) Cheesman), an underutilized and key food security crop in Ethiopia. *Genetic Resources and Crop Evolution, 71*(3), 1159–1176. https://doi.org/10.1007/s10722-023-01683-9

Haile, A., Sagen-Johnsen, S., Kovi, M., Hvoslef-Eide, A., Tesfaye, B., & Rognli, O. (2021). Comparison of different leaf preservation methods to obtain high quality DNA from enset (*Ensete ventricosum* (Welw.) Cheesman), a native and orphan food security crop in Ethiopia. https://doi.org/10.21203/rs.3.rs-786721/v1

Haileslassie, A., Priess, J., Veldkamp, E., & Lesschen, J. (2006). Smallholders' soil fertility management in the central highlands of Ethiopia: Implications for nutrient stocks, balances and sustainability of agroecosystems. *Nutrient Cycling in Agroecosystems, 75*(1–3), 135–146. https://doi.org/10.1007/s10705-006-9017-y

Hassona, M. M. (2023). The role of pseudocereals and nontraditional crops in contributing to food gap in Egypt: A review [Preprint]. Preprints. https://www.preprints.org/manuscript/202309.0908/v1?utm_source=researchgate.net&utm_medium=article

Hrušková, M., Švec, I., & Jurinová, I. (2012). Composite flours-characteristics of wheat/hemp and wheat/teff models. *Food and Nutrition Sciences, 3*(11), 1484–1490. https://doi.org/10.4236/fns.2012.311193

Huyer, S. (2016). Closing the gender gap in agriculture. *Gender Technology and Development, 20*(2), 105–116. https://doi.org/10.1177/0971852416643872

Inglett, G. E., Chen, D., & Liu, S. X. (2015). Functional properties of teff and oat composites. *Food and Nutrition Sciences, 6*(17), 1591–1602. https://doi.org/10.4236/fns.2015.617164

Khatri-Chhetri, A., Regmi, P., Chanana, N., & Aggarwal, P. (2019). Potential of climate-smart agriculture in reducing women farmers' drudgery in high climatic risk areas. *Climatic Change, 158*(1), 29–42. https://doi.org/10.1007/s10584-018-2350-8

Kidane, S., Haukeland, S., Meressa, B., Hvoslef-Eide, A., & Coyne, D. (2021). Planting material of enset (*Ensete ventricosum*), a key food security crop in Southwest Ethiopia, is a key element in the dissemination of plant-parasitic nematode infection. *Frontiers in Plant Science, 12*. https://doi.org/10.3389/fpls.2021.664155

Koch, O., Mengesha, W., Pironon, S., Pagella, T., Ondo, I., Rosa, I., Wilkin, P., & Borrell, J. (2021). Modelling potential range expansion of an underutilised food security crop in Sub-Saharan Africa. *Environmental Research Letters*, *17*(1), 014022. https://doi.org/10.1088/1748-9326/ac40b2

Lee, H. (2018). Teff, a rising global crop: Current status of teff production and value chain. *The Open Agriculture Journal*, *12*(1), 185–193. https://doi.org/10.2174/1874331501812010185

Mengesha, W., Inki, L., & Tolesa, Z. (2022). On-farm diversity of enset (*Ensete ventricosum* (Welw.) Cheesman) landraces, use, and the associated Indigenous knowledge in Adola Rede District, Guji Zone, Oromia, Ethiopia. *Advances in Agriculture*, *2022*, 1–12. https://doi.org/10.1155/2022/7732861

Merchuk-Ovnat, L., Bimro, J., Yaakov, N., Kutsher, Y., Amir-Segev, O., & Reuveni, M. (2020). In-depth field characterization of teff [*Eragrostis tef* (*Zucc.*) *Trotter*] variation: From agronomic to sensory traits. *Agronomy*, *10*(8), 1107. https://doi.org/10.3390/agronomy10081107

Mottaleb, K., & Rahut, D. (2018). Household production and consumption patterns of teff in Ethiopia. *Agribusiness*, *34*(3), 668–684. https://doi.org/10.1002/agr.21550

Murgatroyd, A., Thomas, T., Koo, J., Strzepek, K., & Hall, J. (2024). Building Ethiopia's food security resilience to climate and hydrological change. *Environmental Research Food Systems*, *2*(1), 015008. https://doi.org/10.1088/2976-601x/ad99dd

Nascimento, K., Paes, S., Oliveira, I., Reis, I., & Augusta, I. (2018). Teff: Suitability for different food applications and as a raw material of gluten-free, a literature review. *Journal of Food and Nutrition Research*, *6*(2), 74–81. https://doi.org/10.12691/jfnr-6-2-2

Negassa, T., Meressa, A., Abdissa, N., Degu, S., Addis, G., Debebe, E., Abdisa, N., W/kidan, S., Belitibo, D. B., Ashenef, S., Shanko, W., Zuber, Z., Kumsa, L., Kassahun, M., Assamo, F. T., & Endale, M. (2025). Exploring Indigenous knowledge and practices of the Gurage Community on the biosystematics and utilization of enset landraces for bone fracture and regeneration: The case of Gurage Zone, Central Ethiopia Region. *Frontiers in Pharmacology*, *16*. https://doi.org/10.3389/fphar.2025.1563898

Nespeca, M. G., Vieira, A. L., Júnior, D. S., Neto, J. A. G., & Ferreira, E. C. (2020). Detection and quantification of adulterants in honey by LIBS. *Food Chemistry*, *311*, 125886. https://doi.org/10.1016/j.foodchem.2019.125886

Nuraga, G., Feyissa, T., Tesfaye, K., Biswas, M., Schwarzacher, T., Borrell, J., Wilkin, P., Demissew, S., Tadele, Z., & Heslop-Harrison, J. (2020). The genotypic and genetic diversity of enset (*Ensete ventricosum*) landraces used in traditional medicine is similar to the diversity found in starchy landraces. https://doi.org/10.1101/2020.08.31.27485

Nuraga, G., Feyissa, T., Tesfaye, K., Biswas, M., Schwarzacher, T., Borrell, J., Wilkin, P., Demissew, S., Tadele, Z., & Heslop-Harrison, J. (2022). The genetic diversity of enset (*Ensete ventricosum*) landraces used in traditional medicine is similar to the diversity found in non-medicinal landraces. *Frontiers in Plant Science*, *12*. https://doi.org/10.3389/fpls.2021.756182

Oh, H., Quinlan, R., & Yoder, J. (2022). Crop diversification, impulsivity, and resilience in Ethiopia. *Review of Development Economics*, *26*(4), 2140–2162. https://doi.org/10.1111/rode.12919

Olango, T., Tesfaye, B., Catellani, M., & Pè, M. (2014). Indigenous knowledge, use and on-farm management of enset (*Ensete ventricosum* (Welw.) Cheesman) diversity in Wolaita, Southern Ethiopia. *Journal of Ethnobiology and Ethnomedicine*, *10*(1), 41. https://doi.org/10.1186/1746-4269-10-41

Paudyal, B., Chanana, N., Khatri-Chhetri, A., Sherpa, L., Kadariya, I., & Aggarwal, P. (2019). Gender integration in climate change and agricultural policies: The case of Nepal. *Frontiers in Sustainable Food Systems*, *3*. https://doi.org/10.3389/fsufs.2019.00066

Quinlan, M., Quinlan, R., & Dira, S. (2014). Sidama agro-pastoralism and ethnobiological classification of its primary plant, enset (*Ensete ventricosum*). *Ethnobiology Letters*, *5*, 116–125. https://doi.org/10.14237/ebl.5.2014.222

Quinlan, R., Quinlan, M., Dira, S., Caudell, M., Sooge, A., & Assoma, A. (2015). Vulnerability and resilience of Sidama enset and maize farms in Southwestern Ethiopia. *Journal of Ethnobiology*, *35*(2), 314–336. https://doi.org/10.2993/etbi-35-02-314-336.1

Sahle, M., Subramanian, S., & Saitô, O. (2023). Harnessing insights from indicators-based resilience assessment for enhancing sustainability in Ethiopia's Gurage socio-ecological production landscape. *Environmental Management*, *71*(6), 1269–1287. https://doi.org/10.1007/s00267-023-01794-0

Sankaranarayanan, S., Zhang, Y., Carney, J., Nigussie, Y., Esayas, B., Simane, B., Zaitchik, B., & Siddiqui, S. (2020). What are the domestic and regional impacts of Ethiopia's policy on the export ban of teff? *Frontiers in Sustainable Food Systems, 4.* https://doi.org/10.3389/fsufs.2020.00004

Sirany, T., Tadele, E., Hibistu, T., Kefalew, A., & Reta, H. (2022). Economic viability and use dynamics of the enset food system in Ethiopia: Its implications for food security. *Advances in Agriculture, 2022,* 1–12. https://doi.org/10.1155/2022/5741528

Tekalign, S., Eneyew, A., & Mitiku, F. (2020). Gender roles in the teff value chain in the Borecha District of South-Western Ethiopia: Husband and wife comparisons. *Journal of Agribusiness and Rural Development, 55*(1), 93–105. https://doi.org/10.17306/j.jard.2020.01212

Tsegaye, A., & Struik, P. (2002). Analysis of enset (*Ensete ventricosum*) Indigenous production methods and farm-based biodiversity in major enset-growing regions of Southern Ethiopia. *Experimental Agriculture, 38*(3), 291–315. https://doi.org/10.1017/s0014479702003046

Wang, Y., & Çakır, M. (2021). Welfare impacts of increasing teff prices on Ethiopian consumers. *Agricultural Economics, 52*(2), 195–213. https://doi.org/10.1111/agec.12614

White, O., Biswas, M., Abebe, W., Dussert, Y., Kebede, F., Nichols, R., Buggs, R. J. A., Demissew, S., Woldeyes, F., & Papadopulos, A. (2022). Maintenance and expansion of genetic and trait variation following domestication in a clonal crop. https://doi.org/10.32942/osf.io/3p6h5

Woldeyohannes, A., Desta, E., Fadda, C., Pè, M., & Dell'Acqua, M. (2022). Value of teff (*Eragrostis tef*) genetic resources to support breeding for conventional and smallholder farming: A review. *Cabi Agriculture and Bioscience, 3*(1). https://doi.org/10.1186/s43170-022-00076-9

Worojie, T., & Geremew, T. (2025). The concept of hierarchy in ethnobiological classification: On Kafficho folk botany of enset (*Ensete ventricosum*) in Southwest Ethiopia. *Journal of Ethnobiology, 45*(3), 275–288. https://doi.org/10.1177/02780771251338616

Woyesa, T., & Kumar, S. (2021). "Tree against hunger": Potential of enset-based culinary tourism for sustainable development in rural Ethiopia. *Journal of Cultural Heritage Management and Sustainable Development, 12*(4), 497–512. https://doi.org/10.1108/jchmsd-07-2020-0102

Yemata, G. (2020). Ensete ventricosum: A multipurpose crop against hunger in Ethiopia. *The Scientific World Journal, 2020,* 1–10. https://doi.org/10.1155/2020/6431849

Yemataw, Z., Bekele, A., Blomme, G., Muzemil, S., Tesfaye, K., & Jacobsen, K. (2018). A review of enset [*Ensete ventricosum* (Welw.) Cheesman] diversity and its use in Ethiopia. *Fruits*, *73*(6), 301–309. https://doi.org/10.17660/th2018/73.6.1

Yemataw, Z., Chala, A., Ambachew, D., Studholme, D., Grant, M., & Tesfaye, K. (2017). Morphological variation and inter-relationships of quantitative traits in enset (*Ensete ventricosum* (Welw.) Cheesman) germplasm from South and South-Western Ethiopia. *Plants*, *6*(4), 56. https://doi.org/10.3390/plants6040056

Yemataw, Z., Tesfaye, K., Grant, M., Studholme, D., & Chala, A. (2018). Multivariate analysis of morphological variation in enset (*Ensete ventricosum* (Welw.) Cheesman) reveals regional and clinal variation in germplasm from south and south-western Ethiopia. *Australian Journal of Crop Science*, *12*(12), 1849–1858. https://doi.org/10.21475/ajcs.18.12.12.p1135

Yemataw, Z., Tesfaye, K., Zeberga, A., & Blomme, G. (2016). Exploiting Indigenous knowledge of subsistence farmers for the management and conservation of enset (*Ensete ventricosum* (Welw.) Cheesman) (Musaceae family) diversity on-farm. *Journal of Ethnobiology and Ethnomedicine*, *12*(1), 34. https://doi.org/10.1186/s13002-016-0109-8

Zerssa, G., Feyssa, D., Kim, D., & Eichler-Löbermann, B. (2021). Challenges of smallholder farming in Ethiopia and opportunities by adopting climate-smart agriculture. *Agriculture*, *11*(3), 192. https://doi.org/10.3390/agriculture11030192

CHAPTER 7

Quinoa and Kiwicha in the Andes

Abstract

Quinoa (*Chenopodium quinoa*) and kiwicha (*Amaranthus caudatus*) are native to the Andean highlands and represent two of the most nutrient-dense and culturally embedded crops in Indigenous food systems. Long marginalized under colonial and postcolonial regimes, these crops have re-emerged globally as "superfoods" due to their exceptional amino acid profiles, resilience to abiotic stress, and potential contributions to sustainable agriculture. This chapter explores the coevolutionary relationships between Andean communities and these crops, focusing on traditional cultivation practices, gendered knowledge systems, and ritual significance. It critically examines how Indigenous epistemologies and agrobiodiversity stewardship have historically shaped—and continue to sustain—ecological and nutritional resilience in harsh mountainous environments. Simultaneously, the chapter addresses the challenges of globalization, particularly the commodification and neoliberal appropriation of quinoa and kiwicha, which have often undermined local food security and altered traditional practices. Drawing on interdisciplinary literature and community-based case studies from Bolivia, Peru, and Ecuador, this chapter situates these grains at the intersection of biocultural heritage, food sovereignty, and global market dynamics. It argues that quinoa and kiwicha are not merely nutritionally valuable crops but are also symbolic of broader struggles for autonomy, sustainability, and Indigenous rights. The chapter calls for a revalorization of Indigenous foodways and policy frameworks that uphold agroecological integrity and equitable benefit sharing. In doing so, it contributes to critical conversations on postcolonial agrifood systems, climate adaptation, and decolonial sustainability.

Keywords: Andean food systems, Indigenous knowledge, food sovereignty, biocultural heritage, commodification, agroecology

7.1 Introduction

Quinoa (*Chenopodium quinoa*) and kiwicha (*Amaranthus caudatus*) are integral components of Andean food systems, recognized for their resilience to harsh environmental conditions and their rich nutrient profiles, which include high levels of proteins, vitamins, and essential amino acids (Anaya-González et al., 2022; Ruiz et al., 2013). The incorporation of quinoa and kiwicha into modern diets has surged in recent years, heralded largely as "superfoods" due to their perceived health benefits, including anti-inflammatory and antioxidant properties (Anaya-González et al., 2022; Ng & Wang, 2021). The historical underpinnings of quinoa and kiwicha can be traced back approximately 7,000 years, marking their significance as ancient grains cultivated by pre-Columbian societies in the Andean highlands (Bazile et al., 2016; Winkel et al., 2016). This antiquity underscores their traditional importance, with archaeological evidence suggesting widespread consumption and cultural significance among the Andean peoples (Bazile et al., 2016; Winkel et al., 2016). In contrast, the contemporary resurgence of these grains is viewable through the lens of globalization, where a growing demand for healthy, sustainable foods in international markets has led to a revitalization of interest in quinoa and kiwicha (Jacobsen, 2011; Gamboa et al., 2020).

From sociobiological perspectives, quinoa and kiwicha are regarded as "superfoods" due to their nutrient density and the significant cultural heritage they embody, which is tied to the communities that cultivate them (Jacobsen, 2011; Ng & Wang, 2021). Quinoa is noted for its complete amino acid profile, making it a valuable protein source for vegetarians and vegans, while kiwicha adds to this narrative with its own range of nutritional benefits (Ramos-Pacheco et al., 2025; Ruiz et al., 2013). This portrayal of both grains as essential dietary staples aligns them with the growing trends toward sustainable agriculture and health-conscious eating (Winkel et al., 2016). The complex socioeconomic dynamics surrounding quinoa production in the Andes reveal challenges related to both traditional agricultural practices and modern market demands. As the international popularity of quinoa has increased, local farmers face paradoxical pressures; many are incentivized to sell their produce at high market prices while simultaneously choosing lower-quality, less nutritious foods for personal consumption (Anaya-González et al., 2022; Jacobsen, 2011). This situation not only jeopardizes local food sovereignty but also raises ethical questions about the sustainability and equity of food systems (Winkel et al., 2012).

In the context of environmental change, quinoa exhibits beneficial adaptability that enables it to thrive in a variety of conditions (Pinedo-Taco et al., 2022; Winkel et al., 2016). This resilience fuels the on-going interest in these grains as viable solutions for enhancing food security in a world increasingly affected by climate fluctuations (Ng & Wang, 2021; Winkel et al., 2012). Simultaneously, the academic discourse on quinoa and kiwicha explores genetic diversity and agricultural practices, highlighting the importance of conservation and breeding strategies to optimize yields while preserving the genetic heritage of these ancestral crops (Christensen et al., 2007; Mason et al., 2005). As quinoa and kiwicha gain popularity, it is essential to consider the sustainability of their production. Challenges such as land-use changes, environmental degradation, and socioeconomic disparities must be addressed to ensure that the cultivation of these grains benefits not only the global market but also the local communities that have nurtured them for generations (Winkel et al., 2012). The intersection of traditional knowledge, cultural significance, and modern agricultural science provides a unique framework for understanding the current state and future directions of quinoa and kiwicha production in the Andes (Ruiz et al., 2013; Winkel et al., 2016).

This chapter critically investigates the sociobiological significance of quinoa and kiwicha, two ancient Andean crops that have emerged as globally celebrated "superfoods." Therefore, this chapter aims to illuminate how their adaptive agronomic traits, exceptional nutritional profiles, and embedded cultural meanings have sustained Indigenous livelihoods and agroecosystems in the Andean highlands for millennia. Through an interdisciplinary lens, the chapter examines how traditional cultivation practices, ritual knowledge systems, and gendered stewardship intersect with ecological resilience and food sovereignty. Furthermore, it interrogates the paradoxes of commodification, highlighting the dissonance between global market acclaim and the socioeconomic displacement of smallholder producers. By drawing on recent scholarship and case studies from Peru, Bolivia, and Ecuador, the chapter contributes a grounded perspective to the growing discourse on Indigenous food systems, agroecological transitions, and postcolonial food politics. Ultimately, it positions quinoa and kiwicha as emblematic of broader struggles and possibilities for sustainable development, cultural resurgence, and biocultural conservation in the Anthropocene. This contribution is especially valuable to scholars and policymakers interested in food justice, climate-resilient agriculture, and decolonial knowledge systems.

7.2 Agroecological and Biological Foundations of Quinoa and Kiwicha in the Andes

Chenopodium quinoa and *Amaranthus caudatus* are two prominent crops native to the Andean highlands, recognized for their nutritious profiles and ecological resilience in challenging growing conditions. Their botanical characteristics, adaptive traits, and traditional cultivation techniques reflect a deep connection to the ecological systems they inhabit and the cultural practices of Andean farmers. Quinoa exhibits a notable range of physiological and morphological adaptations that enable its cultivation in diverse environments. As a member of the Amaranthaceae family, quinoa is classified as a pseudocereal due to its seed composition. It can grow to heights of up to 3 meters, with branched inflorescences that may be green, red, or yellow, influencing its marketability and attractiveness (Angeli et al., 2020; El-Serafy et al., 2021). Kiwicha shares similar morphological traits, characterized by a broad leafy structure and colorful seeds, which enhance its adaptability and appeal. Both species possess physiological characteristics that allow them to withstand abiotic stresses such as drought, frost, and salinity (Figures 7.1 and 7.2).

Figure 7.1: Quinoa Plant Showing the Leaves and Inflorescence

Source: Hussain et al. (2021)

Figure 7.2: Kiwicha Plant Showing Leaves and Inflorescence

Source: Habtemariam and Dana (2025)

The resilience of quinoa and kiwicha to adverse environmental conditions is particularly noteworthy; they demonstrate exceptional tolerance to drought and high altitudes. Quinoa's water-use efficiency surpasses that of many traditional crops, enabling it to thrive in arid regions with low fertility (Hinojosa et al., 2018; Vilcacundo & Hernández-Ledesma, 2017). Furthermore, studies indicate that quinoa's unique physiological mechanisms—including the ability to accumulate salt ions—enhance its adaptation to saline soils, a feature also observed in kiwicha (Jiang et al., 2023; Schmöckel et al., 2017). This capacity to grow in marginal soils positions quinoa as a promising candidate for sustainable production systems amid climatic unpredictability (Hinojosa et al., 2021). Adaptation to climate change is essential for agricultural sustainability, and both quinoa and kiwicha have shown resilience. Andean farmers have diversified their cropping systems by employing traditional intercropping and rotation methods, which enhance biodiversity and ecological stability (Ballesteros & Isaza, 2021). These practices maximize the unique properties of both crops while preserving Andean agricultural heritage. Additionally, both quinoa and kiwicha typically thrive at elevations ranging from 2,500 to 4,000 meters above sea level, where few other crops can survive (Galindo-Luján et al., 2024).

Traditional cultivation techniques employed by Andean farmers highlight their sustainable relationship with the land. These practices include the use of native vegetative barriers and organic fertilizers derived from local resources, which bolster soil health and promote biodiversity (Barrientos-Pérez et al., 2023; Hussain et al., 2021). Andean agricultural systems often rely on ancestral knowledge, maintaining crop diversity through the intercropping of quinoa and kiwicha with other native crops such as potatoes and maize. This intercropping enhances soil fertility, reduces pest pressures, and contributes to a balanced agroecosystem (Bazile et al., 2016; Ruiz et al., 2013). Despite the emergence of modern agricultural practices in the region, traditional methods remain vital to Andean agriculture, ensuring the preservation of local ecosystems and cultural identity. The ecosystem services provided by quinoa and kiwicha extend beyond crop yield, encompassing broader environmental and social functions. Cultivation of these crops can lead to improved soil quality, enhanced water retention, and greater resilience against climate shocks (Angeli et al., 2020). The adaptability of quinoa also enables smallholder farmers to respond effectively to market demands and environmental challenges, reiterating the innovative strategies embedded in traditional farming systems (Ballesteros & Isaza, 2021). Quinoa has gained recognition for its nutritional benefits, often labeled a "superfood" due to its high-quality protein content and essential amino acids that surpass those found in most cereal grains (Pathan et al., 2023). Kiwicha is also noted for its protein content and essential fatty acids, making both crops valuable nutritional staples for local populations. These advantages, combined with their adaptability, underscore the potential of both crops to significantly contribute to food security, particularly in regions susceptible to climate variability (Sosa-Zuniga et al., 2017).

Research into the genetic diversity of quinoa and kiwicha highlights their potential for adapting to environmental stresses. Studies have shown that cultivating various local genotypes can enhance resilience against salinity, drought, and extreme temperatures (Ruiz et al., 2013). The conservation and utilization of this genetic diversity through community seed banks and home gardens play a crucial role in sustaining these crops and the livelihoods of farmers across the Andes (Hinojosa et al., 2021; Panuccio et al., 2014). In recent years, global interest in quinoa has expanded its cultivation beyond its native lands, raising concerns over agroecological impacts and the preservation of traditional knowledge (El-Serafy et al., 2021). As demand increases, ensuring sustainable practices in quinoa cultivation is vital to prevent the loss of biodiversity and ecological integrity in

Andean ecosystems (Bazile et al., 2016; Ruiz et al., 2024). Understanding the integration of traditional ecological practices with modern agricultural science is crucial for consolidating a comprehensive approach to quinoa and kiwicha cultivation. Agroecology offers a framework that fosters resilience and sustainability by merging local knowledge with scientific advances in genetic and agronomic research (Altieri et al., 2015). Collaboration between scientists and Andean farmers can develop strategies that enhance the adaptability of these crops, ensuring their continued relevance in a changing world (Ballesteros & Isaza, 2021; Sosa-Zuniga et al., 2017).

Anthropologically, the cultural significance of quinoa and kiwicha in Andean societies is profound. Both crops have been integral to cultural identity, spirituality, and social cohesion among Indigenous communities. Preserving traditional agricultural systems that include these crops helps maintain biodiversity and uphold the cultural heritage that has defined these communities for millennia (Hinojosa et al., 2021; Ruiz et al., 2024).

7.3 Nutritional and Medicinal Value of Quinoa and Kiwicha in the Andes

Quinoa and kiwicha are cultivated in the harsh environmental conditions of the Andes, leading to the development of grains that are versatile in culinary applications and rich in essential macronutrients and micronutrients crucial for human health. Both grains offer a rich profile of proteins, fiber, vitamins, and minerals, making them beneficial components of both traditional and modern diets.

7.3.1 Macronutrient and Micronutrient Profiles

Quinoa is particularly praised for its high protein content, which typically ranges from 8% to 22.5% depending on the variety and processing methods used. This protein is classified as a complete protein because it contains all nine essential amino acids, making it comparable to animal proteins, a rarity among plant-based foods (Pino-Ramos et al., 2023; Vásquez-Ocmín et al., 2022). Kiwicha also displays high protein levels and provides essential amino acids, notably high in lysine compared to traditional grains (Drzewiecki et al., 2003). Both grains are gluten-free, offering safe alternatives for people with celiac disease or gluten sensitivity (Repo-Carrasco-Valencia et al., 2020). In addition to proteins, quinoa and kiwicha are rich in dietary fiber, which supports digestive health and promotes feelings of fullness (Vidaurre-Ruiz et al., 2020). Both grains are predominantly composed of insoluble fiber, beneficial for bowel health, while their soluble fiber also helps maintain healthy cholesterol levels (Vidaurre-Ruiz et al., 2020). Micronutritionally, these pseudocereals are rich in minerals like iron, calcium, potassium, and magnesium, as well as various vitamins, including B vitamins and vitamin E, enhancing their reputation as nutritious foods (García-Ramón et al., 2023; Repo-Carrasco-Valencia et al., 2010).

7.3.2 Comparative Health Benefits

The health benefits associated with quinoa and kiwicha arise from their unique amino acid profiles. Each grain offers a high level of essential amino acids; for example, kiwicha is particularly recognized for its elevated lysine content, making it a valuable protein source for vegetarian and vegan diets (Drzewiecki et al., 2003). Research shows that quinoa has a high protein digestibility rate, which is vital for nutrient absorption (Ortiz-Chura et al., 2018). Saponins found in quinoa, while sometimes regarded as antinutritional due to their potential to hinder the absorption of certain nutrients, also demonstrate beneficial properties such as lowering cholesterol levels and exhibiting anticancer characteristics (Vásquez-Ocmín et al., 2022). Kiwicha has displayed promising antioxidant properties, crucial for combating oxidative stress, which can be enhanced through enzymatic treatments (Chirinos et al., 2020; Paz et al., 2021). The presence of flavonoids and other phenolic compounds in both grains further enhances their antioxidant capabilities, providing significant health benefits (Repo-Carrasco-Valencia et al., 2010).

7.3.3 Role in Traditional Diets and Therapeutic Uses

Quinoa and kiwicha have been staples in the diets of Andean populations for thousands of years. Their significance extends beyond sustenance; these grains are deeply embedded in cultural practices and traditional medicine. They are used in various culinary forms, from soups to fermented beverages, showcasing their versatility (Ramos et al., 2021). Traditionally, these grains are believed to support overall health, often used to aid digestion, promote cardiovascular health, and serve as energy boosters (Paucar-Menacho et al., 2023). Modern research continues to explore the therapeutic uses of quinoa and kiwicha. Their protective effects against oxidative damage indicate potential for inclusion in diets aimed at disease prevention, particularly for chronic conditions such as diabetes and cardiovascular diseases (García-Ramón et al., 2023). Enhanced antioxidant activities, particularly in kiwicha, underscore its potential as a functional food ingredient that contributes to health (Chirinos et al., 2020; Paz et al., 2021). Advancements in food technology have optimized quinoa and kiwicha for various applications, leading to improved products like gluten-free breads and snacks that cater to health-conscious consumers (Dávalos et al., 2023; Vidaurre-Ruiz et al., 2020). Incorporating sprouted grains has shown improvements in nutrient bioavailability, significantly benefiting health outcomes (Paucar-Menacho et al., 2022, 2024). The global recognition of quinoa and kiwicha as "superfoods" has resulted in heightened interest in their applications, leading to increased demand not only for culinary use but also in nutritional supplements and functional foods aimed at enhancing health (Pino-Ramos et al., 2023; Ramos et al., 2021). This rising demand has implications for agricultural practices, sustainability, and the livelihoods of Andean farmers cultivating these crops (Coayla & Bedón, 2020).

7.4 Cultural, Spiritual, and Ritual Significance of Quinoa and Kiwicha in the Andes

Quinoa is not merely sustenance; it embodies historical roots, traditional knowledge, and spiritual connections to the land. Its integration into daily life exemplifies the relationship between agriculture, culture, and spirituality within Andean societies (Table 7.1).

7.4.1 Indigenous Cosmologies and Ritual Uses of Quinoa and Kiwicha

The spiritual dimensions of quinoa and kiwicha are deeply interwoven within the cosmological framework of Andean Indigenous cultures. These crops are often referred to as "mother grains" and are integral to agricultural rituals aimed at ensuring successful harvests. Historically, pre-Inca societies, such as the Tiwanaku and the Incas, revered quinoa, integrating it into religious ceremonies and agricultural calendars. Numerous studies document the role of quinoa in offerings to Pachamama (Mother Earth), where it is presented in rituals aimed at fostering fertility and well-being (Ruiz et al., 2013, 2024). Kiwicha, similarly, is celebrated for its nutritional properties and resilience, making it a staple in many Andean households. It, too, enters various rituals celebrating agricultural cycles, affirming the connection between spiritual and earthly realms (Bedoya-Perales et al., 2018; Ruiz et al., 2013). For these communities, the cycle of planting and

Table 7.1: Indigenous Knowledge and Practices Associated with Quinoa
(*Chenopodium quinoa*) and Kiwicha (*Amaranthus caudatus*) in the Andes

Category	Quinoa (Chenopodium quinoa)	Kiwicha (Amaranthus caudatus)
Traditional Cultivation	Terracing and intercropping on high-altitude fields; use of llama manure as fertilizer	Grown in valley and hillside plots; mixed with maize and potatoes
Seed Selection	Based on grain color, yield, resistance to drought and frost	Selection by panicle size and resistance to lodging
Cultural Significance	Sacred crop of the Inca; used in rituals such as Inti Raymi and agricultural festivals	Symbol of health and fertility; linked to ceremonial offerings
Processing Techniques	Washing and toasting to remove saponins; sun-drying on woven mats	Toasting and grinding seeds into flour; popping seeds for snacks
Food Uses	Cooked whole, made into soups, chicha (fermented drink), or flour	Consumed as porridge, flour for breads and drinks
Medicinal Uses	Used for digestive issues, anti-inflammatory properties	Used for anemia, postpartum recovery, and as a tonic
Storage Practices	Stored in clay pots or woven bags in cool, dry places	Hung in panicles in kitchens or granaries to avoid pest infestation
Knowledge Transmission	Oral transfer through family members and communal work days (mink'a)	Passed down during planting/harvest rituals and women's gatherings
Biodiversity Conservation	Multiple landraces conserved through in situ planting and barter systems	Landraces preserved in home gardens and exchanged in local markets
Adaptation to Climate	Selected for resistance to frost and drought; rotated with legumes	Valued for tolerance to poor soils and shifting rainfall patterns

harvesting is not merely a laborious task but a sacred duty that aligns with a cosmic order guided by ancestral wisdom (Ruiz et al., 2013).

7.4.2 Gendered Knowledge in Seed Preservation and Food Preparation

The knowledge surrounding the preservation of seeds and the preparation of traditional foods from quinoa and kiwicha is often gendered, with women playing a pivotal role. Women in Andean communities are the custodians of agrobiodiversity, holding the knowledge necessary for seed selection and preservation to sustain these culturally significant crops (Flores et al., 2003; Ruiz et al., 2013). This vital practice is rooted in generational knowledge passed down through generations, linking women to both the family and community's food security systems. Studies indicate that women's involvement in agricultural practices brings about a nuanced understanding of how to cultivate these crops effectively, considering the intricate dynamics of local climate conditions and soil health (Flores et al., 2003; Garay & Larrabure, 2011). Their role extends beyond agriculture; they often engage in culinary traditions that utilize quinoa and kiwicha, making them essential actors in the cultural transmission of this knowledge (Flores et al., 2003; Garay & Larrabure, 2011).

7.4.3 Sacred Ceremonies and Agricultural Calendars

In the Andean highlands, sacred ceremonies mark significant points in the agricultural calendar, especially around the sowing and harvesting of quinoa and kiwicha. The Indigenous agricultural calendar, which guides the timing of planting and harvesting, is intricately linked to celestial events and the changing seasons (Stadel, 2020). Events like the Inti Raymi (Festival of the Sun) align with rituals that celebrate and honor the natural cycles upon which crop success depends (Ruiz et al., 2013; Stadel, 2020). Ceremonial practices often include music, dance, and the creation of offerings, reaffirming community bonds and highlighting their shared cultural heritage and values. This blend of agrarian and spiritual practices reflects a holistic worldview in which nature, divinity, and social cohesion are interconnected (Garay & Larrabure, 2011; Ruiz et al., 2013). During these rituals, the cornucopia of offerings typically includes grains, particularly quinoa and kiwicha, reinforcing their significance not only as food sources but also as embodiments of spiritual belief systems. The act of ritualistic offering represents gratitude and a plea for protection over the crops, paying homage to syncretic beliefs that blend Indigenous and colonial influences (Boillat & Berkes, 2013; Ruiz et al., 2024). Furthermore, the understanding of crops' temporal cycles emphasizes a reciprocal relationship with the land. Farmers are not merely extractors of resources but are a part of a broader ecological community, implying a responsibility to nurture and protect these environments (Ruiz et al., 2013).

7.4.4 Social Implications of Quinoa and Kiwicha Cultivation

The immense rise in global demand for quinoa has had substantial socioeconomic implications for communities in the Andes, altering traditional agricultural practices and impacting local food security (Bedoya-Perales et al., 2018; Winkel et al., 2014). This shift reflects a double-edged sword—while the influx of capital offers potential economic benefits, it also threatens ancestral practices and local biodiversity as farmers prioritize marketable varieties over traditional strains (Bedoya-Perales et al., 2018; Winkel et al., 2014). Local communities face challenges such as land degradation, shifts in social structures, and the risk of cultural erosion, leading to a complex interplay between profit motives and the preservation of Indigenous knowledge systems (Brooks, 2016; Winkel et al., 2014). The ongoing tensions reveal the urgent need for frameworks that facilitate sustainable practices while honoring Indigenous rights, ensuring that agricultural expansion does not sideline the cultural significance of these ancient staples (Chalampuente-Flores et al., 2023; Winkel et al., 2014).

7.5 Political Economy and Global Commodification of Quinoa and Kiwicha in the Andes

Historically, both quinoa and kiwicha have been integral to Indigenous diets in the Andes, cultivated for thousands of years by the Quechua and Aymara peoples. However, the commercialization of these crops has prompted significant changes, particularly as quinoa has gained international acclaim for its nutritional benefits and potential as a gluten-free alternative to traditional grains, leading to a surge in global demand and consequent economic implications for local farmers and communities (Seligmann, 2023). The quinoa boom has dramatically altered the agricultural landscape in the Andes, with many smallholder farmers shifting from subsistence farming to cash cropping to capitalize on export markets. This transition, while economically beneficial for some, has raised pressing concerns regarding food security and sovereignty. Researchers argue that the drive toward market-oriented production has undermined traditional growing practices and local food systems, leading to greater dependency on global market fluctuations and affecting food availability for local populations (Torres et al., 2017). The emphasis on quinoa as a cash crop has often diverted resources away from the production of staple foods, resulting in increased food prices and diminished access to diverse diets. For example, Gonzales-Valero highlights that in the Puno region, climatic changes exacerbated by the commercialization of quinoa threaten caloric availability, underscoring the challenges faced by farmers in balancing cash crop production with local food needs (Gonzales-Valero, 2018).

Moreover, elevating quinoa to "superfood" status entails critical discussions about food sovereignty. The push for increased production has raised concerns over the consequences for local farmers' control over their resources, including land and seed varieties. As noted by Seligmann, the commodification of quinoa involves significant tensions: The very attributes that make quinoa valuable on a global scale threaten to undermine Indigenous control over these biogenetic resources. This has prompted calls for reevaluating development policies that encourage export-oriented agriculture at the expense of local needs (Seligmann, 2023). Particularly in Bolivia, Kerssen discusses the challenges of "re-peasantisation" considering the quinoa boom, elucidating how globalization impacts local peasant economies (Kerssen, 2018). Farmers face a dual struggle; on one hand, they have opportunities to enhance their income through export, yet on the other, they encounter pressures from international markets that can dictate terms detrimental to local practices and food systems. These dynamics necessitate careful consideration of strategies that not only support economic viability but also uphold traditional agricultural practices and ensure food access for local communities.

Another layer to the quinoa and kiwicha narrative is the global response to trade liberalization and certification processes that can hinder grassroots efforts within the Andean communities. As global demand rises, the pressure on farmers to conform to certification standards can lead to economic and ecological ramifications. While these certification processes might open new market opportunities, they also impose financial and operational burdens that many small-scale farmers struggle to manage (Seligmann, 2023). Torres et al. underscore that while the quest for food justice is paramount, it often gets overshadowed by the commercial interests that exploit local crops for profit at the expense of Indigenous rights and cultural practices (Torres et al., 2017). In connection with the impacts of global commodification, bioprospecting also raises ethical concerns. As quinoa and kiwicha gain popularity, foreign companies may seek to patent certain genetic traits or processing methods, often without compensating the Indigenous communities that have historically cultivated these crops (Seligmann, 2023). Seligmann illustrates how such practices threaten not only local food systems but also cultural identities tied to these food sources, ultimately calling for frameworks that protect Indigenous rights and enhance communal agency over agricultural practices (Seligmann, 2023).

The nutritional value of quinoa remains a significant aspect of its global appeal. Studies show that quinoa is not only rich in protein but also provides essential nutrients compared to other grains, making it a strategic crop for addressing malnutrition in diverse populations (Villacrés et al., 2022). However, this nutritional advantage must be balanced with concerns about accessibility. As demand grows, prices can increase, potentially excluding lower-income consumers from benefiting from its health properties. Thus, equitable distribution of this "superfood" remains a critical challenge in the context of both local economies and global markets. This underlying tension reveals that while quinoa and kiwicha signify immense potential for economic development in the Andes, their global commodification has multifaceted implications that must be navigated carefully. The future of these crops depends significantly on strategies that encompass sustainable development, respect for local knowledge, and safeguards against biopiracy and market monopolization. Stakeholders, including farmers, policymakers, and consumers, need to engage in dialogues that foster resilience in the face of global market pressures while simultaneously promoting food sovereignty and access for all (Kerssen, 2018; Torres et al., 2017).

The intricate relationship between quinoa and kiwicha's rising status and the historical marginalization of the Andean communities reflects broader issues of power and access within the global food system. The intersection of local practices and global phenomena continues to shape the future of these Indigenous crops, beckoning toward a more inclusive paradigm that values cultural heritage alongside economic opportunity. It is essential that strategies developed in response to these dynamics prioritize the rights and needs of Indigenous communities and contribute to sustainable food systems that ensure access to nourishing food for all (Gonzales-Valero, 2018; Seligmann, 2023).

7.6 Food Sovereignty and Indigenous Stewardship of Quinoa and Kiwicha in the Andes

Food sovereignty is an evolving concept that encompasses the rights of Indigenous populations to define and manage their food systems, influenced by an intricate relationship with their cultural identity, ecological stewardship, and historical legacy. Particularly in the Andes, quinoa and kiwicha serve as focal points within this discussion, highlighting Indigenous responses to commodification, community-driven conservation efforts, and the role of agroecology as forms of resistance and resurgence. The commodification of traditional crops like quinoa and kiwicha presents significant challenges for Indigenous communities. The international surge in demand for these "superfoods" ties them to global market ideologies that often overlook the cultural significance and customary practices associated with their production. Indigenous responses to such commodification have been characterized by critically examining their food practices and asserting their rights in managing these resources, emphasizing a return to traditional agricultural practices that honor cultural heritage and ecological sustainability (Sumner et al., 2023; Whyte, 2018). These responses are rooted in a broader political context where Indigenous food sovereignty is framed as an act of decolonization against the forces of capitalism that have historically marginalized these populations (Grey & Patel, 2014; Jernigan et al., 2023).

Community-based conservation movements have emerged in which Indigenous peoples actively engage in preserving their unique agricultural practices and biodiversity. The revival of Indigenous stewardship practices is essential for maintaining food sovereignty. Initiatives that integrate traditional ecological knowledge into contemporary food production systems help ensure that quinoa and kiwicha are grown sustainably and with cultural respect. By leveraging their ancestral knowledge in modern contexts, Indigenous communities can strengthen their food systems against market pressures and environmental changes (McEachern et al., 2022; Robin, 2019). Such community-led conservation efforts resonate deeply with the ideals of food sovereignty, where the local populace takes charge of defining their food systems and preserving their ecological relationships (Wezel et al., 2020).

Agroecology has emerged as a powerful strategy for Indigenous communities in resisting the pressures associated with industrial agriculture and market-driven food systems. This approach not only bolsters food sovereignty but also enhances community resilience. By employing agroecological principles—such as biodiversity, polyculture, and organic farming techniques—communities can sustainably cultivate quinoa and kiwicha, which are better suited to cope with environmental stressors while producing high nutritional yields (Jernigan et al., 2021). These methods represent a resurgence of Indigenous identity and ecological stewardship as communities redefine their relationship with food and agriculture. Moreover, agroecology is becoming a platform for advocacy, merging cultural reclamation with ecological justice, reinforcing the connections among food rights, environmental health, and social equity (Ain et al., 2023; Jamanca-Gonzales et al., 2024).

The interconnectedness of food sovereignty, agroecology, and community-based conservation highlights the depth of Indigenous stewardship throughout the Andes. Preserving traditional crops is also a form of social cohesion that unites community members against external pressures while nurturing cultural identity through knowledge-sharing and participatory practices (Malli et al., 2023). This reassertion of control over food systems challenges injustices linked to colonial histories and fosters new pathways for health and well-being among Indigenous communities (Abdul et al., 2023). Increasingly, food sovereignty is viewed as an essential component of Indigenous community health strategies. The emphasis on culturally relevant and ecologically sound food production methods is crucial for addressing dietary issues, as communities face food insecurity and diet-related diseases that have proliferated due to the colonial disruption of traditional foodways (Jernigan et al., 2023; Nguyen et al., 2023). Therefore, it is vital to broaden the understanding of food sovereignty beyond mere access to food, illustrating it as an interconnected web of rights and responsibilities that includes environmental stewardship and cultural integrity (Maudrie et al., 2023).

Furthermore, integrating agroecological practices into cultural narratives enables the revitalization of agricultural systems and societal structures that have long been affected by colonization. The revival of traditional farming techniques alongside modern practices showcases the cultural and spiritual connections between people and their lands (Coté, 2016). This relationship is intertwined with the identity of Indigenous peoples, further highlighting the importance of managing native crops like quinoa and kiwicha through cultural and environmental stewardship. Indigenous-led movements for food sovereignty also challenge and critique dominant paradigms concerning food systems. By emphasizing Indigenous rights within the broader framework of food sovereignty, these communities engage in advocacy aimed at reshaping perceptions of agricultural production and food distribution, contesting the prevailing capitalist narratives that frequently marginalize their contributions and perspectives (Jernigan et al., 2023; Nikolaus et al., 2022). This resistance is essential to ensuring that Indigenous knowledge and practices are acknowledged as critical contributions to contemporary sustainability efforts, thereby reestablishing their role in global food systems (Grey & Patel, 2014; Pinedo et al., 2021).

As Indigenous communities continue to reclaim their food sovereignty, incorporating agroecological practices serves as a tool for resistance against commodification and fosters resilience in the face of climate change. By focusing on sustainable practices rooted in traditional knowledge, these communities not only preserve their cultural heritage but also contribute to the broader discourse on sustainable agriculture and environmental conservation globally (Jamanca-Gonzales et al., 2024; Poirier et al., 2024). Ultimately, the quest for food sovereignty among Indigenous peoples acts as a catalyst for redefining food systems, advancing justice, and facilitating a resurgence of cultural practices that honor ancestral lifeways (Malli et al., 2023; Whyte, 2018).

7.7 Challenges and Opportunities of Quinoa and Kiwicha in the Andes

The challenges and opportunities facing quinoa and kiwicha in the Andean region are primarily linked to market volatility, land tenure insecurity, climate change threats, and prospects for co-created research and inclusive value chains. The distinctiveness of these crops, their nutritional benefits, and the socioeconomic context of their production and markets further exacerbate these dynamics. Understanding these factors is crucial for developing effective policies and interventions to ensure the sustainability and profitability of quinoa and kiwicha cultivation in the Andes.

Market volatility presents a significant challenge for quinoa and kiwicha farmers. High demand from international markets, especially in Europe and the United States, has not stabilized prices enough, resulting in fluctuations that affect farmer incomes. Production data show Peru as the leading global producer, controlling approximately 105,000 tons or nearly 58% of global output; however, volatility remains an inherent risk due to changing consumer preferences and market dynamics (Basantes-Morales et al., 2019). This issue is compounded by land tenure insecurity, which deters long-term investments in improved agricultural practices that are essential for boosting yields and ensuring sustainability (Alandia et al., 2020). Many farmers lack formal land ownership, limiting their access to credit and resources, which undermines their ability to adapt to market changes or invest in crop improvement.

Land tenure insecurity intertwines with market volatility, as farmers with secure land rights are more likely to adopt sustainable practices that enhance productivity and reduce vulnerability to market fluctuations. Integrated studies reveal that optimizing agricultural practices, particularly concerning quinoa and kiwicha, is closely linked to addressing land rights issues (Mamani & Velásquez,

2024). Farmers often struggle to navigate local, regional, and international regulations regarding land use, which exacerbates their difficulties in accessing emerging markets (Stöcker et al., 2024). As quinoa and kiwicha gain recognition as "superfoods," the stakes for farmers increase, highlighting the need for robust policies that protect their rights and facilitate their integration into high-value supply chains (Alandia et al., 2020).

Climate change further exacerbates vulnerabilities for farmers cultivating quinoa and kiwicha in the Andes, which are specially adapted to high-altitude environments. Increased temperatures, erratic rainfall patterns, and changing pest dynamics threaten the traditional growing conditions essential for these crops (Alandia et al., 2020). The resilience of Andean agrobiodiversity is at stake, as climate change pressures could lead to a decline in local crop varieties, thereby reducing the genetic diversity essential for adapting to new environmental stresses (Basantes-Morales et al., 2019). Although studies indicate that quinoa exhibits some resilience to certain climate stressors, the long-term viability of this adaptation remains in question without targeted agricultural strategies that focus on both conservation and innovation efforts. Conversely, significant opportunities lie ahead for these crops, particularly through the development of inclusive value chains that integrate fair trade practices, enhancing farmers' market access and improving their economic stability. The rising global demand for quinoa and kiwicha necessitates robust supply chain management and value addition, which can be supported by co-created research initiatives involving local communities, academic institutions, and industry stakeholders. Collaborative efforts in product development, such as gluten-free bread and energy bars enriched with these grains, have shown positive outcomes regarding consumer acceptance and market potential (Aguilar et al., 2020; Paucar-Menacho et al., 2024).

Moreover, inclusive research that engages farmers can empower communities by prioritizing local knowledge and practices, improving adaptation strategies to climate change threats, and ensuring food security within the Andes. Recent findings suggest that co-created research efforts can develop improved agricultural practices and varietals that are both climate-resilient and appealing to global markets, ultimately enhancing the nutritional profile of food products (Repo-Carrasco-Valencia, 2020). Such synergies could alleviate the pressures of climate change on local agriculture while promoting economic and social development for marginalized farming communities (Jamanca-Gonzales et al., 2024). Furthermore, the integration of Andean crops into global health-food markets offers significant opportunities for innovation. The potential therapeutic properties of kiwicha and quinoa, including antioxidant and anti-inflammatory benefits, have captured research interest, leading to the development of novel food products that align with health-conscious trends (Chirinos et al., 2020). This corresponds well with the global interest in bioactive compounds inherent in these grains, which has opened up avenues for their inclusion in various food formulations, from pasta to malted beverages (Ortiz-Chura et al., 2018; Paucar-Menacho et al., 2024).

It is evident that the interplay of market forces, climate challenges, and nutritional benefits establishes quinoa and kiwicha as not only staples of Andean diets but also as strategic components in the global fight against food insecurity. Addressing market volatility and land tenure issues is crucial for the sustainability of quinoa and kiwicha production. Incorporating small-scale farmers into value chains and enhancing their positions through policy support and cooperative frameworks will allow these producers to fully utilize the potential of their crops in a changing world. Exploring modern agricultural techniques while drawing from traditional methods may create a hybrid approach that improves product quality and environmental sustainability. Thus, while the challenges are significant, they are not insurmountable. Stakeholders, including farmers, researchers, and policymakers, must work together to develop comprehensive strategies that leverage the unique attributes of quinoa and kiwicha and the opportunities presented by their expanding global market. By emphasizing sustainability, inclusivity, and innovation, the Andes has substantial potential to emerge as a leading region in the international agricultural landscape, thereby ensuring that quinoa and kiwicha remain foundational elements of dietary health and local economies.

7.8 Sociobiological Reflection on Quinoa and Kiwicha in the Andes

The sociobiological reflection on quinoa and kiwicha in the Andes reveals a complex interplay of agricultural practices, cultural identity, and environmental adaptability that is critical in understanding these native crops' roles within their ecosystems and societies. Quinoa and kiwicha, both classified as pseudocereals, are vital in Andean agriculture due to their high protein content, gluten-free characteristics, and resilience to extreme environmental conditions such as drought and salinity, which have historically shaped the agricultural practices of Andean communities (Bazile et al., 2016; Lutz & Bascuñán-Godoy, 2017; Ruiz et al., 2013).

Research indicates that quinoa possesses significant genetic diversity, with various ecotypes adapted to different agroecological zones in the Andes (Fuentes et al., 2012; Ruiz et al., 2024). This genetic richness is vital for on-farm conservation efforts and highlights the role of Andean farmers in preserving agrobiodiversity through traditional farming practices that emphasize seed exchange and the cultivation of landraces (Fuentes et al., 2012; Huillca et al., 2021). The shifting agricultural paradigms due to increased global demand for quinoa have raised concerns regarding land-use management and the implications of its expansion on local ecosystems and traditional farming methods (Bedoya-Perales et al., 2018; Huillca et al., 2021). As the market for quinoa has expanded beyond its native regions, it has also transformed the socioeconomic dynamics of rural communities, creating both opportunities and challenges (Andrews, 2017; Bedoya-Perales et al., 2018).

The sociocultural aspects of quinoa and kiwicha cultivation reflect deep-rooted traditions within Andean societies. These crops have historically served not only as staple foods but also as integral components of cultural practices and rituals (Andrews, 2017; Suero et al., 2023). Ethnobotanical evidence points to the significant symbolic value attached to quinoa among Indigenous populations, highlighting that it was celebrated as a sacred crop by the Incas (Lutz & Bascuñán-Godoy, 2017; Suero et al., 2023). The ceremonial importance of these crops, contrasted with their commercial value today, raises questions about cultural preservation amid globalization and pressures from commodity markets (Andrews, 2017; Hirich et al., 2021).

In terms of nutritional benefits, both quinoa and kiwicha offer health advantages due to their rich content of essential amino acids, vitamins, and minerals (Ain et al., 2023; Bazile et al., 2016; Ruiz et al., 2013). These characteristics position them as promising solutions for addressing food security challenges, especially in light of climate change and the increasing demand for sustainable agricultural practices. Current research focuses on enhancing the sustainability of quinoa and kiwicha cultivation through improved agricultural techniques, including the use of beneficial microorganisms to promote plant growth and maintain soil health (Castillo et al., 2022). Embracing organic farming practices further contributes to the ecological sustainability of these crops, creating a synergy between traditional knowledge and modern agricultural innovations (Galindo-Luján et al., 2025; Hart et al., 2015). The interplay of ecological adaptation, social significance, and economic viability of quinoa and kiwicha in the Andean region highlights their importance as both cultural symbols and nutritional staples. Understanding this multifaceted relationship not only enriches our appreciation of these crops but also points to critical avenues for future research and sustainable practices that respect traditional knowledge while addressing contemporary agricultural and societal challenges.

7.9 Conclusion

Quinoa and kiwicha represent more than just nutrient-dense crops; they serve as living expressions of Andean biocultural heritage, ecological wisdom, and resistance. This chapter illustrates how Indigenous communities have cultivated and safeguarded these grains through generations of adaptive knowledge, developing intricate agronomic techniques, seed exchange networks, and food systems grounded in reciprocity and ecological balance. These practices have enabled quinoa and kiwicha to thrive in marginal, high-altitude environments, emphasizing their essential role in sustaining both landscapes and livelihoods. However, the global recognition of these grains as "superfoods" has exposed the tensions between local food sovereignty and international market forces. The commodification of quinoa and kiwicha has, in many instances, favored export economies at the expense of traditional access and community autonomy. This shift calls for critical reflection and ethical responsibility from both global consumers and policymakers. Going forward, there is a need for policies and partnerships that protect the cultural integrity, genetic diversity, and equitable value chains of these crops. Strengthening local seed sovereignty, supporting community-led agroecological initiatives, and promoting context-sensitive innovations are essential to ensuring that Indigenous communities remain at the center of the quinoa and kiwicha narratives. In the Anthropocene, where climate stress and food insecurity converge, quinoa and kiwicha offer pathways not only for nutrition and sustainability but also for decolonizing food systems and honoring ancestral knowledge. Their sociobiological significance reaffirms the centrality of Indigenous stewardship in crafting just, resilient, and regenerative food futures.

Chapter Reflection

1. How do teff and enset reflect Ethiopia's sociobiological systems?
2. What role does Indigenous knowledge play in their cultivation and processing?

3. How are women's contributions central yet often overlooked in teff and enset production?
4. How does genetic diversity support climate resilience in these crops?
5. What risks does teff commercialization pose to genetic diversity?
6. How do teff and enset contribute to Ethiopian cultural and spiritual life?
7. What policy gaps hinder sustainable production of teff and enset?
8. Why is a systems-thinking approach important for managing these crops?
9. What global lessons can be learned from Ethiopia's teff and enset systems?
10. Why is urgent action needed to protect these crops and their knowledge systems?

REFERENCES

Abdul, M., Ingabire, A., Lam, C., Bennett, B., Menzel, K., MacKenzie-Shalders, K., & van Herwerden, L. (2023). Indigenous food sovereignty assessment—A systematic literature review. *Nutrition and Dietetics*, *81*(1), 12–27. https://doi.org/10.1111/1747-0080.12813

Aguilar, M., Pastor, A., Domínguez, J., Siché, R., & Barraza-Jáuregui, G. (2020). *Barras energéticas a base de quinua, kiwicha y chía: Características texturales, acústicas y sensoriales*. LACCEI. https://doi.org/10.18687/laccei2020.1.1.377

Ain, Q., Siddique, K., Bawazeer, S., Ali, I., Mazhar, M., Rasool, R., Mubeen, B., Ullah, F., Unar, A., & Jafar, T. (2023). Adaptive mechanisms in quinoa for coping in stressful environments: An update. *PeerJ*, *11*, e14832. https://doi.org/10.7717/peerj.14832

Alandia, G., Rodríguez, J., Jacobsen, S., Bazile, D., & Condori, B. (2020). Global expansion of quinoa and challenges for the Andean region. *Global Food Security*, *26*, 100429. https://doi.org/10.1016/j.gfs.2020.100429

Altieri, M., Nicholls, C., Henao, A., & Lana, M. (2015). Agroecology and the design of climate change-resilient farming systems. *Agronomy for Sustainable Development*, *35*(3), 869–890. https://doi.org/10.1007/s13593-015-0285-2

Anaya-González, R., Cruz, E., Muñoz-Centeno, L., Alarcón, R., León, R., & Carhuaz, R. (2022). Food and medicinal uses of ancestral Andean grains in the districts of Quinua and Acos Vinchos (Ayacucho-Peru). *Agronomy*, *12*(5), 1014. https://doi.org/10.3390/agronomy12051014

Andrews, D. (2017). Race, status, and biodiversity: The social climbing of quinoa. *Culture Agriculture Food and Environment*, *39*(1), 15–24. https://doi.org/10.1111/cuag.12084

Angeli, V., Silva, P., Massuela, D., Khan, M., Hamar, A., Khajehei, F., Graeff-Hönninger, S., & Piatti, C. (2020). Quinoa (*Chenopodium quinoa* willd.): An overview of the potentials of the "golden grain" and socio-economic and environmental aspects of its cultivation and marketization. *Foods*, *9*(2), 216. https://doi.org/10.3390/foods9020216

Ballesteros, J., & Isaza, C. (2021). Adaptation measures to climate change as perceived by smallholder farmers in the Andes. *Journal of Ethnobiology*, *41*(3), 428–446. https://doi.org/10.2993/0278-0771-41.3.428

Barrientos-Pérez, E., Carevic, F., Rodríguez, J., Arenas-Charlín, J., & Delatorre-Herrera, J. (2023). Effect of native vegetative barriers to prevent wind erosion: A sustainable alternative for quinoa (*Chenopodium quinoa* willd.) production. *Agriculture*, *13*(7), 1432. https://doi.org/10.3390/agriculture13071432

Basantes-Morales, E., Alconada, M., & Pantoja, J. (2019). Quinoa (*Chenopodium quinoa* willd.) production in the Andean region: Challenges and potentials. *Journal of Experimental Agriculture International*, *36*(6), 1–18. https://doi.org/10.9734/jeai/2019/v36i630251

Bazile, D., Pulvento, C., Verniau, A., Al-Nusairi, M., Ba, D., Breidy, J., Hassan, L., Mohammed, M. I., Mambetov, O., Otambekova, M., Sepahvand, N. A., Shams, A., Souici, D., Miri, K., & Padulosi, S. (2016). Worldwide evaluations of quinoa: Preliminary results from post international year of quinoa FAO projects in nine countries. *Frontiers in Plant Science*, *7*. https://doi.org/10.3389/fpls.2016.00850

Bedoya-Perales, N., Pumi, G., Mújica, A., Talamini, E., & Padula, A. (2018). Quinoa expansion in Peru and its implications for land use management. *Sustainability*, *10*(2), 532. https://doi.org/10.3390/su10020532

Boillat, S., & Berkes, F. (2013). Perception and interpretation of climate change among Quechua farmers of Bolivia: Indigenous knowledge as a resource for adaptive capacity. *Ecology and Society*, *18*(4), 21. https://doi.org/10.5751/es-05894-180421

Brooks, B. (2016). Using a Susto Symptoms Scale to analyze social well-being in the Andes. *Sociology and Anthropology*, *4*(2), 106–113. https://doi.org/10.13189/sa.2016.040208

Castillo, J., Conde, G., Claros, M., & Ortuño, N. (2022). Diversity of cultivable microorganisms associated with quinoa (*Chenopodium quinoa*) and their potential for plant growth-promotion. *Bionatura, 7*(2), 1–13. https://doi.org/10.21931/rb/2022.07.02.61

Chalampuente-Flores, D., Mosquera-Losada, M., Ron, A., Tapia, C., & Sørensen, M. (2023). Morphological and ecogeographical diversity of the Andean lupine (*Lupinus mutabilis* Sweet) in the high Andean region of Ecuador. *Agronomy, 13*(8), 2064. https://doi.org/10.3390/agronomy13082064

Chirinos, R., Pedreschi, R., Velásquez-Sánchez, M., Aguilar-Galvez, A., & Campos, D. (2020). In vitro antioxidant and angiotensin i-converting enzyme inhibitory properties of enzymatically hydrolyzed quinoa (*Chenopodium quinoa*) and kiwicha (*Amaranthus caudatus*) proteins. *Cereal Chemistry, 97*(5), 949–957. https://doi.org/10.1002/cche.10317

Christensen, S., Pratt, D., Pratt, C., Nelson, P., Stevens, M., Jellen, E., Coleman, C. E., Fairbanks, D. J., Bonifacio, A., & Maughan, P. (2007). Assessment of genetic diversity in the USDA and CIP-FAO international nursery collections of quinoa (*Chenopodium quinoa* willd.) using microsatellite markers. *Plant Genetic Resources, 5*(2), 82–95. https://doi.org/10.1017/s1479262107672293

Coayla, E., & Bedón, Y. (2020). The agro exports of organic native products and environmental security in Peru. *European Journal of Economics and Business Studies, 6*(3), 105. https://doi.org/10.26417/175umi47d

Coté, C. (2016). "Indigenizing" food sovereignty. revitalizing Indigenous food practices and ecological knowledges in Canada and the United States. *Humanities, 5*(3), 57. https://doi.org/10.3390/h5030057

Dávalos, J., Tirado, A., Romero, V., Cisneros-Santos, G., & Gamarra, F. (2023). Structural, thermal and energetic properties of Andean-pseudocereal flours with high nutritional values. *Journal of Thermal Analysis and Calorimetry, 148*(14), 7207–7215. https://doi.org/10.1007/s10973-023-12224-y

Drzewiecki, J., Delgado, E., Haruenkit, R., Pawelzik, E., Martín-Belloso, O., Park, Y., Jung, S. T., Trakhtenberg, S., & Gorinstein, S. (2003). Identification and differences of total proteins and their soluble fractions in some pseudocereals based on electrophoretic patterns. *Journal of Agricultural and Food Chemistry, 51*(26), 7798–7804. https://doi.org/10.1021/jf030322x

El-Serafy, R., El-Sheshtawy, A., El-Razek, U., El-Hakim, A., Hasham, M., Sami, R., Khojah, E., & Al-Mushhin, A. (2021). Growth, yield, quality, and phytochemical behavior of three cultivars of quinoa in response to moringa and azolla extracts under organic farming conditions. *Agronomy*, *11*(11), 2186. https://doi.org/10.3390/agronomy11112186

Flores, H., Walker, T., Guimarães, R., Bais, H., & Vivanco, J. (2003). Andean root and tuber crops: Underground rainbows. *Hortscience*, *38*(2), 161–167. https://doi.org/10.21273/hortsci.38.2.161

Fuentes, F., Bazile, D., Bhargava, A., & Martínez, E. (2012). Implications of farmers' seed exchanges for on-farm conservation of quinoa, as revealed by its genetic diversity in Chile. *The Journal of Agricultural Science*, *150*(6), 702–716. https://doi.org/10.1017/s0021859612000056

Galindo-Luján, R., Pont, L., Jacobo, F., Sanz-Nebot, V., & Benavente, F. (2024). Matrix-assisted laser desorption ionization time-of-flight mass spectrometry combined with chemometrics for protein profiling and classification of boiled and extruded quinoa from conventional and organic crops. *Foods*, *13*(12), 1906. https://doi.org/10.3390/foods13121906

Galindo-Luján, R., Pont, L., Minić, Z., Berezovski, M., Jacobo, F., Sanz-Nebot, V., & Benavente, F. (2025). Comprehensive characterization of raw and processed quinoa from conventional and organic farming by label-free shotgun proteomics. *Journal of Agricultural and Food Chemistry*, *73*(4), 2669–2677. https://doi.org/10.1021/acs.jafc.4c08623

Gamboa, C., Schuster, M., Schrevens, E., & Maertens, M. (2020). Price volatility and quinoa consumption among smallholder producers in the Andes. *Scientia Agropecuaria*, *11*(1), 113–125. https://doi.org/10.17268/sci.agropecu.2020.01.13

Garay, E., & Larrabure, J. (2011). Relational knowledge systems and their impact on management of mountain ecosystems. *Management of Environmental Quality an International Journal*, *22*(2), 213–232. https://doi.org/10.1108/14777831111113392

García-Ramón, F., Sotelo-Méndez, A., Alvarez-Chancasanampa, H., Norabuena, E., Sumarriva, L., Yachi, K., Huamán, T. G., Vega, M. N., & Cornelio-Santiago, H. (2023). Influence of Peruvian Andean grain flours on the nutritional, rheological, physical, and sensory properties of sliced bread. *Frontiers in Sustainable Food Systems*, *7*. https://doi.org/10.3389/fsufs.2023.1202322

Gonzales-Valero, W. (2018). Hazards to food caloric availability and coverage per capita due to climate change in the Puno Region, Peruvian Altiplano: Challenges in food security and sovereignty. *Food and Energy Security*, *7*(2). https://doi.org/10.1002/fes3.134

Grey, S., & Patel, R. (2014). Food sovereignty as decolonization: Some contributions from Indigenous movements to food system and development politics. *Agriculture and Human Values*, *32*(3), 431–444. https://doi.org/10.1007/s10460-014-9548-9

Habtemariam, T. H., & Dana, B. K. (2025). Nutritional and mineral composition of *Amaranthus caudatus* leaves in Wolaita Zone, Southern Ethiopia. *Journal of Food Composition and Analysis*, *144*, 107641. https://doi.org/10.1016/j.jfca.2025.107641

Hart, A., McMichael, P., Milder, J., & Scherr, S. (2015). Multi-functional landscapes from the grassroots? The role of rural producer movements. *Agriculture and Human Values*, *33*(2), 305–322. https://doi.org/10.1007/s10460-015-9611-1

Hinojosa, L., González, J., Barrios-Masias, F., Fuentes, F., & Murphy, K. (2018). Quinoa abiotic stress responses: A review. *Plants*, *7*(4), 106. https://doi.org/10.3390/plants7040106

Hinojosa, L., Leguizamo, A., Carpio, C., Muñoz, D., Mestanza, C., Ochoa, J., Castillo, C., Murillo, A., Villacréz, E., Monar, C., Pichazaca, N., & Murphy, K., & Murphy, K. (2021). Quinoa in Ecuador: Recent advances under global expansion. *Plants*, *10*(2), 298. https://doi.org/10.3390/plants10020298

Hirich, A., Rafik, S., Rahmani, M., Fetouab, A., Azaykou, F., Filali, K., Ahmadzai, H., Jnaoui, Y., Soulaimani, A., Moussafir, M., El Gharous, M., Karboune, S., Sbai, A., & Choukr-Allah, R. (2021). Development of quinoa value chain to improve food and nutritional security in rural communities in Rehamna, Morocco: Lessons learned and perspectives. *Plants*, *10*(2), 301. https://doi.org/10.3390/plants10020301

Huillca, J., Río, B., & Álvarez, A. (2021). Quinoa expansion in Peruvian departments and land use change before the health crisis. *Agricultural Sciences*, *12*(8), 827–843. https://doi.org/10.4236/as.2021.128053

Hussain, M., Farooq, M., Syed, Q. A., Ishaq, A., Al-Ghamdi, A. A., & Hatamleh, A. A. (2021). Botany, nutritional value, phytochemical composition and biological activities of quinoa. *Plants*, *10*(11), 2258. https://doi.org/10.3390/plants10112258

Jacobsen, S. (2011). The situation for quinoa and its production in Southern Bolivia: From economic success to environmental disaster. *Journal of Agronomy and Crop Science*, *197*(5), 390–399. https://doi.org/10.1111/j.1439-037x.2011.00475.x

Jamanca-Gonzales, N., Ocrospoma-Dueñas, R., Eguilas-Caushi, Y., Padilla-Fabian, R., & Silva-Paz, R. (2024). Technofunctional properties and rheological behavior of quinoa, kiwicha, wheat flours and their mixtures. *Molecules, 29*(6), 1374. https://doi.org/10.3390/molecules29061374

Jernigan, V., Maudrie, T., Nikolaus, C., Benally, T., Johnson, S., Teague, T., Mayes, M., Jacob, T., & Taniguchi, T. (2021). Food sovereignty indicators for Indigenous community capacity building and health. *Frontiers in Sustainable Food Systems, 5*. https://doi.org/10.3389/fsufs.2021.704750

Jernigan, V., Nguyen, C. J., Maudrie, T. L., Demientieff, L. X., Black, J. C., Wilbur, R. E., Mortenson, R., Clyma, K. R., Lewis, M., & Lopez, S. (2023). Food sovereignty and health: A conceptual framework to advance research and practice. *Health Promotion Practice, 24*(6), 1070–1074. https://doi.org/10.1177/15248399231190367

Jernigan, V., Taniguchi, T., Nguyen, C., London, S., Henderson, A., Maudrie, T. L., Blair, S., Clyma, K. R., Lopez, S. V., & Jacob, T. (2023). Food systems, food sovereignty, and health: Conference shares linkages to support Indigenous community health. *Health Promotion Practice, 24*(6), 1109–1116. https://doi.org/10.1177/15248399231190360

Jiang, Y., Yasir, M., Cao, Y., Hu, L., Yan, T., Zhu, S., & Lu, G. (2023). Physiological and biochemical characteristics and response patterns of salinity stress responsive genes (SSRGS) in wild quinoa *(Chenopodium quinoa* L.). *Phyton, 92*(2), 399–410. https://doi.org/10.32604/phyton.2022.022742

Kerssen, T. (2018). Food sovereignty and the quinoa boom: Challenges to sustainable re-peasantisation in the Southern Altiplano of Bolivia. In *Food sovereignty* (pp. 59–77). Routledge. https://doi.org/10.4324/9781315227580-4

Lutz, M., & Bascuñán-Godoy, L. (2017). *The revival of quinoa: A crop for health*. InTech. https://doi.org/10.5772/65451

Malli, A., Monteith, H., Hiscock, C., Smith, E., Fairman, K., Galloway, T., & Mashford-Pringle, A. (2023). Impacts of colonization on Indigenous food systems in Canada and the United States: A scoping review. *BMC Public Health, 23*(1). https://doi.org/10.1186/s12889-023-16997-7

Mamani, A., & Velásquez, R. (2024). Environmental sustainability of the quinoa *(Chenopodium quinoa)* production chain in the Inclán-Tacna District, Peru. *Revista De Gestão Social E Ambiental, 18*(9), e07094. https://doi.org/10.24857/rgsa.v18n9-130

Mason, S. L., Stevens, M. R., Jellen, E. N., Bonifacio, A., Fairbanks, D. J., Coleman, C. E., McCarty, R. R., Rasmussen, A. G., & Maughan, P. J. (2005). Development and use of microsatellite markers for germplasm characterization in quinoa (*Chenopodium quinoa* willd.). *Crop Science*, *45*(4), 1618–1630. https://doi.org/10.2135/cropsci2004.0295

Maudrie, T., Nguyen, C., Wilbur, R., Mucioki, M., Clyma, K., Ferguson, G., & Jernigan, V. (2023). Food security and food sovereignty: The difference between surviving and thriving. *Health Promotion Practice*, *24*(6), 1075–1079. https://doi.org/10.1177/15248399231190366

McEachern, L., Yessis, J., Zupko, B., Yovanovich, J., Valaitis, R., & Hanning, R. (2022). Learning circles: An adaptive strategy to support food sovereignty among First Nations Communities in Canada. *Applied Physiology Nutrition and Metabolism*, *47*(8), 813–825. https://doi.org/10.1139/apnm-2021-0776

Ng, C., & Wang, M. (2021). The functional ingredients of quinoa (Chenopodium quinoa) and physiological effects of consuming quinoa: A review. *Food Fron*, *2*(3), 329–356. https://doi.org/10.1002/fft2.109

Nguyen, C., Wilbur, R., Henderson, A., Sowerwine, J., Mucioki, M., Sarna-Wojcicki, D., Ferguson, G. L., Maudrie, T. L., Moore-Wilson, H., Wark, K., & Jernigan, V. (2023). Framing an Indigenous food sovereignty research agenda. *Health Promotion Practice*, *24*(6), 1117–1123. https://doi.org/10.1177/15248399231190362

Nikolaus, C., Johnson, S., Benally, T., Maudrie, T., Henderson, A., Nelson, K., Lane, T., Segrest, V., Ferguson, G. L., Buchwald, D., Jernigan, V. B. B., & Sinclair, K. (2022). Food insecurity among American Indian and Alaska Native People: A scoping review to inform future research and policy needs. *Advances in Nutrition*, *13*(5), 1566–1583. https://doi.org/10.1093/advances/nmac008

Ortiz-Chura, A., Pari-Puma, R., Rodríguez-Huanca, F., Cerón-Cucchi, M., & Araníbar, M. (2018). Apparent digestibility of dry matter, organic matter, protein and energy of Native Peruvian feedstuffs in juvenile rainbow trout (*Oncorhynchus mykiss*). *Fisheries and Aquatic Sciences*, *21*(1), 32. https://doi.org/10.1186/s41240-018-0111-2

Panuccio, M., Jacobsen, S., Akhtar, S., & Muscolo, A. (2014). Effect of saline water on seed germination and early seedling growth of the halophyte quinoa. *Aob Plants*, *6*, plu047. https://doi.org/10.1093/aobpla/plu047

Pathan, S., Ndunguru, G., Clark, K., & Ayele, A. (2023). Yield and nutritional responses of quinoa (*Chenopodium quinoa* willd.) genotypes to irrigated, rainfed, and drought-stress environments. *Frontiers in Sustainable Food Systems*, *7*. https://doi.org/10.3389/fsufs.2023.1242187

Paucar-Menacho, L., Guzmán, J., Símpalo-López, W., Martínez, W., & Martínez-Villaluenga, C. (2023). Enhancing nutritional profile of pasta: The impact of sprouted pseudocereals and cushuro on digestibility and health potential. *Foods*, *12*(24), 4395. https://doi.org/10.3390/foods12244395

Paucar-Menacho, L., Símpalo-López, W., Martínez, W., Paredes, L., & Martínez-Villaluenga, C. (2022). Improving nutritional and health benefits of biscuits by optimizing formulations based on sprouted pseudocereal grains. *Foods*, *11*(11), 1533. https://doi.org/10.3390/foods11111533

Paucar-Menacho, L., Símpalo-López, W., Martínez, W., Paredes, L., Martínez-Villaluenga, C., & Schmiele, M. (2024). Optimization of rheological properties of bread dough with substitution of wheat flour for whole grain flours from germinated Andean pseudocereals. *Ciência Rural*, *54*(11). https://doi.org/10.1590/0103-8478cr20220402

Paz, S., Martínez-López, A., Villanueva, Á., Pedroche, J., Millán, F., & Millán-Linares, M. (2021). Identification and characterization of novel antioxidant protein hydrolysates from kiwicha (*Amaranthus caudatus* L.). *Antioxidants*, *10*(5), 645. https://doi.org/10.3390/antiox10050645

Pinedo, S., Escobar, L., & Neufeld, H. (2021). *The white/wiphala paper on Indigenous peoples' food systems*. FAO. https://doi.org/10.4060/cb4932en

Pinedo-Taco, R., Gomez-Pando, L., & Anderson-Berens, D. (2022). Production sustainability index of organic quinoa (*Chenopodium quinoa* willd.) in the Inter-Andean Valleys of Peru. *Tropical and Subtropical Agroecosystems*, *25*(2). https://doi.org/10.56369/tsaes.3925

Pino-Ramos, L., Laurie, V., Gómez-Plaza, E., & Bautista-Ortín, A. (2023). Effect of fining with new plant proteins on the aroma composition, phenolic compounds, and color of a Monastrell wine. *Bio Web of Conferences*, *68*, 02012. https://doi.org/10.1051/bioconf/20236802012

Poirier, B., Soares, G., Neufeld, H., Hedges, J., Sethi, S., & Jamieson, L. (2024). Conceptualising the relationships between food sovereignty, food security and oral health among global Indigenous communities: A scoping review. *Public Health Nutrition*, *27*(1), e147. https://doi.org/10.1017/s1368980024001198

Ramos, E., Coles, P., Chavez, M., & Hazen, B. (2021). Measuring agri-food supply chain performance: Insights from the Peruvian kiwicha industry. *Benchmarking an International Journal*, *29*(5), 1484–1512. https://doi.org/10.1108/bij-10-2020-0544

Ramos-Pacheco, B., Ligarda-Samanez, C., Choque-Quispe, D., Choque-Quispe, Y., Solano-Reynoso, A., Choque-Quispe, K., Palomino-Rincón, H., Taipe-Pardo, F., Peralta-Guevara, D. E., Moscoso-Moscoso, E., Diaz-Barrera, Y., & Agreda-Cerna, H. W. (2025). Study of the physical–chemical, thermal, structural, and rheological properties of four high Andean varieties of germinated Chenopodium quinoa. *Polymers*, *17*(3), 312. https://doi.org/10.3390/polym17030312

Repo-Carrasco-Valencia, R. (2020). Nutritional value and bioactive compounds in Andean ancient grains. *Proceedings*, *53*(1), 1. https://doi.org/10.3390/proceedings2020053001

Repo-Carrasco-Valencia, R., E.C., Binaghi, M., Greco, C., & Ferrer, P. (2010). Effects of roasting and boiling of quinoa, kiwicha and kañiwa on composition and availability of minerals in vitro. *Journal of the Science of Food and Agriculture*, *90*(12), 2068–2073. https://doi.org/10.1002/jsfa.4053

Repo-Carrasco-Valencia, R., Hellström, J., Pihlava, J., & Mattila, P. (2010). Flavonoids and other phenolic compounds in Andean Indigenous grains: Quinoa (*Chenopodium quinoa*), kañiwa (*Chenopodium pallidicaule*) and kiwicha (*Amaranthus caudatus*). *Food Chemistry*, *120*(1), 128–133. https://doi.org/10.1016/j.foodchem.2009.09.087

Repo-Carrasco-Valencia, R., Vidaurre-Ruiz, J., & Luna-Mercado, G. (2020). Development of gluten-free breads using Andean native grains quinoa, kañiwa, kiwicha and tarwi. *Proceedings*, *53*(1), 15. https://doi.org/10.3390/proceedings2020053015

Robin, T. (2019). Our hands at work: Indigenous food sovereignty in Western Canada. *Journal of Agriculture Food Systems and Community Development*, *9*, 1–15. https://doi.org/10.5304/jafscd.2019.09b.007

Ruiz, F., Villanueva, A., & Bazile, D. (2024). Chorematic modeling to represent dynamics in the quinoa agroecosystems in Peru. *Plos One*, *19*(4), e0300464. https://doi.org/10.1371/journal.pone.0300464

Ruiz, K., Biondi, S., Oses, R., Acuña-Rodríguez, I., Antognoni, F., Martinez-Mosqueira, E., Coulibaly, A., Canahua-Murillo, A., Pinto, M., Zurita-Silva, A., Bazile, D., Jacobsen, S.-E., & Molina-Montenegro, M. (2013). Quinoa biodiversity and sustainability for food security under climate change: A review. *Agronomy for Sustainable Development*, *34*(2), 349–359. https://doi.org/10.1007/s13593-013-0195-0

Schmöckel, S., Lightfoot, D., Razali, R., Tester, M., & Jarvis, D. (2017). Identification of putative transmembrane proteins involved in salinity tolerance in Chenopodium quinoa by integrating physiological data, RNASEQ, and SNP analyses. *Frontiers in Plant Science, 8.* https://doi.org/10.3389/fpls.2017.01023

Seligmann, L. J. (2023). Food sovereignty, food security, and sustainability. In L. J. Seligmann (Ed.), *Quinoa: Food politics and Agrarian life in the Andean highlands* (pp. 103–116). University of Illinois Press. https://doi.org/10.5622/illinois/9780252044793.003.0006

Sosa-Zuniga, V., Brito, V., Fuentes, F., & Steinfort, Ú. (2017). Phenological growth stages of quinoa (*Chenopodium quinoa*) based on the BBCH Scale. *Annals of Applied Biology, 171*(1), 117–124. https://doi.org/10.1111/aab.12358

Stadel, C. (2020). Horizontal and vertical archipelagoes of agriculture and rural development in the Andean realm. In M. J. Bastante-Ceca, J. L. Fuentes-Bargues, L. Hufnagel, F.-C. Mihai, & C. Iatu (Eds.), *Sustainability assessment at the 21st century.* IntechOpen. https://doi.org/10.5772/intechopen.86841

Stöcker, N., Hernández, H., & Torrico-Albino, J. (2024). The Bolivian organic quinoa in the fairtrade market, implications for the weakest link in the value chain: Small-scale farmers. *Tropical and Subtropical Agroecosystems, 27*(3). https://doi.org/10.56369/tsaes.5814

Suero, A., Goldstein, P., Guedes, J., & Sitek, M. (2023). Homeland food traditions in the Tiwanaku Colonies: Quinoa and amaranthaceae cultivation in the middle horizon (ad 600–1100) Locumba Valley, Peru. *Latin American Antiquity, 35*(4), 927–945. https://doi.org/10.1017/laq.2023.46

Sumner, J., McMurtry, J., & Tarhan, D. (2023). Growing community sustenance: The social economy as a route to Indigenous food sovereignty. *Canadian Journal of Nonprofit and Social Economy Research, 14*(S1). https://doi.org/10.29173/cjnser535

Torres, L., Salas, J., & Lafosse, H. (2017). Commercial opportunities for productive development of the Quinua. A reflection from the security, sovereignty and food justice. *Cooperativismo and Desarrollo, 25*(111). https://doi.org/10.16925/co.v25i111.1877

Vásquez-Ocmín, P. G., Marti, G., Gadéa, A., Cabanac, G., Vásquez-Briones, J. A., Casavilca-Zambrano, S., Ponts, N., Jargeat, P., Haddad, M., & Bertani, S. (2022). Metabotyping of Andean pseudocereals and characterization of emerging mycotoxins. https://doi.org/10.1101/2022.06.23.497323

Vidaurre-Ruiz, J., Salas-Valerio, F., Schöenlechner, R., & Repo-Carrasco-Valencia, R. (2020). Rheological and textural properties of gluten-free doughs made from Andean grains. *International Journal of Food Science and Technology*, *56*(1), 468–479. https://doi.org/10.1111/ijfs.14662

Vilcacundo, R., & Hernández-Ledesma, B. (2017). Nutritional and biological value of quinoa (*Chenopodium quinoa* willd.). *Current Opinion in Food Science*, *14*, 1–6. https://doi.org/10.1016/j.cofs.2016.11.007

Villacrés, E., Quelal, M., Galarza, S., Iza, D., & Silva, E. (2022). Nutritional value and bioactive compounds of leaves and grains from quinoa (*Chenopodium quinoa* willd.). *Plants*, *11*(2), 213. https://doi.org/10.3390/plants11020213

Wezel, A., Herren, B., Kerr, R., Barrios, E., Gonçalves, A., & Sinclair, F. (2020). Agroecological principles and elements and their implications for transitioning to sustainable food systems. A review. *Agronomy for Sustainable Development*, *40*(6), 40. https://doi.org/10.1007/s13593-020-00646-z

Whyte, K. (2018). Food sovereignty, justice, and Indigenous peoples: An Essay on Settler Colonialism and Collective Continuance. In A. Barnhill, M. Budolfson, & T. Doggett (Eds.), *The Oxford handbook of food ethics* (pp. 345–366). Oxford University Press. https://doi.org/10.1093/oxfordhb/9780199372263.013.34

Winkel, T., Álvarez-Flores, R., Bertero, D., Cruz, P., Castillo, C., Joffre, R., Parada, S. P., & Tonacca, L. (2014). Calling for a reappraisal of the impact of quinoa expansion on agricultural sustainability in the Andean Highlands. *Idesia (Arica)*, *32*(4), 95–100. https://doi.org/10.4067/s0718-34292014000400012

Winkel, T., Bertero, H., Bommel, P., Bourliaud, J., Lazo, M., Cortés, G., Gasselin, P., Geerts, S., Joffre, R., Léger, F., Martinez Avisa, B., Rambal, S., Rivière, G., Tichit, M., Tourrand, J. F., Vassas Toral, A., Vacher, J. J., Vieira Pak, M., & Pak, M. (2012). The sustainability of quinoa production in Southern Bolivia: From misrepresentations to questionable solutions. Comments on Jacobsen (2011, J. Agron. Crop sci. 197: 390–399). *Journal of Agronomy and Crop Science*, *198*(4), 314–319. https://doi.org/10.1111/j.1439-037x.2012.00506.x

Winkel, T., Bommel, P., Chevarría-Lazo, M., Cortés, G., Del Castillo, C., Gasselin, P., Léger, F., Nina-Laura, J.-P., Rambal, S., Tichit, M., Tourrand, J.-F., Vacher, J.-J., Vassas-Toral, A., Vieira-Pak, M., & Joffre, R. (2016). Panarchy of an Indigenous agroecosystem in the globalized market: The quinoa production in the Bolivian Altiplano. *Global Environmental Change*, *39*, 195–204. https://doi.org/10.1016/j.gloenvcha.2016.05.007

CHAPTER 8

Cassava and Yam in West and Central Africa

Abstract

Cassava (*Manihot esculenta*) and yam (*Dioscorea* spp.) are essential tuber crops in West and Central Africa, vital for agroecological resilience, nutrition, and cultural identity. This chapter provides a sociobiological synthesis of their domestication histories, genetic diversity, nutritional properties, toxicological management, and cultural significance. It also explore cassava's South American origins and rapid spread across African landscapes, emphasizing its drought tolerance and yield stability, while yam's native domestication emphasizes its deep cultural embeddedness, particularly in ceremonial and ritual practices. The chapter investigates the biochemical challenges presented by cassava's cyanogenic compounds and the sophisticated ethnoscientific processing methods—fermentation, soaking, grating, and drying—that ensure food safety and nutritional bioavailability. Nutritionally, while cassava primarily serves as an energy source, yam offers higher protein content, fiber, and essential micronutrients, contributing to dietary diversification and resilience. Gendered labor divisions highlight women's crucial roles in seed selection, processing, and intergenerational knowledge transmission, positioning them as guardians of agrobiodiversity and food sovereignty. The socioecological functions of intercropping systems, soil stabilization, and climate resilience are assessed, alongside threats from monoculture expansion, genetic erosion, and market-driven commodification. The chapter advocates for integrative approaches that blend Indigenous knowledge with modern breeding and policy frameworks to strengthen sustainable cassava and yam systems. In doing so, it contributes to broader discussions on food security, environmental health, and the sociobiology of Indigenous foods amid global change.

Keywords: cyanogenic compounds, food safety, traditional knowledge, Central Africa, agrobiodiversity, tuber systems

8.1 Introduction

Cassava (*Manihot esculenta*) and yam (*Dioscorea* spp.) are vital staple tuber crops in the agricultural and nutritional landscapes of West and Central Africa. Their cultivation is deeply integrated into the cultural fabric of the region, with a history that extends back centuries. Both cassava and yam are essential not only in dietary practices but also in socioeconomic structures, serving as primary sources of carbohydrates for millions of people. These tubers are critical in addressing food security challenges faced within the socioecological frameworks of the region (Alene et al., 2015; Tizé et al., 2021).

Cassava has played a key role in alleviating food insecurity across tropical and subtropical regions, primarily due to its resilience to drought and poor soil conditions. The crop was historically introduced to Africa from South America, where it remarkably adapted to diverse agroecologies, resulting in the emergence of numerous local varieties that perform well under various climatic conditions (Guo, 2020; Tizé et al., 2021). Cassava mosaic disease (CMD) and cassava brown streak disease (CBSD) pose significant threats to its cultivation, necessitating genetic research and the development of resistant cultivars (Alicai et al., 2019; Houngue et al., 2019; Imarhiagbe et al. 2024; Yadav et al., 2011). The role of yam, which is primarily Indigenous to Africa, is similarly significant; it is linked to cultural practices and is often featured in festive meals and rites of passage (Alene et al., 2015; Guo, 2020).

Culturally, yam is not only a dietary staple but also a symbol of prosperity and social status. Its annual festivals in several West African communities underscore its importance in agricultural traditions and regional identity (Guo, 2020; Opabode, 2014). Cultivation methods for these crops depend on local practices that have evolved over time, weaving agricultural knowledge into community rituals and cultural significance (Tizé et al., 2021). The relationship between these practices and sustainable agriculture enhances both food security and agricultural resilience, as both crops rank among the top sources of calories, providing essential nutrients for the region's growing populations (Anikwe & Ikenganyia, 2018; Babatunde et al., 2022).

Food security in West and Central Africa is increasingly threatened by population growth, climate variability, and emerging pests and diseases. The hardiness of these tubers is crucial in mitigating these challenges. For instance, cassava is known for its ability to thrive in marginal conditions and yield year-round harvests, significantly contributing to food availability and accessibility

(Anıkwe & Ikenganyia, 2018; Hassall et al., 2024). Additionally, agricultural practices involving yam highlight its adaptability and critical role in local economies, promoting dietary diversity (Alene et al., 2015; Anikwe & Ikenganyia, 2018; Guo, 2020). As food systems continue to evolve, the adaptability of these tubers presents a sustainable pathway to enhancing food sovereignty and resilience (Tizé et al., 2021). The increasing cultivation of cassava in several African nations, especially Nigeria, reaffirms its status as a key food crop necessary for national food security (Godding et al., 2023; Guo, 2020). In recent decades, research has focused on improving cassava yields through better agricultural practices and biotechnological innovations (Guo, 2020; Yadav et al., 2011). Intercropping yam and cassava is a common method that diversifies food production, improves soil fertility, and reduces pest issues (Hassall et al., 2024; Legg et al., 2014). Despite the importance of these crops, they face substantial threats from diseases and pests that could undermine entire harvests, emphasizing the urgent need for integrated pest management strategies and disease resistance research (Alicai et al., 2019; Legg et al., 2014; Nwezeobi et al., 2020a, 2020b).

As agricultural systems in West and Central Africa continue to evolve, integrating traditional knowledge with modern agricultural science becomes crucial. This integration can enhance crop resilience, mitigate risks associated with climate change, and improve the livelihoods of millions of smallholder farmers (Egbebiyi et al., 2023; Halake & Chinthapalli, 2020; Opabode, 2014). Developing improved varieties of cassava and yam that exhibit better disease resistance and higher yields could significantly improve food security and nutritional levels in the region (Alene et al., 2015; Babatunde et al., 2022; Mombo et al., 2016). Moreover, the significance of these tubers extends beyond crop yield; they contribute to the socioeconomic stability of communities, reinforcing local identities and promoting sustainable agricultural practices (Godding et al., 2023; Montemayor et al., 2014; Opabode, 2014).

This chapter aims to provide a comprehensive sociobiological analysis of cassava and yam within West and Central African societies, emphasizing their ecological, nutritional, and cultural significance. Drawing on ethnobotanical, nutritional, and anthropological perspectives, the chapter explores how these tubers are not only dietary staples but also integral to traditional ecological knowledge systems, gendered practices, and community resilience. By examining Indigenous detoxification methods for managing cyanogenic compounds in cassava and traditional processing and culinary techniques associated with yam, the chapter highlights how local knowledge systems serve as embedded food safety mechanisms evolved through generations of environmental interaction. It also investigates the gendered dimensions of cultivation and knowledge transmission, positioning women as central actors in maintaining tuber-based agrobiodiversity. The chapter contributes to the broader discourse on food sovereignty, Indigenous science, and sustainable agriculture by illuminating the threats posed by commercial monocultures and commodification while underscoring the resilience and adaptive strategies inherent in traditional tuber systems. In doing so, this work enriches sociobiological scholarship by bridging ecological function, cultural heritage, and nutritional science within a framework rooted in African realities.

8.2 Origins, Domestication, and Distribution of Cassava and Yam in West and Central Africa

The origins, domestication, and distribution of key root crops such as *Dioscorea* spp. (Figures 8.1 and 8.2) and *Manihot esculenta* (Figure 8.3) in West and Central Africa represent a significant area of agricultural research, particularly given the region's socioeconomic reliance on these crops. Archaeobotanical and ethnobotanical evidence highlights early cultivation practices, demonstrating a complex interplay of environmental conditions, human migration, and trade that facilitated the spread of these tubers. Archaeobotanical evidence suggests that cassava and yam were among the earliest cultivated plants in West Africa. Specifically, yam is believed to have originated in present-day Nigeria and Cameroon, while cassava was domesticated in South America before being brought to Africa. The domestication of these species seems to have originated in diverse environments, particularly in areas characterized by rich biodiversity. A range of studies confirms that the early cultivation of yam, specifically *Dioscorea rotundata*, was concentrated in the forest zones of West Africa; this is supported by the biocultural heritage of Indigenous yams and their cultivation practices, which have evolved over generations (Condé et al., 2024). Furthermore, palynological studies indicate that cultivators adapted to changes in climate and ecosystems, which enhanced their crop yields through selective breeding and cultivation techniques developed over centuries (Abajue & Ogwu 2024; Chia, 2023).

Genetic studies illustrate the diversity of yam cultivars in the region, with a significant number of landraces still cultivated. For instance, Dioscorea rotundata, commonly referred to as white Guinea yam, displays remarkable agromorphological diversity as revealed through SNP marker analysis, highlighting the extensive traditional knowledge embedded in local agricultural practices (Agre et al., 2021). Such genetic diversity is crucial for developing resilient cultivars,

Figure 8.1: Health Leaves of Dioscorea Species

Source: Adapted from Krüger et al. (2024)

enabling adaptation to various environmental stresses faced by farmers across the region (Egboduku et al., 2024a, 2024b). The importance of local cultivars in cassava and yam breeding programs is paramount, as they represent essential genetic reservoirs vital for improving yields and resilience to biotic and abiotic stress (Alene et al., 2015).

Figure 8.2: Tubers of Dioscorea Species

The ethnobotanical narratives surrounding cassava highlight its rapid spread from its center of origin in South America to African coastal regions, driven by trade and human migration patterns (Ogwu & Kulkarni, 2024). Initially introduced by Portuguese traders in the sixteenth century, cassava quickly became integrated into local diets and agricultural systems (Chia, 2023). This adaptability, along with the tuber's drought resistance and high caloric yield per hectare, facilitated its rise to prominence in Africa, particularly in areas with challenging agricultural conditions. Such dynamics illustrate the ongoing interaction between culture,

Figure 8.3: Cassava Tubers and Leaves

Source: Olukanni and Olatunji (2018)

environment, and economy, where yam and cassava play essential roles in rural livelihoods (Condé et al., 2024). There are notable regional variations in the species and cultivars of yams grown across West and Central Africa. For example, the diversity in local practices and climatic conditions leads to a multitude of cultivars adapted to specific environments, showcasing a local agricultural wisdom that reflects both cultural identity and ecosystem relationships (Agre et al., 2021). Indeed, Dioscorea alata, or water yam, is predominantly cultivated in Nigeria, while Dioscorea rotundata remains dominant in Benin, indicating a need to understand the specific ecological niches occupied by these different species (Alene et al., 2015). The food security implications of this variability are significant, as crop diversification is particularly crucial for resilience against climate change and food scarcity.

Interactions among human migration, trade, and tuber spread further contextualize the patterns of yam and cassava cultivation (Agoh et al. 2024). Historical accounts and modern analyses reveal that migrations, particularly of various ethnic groups within West Africa, have led to the dissemination of knowledge regarding crop cultivation techniques and the introduction of diverse cultivars across regions. The trade networks established along coastal areas and rivers served as conduits for these root crops, enhancing agricultural diversity as farmers encountered new planting and working techniques from different cultures (Alene et al., 2015). Moreover, trade not only facilitated the exchange of plant materials, but it also resulted in sharing agricultural practices, fostering innovation and resilience in farming communities (Condé et al., 2024). Intertwined with these migrations are the ecological shifts and adaptations that have subsequently shaped agricultural practices. The spread of invasive species and changing climatic conditions necessitated adaptive strategies among farmers. Over time, this evolving agricultural landscape has witnessed a transition from traditional farming methods to more intensive practices involving synthetic fertilizers and herbicides, raising concerns about sustainability and ecological integrity in contemporary yam and cassava production systems in Guinea (Condé et al., 2024). Such changes emphasize the need for research and development in sustainable agricultural practices that honor traditional knowledge while incorporating modern techniques.

8.3 Nutritional Value and Functional Roles of Cassava and Yam in West and Central Africa

Cassava and yam are two staple crops of great socioeconomic significance in West and Central Africa. Their nutritional profiles, functional roles, and contributions to dietary diversification are crucial for food security in the region. This synthesis examines the comparative nutritional profiles of cassava and yam, their multifunctional uses, and their roles in enhancing dietary resilience.

8.3.1 Comparative Nutritional Profiles of Cassava and Yam

Cassava is primarily recognized for its high carbohydrate content, which ranges from 30% to 40% of dry weight, mainly in the form of starch. This high caloric density makes it a significant energy source for many communities throughout sub-Saharan Africa (Meilawaty & Kusumawardani, 2016; Mohidin et al., 2023). In addition to carbohydrates, cassava also provides essential micronutrients, although in lower quantities compared to other tubers (Meilawaty & Kusu-mawardani, 2016; Mohidin et al., 2023). Nutritional analyses have revealed that cassava contains limited protein (approximately 1%–2%) and moderate amounts of vitamin C and B vitamins, including thiamine and riboflavin (Mohidin et al., 2023; Opoku-Agyemang et al., 2024). However, its roots may also harbor toxic compounds such as cyanogenic glycosides, depending on the variety and processing methods used (Kosoe & Ogwu, 2024; Mohidin et al., 2023; Zhang et al., 2022). In contrast, yam offers a more balanced nutritional profile, contain-ing approximately 12%–15% protein and higher levels of micronutrients, such as potassium, magnesium, and vitamins A and C (Li et al., 2017; Zhang et al., 2022). Studies have indicated that yam also provides significant fiber, which is crucial for digestive health; it typically has a fiber content of around 4%–5% of its dry weight (Arise et al., 2023; Zhang et al., 2022). Moreover, the amino acid composition of yam is more favorable compared to cassava, with higher levels of essential amino acids that are critical for human health (Arise et al., 2023). This difference allows yam to play a more significant role in supporting overall nutrition, contrasting with cassava's function primarily as a calorie source.

8.3.2 Functional Uses Beyond Calories

Beyond their caloric contributions, cassava and yam serve various functional roles in cultural practices, medicine, and cuisines throughout West and Central Africa. Both crops are often utilized in ceremonial applications and traditional dishes, forming an integral part of the cultural heritage (Chandrasekara & Kumar, 2016; Eijck et al., 2012). For instance, yam is celebrated during various festivals, signifying prosperity and harvest, while cassava is transformed into a wide range of traditional dishes, from fufu to gari, showcasing its versatility (Chandrasekara & Kumar, 2016; Eijck et al., 2012). Medicinal applications also feature prominently in the use of both crops. Cassava leaves have been traditionally employed in herbal medicine for their anti-inflammatory properties and are known for their high content of flavonoids and other phytochemicals beneficial for health (Cai et al., 2025; Okoro, 2020). These leaves, though of-ten underutilized in food production, can be consumed and have been shown to possess analgesic and anti-inflammatory effects, making them useful in managing conditions such as periodontitis and rheumatism (Cai et al., 2025;

Okoro, 2020). Meanwhile, yam has been associated with improved reproductive health and has been traditionally used in treating various ailments, including gastrointestinal disorders (Zhang et al., 2022). Furthermore, research highlights that extracts from both cassava and yam can modulate important biological pathways, making them useful in the development of nutraceuticals and functional foods (Eijck et al., 2012; Opoku-Agyemang et al., 2024). For example, the antidiabetic properties of cassava have been investigated due to its effect on glucose metabolism and its antioxidant potential, while yam's role in hormonal balance and fertility treatments has drawn attention (Cai et al., 2025; Okoro, 2020).

8.3.3 Role in Dietary Diversification and Resilience

The cultivation of cassava and yam plays a vital role in dietary diversification and food security in West and Central Africa. As staple foods, they provide caloric sustenance while supporting the agricultural economy of rural communities (Li et al., 2017; Meilawaty & Kusumawardani, 2016). Their ability to grow in varied conditions makes them resilient crops, better suited to withstand the erratic climate patterns increasingly faced in the region due to climate change (Li et al., 2017; Mohidin et al., 2023). Cassava, in particular, exhibits notable drought resistance, making it an essential crop for ensuring food availability even in adverse conditions (Phan et al., 2024; Wang et al., 2023). Incorporating cassava and yam into local diets enhances nutritional diversity. This can be critical for combating malnutrition, especially in regions where protein sources may be limited (Li et al., 2017; Okoro, 2020). The cultivation and utilization of these tubers support smallholder farmers' livelihoods and local economies by creating job opportunities and fostering sustainable agricultural practices (Wang et al., 2023). When viewed through the lens of dietary resilience, the inclusion of diverse food sources, including yam and cassava, empowers communities to better adapt to economic strains and climate-induced challenges (Li et al., 2017; Opoku-Agyemang et al., 2024). Moreover, the value-added potential of cassava and yam through processing into flour, snack products, and bioethanol underlines their versatility and importance beyond basic food crop status (Li et al., 2017; Wang et al., 2023). Producing diverse products can pave the way for food security and create new market opportunities for rural farmers. Through these perspectives, it is evident that both cassava and yam not only sustain nutritional needs but also enhance socioeconomic stability within West and Central African communities.

8.4 Toxicology, Processing, and Safety of Cassava and Yam in West and Central Africa

The processing and safety of cassava (Manihot esculenta) and yam in West and Central Africa present significant public health challenges, primarily due to the presence of cyanogenic glucosides in cassava, which can lead to acute toxicity if not properly managed. The consumption of improperly processed cassava can result in chronic cyanide poisoning, manifesting as conditions such as konzo, a neurological disease predominantly affecting communities that rely on cassava as a staple food (Montagnac et al., 2008; Nizzy & Ogwu, 2024; Onabolu et al., 2001). To mitigate these risks, traditional detoxification methods such as fermentation, drying, soaking, and other techniques are vital for rendering cassava safe for consumption.

Cyanogenic glucosides, specifically linamarin and lotaustralin, are found in varying concentrations across different cassava varieties (Egboduku et al., 2024a). The mean cyanogen content in cassava food samples collected from Nigerian markets has indicated that a significant proportion of the samples exceeded the recommended safety limit established by the Food and Agricultural Organization (Onabolu et al., 2001). This highlights the need for effective processing methods to reduce cyanogenic content and ensure food safety. Fermentation, for example, has been shown to significantly lower the levels of cyanogenic compounds, making the product safer for human consumption while also improving its nutritional content (Abiodun et al., 2020; Montagnac et al., 2008; Nainggolan et al., 2024). Various processing methods, such as soaking or drying, can be employed to detoxify cassava effectively. For instance, Airaodion et al. (2019) noted that processing cassava into garri involves a method of soaking and fermenting that not only helps in cyanide reduction but also enhances the product's nutritional profile by improving digestibility and the availability of micronutrients. Similarly, studies have shown that traditional methods like sun-drying can remove a substantial percentage of antinutrients, thereby improving overall nutritional quality (Nambisan, 2011).

In addition to their health risks, the cultural aspects of cassava processing encompass deep-rooted knowledge systems and practices that are transmitted through generations. Ethnoscientific knowledge regarding safe preparation techniques is critical in these communities, as local practices often dictate the safety and efficacy of cassava processing methods. For instance, studies demonstrate that community education on appropriate cassava handling can lead to improved food safety practices (Bokundabi et al., 2023). These cultural norms emphasize safe preparation practices and the transmission of knowledge regarding the detoxification of cassava, which are essential for community health. Moreover, it is important to examine the role of child education concerning food preparation and safety. Educational initiatives aimed at younger generations can influence future practices within households, promoting a culture of safety from an early age. Proper educational frameworks can empower children with the necessary knowledge of food safety, enhancing community resilience against foodborne illnesses (Aquino et al., 2021; Hawashi et al., 2020). Observations from various studies indicate that food handlers and caregivers, who possess adequate training in food safety, exhibit more reliable practices in food preparation compared to those without such training (Husain et al., 2016). Cultural norms and practices notably impact how communities perceive and implement food safety measures. Many households rely on traditional methods that have been adapted to local circumstances, but these methods may not always align with modern food safety standards. Studies indicate that when communities are introduced to improved practices through engagement and education, there is a marked change in attitudes toward food preparation, ultimately leading to safer consumption of foods such as cassava (Ayetigbo et al., 2018; Nainggolan et al., 2024).

Community-based interventions focusing on enhancing the knowledge and skills of individuals involved in cassava processing represent an essential avenue for reducing food safety risks. Empowering food handlers through targeted training has shown positive results; however, the sustainability of such methods hinges on addressing structural barriers and ensuring that knowledge is effectively transmitted (Bokundabi et al., 2023). For food safety initiatives to be effective, they must consider the sociocultural context of the communities they aim to serve. The integration of local knowledge with scientific practices can produce better outcomes in terms of food safety and public health. For instance, the use of local fermentation techniques is not only culturally accepted but also environmentally friendly and economically feasible, highlighting the need for approaches that harmonize traditional wisdom with contemporary food science (Hawashi et al., 2020; Montagnac et al., 2008).

8.5 Culinary Ethnoscience of Cassava and Yam in West and Central Africa

The culinary ethnoscience of cassava and yam in West and Central Africa presents a rich tapestry of traditional cooking techniques, ingredients, and dishes that vary markedly across geographical and ethnic boundaries. These two staples are cornerstones of the diet in this region, and they embody cultural identity and traditional practices tied deeply to local agroecological conditions. The cultivation and consumption of these tubers are rooted in historical practices that have been preserved and adapted through generations, illustrating their critical role in food security and social cohesion. Yam, particularly the species Dioscorea rotundata, dominates the culinary landscape in many parts of West Africa, accounting for over 90% of global yam production (Morse, 2021; Sugihara et al., 2020). Traditional culinary techniques dictate that yam is often prepared through boiling, steaming, or pounding, resulting in variations of dishes such as pounded yam (Iyan), yam porridge (Asaro), and yam fries. These dishes are not merely food; they reflect the social traditions and communal practices inherent to many local cultures (Barlagne et al., 2016). In Nigeria, for instance, pounded yam is a staple often served during social gatherings and ceremonies, illustrating its significance beyond mere sustenance and into the realms of cultural identity (Olagunju-Yusuf et al., 2019).

Cassava is another vital tuber that is similarly integrated into the culinary practices of West and Central Africa. It is often processed into various forms, such as garri, fufu, and cassava flour, reflecting a culinary adaptability that allows it to serve multiple functions within the diet (Ferraro et al., 2015). The fermentation process used to produce garri not only enhances its flavor but also helps in detoxifying harmful compounds like cyanogens, demonstrating sophisticated knowledge about food safety among local populations (Padhan & Panda, 2020). Variations in the usage and preparation of cassava and yam are greatly influenced by ethnic group preferences and regional geography. In Southern Nigeria, the Yoruba people have distinct ways of preparing yam that incorporate local spices and flavors, while the Ibibio group in the southeastern region may prefer different varieties of yam and alternative preparation methods. Such culinary diversity also extends to cassava, where different communities have developed unique dishes that reflect local ingredients and cultural practices, thus preserving their heritage and fostering a sense of community and belonging.

Texture and taste play a pivotal role in how dishes made from yam and cassava are perceived and enjoyed. The sensory attributes of yam are particularly important; the desired smoothness of pounded yam or the right degree of firmness in boiled yam can dictate its popularity (Otegbayo et al., 2023). In a focus group study by Barlagne et al. (2016), consumer preferences highlighted the importance of taste and texture in the acceptability of fresh yam products, emphasizing that personal and cultural factors significantly influence these preferences. Similarly, the culinary treatment of cassava often seeks to balance its starchy texture with various flavors, leading to innovative dishes that cater to local palates (Ferraro et al., 2015). Local preferences also dictate agricultural practices surrounding cassava and yam cultivation. For instance, specific cultivars are favored in certain areas because they are better suited to local soil conditions and climate, impacting yield and the culinary characteristics of the tubers (Asiedu & Sartie, 2010). The integration of modern agricultural techniques with traditional practices in the production of these tubers aims to enhance food security while maintaining cultural heritage (Danquah et al., 2022). Knowledge about the nutritional value of these tubers guides community practices in their consumption, thus reinforcing the connection between food and health (Olagunju-Yusuf et al., 2019).

Recognition of the economic significance of yam and cassava contributes to community resilience. These tubers serve as vital food sources and key components of local economies, supporting livelihoods through market sales and generating income for farmers (Pouya et al., 2022). The dual role of these tubers as staples and economic commodities fosters agroeconomic stability within communities, illustrating how traditional knowledge and contemporary practices intersect (Adewale & Nnamani, 2022). The impact of climate change represents a significant challenge to yam and cassava cultivation, prompting researchers and practitioners to adapt their methods for cultivation and processing (Condé et al., 2024). Sustainable practices that align with traditional knowledge are essential for enhancing the resilience of yam and cassava farming systems, ensuring these essential crops continue to fulfill their vital roles (Kiba et al., 2020). As traditional recipes evolve alongside modern techniques, the culinary landscapes of West and Central Africa remain dynamic, characterized by a continuous interplay of heritage and innovation. The relationship between food, culture, and identity cannot be overstated in the context of yam and cassava. These tubers are enmeshed in local traditions and practices, serving as markers of cultural heritage (Asiedu & Sartie, 2010). Whether through ceremonial preparations or everyday meals, the culinary use of these crops reflects broader themes of community, belonging, and continuity of practice, shaping the identity of various ethnic groups across West and Central Africa.

8.6 Indigenous Knowledge Systems and Gendered Practices of Cassava and Yam in West and Central Africa

The complex interrelationship between Indigenous knowledge systems and gendered practices surrounding the cultivation, preparation, and selection of tubers such as cassava and yam in West and Central Africa is a key area of research (Table 8.1). Women play an instrumental role as custodians of these practices. Their contributions encompass not only the physical aspects of cultivation and preparation but also significant responsibilities such as seed selection, which has profound implications for agricultural sustainability and food sovereignty within their communities.

Women in many African cultures have been recognized as primary custodians of tuber cultivation, spanning generations. They participate in activities that include selecting healthy seed tubers and essential farming practices such as planting, weeding, and harvesting. Evidence shows that women's involvement in cassava production generates economic benefits, indicating that their expertise is fundamental to the success of cassava farming (Immanuel et al., 2024). This expertise encompasses traditional knowledge systems that guide the selection of quality tubers based on their physical characteristics, which are often passed down through generations. Furthermore, these practices are typically supported by detailed understanding of environmental conditions and tuber genetics (Chaïr et al., 2010).

Intergenerational knowledge transfer plays a pivotal role in maintaining Indigenous practices. Effective practices for intergenerational transfer often involve tacit and explicit knowledge-sharing mechanisms, including oral storytelling and communal planting activities, where younger generations learn through participation. This method ensures the survival of agricultural techniques and reinforces cultural values related to land and food. Intergenerational learning is crucial for adapting to changing climatic conditions and market demands while maintaining biodiversity, especially in yam and cassava cultivation (Rupčić, 2018). Studies underscore the significance of women as central figures in this knowledge transfer process, enriching the preservation and adaptation of traditional practices over time (Rupčić, 2018). The gendered division of labor is evident in farming and food preparation associated with these tubers. Women typically undertake most responsibilities regarding cultivation and preparation, while men may be more involved in commercial aspects or cash crop production (Ogwu et al., 2024). These gender roles are deeply ingrained within community structures and significantly shape agricultural decision-making and organization (Cavallari et al., 2016). Such dynamics not only impact the economic status of households but also food security, as women's contributions can enhance the nutritional quality and availability of food (Cavallari et al., 2016; Immanuel et al., 2024). The communal organization of labor further emphasizes the collective nature of tuber cultivation and food preparation practices among women, fostering solidarity

Table 8.1: Indigenous Knowledge Systems for Cassava and Yam in West and Central Africa

Knowledge Area	Cassava	Yam
Crop Choice	Chosen for bitterness (for safety) and drought tolerance	Selected by size, health, and past harvest quality
Land Preparation	Mounds or ridges made, ash added to improve the soil	Raised beds or ridges based on the local land shape
When to Plant	Based on local calendars, the moon, and the rains	Based on traditions and festivals
Tools Used	Hoes and cutlasses to space and dig roots	Digging sticks and stakes to hold up vines
Weeds and Pests	Grown with maize/legumes; ash or pepper used to scare pests	Hand weeding, crop rotation, and mixed planting used
How It's Processed	Fermented (e.g., for gari, fufu) to remove toxins	Roasted, boiled, or pounded into flour
Storage Methods	Kept in underground pits, baskets with leaves, or sun-dried	Stored in special yam barns with stakes and ropes
Cultural Use	Used in ceremonies, feasts, and women's cooperatives	Celebrated in yam festivals and used in rites of passage
Gender Knowledge	Women lead the processing and selling	Men grow yams; women also help in processing
Local Naming	Called "sweet" or "bitter"; also known for healing vs. food use	Names describe look, use, and history
Environmental Fit	Grown with fallow periods to keep the soil healthy	Local knowledge matches yams to the right soil
Spiritual Beliefs	Spirits believed to protect fields or punish wrong harvesting	First yams are offered to the gods or ancestors before anyone eats

and support networks. These networks are vital during times of crisis, such as extreme weather events or market fluctuations, as they facilitate resource sharing and collective action. Thus, women's roles as custodians extend beyond individual contributions; they are integral to a larger socioeconomic network that sustains local agricultural systems. Understanding these labor divisions is essential for grasping the sociocultural context of tuber cultivation and its evolution to meet both local and global food demands (Cavallari et al., 2016; Chaïr et al., 2010).

Moreover, the significance of cassava and yam in West and Central Africa extends beyond their roles as food staples. These tubers embody cultural identity and heritage, frequently featuring in local cuisines and ceremonies. By preserving traditional cultivation and preparation methods, women safeguard cultural values inherent to their communities. Their roles as knowledge custodians contribute to broader discussions on food sovereignty, where control over food systems is crucial for sustainable development and community resilience (Cavallari et al., 2016; Immanuel et al., 2024). When considering the future, sustainable practices surrounding yam and cassava cultivation must adapt to challenges such as climate change and economic pressures. Integrating women's knowledge into modern agricultural practices can provide innovative solutions that honor traditional wisdom while embracing advancements. For example, initiatives aimed at improving seed diversity and soil health can greatly benefit from the extensive knowledge women possess regarding local ecosystems and plant varieties (Cavallari et al., 2016; Rupčić, 2018).

Community-based programs that focus on empowering women in agricultural research and policy frameworks are essential. By recognizing and elevating women's roles within these systems, stakeholders can create more equitable structures that enhance food security and promote sustainable agricultural practices. This empowerment can involve education on sustainable practices and participation in decision-making about land use and crop selection (Cavallari et al., 2016; Rupčić, 2018).

8.7 Socioecological and Agroecological Roles of Cassava and Yam in West and Central Africa

Cassava and yam hold vital socioecological roles in West and Central Africa, significantly contributing to ecological balance, climate adaptation, and resilience against drought conditions. This section examines how these tubers influence ecological dynamics, food security, and agricultural biodiversity across different cultivation practices, intercropping systems, and social frameworks within these regions. Cassava and yam promote ecological balance by enhancing soil retention and promoting soil health. The cultivation of these crops helps stabilize soil structures, which is particularly significant in areas prone to erosion and nutrient leaching. Both crops possess extensive root systems that anchor soil and improve its organic content, promoting nutrient cycling (Ogwu et al. 2024; Oshunsanya & Nwosu, 2018; Villarino et al., 2020). Research highlights that cassava's adaptability enables it to flourish in marginal soils, potentially restoring degraded land through efficient nutrient uptake (Villarino et al., 2020). These qualities align with sustainable farming practices that mitigate soil erosion and prevent nutrient runoff, providing a dual benefit of preserving land while feeding local communities.

In addition to soil retention, intercropping with cassava and yam can bolster agricultural productivity and increase ecological diversity. Intercropping systems utilizing yam and cassava are recognized for their ability to enhance land use efficiency and provide a buffer against pests and diseases (Tchabi et al., 2008, 2009). Diversifying crop systems can lead to better soil fertility by varying nutrient extraction and accumulation patterns, which further supports farming sustainability in regions with sparse biodiversity (Condé et al., 2024). Such systems enable farmers to benefit from both crops while minimizing the risks associated with monocultures, which are often prone to pest infestations and soil nutrient depletion. The role of cassava and yam extends beyond ecological functions to address climate adaptation and resilience. Both crops are well-suited for cultivation under conditions of limited water availability and fluctuating weather patterns. Cassava, in particular, is noted for its drought resistance, making it an attractive option for resource-poor farmers facing climate variability (Mtunguja et al., 2019). In areas where conventional crops fail, cassava has been pivotal in maintaining food security, providing consistent yields under challenging climatic conditions. Furthermore, yam's deep-rooting systems enable it to access subsurface moisture, thus reinforcing farmers' resilience to drought conditions (Tchabi et al., 2008, 2009).

Despite the many benefits of cassava and yam, monocropping poses significant threats to biodiversity in agricultural landscapes. The increased focus on a single crop can lead to the loss of genetic diversity, making farmland more susceptible to diseases and pests (Mota et al., 2017; Thresh & Cooter, 2005). The introduction of high-yielding, uniform varieties can exacerbate this problem, as it displaces diverse local landraces traditionally cultivated for their resilience and nutritional value. Such practices not only threaten ecosystems but also diminish the cultural heritage associated with diverse agricultural systems (Condé et al., 2024). As farmers shift toward monocultures, the ecological balance can be disrupted, leading to soil degradation, habitat loss, and overall reduced biodiversity. Moreover, the environmental implications of extensive monocropping practices in cassava and yam cultivation are concerning. Recent studies indicate that increased reliance on chemical inputs and reduced traditional farming practices linked to biodiversity can lead to soil health deterioration and increased greenhouse gas emissions (Bilong et al., 2022; Kouassi et al., 2023). The historical interdependence of local communities with their agricultural systems is being undermined as they transition toward more industrialized farming methods that favor short-term profits over long-term sustainability. This shift signals a pressing need to cultivate awareness regarding the benefits of ecological farming methods, which can include intercropping and agroforestry to restore soil health and enhance biodiversity.

In response to these challenges, innovative strategies to capitalize on the agroecological potential of yam and cassava are gaining momentum. Sustainable initiatives aim to integrate traditional practices with modern agricultural advancements to create resilient farming communities (Mtunguja et al., 2019; Villarino et al., 2020). By developing agroecological methods that harness the natural interactions between these crops and their environment, it is possible to promote synergy within farming systems that bolsters productivity without compromising biodiversity. For instance, employing organic fertilizers and practicing crop rotation can enhance soil fertility and health while mitigating environmental pressures associated with intensive agricultural practices (Bilong et al., 2022). Adopting intercropping practices with yam and cassava can be a transformative step toward climate-resilient agriculture. A synergistic approach that incorporates crop diversity along with sustainable land use can lead to increased yields and improved soil quality while fostering ecological stability (Bilong et al., 2022; Kouassi et al., 2023). Farmers can benefit from this approach by enhancing their livelihoods and securing food sovereignty, which is increasingly critical as climate change impacts become more severe.

8.8 Threats to Cassava and Yam in West and Central Africa

The examination of threats, transformations, and commodification of cassava and yam in West and Central Africa intricately weaves together issues of globalization, improved agricultural varieties, and cash cropping practices. This industry shift is particularly critical when analyzing the socioeconomic landscape and the implications for food sovereignty in these regions. Globalization has significantly influenced agricultural practices, promoting the standardization of crop varieties in response to increasing demand for cash crops. Improved varieties of cassava, such as those resistant to CMD and CBSD, have been developed, which is crucial given the historical prevalence of these diseases in Africa. Breeding programs have emphasized the development of cultivars that demonstrate resistance, thereby increasing yield potential in affected areas (Odedina et al., 2020; Soro et al., 2021). Such improvements in crop genetics have the capability to enhance food security; however, they simultaneously invoke pressures on Indigenous varieties, which are often marginalized in favor of these commercial strains (Table 8.2).

The introduction of improved cassava varieties can lead to monoculture practices, which contribute to a significant loss of biodiversity. Studies reveal that while hundreds of cassava varieties exist in local communities, the trend is toward the adoption of a few elite varieties that meet commercial requirements (Ogwok et al., 2015; Oluwole et al., 2007). This narrowing genetic pool raises concerns about vulnerability to pests and diseases, along with the potential extinction of traditional landraces. Research indicates that landraces can harbor unique traits that confer resilience to environmental stresses (Kawuki et al., 2013; Onyango et al., 2024). The genetic dilution caused by shifting to improved varieties not only threatens agricultural diversity but also compromises the traditional agricultural knowledge systems that have sustained local communities for generations (Anjanappa et al., 2017).

Table 8.2: Threats to Transformations and Commodification of Cassava and Yam in West and Central Africa

Threat	Cassava	Yam
Ecological Threats	Soil nutrient depletion from monoculture; susceptibility to diseases like cassava mosaic virus and bacterial blight	Land degradation from shifting cultivation; pests like yam beetles and nematodes
Climate-Related Stressors	Drought and erratic rainfall patterns affecting root bulking	Altered rainfall impacting tuber formation and traditional planting calendars

Threat	Cassava	Yam
Cultural Transformations	Decline in traditional fermentation and processing knowledge due to urbanization	Loss of yam festivals and rituals (e.g., New Yam Festival) in diaspora and urban contexts
Labor Shifts	Mechanization in processing; youth migration from rural areas	Reduced youth interest in labor-intensive yam cultivation
Gender Roles	Women dominate processing and marketing (gari, fufu); feminization of value chains	Men traditionally control yam fields; women increasingly involved in marketing
Land Tenure and Access	Land pressure and commodification of farmlands affect cassava expansion	Inheritance systems restrict access for women and youth in yam-growing communities
Market Integration	Growth of cassava-based products (e.g., starch, ethanol) into formal markets	High-value export potential of ware and seed yams; pricing volatility affects local access
Policy and Subsidies	Cassava promoted in national food security policies (e.g., Nigeria's Cassava Transformation Agenda)	Limited support for yam preservation; inadequate storage infrastructure
Value Addition & Innovation	Fortification (biofortified cassava), industrial uses (glue, flour)	Seed yam innovations (aeroponics, minisetts); emergence of yam flour products
Cultural Commodification	Traditional uses being rebranded for niche markets (e.g., organic gari)	Ritual significance diluted in commercial contexts (e.g., yams as luxury items for festivals)

The challenge of balancing commercialization with food sovereignty emerges prominently in this discourse. Cash cropping, motivated by global market demands, often prioritizes export-oriented production over local food needs. This transactional approach can undermine farmers' ability to grow food for subsistence, creating an imbalance where food becomes a commodity rather than a basic right (Han et al., 2016; Laya et al., 2018). As farmers increasingly adopt cash crops like cassava, they may neglect other crops essential for dietary diversity and nutritional security (Ano et al., 2021). Moreover, reliance on these commercial varieties diminishes the resilience of local food systems in the face of climatic changes or market fluctuations. Tensions surrounding this commercialization paradigm are further exacerbated by the socioeconomic conditions faced by smallholder farmers, who may be trapped in a cycle of debt and dependency (Kaweesi et al., 2014). Enhanced crop varieties may promise higher yields, but the input costs associated with growing these improved crops can outweigh the financial benefits, especially when global commodity prices fluctuate (Anjanappa et al., 2017; Masinde et al., 2018). Consequently, the promise of improved food security through commercialization may yield the opposite effect, fostering vulnerability and dependence among local farmers.

In addition, the introduction of improved varieties must be accompanied by strategies that safeguard local agricultural biodiversity. It is vital to combine modern breeding techniques with the preservation of traditional varieties through community seed banks and participatory breeding programs, ensuring that farmers retain access to diverse genetic resources that can adapt to changing agricultural conditions (Ano et al., 2021; Kawuki et al., 2016). Educating farmers about the value of diversity and traditional agricultural practices is essential to mitigate the impacts of market-driven agricultural policies that ignore local needs (Sadiku et al., 2019). Funding models for agrarian research should prioritize sustainable practices that include the participation of local communities, fostering a sense of ownership and responsibility toward preserving biodiversity (Izah & Ogwu 2023; Ogwu & Izah 2023; Prasangika et al., 2008). Agricultural policies should seek a balance between developmental goals and ecological sustainability, ensuring that local farming practices are respected and integrated into broader economic frameworks. Collaborative efforts to empower farmers through training on sustainable farming practices and investment in local food systems can facilitate resilience against market shocks and environmental changes (Emmanuel et al., 2019; Soro et al., 2021).

8.9 Sociobiology Reflection on Cassava and Yam in West and Central Africa

The socioeconomic significance of cassava and yam in West and Central Africa is deeply intertwined with their roles in food security, cultural practices, and regional agricultural economies. Both crops serve as essential dietary staples; cassava is a vital source of carbohydrates for millions, while yam holds substantial cultural and economic relevance, particularly in countries like Nigeria, where it is widely cultivated and integrated into traditional diets (Awoyale et al., 2020). Indeed, yam is recognized as the second most important tuber crop in Africa, with its production largely concentrated in West African nations, underscoring its cultural and socioeconomic importance (Awoyale et al., 2020).

In terms of agricultural output, cassava stands out due to its resilience and high yield potential under suboptimal conditions, which is critical for food security in many rural subsistence communities facing challenges such as climate change and pest diseases. Cassava cultivation has expanded significantly, demonstrating adaptability to various environmental conditions and yielding reasonable outputs even in poor soils, with some reports indicating yields averaging around 10 t/ha (Parmar et al., 2017). This adaptability is vital as the region grapples with the ramifications of rapid population growth and dwindling agricultural productivity (Ozimati et al., 2022). Efforts to breed disease-resistant cassava varieties have been pivotal in enhancing yields and securing the livelihoods of farmers across sub-Saharan Africa (Beyene et al., 2017; Okogbenin et al., 2012).

Disease resistance is indeed a pressing concern, particularly with threats posed by CMD and CBSD. These viral diseases, transmitted by whiteflies, have caused significant economic losses and food insecurity in the region (Alicai et al., 2019; Beyene et al., 2017). For instance, CMD has led to root losses of up to 100% in highly susceptible cultivars, emphasizing the importance of developing and disseminating genetically resistant varieties (Houngue et al., 2019). The need for such varieties is critical, especially as environmental conditions become more favorable for the vectors responsible for their spread (Nwezeobi et al., 2020a, 2020b). Conversely, yam production also faces challenges, though not to the same extent as cassava, with its cultural significance transcending nutritional value and playing a role in rituals and social gatherings, particularly in Nigerian communities (Awoyale et al., 2020). However, research on yam is less prevalent compared to cassava, underscoring a gap that needs addressing in agricultural practices and research priorities in the region.

8.10 Conclusion

Cassava and yam cultivation in West and Central Africa offers a compelling lens into the coevolution of human societies and tuber-based agroecosystems. These crops, far from being mere dietary staples, encapsulate a wealth of ecological knowledge, biochemical management, and cultural meaning that has been fine-tuned over generations. The processing of bitter cassava to neutralize cyanogenic compounds exemplifies a remarkable biocultural adaptation—one that marries food safety with culinary tradition and underscores the scientific rigor embedded within Indigenous knowledge systems. Similarly, yam production and its associated rituals illuminate how agriculture, identity, and social organization are deeply intertwined. This chapter has shown that traditional processing methods are not only effective in managing toxicological risks but also serve as socioecological safeguards that ensure long-term food and environmental security. Moreover, the gendered and intergenerational transmission of knowledge related to tuber cultivation highlights the importance of inclusive approaches to food systems research and policy. However, contemporary pressures such as land-use change, climate variability, and market-driven agricultural transformation pose serious threats to these systems. In response, there is an urgent need for policies that bridge scientific innovation with Indigenous expertise to support resilient, culturally grounded food systems. Future research should prioritize participatory frameworks that validate local practices while enhancing ecological sustainability and nutritional outcomes. Recognizing the sociobiological value of cassava and yam not only enriches academic discourse but also empowers communities to assert sovereignty over their food, health, and ecological futures.

Chapter Reflection

1. How did cassava and yam originate and become central to the diets of West and Central African communities?
2. In what ways do cassava and yam differ nutritionally, and how do these differences impact food security?
3. What roles do women play in the cultivation, processing, and preservation of cassava and yam?
4. How do Indigenous detoxification techniques ensure cassava safety despite its cyanogenic compounds?

5. How have intergenerational knowledge systems contributed to sustaining cassava and yam cultivation?
6. What are the agroecological benefits of intercropping cassava and yam in terms of soil health and resilience?
7. How do cultural festivals and ceremonies reflect the symbolic importance of yam and cassava?
8. What are the main risks associated with the commercialization and mono-culture expansion of these tubers?
9. How is climate change affecting cassava and yam production in West and Central Africa?
10. Why is it important to integrate traditional knowledge with scientific in-novation to ensure the future sustainability of cassava and yam systems?

REFERENCES

Abajue, M. C., & Ogwu, M. C. (2024). Morphological features, physicochem-ical and microbial characterization of cassava, and its products in the tropics. In M. C. Ogwu, S. C. Izah, A. A. Cunha Alves, & S. C. Babu (Eds.), *Plant biology, sustainability and climate change, sustainable cassava* (pp. 19–38). Academic Press. https://doi.org/10.1016/B978-0-443-21747-0.00023-0

Abiodun, O., Ayano, B., & Amanyunose, A. (2020). Effect of fermentation periods and storage on the chemical and physicochemical properties of biofort-ified cassava gari. *Journal of Food Processing and Preservation*, *44*(12), e14958. https://doi.org/10.1111/jfpp.14958

Adewale, B., & Nnamani, C. (2022). Introduction to food, feed, and health wealth in African yam bean, a locked-in African Indigenous tuberous legume. *Frontiers in Sustainable Food Systems*, *6*. https://doi.org/10.3389/fsufs.2022.726458

Agoh, E. C., Ogwu, M. C., Chukwuemeka, O. S., & Ekeledo, P. I. (2024). Environmental and human health effects of cassava processing and processing waste. In M. C. Ogwu, S. C. Izah, A. A. Cunha Alves, & S. C. Babu (Eds.), *Plant biology, sustainability and climate change, sustainable cassava* (pp. 203–219). Academic Press. https://doi.org/10.1016/B978-0-443-21747-0.00001-1

Agre, P., Dassou, A. G., Loko, L. E. Y., Idossou, R., Dadonougbo, E., Gbaguidi, A., Mondo, J. M., Muyideen, Y., Adebola, P. O., Asiedu, R., Dansi, A. A., & Asfaw,

A. (2021). Diversity of white guinea yam (*Dioscorea rotundatapoir.*) cultivars from benin as revealed by agro-morphological traits and SNP markers. *Plant Genetic Resources, 19*(5), 437–446. https://doi.org/10.1017/s1479262121000526

Airaodion, A., Airaodion, E., Ogbonnaya, E., Ogbuagu, E., & Ogbuagu, U. (2019). Nutritional and anti–nutritional evaluation of garri processed by traditional and instant mechanical methods. *Asian Food Science Journal, 9*(4), 1–13. https://doi.org/10.9734/afsj/2019/v9i430021

Alene, A., Abdoulaye, T., Rusike, J., Manyong, V., & Walker, T. (2015). The effectiveness of crop improvement programmes from the perspectives of varietal output and adoption: Cassava, cowpea, soybean and yam in sub-Saharan Africa and maize in West and Central Africa. In T. S. Walker & J. Alwang (Eds.), *Crop improvement, adoption, and impact of improved varieties in food crops in sub-Saharan Africa* (pp. 74–122). CABI. https://doi.org/10.1079/9781780644011.0074

Alicai, T., Szyniszewska, A. M., Omongo, C. A., Abidrabo, P., Okao-Okuja, G., Baguma, Y., Ogwok, E., Kawuki, R., Esuma, W., Tairo, F., Bua, A., Legg, J. P., Stutt, R. O. J. H., Godding, D., Sseruwagi, P., Ndunguru, J., & Gilligan, C. (2019). Expansion of the cassava brown streak pandemic in Uganda revealed by annual field survey data for 2004 to 2017. *Scientific Data, 6*(1), 327. https://doi.org/10.1038/s41597-019-0334-9

Anikwe, M., & Ikenganyia, E. (2018). Ecophysiology and production principles of cassava (manihot species) in southeastern Nigeria. In V. Waisundara (Ed.), *Cassava.* InTech. https://doi.org/10.5772/intechopen.70828

Anjanappa, R., Mehta, D., Okoniewski, M., Szabelska-Berȩsewicz, A., Gruissem, W., & Vanderschuren, H. (2017). Early transcriptome analysis of the brown streak virus–cassava pathosystem provides molecular insights into virus susceptibility and resistance. https://doi.org/10.1101/100552

Anjanappa, R., Mehta, D., Okoniewski, M., Szabelska-Berȩsewicz, A., Gruissem, W., & Vanderschuren, H. (2017). Molecular insights into cassava brown streak virus susceptibility and resistance by profiling of the early host response. *Molecular Plant Pathology, 19*(2), 476–489. https://doi.org/10.1111/mpp.12565

Ano, C., Ochwo-Ssemakula, M., Ibanda, A., Ozimati, A., Gibson, P., Onyeka, J., Njoku, D., Egesi, C., & Kawuki, R. (2021). Cassava brown streak disease response and association with agronomic traits in elite Nigerian cassava cultivars. *Frontiers in Plant Science, 12.* https://doi.org/10.3389/fpls.2021.720532

Aquino, H., Yap, T., Lacap, J., Tuazon, G., & Flores, M. (2021). Food safety knowledge, attitudes, practices and training of fast-food restaurant food handlers: A moderation analysis. *British Food Journal, 123*(12), 3824–3840. https://doi.org/10.1108/bfj-01-2021-0026

Arise, A., Esan, O., & Famakinde, T. (2023). Amino acid, pasting and sensory properties of "poundo" yam enriched with fermented bambara groundnut flour. *Ceylon Journal of Science, 52*(1), 83. https://doi.org/10.4038/cjs.v52i1.8107

Asiedu, R., & Sartie, A. (2010). Crops that feed the world 1. yams. *Food Security, 2*(4), 305–315. https://doi.org/10.1007/s12571-010-0085-0

Awoyale, W., Oyedele, H., & Maziya-Dixon, B. (2020). Correlation of the sensory attributes of thick yam paste (amala) and the functional and pasting properties of the flour as affected by storage periods and packaging materials. *Journal of Food Processing and Preservation, 44*(10), e14732. https://doi.org/10.1111/jfpp.14732

Ayetigbo, O., Latif, S., Abass, A., & Müller, J. (2018). Comparing characteristics of root, flour and starch of biofortified yellow-flesh and white-flesh cassava variants, and sustainability considerations: A review. *Sustainability, 10*(9), 3089. https://doi.org/10.3390/su10093089

Babatunde, O., Salami, A., & Oriola, E. (2022). Variation in cassava yield cultivated on ferralsols and ferruginous soils in Kwara State, Nigeria. *International Journal of Multidisciplinary Studies, 9*(2), 1. https://doi.org/10.4038/ijms.v9i2.160

Barlagne, C., Cornet, D., Blazy, J., Diman, J., & Ozier-Lafontaine, H. (2016). Consumers' preferences for fresh yam: A focus group study. *Food Science & Nutrition, 5*(1), 54–66. https://doi.org/10.1002/fsn3.364

Beyene, G., Chauhan, R., Ilyas, M., Wagaba, H., Fauquet, C. M., Miano, D., Alicai, T., & Taylor, N. (2017). A virus-derived stacked RNAi construct confers robust resistance to cassava brown streak disease. *Frontiers in Plant Science, 7.* https://doi.org/10.3389/fpls.2016.02052

Bilong, E. G., Abossolo-Angue, M., Ajebesone, F. N., Anaba, B. D., Madong, B. A., Nomo, L. B., & Bilong, P. (2022). Improving soil physical properties and cassava productivity through organic manures management in the Southern Cameroon. *Heliyon, 8*(6), e09570. https://doi.org/10.1016/j.heliyon.2022.e09570

Bokundabi, G., Haskins, L., Horwood, C., Kuwa, C., Mutombo, P. B., John, V. M., Mapatano, M. A., & Banea, J. (2023). When knowledge is not enough:

Barriers to recommended cassava processing in resource-constrained Kwango, Democratic Republic of Congo. *Journal of Public Health in Africa, 14*(5), 5. https://doi.org/10.4081/jphia.2023.2052

Cai, J., Wenli, Z., Xue, J., Ma, Y., Li, K., Zhang, L., Aluko, O. O., Chen, S., Luo, X., & An, F. (2025). A comprehensive analysis of chemical composition and anti-inflammatory effects of cassava leaf extracts in two varieties in *Manihot esculenta* crantz. *International Journal of Molecular Sciences, 26*(9), 4140. https://doi.org/10.3390/ijms26094140

Cavallari, J., Ahuja, M., Dugan, A. G., Meyer, J. D., Simcox, N., Wakai, S., & Garza, J. L. (2016). Differences in the prevalence of musculoskeletal symptoms among female and male custodians. *American Journal of Industrial Medicine, 59*(10), 841–852. https://doi.org/10.1002/ajim.22626

Chaïr, H., Cornet, D., Deu, M., Baco, M., Agbangla, A., Duval, M., & Noyer, J. (2010). Impact of farmer selection on yam genetic diversity. *Conservation Genetics, 11*(6), 2255–2265. https://doi.org/10.1007/s10592-010-0110-z

Chandrasekara, A., & Kumar, T. (2016). Roots and tuber crops as functional foods: A review on phytochemical constituents and their potential health benefits. *International Journal of Food Science, 2016*, 1–15. https://doi.org/10.1155/2016/3631647

Chia, R. (2023). *Agriculture, West African forests* (pp. 1–8). Wiley. https://doi.org/10.1002/9781119399919.eahaa00473

Condé, N., Burton, G., Touré, M., Gori, B., Cheek, M., Magassouba, S., Wilkin, P., Couch, C., & Ryan, P. (2024). The biocultural heritage and changing role of Indigenous yams in the Republic of Guinea, West Africa. *Plants People Planet, 7*(3), 719–733. https://doi.org/10.1002/ppp3.10498

Danquah, E. O., Danquah, F. O., Frimpong, F., Dankwa, K. O., Weebadde, C. K., Ennin, S. A., Asante, M. O. O., Brempong, M. B., Dwamena, H. A., Addo-Danso, A., Nyamekye, D. R., Akom, M., & Opoku, A. (2022). Sustainable intensification and climate-smart yam production for improved food security in West Africa: A review. *Frontiers in Agronomy, 4*. https://doi.org/10.3389/fagro.2022.858114

Egbebiyi, T., Lennard, C., Pinto, I., Odoulami, R., Wiolski, P., Tilmes, S., & EGBEBIYI, T. S. (2023). *How will solar radiation modification affect cropland suitability in West Africa?*. EGU23-703. https://doi.org/10.5194/egusphere-egu23-703

Egboduku, W. O., Egboduku, T., Golohor, O. M., Imarhiagbe, O., & Ogwu, M. C. (2024a). Cassava as raw material for sustainable bioeconomy development. In M. C. Ogwu, S. C. Izah, A. A. Cunha Alves, & S. C. Babu (Eds.), *Plant biology, sustainability and climate change, sustainable cassava* (pp. 57–73). Academic Press. https://doi.org/10.1016/B978-0-443-21747-0.00022-9

Egboduku, W. O., Ogwu, M. C., Egboduku, T., Enujeke, E. C., Edokpaiawe, S., & Golohor, O. M. (2024b). Sustainable cassava processing techniques to eliminate cyanogenic glycosides. In M. C. Ogwu, S. C. Izah, A. A. Cunha Alves, & S. C. Babu (Eds.), *Plant biology, sustainability and climate change, sustainable cassava* (pp. 379–394). Academic Press. https://doi.org/10.1016/B978-0-443-21747-0.00007-2.

Eijck, J., Smeets, E., & Faaij, A. (2012). The economic performance of jatropha, cassava and eucalyptus production systems for energy in an East African smallholder setting. *GCB Bioenergy*, *4*(6), 828–845. https://doi.org/10.1111/j.1757-1707.2012.01179.x

Emmanuel, C., Inthuja, A., & Keshiga, A. (2019). Status of cassava cultivation in Jaffna Peninsula and detection of cassava mosaic disease causing agent. *Vingnanam Journal of Science*, *14*(2), 1. https://doi.org/10.4038/vingnanam.v14i2.4150

Ferraro, V., Piccirillo, C., Tomlins, K., & Pintado, M. (2015). Cassava (*Manihot esculenta Crantz*) and yam (*Dioscorea* spp.) crops and their derived foodstuffs: Safety, security and nutritional value. *Critical Reviews in Food Science and Nutrition*, *56*(16), 2714–2727. https://doi.org/10.1080/10408398.2014.922045

Godding, D., Stutt, R., Alicai, T., Abidrabo, P., Okao-Okuja, G., & Gilligan, C. (2023). Developing a predictive model for an emerging epidemic on cassava in sub-Saharan Africa. *Scientific Reports*, *13*(1), 12603. https://doi.org/10.1038/s41598-023-38819-x

Guo, Y. (2020). Closing yield gap of cassava for food security in West Africa. *Nature Food*, *1*(6), 325. https://doi.org/10.1038/s43016-020-0107-9

Halake, N., & Chinthapalli, B. (2020). Fermentation of traditional African cassava-based foods: Microorganisms role in nutritional and safety value. *Journal of Experimental Agriculture International*, *42*(9), 56–65. https://doi.org/10.9734/jeai/2020/v42i930587

Han, B., Fu, L., Zhang, D., He, X., Chen, Q., Peng, M., & Zhang, J. (2016). Interspecies and intraspecies analysis of trehalose contents and the biosynthesis pathway gene family reveals crucial roles of trehalose in osmotic-stress tolerance in cassava. *International Journal of Molecular Sciences, 17*(7), 1077. https://doi.org/10.3390/ijms17071077

Hassall, K., Chávez, V., Sint, H., Helps, J., Abidrabo, P., Okao-Okuja, G., Eboulem, R. G., Amoakon, W. J.-L., Otron, D. H., & Szyniszewska, A. (2024). Validating a cassava production spatial disaggregation model in sub-Saharan Africa. *Plos One, 19*(11), e0312734. https://doi.org/10.1371/journal.pone.0312734

Hawashi, M., Widjaja, T., & Gunawan, S. (2020). Solid-state fermentation of cassava products for degradation of anti-nutritional value and enrichment of nutritional value. In R. M. Martínez-Espinosa (Ed.), *New advances on fermentation processes*. IntechOpen. https://doi.org/10.5772/intechopen.87160

Houngue, J., Pita, J., Ngalle, H., Zandjanakou-Tachin, M., Kuate, A., Cacaï, G., Bell, J. M., & Ahanhanzo, C. (2019). Response of cassava cultivars to African cassava mosaic virus infection across a range of inoculum doses and plant ages. *Plos One, 14*(12), e0226783. https://doi.org/10.1371/journal.pone.0226783

Husain, N., Muda, W., Jamil, N., Hanafi, N., & Rahman, R. (2016). Effect of food safety training on food handlers' knowledge and practices. *British Food Journal, 118*(4), 795–808. https://doi.org/10.1108/bfj-08-2015-0294

Imarhiagbe, O., Ogwu, M. C., Ikponmwosa, B. O., Mukah, F. E., Akemu, S. E., & Ohiaba, E. E. (2024). Threats to cassava cultivation, production, and processing: Global status and sustainable management strategies. In M. C. Ogwu, S. C. Izah, A. A. Cunha Alves, & S. C. Babu (Eds.), *Plant Biology, sustainability and climate change, sustainable cassava* (pp. 75–97). Academic Press. https://doi.org/10.1016/B978-0-443-21747-0.00016-3.

Immanuel, S., Jaganathan, D., & Prakash, P. (2024). Gender analysis and empowerment of women and men in cassava (*Manihot esculenta*) production in Kerala. *Journal of Horticultural Sciences, 19*(1). https://doi.org/10.24154/jhs.v19i1.2646

Izah, S. C., & Ogwu, M. C. (2023). *Sustainable utilization and conservation of Africa's biological resources and environment*. Sustainable Development and Biodiversity Series (1st ed., Vol. 32, 691p.) (Ramawat K.G.). Springer Nature. https://doi.org/10.1007/978-981-19-6974-4

Kaweesi, T., Kawuki, R., Kyaligonza, V., Baguma, Y., Tusiime, G., & Ferguson, M. (2014). Field evaluation of selected cassava genotypes for cassava brown streak disease based on symptom expression and virus load. *Virology Journal*, *11*(1), 216. https://doi.org/10.1186/s12985-014-0216-x

Kawuki, R. S., Herselman, L., Labuschagne, M. T., Nzuki, I., Ralimanana, I., Bidiaka, M., Kanyange, M. C., Gashaka, G., Masumba, E., Mkamilo, G., Gethi, J., Wanjala, B., Zacarias, A., Madabula, F., & Ferguson, M. (2013). Genetic diversity of cassava (*Manihot esculenta Crantz*) landraces and cultivars from Southern, Eastern and Central Africa. *Plant Genetic Resources*, *11*(2), 170–181. https://doi.org/10.1017/s1479262113000014

Kawuki, R., Kaweesi, T., Esuma, W., Pariyo, A., Kayondo, S., Ozimati, A., Kyaligonza, V., Abaca, A., Orone, J., Tumuhimbise, R., Nuwamanya, E., Abidrabo, P., Amuge, T., Ogwok, E., Okao, G., Wagaba, H., Adiga, G., Alicai, T., Omongo, C., … Baguma, Y. (2016). Eleven years of breeding efforts to combat cassava brown streak disease. *Breeding Science*, *66*(4), 560–571. https://doi.org/10.1270/jsbbs.16005

Kiba, D. I., Hgaza, V. K., Aighewi, B., Aké, S., Barjolle, D., Bernet, T., Diby, L. N., Ilboudo, L. J., Nicolay, G., Oka, E., Ouattara, F. Y., Pouya, N., Six, J., & Frossard, E., & Frossard, E. (2020). A transdisciplinary approach for the development of sustainable yam (*Dioscorea* sp.) production in West Africa. *Sustainability*, *12*(10), 4016. https://doi.org/10.3390/su12104016

Kosoe, E. A., & Ogwu, M. C. (2024). Sustainable cassava processing: Processes, techniques, wastes, and waste streams. In M. C. Ogwu, S. C. Izah, A. A. Cunha Alves, & S. C. Babu (Eds.), *Plant biology, sustainability and climate change, sustainable cassava* (pp. 257–272). Academic Press. https://doi.org/10.1016/B978-0-443-21747-0.00011-4

Kouassi, K., Kouadio, Y., Kouassi, K., N'dri, Y., & Amani, N. (2023). Impacts of storage practices on the physical, culinary and sensory quality of kponan yam (*Dioscorea cayenensis-rotundata*) from côte d'ivoire during storage. *Journal of the Science of Food and Agriculture*, *104*(4), 2023–2029. https://doi.org/10.1002/jsfa.13093

Krüger, D., Weng, A., & Baecker, D. (2024). Investigating the discoloration of leaves of *Dioscorea polystachya* using developed atomic absorption spectrometry methods for manganese and molybdenum. *Molecules*, *29*(16), 3975. https://doi.org/10.3390/molecules29163975

Laya, A., Koubala, B., Habiba, K., & Nukenine, E. (2018). Effect of harvest period on the proximate composition and functional and sensory properties of gari produced from local and improved cassava (*Manihot esculenta*) varieties. *International Journal of Food Science, 2018*, 1–15. https://doi.org/10.1155/2018/6241035

Legg, J., Shirima, R., Tajebe, L., Guastella, D., Boniface, S., Jeremiah, S., Nsami, E., Chikoti, P., & Rapisarda, C. (2014). Biology and management of bemisia whitefly vectors of cassava virus pandemics in Africa. *Pest Management Science, 70*(10), 1446–1453. https://doi.org/10.1002/ps.3793

Li, S., Cui, Y., Zhou, Y., Luo, Z., Liu, J., & Zhao, M. (2017). The industrial applications of cassava: Current status, opportunities and prospects. *Journal of the Science of Food and Agriculture, 97*(8), 2282–2290. https://doi.org/10.1002/jsfa.8287

Masinde, E., Mkamillo, G., Ogendo, J., Hillocks, R., Mulwa, R., Kimata, B., & Maruthi, M. N. (2018). Genotype by environment interactions in identifying cassava (*Manihot esculenta Crantz*) resistant to cassava brown streak disease. *Field Crops Research, 215*, 39–48. https://doi.org/10.1016/j.fcr.2017.10.001

Meilawaty, Z., & Kusumawardani, B. (2016). Effect of cassave leaf flavonoid extract on tnf-α expressions in rat models suffering from periodontitis. *Dental Journal (Majalah Kedokteran Gigi), 49*(3), 137. https://doi.org/10.20473/j.djmkg.v49.i3.p137-142

Mohidin, S., Moshawih, S., Hermansyah, A., Ikmal, A., Shafqat, N., & Ming, L. (2023). Cassava (*Manihot esculenta* Crantz): A systematic review for the pharmacological activities, traditional uses, nutritional values, and phytochemistry. *Journal of Evidence-Based Integrative Medicine, 28*. https://doi.org/10.1177/2515690x231206227

Mombo, S., Dumat, C., Shahid, M., & Schreck, E. (2016). A socio-scientific analysis of the environmental and health benefits as well as potential risks of cassava production and consumption. *Environmental Science and Pollution Research, 24*(6), 5207–5221. https://doi.org/10.1007/s11356-016-8190-z

Montagnac, J., Davis, C., & Tanumihardjo, S. (2008). Processing techniques to reduce toxicity and antinutrients of cassava for use as a staple food. *Comprehensive Reviews in Food Science and Food Safety, 8*(1), 17–27. https://doi.org/10.1111/j.1541-4337.2008.00064.x

Montemayor, S., Dellapé, P., & Melo, M. (2014). Predicting the potential invasion suitability of regions to cassava lacebug pests (Heteroptera: tingidae: *Vatiga* spp.). *Bulletin of Entomological Research, 105*(2), 173–181. https://doi.org/10.1017/s0007485314000856

Morse, S. (2021). The role of plant health in the sustainable production of seed yams in Nigeria: A challenging nexus between plant health, human food security, and culture. *Plant Pathology, 71*(1), 43–54. https://doi.org/10.1111/ppa.13409

Mota, A., Lima, P., Silva, D., Abreu, V., Freitas, E., & Pereira, A. (2017). Internal quality of eggs coated with cassava and yam starches. *Revista Brasileira De Ciências Agrárias—Brazilian Journal of Agricultural Sciences, 12*(1), 47–50. https://doi.org/10.5039/agraria.v12i1a5420

Mtunguja, M., Beckles, D., Laswai, H., Ndunguru, J., & Sinha, N. (2019). Opportunities to commercialize cassava production for poverty alleviation and improved food security in Tanzania. *African Journal of Food Agriculture Nutrition and Development, 19*(1), 13928–13946. https://doi.org/10.18697/ajfand.84.blfb1037

Nainggolan, E., Banout, J., & Urbanová, K. (2024). Recent trends in the pre-drying, drying, and post-drying processes for cassava tuber: A review. *Foods, 13*(11), 1778. https://doi.org/10.3390/foods13111778

Nambisan, B. (2011). Strategies for elimination of cyanogens from cassava for reducing toxicity and improving food safety. *Food and Chemical Toxicology, 49*(3), 690–693. https://doi.org/10.1016/j.fct.2010.10.035

Nizzy, A. M., & Ogwu, M. C. (2024). Valorization of cassava processing by-products into biofuel for a sustainable environment. In M. C. Ogwu, S. C. Izah, A. A. Cunha Alves, & S. C. Babu (Eds.), *Plant biology, sustainability and climate change, sustainable cassava* (pp. 291–309). Academic Press. https://doi.org/10.1016/B978-0-443-21747-0.00010-2

Nwezeobi, J., Onyegbule, O., Nkere, C., Onyeka, J., Brunschot, S., Seal, S., & Colvin, J. (2020a). Cassava whitefly species in Eastern Nigeria and the threat of vector-borne pandemics from East and Central Africa. *Plos One, 15*(5), e0232616. https://doi.org/10.1371/journal.pone.0232616

Nwezeobi, J., Onyegbule, O., Nkere, C., Onyeka, J., Brunschot, S., Seal, S., & Colvin, J. (2020b). Protocol for whitefly genetic analysis v1. https://doi.org/10.17504/protocols.io.bd49i8z6

Odedina, S., Ajayi, A., & Awoyemi, S. (2020). Assessing yield responses of four improved cassava varieties in Akure, Nigeria. *International Journal of Food Science and Agriculture*, *4*(1). https://doi.org/10.26855/ijfsa.2020.03.002

Ogwok, E., Alicai, T., Rey, M., Beyene, G., & Taylor, N. (2015). Distribution and accumulation of cassava brown streak viruses within infected cassava (*Manihot esculenta*) plants. *Plant Pathology*, *64*(5), 1235–1246. https://doi.org/10.1111/ppa.12343

Ogwu, M. C., & Izah, S. C. (2023). *One health implications of agrochemicals and their sustainable alternatives*. Sustainable Development and Biodiversity Series (1st ed., Vol. 34, 826p.) (Ramawat K. G.). Springer Nature. https://doi.org/10.1007/978-981-99-3439-3

Ogwu, M. C., Izah, S. C., Alves, A. A. C., & Babu, S. (2024). *Sustainable cassava: Strategies from production through waste management*. Elsevier.

Ogwu, M. C., & Kulkarni, S. (2024). Sustainable cassava: An overview. In M. C. Ogwu, S. C. Izah, A. A. Cunha Alves, & S. C. Babu (Eds.), *Plant biology, sustainability and climate change, sustainable cassava* (pp. 1–15). Academic Press. https://doi.org/10.1016/B978-0-443-21747-0.00024-2.

Ogwu, M. C., Odozi, I. P., Ahonsi, O. C., Uleanya, K. O., & Odozi, E. B. (2024). Organic acid production from cassava. In M. C. Ogwu, S. C. Izah, A. A. Cunha Alves, & S. C. Babu (Eds.), *Plant biology, sustainability and climate change, sustainable cassava* (pp. 395–418). Academic Press. https://doi.org/10.1016/B978-0-443-21747-0.00009-6

Okogbenin, E., Egesi, C. N., Olasanmi, B., Ogundapo, O., Kahya, S., Hurtado, P., Marin, J., Akinbo, O., Mba, C., Gomez, H., de Vicente, C., Baiyeri, S., Uguru, M., Ewa, F., & Fregene, M. (2012). Molecular marker analysis and validation of resistance to cassava mosaic disease in elite cassava genotypes in Nigeria. *Crop Science*, *52*(6), 2576–2586. https://doi.org/10.2135/cropsci2011.11.0586

Okoro, I. (2020). Two extracts from manihot esculenta leaves efficiently inhibit α-glucosidase and α-amylase: A new approach for the management of diabetes. *Iranian Journal of Toxicology*, *14*(3), 131–138. https://doi.org/10.32598/ijt.14.3.583.3

Olagunju-Yusuf, O., Adebowale, A., Sobukola, O., & Sanni, L. (2019). The optimization of production of instant pounded yam flour using cultivars of white yam (*Dioscorea rotundata*). *Asian Food Science Journal, 13*(4), 1–9. https://doi.org/10.9734/afsj/2019/v13i430113

Olukanni, D. O., & Olatunji, T. O. (2018). Cassava waste management and biogas generation potential in selected local government areas in Ogun State, Nigeria. *Recycling, 3*(4), 58. https://doi.org/10.3390/recycling3040058

Oluwole, O., Onabolu, A., Mtunda, K., & Mlingi, N. (2007). Characterization of cassava (*Manihot esculenta Crantz*) varieties in Nigeria and Tanzania, and farmers' perception of toxicity of cassava. *Journal of Food Composition and Analysis, 20*(7), 559–567. https://doi.org/10.1016/j.jfca.2007.04.004

Onabolu, A., Oluwole, O., Bokanga, M., & Rosling, H. (2001). Ecological variation of intake of cassava food and dietary cyanide load in Nigerian communities. *Public Health Nutrition, 4*(4), 871–876. https://doi.org/10.1079/phn2001127

Onyango, E., Kituyi, S., Hunja, C., Kimatu, J., & Nyaboga, E. (2024). First time morphological and molecular isolation of epicoccum sorghinum pathogenicity in the cassava brown leaf spot disease in Kenya. https://doi.org/10.21203/rs.3.rs-4228831/v1

Opabode, J. (2014). Influence of type and age of primary somatic embryo on secondary and cyclic somatic embryogenesis of cassava (*Manihot esculenta Crantz*). *British Biotechnology Journal, 4*(3), 254–269. https://doi.org/10.9734/bbj/2014/3624

Opoku-Agyemang, F., Amissah, J., Owusu-Nketia, S., Ofori, P., & Notaguchi, M. (2024). Optimization of cassava (*Manihot esculenta Crantz*) grafting technique to enhance its adoption in cassava cultivation. *Methodsx, 13*, 102904. https://doi.org/10.1016/j.mex.2024.102904

Oshunsanya, S., & Nwosu, N. (2018). *Soil-water-crop relationship: A case study of cassava in the tropics*. InTech. https://doi.org/10.5772/intechopen.71968

Otegbayo, B., Oluyinka, O., Tanimola, A., Bisi, F., Ayomide, A., Tomilola, B., Madu, T., Okoye, B., Chijioke, U., Ofoeze, M., Alamu, E. O., Adesokan, M., Ayetigbo, O. Bouniol, A., DJibril-Mousa, I., Adinsi, L., Akissoe, N., Cornet, D., Agre, P., ... Maziya-Dixon, B. (2023). Food quality profile of pounded yam and implications for yam breeding. *Journal of the Science of Food and Agriculture, 104*(8), 4635–4651. https://doi.org/10.1002/jsfa.12835

Ozimati, A., Esuma, W., Manze, F., Iragaba, P., Kanaabi, M., Ano, C., Egesi, C., & Kawuki, R. (2022). Utility of Ugandan genomic selection cassava breeding populations for prediction of cassava viral disease resistance and yield in West African clones. *Frontiers in Plant Science*, *13*. https://doi.org/10.3389/fpls.2022.1018156

Padhan, B., & Panda, D. (2020). Potential of neglected and underutilized yams (*Dioscorea spp.*) for improving nutritional security and health benefits. *Frontiers in Pharmacology*, *11*. https://doi.org/10.3389/fphar.2020.00496

Parmar, A., Sturm, B., & Hensel, O. (2017). Crops that feed the world: production and improvement of cassava for food, feed, and industrial uses. *Food Security*, *9*(5), 907–927. https://doi.org/10.1007/s12571-017-0717-8

Phan, T. Q., Wang, S., Nguyen, T. H., Nguyen, T. H., Pham, T. H. T., Đoàn, M. D., Tran, T. H. T., Ngo, V. A., Nguyen, A. D., & Nguyen, V. B. (2024). Using cassava starch processing by-product for bioproduction of 1-hydroxyphenazine: A novel fungicide against fusarium oxysporum. *Recycling*, *9*(1), 12. https://doi.org/10.3390/recycling9010012

Pouya, N., Hgaza, V., Kiba, D., Bomisso, L., Aighewi, B., Aké, S., & Frossard, E. (2022). Water yam (*Dioscorea alata* L.) growth and tuber yield as affected by rotation and fertilization regimes across an environmental gradient in West Africa. *Agronomy*, *12*(4), 792. https://doi.org/10.3390/agronomy12040792

Prasangika, H., Salim, N., & Razak, M. (2008). Evaluation of susceptibility of cassava germplasm to cassava mosaic disease. *Journal of the National Science Foundation of Sri Lanka*, *36*(1), 99. https://doi.org/10.4038/jnsfsr.v36i1.137

Rupčić, N. (2018). Intergenerational learning and knowledge transfer – challenges and opportunities. *The Learning Organization*, *25*(2), 135–142. https://doi.org/10.1108/tlo-11-2017-0117

Sadiku, B., Kemabonta, K., & Makanjuola, W. (2019). Reproductive performance of the larger grain borer *Prostephanus truncatus* (Horn) (Coleoptera: Bostrichidae) on three different food hosts. *Nigeria Journal of Entomology*, *35*(1), 19–30. https://doi.org/10.36108/nje/9102/53.01.30

Soro, M., Somé, K., Tiendrébéogo, F., Pita, J. S., Romba, R., Néya, B. J., & Daouda, K. (2021). Evaluation of ten cassava varieties for resistance to cassava mosaic disease in Burkina Faso. Universal *Journal of Agricultural Research*, *9*(6), 266–276. https://doi.org/10.13189/ujar.2021.090605

Sugihara, Y., Darkwa, K., Yaegashi, H., Natsume, S., Shimizu, M., Abe, A., Hirabuchi, A., Ito, K., Oikawa, K., Tamiru-Oli, M., Ohta, A., Matsumoto, R., Agre, P., De Koeyer, D., Pachakkil, B., Yamanaka, S., Muranaka, S., Takagi, H., White, B., … Terauchi, R. (2020). Genome analyses reveal the hybrid origin of the staple crop white guinea yam (*Dioscorea rotundata*). *Proceedings of the National Academy of Sciences, 117*(50), 31987–31992. https://doi.org/10.1073/pnas.2015830117

Tchabi, A., Burger, S., Coyne, D., Hountondji, F., Lawouin, L., Wiemken, A., & Oehl, F. (2009). Promiscuous arbuscular mycorrhizal symbiosis of yam (*Dioscorea* spp.), a key staple crop in West Africa. *Mycorrhiza, 19*(6), 375–392. https://doi.org/10.1007/s00572-009-0241-6

Tchabi, A., Coyne, D., Hountondji, F., Lawouin, L., Wiemken, A., & Oehl, F. (2008). Arbuscular mycorrhizal fungal communities in Sub-Saharan savannas of Benin, West Africa, as affected by agricultural land use intensity and ecological zone. *Mycorrhiza, 18*(4), 181–195. https://doi.org/10.1007/s00572-008-0171-8

Thresh, J., & Cooter, R. (2005). Strategies for controlling cassava mosaic virus disease in Africa. *Plant Pathology, 54*(5), 587–614. https://doi.org/10.1111/j.1365-3059.2005.01282.x

Tizé, I., Kuate Fotso, A., Nukenine, E. N., Masso, C., Ngome, F. A., Suh, C., Lendzemo, V. W., Nchoutnji, I., Manga, G., Parkes, E., Kulakow, P., Kouebou, C., Fiaboe, K. K. M., & Hanna, R. (2021). New cassava germplasm for food and nutritional security in Central Africa. *Scientific Reports, 11*(1). https://doi.org/10.1038/s41598-021-86958-w

Villarino, M., Silva, M., López-Lavalle, L., & Castro-Nuñez, A. (2020). "Rambo root" to the rescue: How a simple, low-cost solution can lead to multiple sustainable development gains. *Conservation Science and Practice, 3*(2), e320. https://doi.org/10.1111/csp2.320

Wang, S., Li, R., Zhou, Y., Fernie, A., Ding, Z., Zhou, Q., Che, Y., Yao, Y., Liu, J., Wang, Y., Hu, X., & Guo, J. (2023). Integrated characterization of cassava (*Manihot esculenta*) pectin methylesterase (*MePME*) genes to filter candidate gene responses to multiple abiotic stresses. *Plants, 12*(13), 2529. https://doi.org/10.3390/plants12132529

Yadav, J. S., Ogwok, E., Wagaba, H., Patil, B. L., Bagewadi, B., Alicai, T., Gaitan-Solis, E., Taylor, N. J., & Fauquet, C. M. (2011). Rnai-mediated resistance to cassava brown streak Uganda virus in transgenic cassava. *Molecular Plant Pathology*, *12*(7), 677–687. https://doi.org/10.1111/j.1364-3703.2010.00700.x

Zhang, C., Ketnawa, S., Thuengtung, S., Cai, Y., Qin, W., & Ogawa, Y. (2022). Simulated in vitro digestive characteristics of raw yam tubers in Japanese diet: Changes in protein profile, starch digestibility, antioxidant capacity and microstructure. *Foods*, *11*(23), 3892. https://doi.org/10.3390/foods11233892

CHAPTER 9

Maize in Mesoamerican Civilizations

Abstract

Maize (*Zea mays*) occupies an unparalleled place in the cultural, spiritual, and ecological lifeworlds of Mesoamerican civilizations. This chapter explores the coevolutionary relationship between maize and Indigenous societies of the region, emphasizing its sociobiological relevance from pre-Columbian times to the contemporary era. Drawing from interdisciplinary sources in ethnobotany, molecular biology, anthropology, and agroecology, the chapter examines maize's centrality in shaping cosmological belief systems, such as the Mayan Popol Vuh, and its embeddedness in daily life, cuisine, and rituals. The milpa system—an Indigenous polycultural farming practice based on maize, beans, and squash—is analyzed as a model of agrobiodiversity, ecological resilience, and cultural continuity. Through the lens of sociobiology, maize is portrayed not merely as a staple crop but as a vehicle of intergenerational knowledge transfer, gendered labor, and territorial stewardship. In addition to honoring the deep historical roots of maize, the chapter addresses contemporary pressures, including the commodification of maize, biopiracy, loss of native landraces, and the contested introduction of genetically modified organisms. It highlights Indigenous-led movements to reclaim seed sovereignty and restore traditional practices in the face of agro-industrial disruption. By centering maize as a cultural keystone species, the chapter contributes to broader discourses on food justice, decolonial sustainability, and the epistemic value of Indigenous knowledge systems in addressing global ecological challenges. Ultimately, this work underscores maize's enduring role as both a biological and cultural seed of life in Mesoamerican landscapes.

Keywords: milpa system, food sovereignty, Indigenous knowledge, agrobiodiversity, cosmology, sociobiology

9.1 Introduction

Maize (*Zea mays*) serves as a significant sociobiological cornerstone of Mesoamerican civilizations, intricately woven into their cultural, economic, and environmental fabric. The cultivation of maize facilitated the development of complex societies by providing a reliable food source, thereby paving the way for population growth, social stratification, and cultural innovations. The commitment to maize-based food production within these early civilizations has been substantiated by both archaeological evidence and isotopic analyses, which indicate the grain's status as a staple in their diets and agricultural systems (Kennett et al., 2020). Early isotopic studies demonstrate that reliance on maize played a pivotal role in establishing robust agricultural networks, which led to advancements in agricultural techniques, including forest management practices that incorporated the sustainable use of diverse crop species, including maize (Kennett et al., 2020).

From the historical context of maize domestication, it is understood that maize originated in the Balsas River Valley of Mexico over 9,000 years ago, marking a period of significant agricultural revolution (Kistler et al., 2018; van Heerwaarden et al., 2010). This domestication process involved the selective breeding of teosinte, its wild ancestor, influenced by early Mesoamerican populations. The switch from wild to domesticated maize represented not only an agronomic transformation but also acted as a catalyst for demographic shifts, as growing populations began forming sedentary settlements that led to the development of complex societal structures (Kistler et al., 2018; van Heerwaarden et al., 2010). Recent studies indicate that between 4,300 and 2,500 years ago, significant genetic diversity within maize was enhanced due to ancient gene flow from South America, which also aided its adaptation and the evolution of agricultural practices across Mesoamerica (Kistler et al., 2020).

Ecologically, maize has shown remarkable adaptability, thriving in various environments from humid lowlands to arid highlands. This versatility can be attributed to its genetic diversity, which has allowed it to withstand numerous climatic challenges, thereby ensuring agricultural resilience (van Heerwaarden et al., 2010). The ecological impacts of maize cultivation have been profound, contributing to deforestation and soil degradation in the pursuit of agricultural expansion (Kennett et al., 2020). The adoption of nixtamalization—a cultural practice dating back to pre-Columbian societies—has increased the nutritional value of maize and made it more palatable, thereby reinforcing its role as a food staple (Serrano-Gamboa et al., 2019). This process exemplifies the intricate relationship between culture and agricultural practice, where innovations have contributed to diet and sustenance, crucially supporting evolving societal norms.

The discussion of maize within the historiography of Mesoamerica is critical to understanding the rise and fall of various civilizations in the region. Mesoamerican landscapes were shaped by human intervention through maize cultivation, leading to significant evolutionary trajectories in agricultural practices over millennia. The complex interplay of climatic changes and agricultural choices, such as crop selection and planting strategies, not only influenced local diets but also spread across broader food systems, as observed in trade networks that facilitated crop exchange between Mesoamerican and Andean cultures (Mir et al., 2013; Therrell et al., 2006). This diffusion fostered genetic exchange and agricultural innovation, establishing maize as a pivotal element of transcontinental agrarian traditions (Ogwu et al., 2024). Over centuries, the cultivation of maize has transformed from subsistence farming practices to organized, large-scale agricultural systems as populations increased. This shift was driven by various factors, including technological advancements and changes in social structure, leading to evolutionary changes in maize breeding and cultivation strategies (Kennett et al., 2017; Omoigui et al., 2016). Additionally, archaeological findings corroborate that maize productivity increased as civilizations transitioned toward more complex societal organizations, enabling the establishment of political and economic power structures (Kennett et al., 2017). Indeed, the sociopolitical landscape of Mesoamerica was significantly influenced by agricultural practices, particularly those revolving around maize cultivation.

The adoption of maize-based agriculture impacted demographic trends and triggered transformations in land use and the ecological environment. Such shifts led to diverse agricultural systems that embraced intercropping and crop rotation, primarily designed to maximize yields while maintaining soil health and environmental balance (Bedoya et al., 2017; Therrell et al., 2006). A hallmark of Mesoamerican agriculture, the milpa system, which integrates maize with other crops, exemplifies a sustainable approach that facilitated agricultural resilience against climate variability (Castillo & Rivera-García, 2022). This method underscores the depth of ecological knowledge possessed by Indigenous farmers and their ability to innovate in response to environmental challenges. The sociocultural significance of maize continues to resonate in contemporary society, underpinning economic and dietary practices in Mesoamerica and beyond. Today, maize remains a staple in various forms and plays a crucial role in traditional diets, regional cuisines, and food security initiatives (Bedoya et al., 2017). Cultural festivities, rituals, and culinary traditions perpetuate maize's historical legacy, affirming its status as a symbol of resilience and identity for many communities. Efforts to conserve and protect traditional maize varieties further reflect the ongoing acknowledgment of its importance to biodiversity and cultural heritage, showcasing the intertwined fates of agriculture, culture, and ecology (Kistler et al., 2020).

This chapter aims to illuminate the multifaceted role of maize (*Zea mays*) as a sociobiological and cultural cornerstone of Mesoamerican civilizations. The chapter explores the deep coevolutionary relationship between maize and human societies in this region by weaving together archaeological, ecological, ethnographic, and genetic evidence. It investigates how maize has served as a vital source of sustenance and shaped cosmologies, gender roles, and agroecological systems such as the milpa. Particular attention is given to the cultural symbolism of maize in origin myths, its spiritual centrality in Indigenous rituals, and the role of women in sustaining maize-based knowledge systems and culinary practices. In doing so, the chapter contributes to a broader understanding of how food systems function as dynamic socioecological constructs. The chapter further engages with contemporary challenges, including the commodification of maize, the erosion of native landraces, and the implications of industrial agriculture and biotechnology. It highlights Indigenous-led efforts to reclaim food sovereignty, conserve agrobiodiversity, and restore ecological integrity through traditional knowledge systems. Overall, this chapter contributes to the book's broader goal of centering Indigenous foodways as resilient and instructive frameworks for navigating sustainability, cultural identity, and food justice in the Anthropocene. Through a sociobiological lens, it underscores maize not merely as a crop but as a living legacy of resistance, adaptation, and relational knowledge in Mesoamerican societies.

9.2 Origins and Domestication of Maize in Mesoamerican

The origins and domestication of maize are pivotal topics in understanding agricultural development in Mesoamerica, a region recognized for its early agricultural advances. Maize evolved from its wild ancestor, teosinte, through a complex interplay of genetic alterations and human intervention, as evidenced by archaeological findings and genomic data (Figure 9.1). This thorough examination encompasses three principal aspects: the genetic and archaeological evidence of

Figure 9.1: Comparison of Wild Grass Teosinte (*Zea mays* ssp. *parviglumis*) that Was Domesticated into Modern Maize (*Z. mays* ssp. *mays*)

Source: Pathirana and Carimi (2022)

maize evolution from teosinte, the contributions of early Mesoamerican agriculturalists, and the timeline and geographical expansion of maize domestication.

Genetic evidence reveals that maize originated from the wild grass teosinte, specifically *Zea mays* ssp. *parviglumis*. Studies employing multilocus microsatellite genotyping indicate that maize underwent a single domestication event approximately 9,000 years ago in southwestern Mexico, with evolutionary adaptations influenced by early agriculturalists (Matsuoka et al., 2002). This study illustrates that various morphological traits, such as the transition from small teosinte ears to the larger ears of maize, were influenced by selective breeding practices by Mesoamerican farmers. Furthermore, genomic analyses have identified multiple genes associated with notable phenotypes in maize, suggesting that while certain traits can be traced to specific loci, many others involve complex interactions among various genes (Doebley, 2004; Stitzer & Ross-Ibarra, 2018).

Archaeological evidence corroborates the genetic findings, particularly regarding the timeframe and geographical context of maize domestication. Piperno et al. (2009) provide some of the earliest evidence for maize in the Central Balsas River Valley, dating back approximately 9,000 years, supporting the conclusion that maize cultivation began during a significant climatic change in the Holocene epoch. Furthermore, genome sequencing of a maize cob dated to over 5,000 years ago affirms the long history of maize cultivation and indicates that many domestication traits were established by that time (Ramos-Madrigal et al., 2016). The archaeological record is rich with indications of maize's presence, including starch grains and phytoliths, supporting its early integration into Mesoamerican diets. This evidence illustrates the convergence of biological evolution and human cultivation practices that led to the domestication of maize.

Contributions from early Mesoamerican agriculturalists were instrumental in the development of maize as a staple crop. As agricultural communities emerged, these farmers selected desirable traits in maize, resulting in increased kernel size, improved yields, and enhanced taste. This agricultural innovation coincided with a gradual shift toward sedentary lifestyles, allowing for the establishment of complex societies and the eventual growth of civilizations such as the Maya and Aztecs, whose cultures remained heavily dependent on maize as a dietary staple (Zarrillo et al., 2008). These communities employed sophisticated cultivation techniques, including crop rotation and selective breeding, to enhance maize varieties suited for various environments across Mesoamerica. Understanding the timeline of maize domestication and its geographical expansion is essential for grasping its role in Mesoamerican history. Genetic studies indicate that following its domestication around 9,000 years ago, maize spread throughout Central America by approximately 7,500 years ago and reached South America by around 6,500 years ago (Kistler et al., 2018). This spread was not straightforward; rather, it involved various migration and adaptation processes, as suggested by the expansion of maize landraces and the genetic introgression observed between maize and local teosinte populations. The complex interactions with local wild relatives enriched the genetic diversity of maize, creating a varied array of landraces that showcased different traits adapted to specific environments (Kistler et al., 2020; van Heerwaarden et al., 2010). The eventual introduction of maize to diverse ecological zones allowed it to become a crucial agricultural product across much of the Americas.

9.3 Myth, Identity, and the Sacredness of Maize in Mesoamerican Civilizations

The profound significance of maize in Mesoamerican civilizations is intricately woven into their mythologies, cosmologies, agricultural practices, and spiritual beliefs. *Zea mays* was not only a staple crop but also symbolically connected to the identities of various Mesoamerican cultures, particularly the Maya and the Aztecs. This sacred grain is prominent in the Popol Vuh, the sacred text of the Maya, which recounts the creation stories that define their worldview. In the Popol Vuh, it is said that the gods created humans from maize, emphasizing its divine importance and central role in the Mesoamerican understanding of life and identity (VanDerwarker & Kruger, 2012). Moreover, the cultivation of maize and its integration into societal rituals highlight its symbolic status. Maize was central to numerous festivals, including those that celebrated the harvest and honored deities associated with agriculture (VanDerwarker & Kruger, 2012). These events often included rituals that showcased offerings of maize, reflecting both gratitude and reverence toward the gods for their blessings. This ritualistic aspect is echoed in the archaeological evidence from various sites, indicating that maize consumption was intricately linked to feasting rituals and cultural expressions (VanDerwarker & Kruger, 2012). For instance, evidence from Olmec sites suggests that political power dynamics involved communal maize consumption, thereby reinforcing social bonds and identity (VanDerwarker & Kruger, 2012).

In terms of its agricultural significance, maize served as the foundation of Mesoamerican diets, offering essential calories and nutrients. Studies show that maize accounted for nearly 70% of the daily caloric intake in rural Mexico, highlighting its role in sustaining the population (Therrell et al., 2006). The agricultural practices associated with maize included sophisticated techniques like crop rotation and the use of organic fertilizers, which improved soil health and crop yields. These practices not only demonstrate a deep understanding of environmental stewardship but also reflect the communal nature of farming in Mesoamerican societies (Kennett et al., 2010). The collaborative process of maize cultivation fostered social cohesion, as communities came together to plant, tend, and harvest their staple crop. On a genetic and ecological level, the domestication and spread of maize illustrate a complex interplay between environmental conditions and agricultural innovation (Figure 9.2). Research indicates that Mesoamerican maize originated from wild teosinte in the region and underwent significant adaptation to various ecological zones, showcasing the ingenuity of ancient farmers (Smith, 2005). These adaptations were vital as Mesoamerica transitioned from foraging to agriculture. The geographical spread of maize to areas such as the Southwestern United States and the Andes is well-documented, highlighting maize's essential role in the development of regional food systems and cultural identities throughout the Americas (Larrañaga et al., 2020; Merrill et al., 2009). The trade networks that emerged were not only economic; they

Figure 9.2: Migration Route of Maize Out of the Center of Origin

also represented cultural exchanges through which maize facilitated the sharing of agricultural practices and cosmologies (Kennett et al., 2010).

Blue central marker: The core region of maize origin in present-day southern Mexico. From this center, maize spread through diverse ecological and cultural landscapes across the Americas via multiple routes:

- **Red arrows** depict early expansion southward through Central America.
- **Blue arrows** show dispersal into South America, particularly the Andean and Amazonian regions.

- **Green arrows** indicate widespread diffusion throughout northern South America, the Caribbean, and eventually into the eastern and southern parts of the continent.
- **Dashed black and orange lines** may represent secondary migration routes or cultural trade and exchange corridors facilitating crop diffusion.

Source: Bedoya et al. (2017).

The mystical dimension of maize is further emphasized by its ties to life, death, and rebirth within Mesoamerican cosmologies. For instance, the cyclical nature of maize cultivation—sowing, growing, harvesting, and re-sowing—mirrored concepts of creation and renewal that were integral to Indigenous spiritual beliefs. This cyclical farming process symbolized the life cycles of the gods and the earth, resulting in maize being revered as sacred (Bedoya et al., 2017). Maize-related rituals often incorporated elements of myth and oral tradition, reinforcing its role as a cultural anchor that formed narratives of origin, identity, and community among the Mesoamericans (VanDerwarker & Kruger, 2012). As societies evolved, the cultivation and consumption of maize remained a vital aspect of community identity. Contemporary studies emphasize that modern Mesoamerican societies still view maize as crucial not only for sustenance but also for cultural heritage and identity. The recipes, culinary traditions, and agricultural methods that have thrived over centuries encapsulate a profound respect for this crop (Salazar et al., 2016). Particularly among Indigenous communities, traditional practices surrounding maize are essential to cultural progressions, serving as a vital thread that ties past generations to their descendants (Salazar et al., 2016).

The sacredness of maize in Mesoamerican civilizations emphasizes the importance of agriculture as a cultural and spiritual anchor. The multifaceted role of maize as a food source, a ritual object, and a symbol of creation demonstrates how integral it was to Mesoamerican identities. The narratives embedded within the *Popol Vuh* and the continuing practices of maize cultivation echo the resilience and adaptability of these cultures, reinforcing a legacy that remains relevant today (VanDerwarker & Kruger, 2012). This enduring reverence highlights maize not only as a nutritional foundation but also as a living symbol of a community's history, tradition, and spiritual beliefs. In synthesis, documenting the influences of maize on Mesoamerican civilizations reveals an intricate tapestry of interactions among myth, identity, and spirituality. Maize facilitated the development of agrarian societies that fostered social complexity, cultural richness, and spiritual depth (Murray et al., 2024; VanDerwarker & Kruger, 2012). The agricultural and ritualistic practices surrounding this sacred grain illustrate how Mesoamerican civilizations utilized maize for survival and as a potent symbol that shaped their worldviews and connected them with the divine and the earth. The potency of maize as a life-giving force and a cultural fulcrum in Mesoamerican life accentu-ates its centrality in historical contexts and contemporary discussions surround-ing Indigenous identities and food sovereignty. Recognizing such significance necessitates a greater understanding and appreciation of maize's role as a key element that intertwined the fabric of Mesoamerican cultures throughout history and into the present (Salazar et al., 2016).

9.4 The Milpa System (AKA Three Sisters): Agrobiodiversity and Ecological Balance of Maize Mesoamerican Civilizations

The milpa system, a traditional Mesoamerican agricultural practice, is a complex and sustainable polyculture primarily composed of maize, beans, and squash. This system is characterized by the synergistic coexistence of these three crops, often called "the Three Sisters," which mutually support each other's growth and productivity. Evidence suggests that intercropping maize, beans (*Phaseolus vulgaris*), and squash (*Cucurbita* spp.) enhances total agricultural output and resilience in variable environmental conditions (Ebel et al., 2017; Maravillas et al., 2019). The integrated cultivation approach minimizes soil disturbance through minimal or zero tillage, eliminates the need for extensive irrigation, and optimizes land use on sloped and diverse terrains—an invaluable attribute given the climatic challenges faced in Mesoamerica (Ebel et al., 2017; Ikhajiagbe et al., 2021a, 2021b).

The structural framework of the milpa contains a rich interdependence among its components. Maize, being a tall crop, provides shade for beans and squash, which are generally ground-dwelling plants. The beans, in turn, utilize nitrogen-fixing capabilities to enhance soil fertility, benefiting the maize. Furthermore, squash is a natural mulch, and its broad leaves provide ground cover that suppresses weeds and retains moisture (Liao et al., 2024). This intricate relationship exemplifies the resource efficiency inherent within polycultures. Studies have shown that these crops, when grown together, can achieve yields on a land-equivalent basis that surpass those of monocultures, suggesting spatial and functional complementarity among root systems that optimize nutrient uptake (Postma & Lynch, 2012; "Review for 'Facilitation and Biodiversity Ecosystem Function (BEF) relationships in crop production systems and their role in sustainable farming'," 2020).

Ecologically, the milpa system plays a significant role in nutrient cycling, pest management, and overall soil health. The varying root structures of maize, beans, and squash interact with soil biota in distinct ways, resulting in a diversity of microbial communities that enhance soil health (Maravillas et al., 2019; Postma & Lynch, 2012). For example, the presence of specific bacteria such as Pseudomonas fluorescens in these systems can change root-associated communities, improving the resilience of maize crops within the milpa model (Rojas-Sánchez et al., 2023). Furthermore, having multiple plant species close together disrupts the lifecycles of various pests, with findings suggesting that the polyculture system experiences lower rates of pest infestation compared to monocultures, indicating a form of integrated pest management (Liao et al., 2024; Pennisi, 2024).

Another significant aspect of the milpa system is its role in knowledge transmission among agricultural communities. Indigenous populations have cultivated these crops cooperatively for centuries, with local knowledge being pivotal in selecting crop varieties and understanding agronomic practices that sustain ecosystem balance. Evidence highlights that farmers within the milpa system actively engage in seed exchange networks that bolster genetic diversity and crop resilience (Llamas-Guzmán et al., 2022). Frequencies of crop variety exchanges correlate positively with local agricultural practices, emphasizing the importance of community involvement in enhancing biodiversity (Llamas-Guzmán et al., 2022). This dynamic helps maintain a diverse gene pool, allowing adaptation to changing environmental conditions and agricultural demands (Kamakaula et al., 2024). The cultural context underpinning the milpa is as important as its ecological benefits. The knowledge of intercropping techniques and cultural practices is typically handed down through generations, creating a reservoir of traditional ecological knowledge that remains relevant in contemporary agronomy (Kamakaula et al., 2024; Ulanowska & Siennicka, 2019). The integration of local experiences and scientific understanding facilitates practices that are both ecologically sound and economically viable. Moreover, by embracing Indigenous practices, modern agriculture can learn valuable lessons in sustainability, resilience, and environmental stewardship, suggesting a pathway for future agronomic innovations (Kamakaula et al., 2024).

9.5 Gender, Labor, and Maize Knowledge in Mesoamerican

In the context of Mesoamerican agriculture, particularly concerning maize, gender roles have been crucial in shaping practices such as seed selection, planting, harvesting, and food preparation. Women play a significant role in these areas, often serving as both the primary cultivators and processors of maize. Historical and anthropological assessments reveal that gendered labor divisions are deeply rooted in social and cultural norms that dictate who performs specific agricultural tasks, reflecting a long-standing tradition of women's involvement in food production within Mesoamerican communities (Howard, 2006; Sachs et al., 2020). For example, women are frequently responsible for selecting the seeds that will be planted, a task that requires knowledge of plant varieties and experience with local soils and climatic conditions (López-Ridaura et al., 2021; Sachs et al., 2020). Their role in this process is critical, as the success of maize cultivation often relies on selecting the appropriate seed varieties that align with traditional ecological knowledge.

Furthermore, during the planting and harvesting phases, women contribute significantly not only to the physical labor involved but also to the intergenerational transfer of knowledge regarding crop management and cultivation techniques. This knowledge-sharing among women and across generations is vital for maintaining agricultural practices, particularly in a system where maize is frequently intercropped with other plants in a "milpa" system (Aguirre-Noyola et al., 2021; López-Ridaura et al., 2021). This agroecological practice combines maize, beans, and squash, providing both nutritional diversity and ecological advantages such as pest control and improved soil fertility (Aguirre-Noyola et al., 2021; López-Ridaura et al., 2021). Women's roles in these systems are rooted in their control over domestic resources and food production, which strengthens their authority within the household and community (Howard, 2006; Sachs et al., 2020).

Women play a crucial role in preserving culinary traditions, particularly in tasks such as making tortillas and nixtamalizing maize. The nixtamalization process—where maize kernels are soaked and cooked in an alkaline solution—improves the nutritional value and flavor of the maize, which is essential to traditional dishes throughout Mesoamerica (Preston-Werner, 2008; Sachs et al., 2020). This culinary practice symbolizes women's roles not only as producers but also as cultural custodians, passing down culinary wisdom and traditions to future generations.

The importance of their work goes beyond simple food preparation; it involves cultural identity and heritage, which are vital aspects of Mesoamerican societies (Howard, 2006; Preston-Werner, 2008; Sachs et al., 2020).

The gender dynamics associated with maize agriculture also reflect broader socioeconomic conditions. Women's labor in maize cultivation often complements their roles in the domestic sphere, thus perpetuating traditional gender roles despite external pressures such as economic changes and migration (Radel et al., 2010). Studies indicate that as men migrate for labor, women take on increased agricultural responsibilities at home, which can alter labor dynamics and necessitate a reevaluation of resources allocated to maize production (Atreya & Gartaula, 2022; Radel et al., 2010). This shift highlights the resilience and adaptability of women in maintaining maize cultivation, despite challenges such as labor shortages and economic constraints arising from male outmigration (Gray, 2009; Radel et al., 2010).

Intergenerational knowledge transfer is fundamental in these contexts, reinforcing the importance of women's roles in agricultural education. Women are often the key educators in their families regarding traditional maize cultivation practices and local biodiversity (Mucheru-Muna et al., 2021; Sachs et al., 2020). This transmission of knowledge is crucial not only for sustaining agricultural practices but also for preserving the cultural significance of maize within Mesoamerican identity. As maize is not merely a crop but a symbol of life and sustenance, the role of women in educating their children about its cultivation and preparation is vital (Mucheru-Muna et al., 2021; López-Ridaura et al., 2021).

9.6 Political Economy of Maize in Mesoamerican

The political economy of maize in Mesoamerica encompasses a vast historical narrative that begins with its domestication over 6,000 years ago in central Mexico. This crop not only shaped agricultural landscapes but also significantly influenced social, political, and economic structures in Mesoamerican societies (Table 9.1). Early civilizations, including the Olmecs and Maya, developed sophisticated farming techniques that relied on maize as a staple food and a central element of their tribute systems and empire economies. Tribute systems were intricate mechanisms of political control through which maize served as a vital commodity exchanged for power, resources, and status among various factions (Chávez & Spence, 2012; Lesure et al., 2021). The significance of

Table 9.1: Influence of Political Economy on Maize
Sociobiology in Mesoamerica

Dimension	Influence on Maize in Mesoamerica
Colonial Policies	Spanish colonial rule imposed tribute systems requiring maize, restructuring Indigenous food systems.
Land Tenure and Reforms	Enclosure of communal lands (ejidos) and hacienda systems disrupted traditional maize farming.
Trade and Globalization	Integration into global commodity markets led to dependency on maize exports or imports.
Subsidies and Price Controls	Government subsidies (e.g., through CONASUPO in Mexico) once stabilized maize prices, later withdrawn under neoliberal reforms.
Agroindustrialization	Expansion of agroindustry led to hybrid and GM maize cultivation, reducing biodiversity and local seed use.
NAFTA and Free Trade	NAFTA enabled cheap U.S. maize imports, undermining smallholder farmers and threatening local varieties.
Migration and Labor Shifts	Outmigration from rural areas reduced labor for traditional maize farming and weakened seed-saving culture.
Seed Policy and Intellectual Property Rights	Enforcement of intellectual property rights discouraged seed sharing and criminalized native seed saving.
Food Sovereignty Movements	Political resistance has emerged defending Indigenous maize, promoting milpa systems and seed autonomy.
Cultural-Political Identity	Maize symbolizes Indigenous identity; its politicization reflects broader struggles for autonomy and land rights.

maize in sustaining population growth and fostering complex societies cannot be overstated: As maize yields increased due to advancements in agricultural technology and diversified cultivation practices like milpa and chinampas, so too did social stratification and political power dynamics (Lesure et al., 2021).

The colonization of Mesoamerica fundamentally changed the traditional systems surrounding maize cultivation and its societal role. After the arrival of Spanish colonizers, Indigenous agricultural practices experienced significant disruptions. The introduction of European farming methods and crops not only transformed local diets and farming techniques but also imposed a colonial economy that marginalized native maize varieties and traditional agricultural practices (Bedoya et al., 2017; García-Castañeda, 2022). Moreover, the colonizers enforced new land ownership patterns, using maize primarily for export rather than its historic role in local sustenance, thereby undermining Indigenous agricultural sovereignty. The resulting impact on maize sovereignty can be viewed as both a food security issue and a cultural crisis, as the crop holds deep cultural significance for Indigenous communities, serving as a symbol of identity, heritage, and cosmology (Bedoya et al., 2017; García-Castañeda, 2022).

In contemporary times, the political economy of maize continues to be influenced by complex dynamics involving genetically modified organisms (GMOs), trade policies, and the movement toward food justice. The introduction and commercialization of GMO maize have generated significant tensions in Mesoamerican societies, where Indigenous maize varieties are seen as part of cultural heritage and biodiversity (Bedoya et al., 2017; Walker et al., 2024). There is a palpable fear that the prevalence of GMOs could result in the genetic erosion of native maize varieties, which are inherently adapted to local environments and have the potential to enhance food security under changing climate conditions (Magaña-Cerino et al., 2019). Furthermore, trade policies developed under the modern agricultural regime often favor large-scale agricultural enterprises, which typically displace smallholder farmers and worsen inequalities in access to local resources and markets (García-Castañeda, 2022).

The pursuit of food justice emerges as a critical response to these issues, advocating for the rights of Indigenous peoples to maintain control over their traditional agricultural practices and crops, including maize. Grassroots movements increasingly seek recognition and preservation of traditional maize varieties through sustainable agricultural practices that prioritize ecological balance, cultural significance, and local food sovereignty (Lesure et al., 2021). These movements highlight the urgent need for policy reforms that acknowledge the importance of maize as not only a food source but also a historical and cultural staple, essential for maintaining the social fabric of Mesoamerican societies. Furthermore, the implications of maize in Mesoamerican history extend beyond agricultural practices; it influences linguistic, social, and political landscapes, as seen in the integration of maize-related vocabulary into the

Proto-Uto-Aztecan languages (Hill, 2008). The rich interplay between language and agriculture underpins a shared cultural heritage that transcends temporal and spatial boundaries within Mesoamerican societies. Cultural narratives surrounding maize and its embodiments in mythology further contribute to its status as a pivotal component of identity for various ethnic groups across the region (Read & González, 2002).

9.7 Resilience and Revival Movements of Maize in Mesoamerican

Maize remains a staple crop, deeply rooted in Mesoamerica's agricultural and cultural landscape, yet it has faced numerous challenges, particularly with the rise of GMOs and climate change. Through Indigenous-led initiatives and community-driven engagement, various movements focus on revitalizing resilience in maize landraces, ensuring ecological stability while preserving cultural heritage. These revival movements are supported by comprehensive strategies, ranging from seed banking to agroecological practices and educational outreach. Indigenous communities in Mesoamerica have historically managed maize's genetic diversity, selecting traits that adapt to local environmental conditions over thousands of years. The genetic composition and local adaptation of these maize landraces encompass a wide range of flowering traits and drought-tolerance features that reflect their robustness against various climatic stressors. Arteaga et al. highlight that genomic variation in recently collected maize landraces from Mexico demonstrates how these landraces are tailored for specific altitudinal and ecological conditions, underscoring their value in contemporary climate resilience efforts (Arteaga et al., 2016). Moreover, Worthington et al. state that the shrinking geographic and environmental diversity due to increased migration and agricultural practices threatens these genetic pools, necessitating focused conservation strategies (Worthington et al., 2012).

Community seed banks are crucial for preserving and revitalizing these genetic resources. These banks not only store seeds but also function as centers for education and local knowledge sharing. Community-led initiatives promote the exchange of information on best practices in maize cultivation, directly impacting local agricultural sustainability (Carvalho et al., 2012). Keleman et al. indicate that such efforts can lead to improved farming practices aligned with the local cultural values associated with maize cultivation (Keleman et al., 2009). The establishment of community seed banks aligns with research highlighting the importance of genetic diversity in farming systems, thereby enhancing food security and ecological resilience. Bellon et al. emphasize that traditional maize seed systems are vulnerable to climate change, yet ongoing efforts to conserve these systems underscore their significant role in sustaining agricultural biodiversity (Bellon et al., 2011). Local adaptation, including phenological variation in maize flowering and maturation, enables farmers to utilize strategies that mitigate drought risk and enhance overall resilience (Zimmerer, 2013). Education plays a pivotal role in these revival movements. Outreach programs that educate community members about agroecology promote a deeper understanding of sustainable farming and the cultural significance of maize landraces. Such educational initiatives empower Indigenous populations to actively manage their agricultural heritage while adapting to modern challenges. Research indicates that knowledge transfer among farmers regarding landrace management can significantly boost crop resilience against biotic and abiotic stressors (Londero et al., 2020). Furthermore, community workshops and training can foster innovative approaches to maize cultivation, enabling farmers to adapt to changing climatic conditions while maintaining a connection to their cultural heritage (Sharma et al., 2010).

The role of agroecology in maize production contributes to ecological resurgence movements. Agroecology, as a holistic agricultural approach, emphasizes biodiversity, sustainability, and the adaptive capacities of local farming systems. As Janzen et al. (2022) illustrate, the adaptation of maize landraces to various environmental stresses affirms their potential for breeding programs aimed at enhancing resilience and decreasing dependency on chemical inputs. This transition toward agroecological practices supports sustainable farming and aligns with Indigenous knowledge systems, reinforcing the cultural and ecological identity of Mesoamerican communities. However, the phenomenon of cross-pollination between GM maize and Indigenous landraces poses challenges that require careful management. The complexity of pollen-mediated gene flow underscores the necessity of effective coexistence strategies among diverse maize production systems (Baltazar et al., 2015). Promoting organic practices and adhering to strict ecological guidelines can mitigate the risk of contamination while enhancing the integrity of native maize varieties (Munarini & Nodari, 2021). New pedagogical models addressing these challenges can further strengthen community resilience against external pressures, fostering a cycle of cultural and ecological revival.

Despite advances in preserving maize landraces, challenges persist, including socioeconomic factors such as land access and the impact of global agricultural policies (Ranum et al., 2014). Research shows that socioeconomic disparities among farmers can limit their ability to engage with revival movements

(Alvarado-Ramírez et al., 2024). Systematic reviews of genetic diversity among maize populations indicate that acknowledging sociohistorical contexts is crucial for understanding the factors that influence local farming practices and their implications for maize conservation initiatives (Orozco–Ramírez et al., 2016). Therefore, establishing a supportive policy framework that values traditional agricultural practices alongside modern scientific approaches is essential for enhancing resilience in maize landraces.

9.8 Sociobiological Significance of Maize in the Anthropocene

The sociobiological significance of maize in the Anthropocene can be profoundly understood through its dual roles as a symbol of resilience and its potential in traditional systems for future food security. Historically rooted in Mesoamerican cultures, maize has played a pivotal role not only as a staple food but also as a cultural and spiritual element, illustrating its multifaceted importance in times of climatic and social change. Maize is inherently linked to resilience against climate variability and socioeconomic transitions, which is particularly relevant in the context of the Anthropocene. It has adapted remarkably over centuries to various environmental stresses, largely thanks to traditional agricultural practices such as the milpa system. This system fosters polyculture, enabling maize to thrive alongside other crops like beans and squash, enhancing biodiversity, soil health, and food security under fluctuating climate conditions (Birol et al., 2009). In regions like Oaxaca, farmers have historically narrated the adaptation of cropping systems to climatic adversities, highlighting that such resilience is crucial to addressing recent climatic challenges, including temperature increases and altered rainfall patterns (Rogé & Astier, 2015).

Moreover, the adaptation strategies of maize result from genetic diversity and complex relationships with its microbial community and endophytes, which help the plant acclimatize to abiotic stresses such as drought (González-Mendoza et al., 2023; Vwioko et al., 2018). Studies have shown that beneficial bacteria associated with maize play crucial roles in nutrient acquisition and stress tolerance, further enhancing its resilience (Eskikoy & Kutlu, 2024). In traditional agroecosystems like milpas, a dynamic microbiome boosts maize viability, reflecting an interdisciplinary approach that combines biology, agriculture, and ecology to ensure sustainability (Gastélum et al., 2022). Emerging from traditional practices is the recognition that biodiversity within crops significantly enhances food security. Varieties of Indigenous maize exhibit unique traits that provide advantages in specific environmental conditions. Cultivating these landraces preserves agricultural diversity, acting as a buffer against pests and diseases, which is vital for sustaining food supplies amid environmental threats (Birol et al., 2009). Furthermore, farmers' preferences for maintaining diverse crop species reflect deep-rooted cultural practices that integrate agricultural management with ecological sustainability (Birol et al., 2009).

In light of the ongoing global challenges compounded by climate change and socioeconomic upheaval, these traditional systems provide a framework for future food security strategies. For instance, contemporary engagement with intercropping maize and legumes like cowpeas has demonstrated beneficial effects on yields, enhancing food security in mixed culture systems (Makoi, 2019; Ogwu et al., 2014). The cooperative growth dynamics of maize with other crops can mitigate competition for resources while promoting sustainable practices necessary as agricultural demands increase (Makoi, 2019). It is also important to consider how maize's adaptability can be harnessed through modern agricultural biotechnology, ensuring it remains both an essential resource and a culturally significant symbol. For example, transgenic maize varieties have been developed to resist various stresses, potentially enhancing yield and adaptability to climate fluctuations (Yassitepe et al., 2021). However, it is critical to balance these biotechnological advancements with traditional practices to prevent disrupting local ecosystems and cultural heritage (Gillani et al., 2021).

Additionally, the marginalization of traditional maize varieties in favor of monocultures could jeopardize both biodiversity and cultural significance. Empowering local farming communities to preserve their heritage varieties while adopting scientific advances fosters an integrated approach that promotes both ecological resilience and cultural sustainability (Rodríguez et al., 2017). Furthermore, farmers' insights into climate trends and adaptive practices are vital in informing policies aimed at climate adaptation, ensuring that food security measures are context-specific (Rodríguez et al., 2017). Balancing modern agricultural practices with Indigenous knowledge is thus a cornerstone of resilience in maize cultivation. As maize plays a central role in socioeconomic frameworks across various cultures, ongoing efforts must focus on integrating traditional knowledge with scientific insights to advance agricultural strategies that are adaptive, sustainable, and culturally respectful. The significance of maize transcends its role as a food source; it embodies the intersection of history, technology, and ecology, making it a pivotal element in discussions of resilience and food security.

9.9 Conclusion

Maize is more than a crop; it serves as the epistemological, ecological, and spiritual cornerstone of Mesoamerican civilizations. From its mythic origins in Indigenous cosmologies to its central role in sustaining livelihoods through the milpa system, maize exemplifies a profound coevolution between humans and plants. This chapter illuminates how this relationship is deeply sociobiological—anchored in ecological interdependence, ritual meaning, and generational knowledge. Maize cultivation practices, including intercropping and landrace selection, reflect sophisticated ecological intelligence that supports biodiversity and cultural identity. However, the vitality of maize-based systems faces modern threats. Industrial agriculture, genetic erosion, land dispossession, and climate change pose serious risks to the resilience of traditional foodways. These pressures undermine environmental sustainability and marginalize the Indigenous epistemologies that have safeguarded maize diversity for millennia. The commodification of maize and the introduction of GMOs have further complicated the social and political landscape, challenging local autonomy and food sovereignty. Looking ahead, the sociobiological study of maize offers critical insights for navigating these challenges. Recognizing maize as a biocultural heritage asset highlights the need for inclusive policy frameworks that respect Indigenous land rights, promote agroecological sustainability, and support seed sovereignty movements. Future research should deepen the dialogue between molecular science and traditional knowledge, advocating for coproduced solutions that sustain both biodiversity and cultural resilience. In the age of ecological crisis, the story of maize in Mesoamerica serves as a testament to the enduring power of Indigenous systems to nourish both people and the planet.

Chapter Reflection

1. How did the domestication of maize contribute to the development of complex societies in Mesoamerica?
2. What makes the milpa system an example of sustainable agriculture and ecological balance?
3. How is maize portrayed in Mesoamerican mythologies, particularly in relation to identity and spirituality?
4. In what ways did colonial rule disrupt traditional maize cultivation and Indigenous food systems?

5. How do gender roles shape seed selection, farming practices, and food preparation in maize-based societies?

6. Why is the process of nixtamalization important for both the nutritional value and cultural significance of maize?

7. What challenges arise from the introduction of genetically modified maize into traditional agricultural systems?

8. How do community seed banks help preserve the genetic diversity of maize and support local food sovereignty?

9. How does agroecology blend traditional farming knowledge with scientific approaches to ensure maize sustainability?

10. Why is maize considered both a biological and cultural cornerstone in the context of current global environmental challenges?

REFERENCES

Aguirre-Noyola, J., Rosenblueth, M., Santiago-Martínez, M., & Martínez-Romero, E. (2021). Transcriptomic responses of *Rhizobium phaseoli* to root exudates reflect its capacity to colonize maize and common bean in an intercropping system. *Frontiers in Microbiology*, *12*. https://doi.org/10.3389/fmicb.2021.740818

Alvarado-Ramírez, E., Elghandour, M., Rivas-Jacobo, M., Calabrò, S., Vastolo, A., Cutrignelli, M., Hernández-Ruiz, P. E., Figueroa-Pacheco, E. B., & Salem, A. (2024). Influence of genotype and anaerobic fermentation on in vitro rumen fermentation characteristics and greenhouse gas production of whole-plant maize. *Fermentation*, *10*(1), 42. https://doi.org/10.3390/fermentation10010042

Arteaga, M., Moreno-Letelier, A., Mastretta-Yanes, A., Vázquez-Lobo, A., Breña-Ochoa, A., Moreno-Estrada, A., Eguiarte, L. E., & Piñero, D. (2016). Genomic variation in recently collected maize landraces from Mexico. *Genomics Data*, *7*, 38–45. https://doi.org/10.1016/j.gdata.2015.11.002

Atreya, K., & Gartaula, H. (2022). Changing gender role declines maize yield, but remittances offset: Findings from migrant households in the Central Himalayas, Nepal. *Outlook on Agriculture*, *51*(2), 247–259. https://doi.org/10.1177/00307270221097984

Baltazar, B., Espinoza, L., Banda, A., Martínez, J., Tiznado, J., García, J., Gutiérrez, M. A., Rodríguez, J. L. G., Díaz, O. H., Horak, M. J., Martínez, J. I. M., Schapaugh, A. W., Stojšin, D., Montes, H. R. U., & García, F. (2015). Pollen-mediated gene flow in maize: Implications for isolation requirements and co-existence in Mexico, the center of origin of maize. *Plos One, 10*(7), e0131549. https://doi.org/10.1371/journal.pone.0131549

Bedoya, C., Dreisigacker, S., Hearne, S., Franco, J., Mir, C., Prasanna, B. M., Taba, S., Charcosset, A., & Warburton, M. L. (2017). Genetic diversity and population structure of native maize populations in Latin America and the Caribbean. *Plos One, 12*(4), e0173488. https://doi.org/10.1371/journal.pone.0173488

Bellon, M. R., Hodson, D., & Hellin, J. (2011). Assessing the vulnerability of traditional maize seed systems in Mexico to climate change. *Proceedings of the National Academy of Sciences, 108*(33), 13432–13437. https://doi.org/10.1073/pnas.1103373108

Birol, E., Villalba, E., & Smale, M. (2009). Farmer preferences for milpa diversity and genetically modified maize in Mexico: A latent class approach. *Environment and Development Economics, 14*(4), 521–540. https://doi.org/10.1017/s1355770x08004944

Carvalho, M., Bebeli, P., Bettencourt, E., Costa, G., Dias, S., Dos Santos, T. M. M., & Ślaski, J. J. (2012). Cereal landraces genetic resources in worldwide GeneBanks. A review. *Agronomy for Sustainable Development, 33*(1), 177–203. https://doi.org/10.1007/s13593-012-0090-0

Castillo, R., & Rivera-García, R. (2022). Biophysical and biocultural upheavals in Mesoamerica, a conservation perspective: Mountains, maize-milpa, and globalization. *Frontiers in Forests and Global Change, 5*. https://doi.org/10.3389/ffgc.2022.763009

Chávez, S., & Spence, M. (2012). Interaction among the complex societies of classic-period Mesoamerica. In D. L. Nichols (Ed.), The *Oxford handbook of Mesoamerican archaeology* (pp. 283–298). Oxford University Press. https://doi.org/10.1093/oxfordhb/9780195390933.013.0020

Doebley, J. (2004). The genetics of maize evolution. *Annual Review of Genetics, 38*(1), 37–59. https://doi.org/10.1146/annurev.genet.38.072902.092425

Ebel, R., Cárdenas, J., Miranda, F., & González, J. (2017). Manejo orgánico de la milpa: Rendimiento de maíz, frijol y calabaza en monocultivo y policultivo. *Revista Terra Latinoamericana, 35*(2), 149. https://doi.org/10.28940/terra.v35i2.166

Eskikoy, G., & Kutlu, İ. (2024). Inter-subspecies diversity of maize to drought stress with physio-biochemical, enzymatic and molecular responses. *PeerJ, 12*, e17931. https://doi.org/10.7717/peerj.17931

García-Castañeda, V. (2022). Tecnodiversidad y maíz. sugerencias para la búsqueda de una cosmotécnica mesoamericana / techno-diversity and maize. suggestions for the search of a Mesoamerican cosmotechnics. *Technophany a Journal for Philosophy and Technology, 1*(1), 219–261. https://doi.org/10.54195/technophany.12656

Gastélum, G., Aguirre-von-Wobeser, E., Torre, M., & Rocha, J. (2022). Interaction networks reveal highly antagonistic endophytic bacteria in native maize seeds from traditional milpa agroecosystems. *Environmental Microbiology, 24*(11), 5583–5595. https://doi.org/10.1111/1462-2920.16189

Gillani, S., Rasheed, A., Majeed, Y., Tariq, H., & Peng, Y. (2021). Recent advancements on use of CRISPR/Cas9 in maize yield and quality improvement. *Notulae Botanicae Horti Agrobotanici Cluj-Napoca, 49*(3), 12459. https://doi.org/10.15835/nbha49312459

González-Mendoza, V., Torre, M., & Rocha, J. (2023). Plant-growth promoting endophytic bacteria and their role for maize acclimatation to abiotic stress. In M. Oliveira & A. Fernandes-Silva (Eds.), *Abiotic stress in plants - Adaptations to climate change*. IntechOpen. https://doi.org/10.5772/intechopen.109798

Gray, C. (2009). Rural out-migration and smallholder agriculture in the Southern Ecuadorian Andes. *Population and Environment, 30*(4–5), 193–217. https://doi.org/10.1007/s11111-009-0081-5

Hill, J. (2008). Otomanguean loan words in Proto-Uto-Aztecan maize vocabulary?. In J. D. Bengtson (Ed.), *Hot pursuit of language in prehistory* (pp. 309–320). John Benjamins Publishing Company. https://doi.org/10.1075/z.145.23hill

Howard, P. (2006). Gender and social dynamics in swidden and homegardens in Latin America. In B. M. Kumar & P. K. R. Nair (Eds.), *Tropical homegardens*. Advances in agroforestry (pp. 159–182). Springer. https://doi.org/10.1007/978-1-4020-4948-4_10

Ikhajiagbe, B., Igiebor, F. A., & Ogwu, M. C. (2021a). Growth and yield performances of rice (*Oryza sativa* var. nerica) after exposure to biosynthesized nanoparticles. *Bulletin of the National Research Centre, 45*, 62. https://doi.org/10.1186/s42269-021-00508-y

Ikhajiagbe, B., Ogwu, M. C., Fawehinmi, F. O., & Adekunle, I. J. (2021b). Comparative growth responses of *Amaranthus* [L.] species in humus and ferruginous ultisols using plant growth promoting Rhizobacteria (*Pseudomonas* species). *South African Journal of Botany, 137*, 10–18. https://doi.org/10.1016/j.sajb.2020.09.029

Janzen, G. M., Aguilar-Rangel, M. R., Cíntora-Martínez, C., Blöcher-Juárez, K. A., González-Segovia, E., Studer, A. J., Runcie, D. E., Flint-Garcia, S. A., Rellán-Álvarez, R., Sawers, R. J. H., & Hufford, M. (2022). Demonstration of local adaptation in maize landraces by reciprocal transplantation. *Evolutionary Applications*, *15*(5), 817–837. https://doi.org/10.1111/eva.13372

Kamakaula, Y., Amruddin, A., Demmanggasa, Y., Saprudin, S., & Nugroho, R. J. (2024). The role of local knowledge in natural resources conservation: An environmental anthropological perspective in traditional agriculture. Global International Journal of Innovative Research, 1(2), 97–106. https://pdfs. semanticscholar.org/edcd/dfa4ac839fb51c3a6512703cb52cf6222a8b.pdf

Keleman, A., Hellin, J., & Bellon, M. (2009). Maize diversity, rural development policy, and farmers' practices: Lessons from Chiapas, Mexico. *Geographical Journal*, *175*(1), 52–70. https://doi.org/10.1111/j.1475-4959.2008 .00314.x

Kennett, D. J., Piperno, D. R., Jones, J. G., Neff, H., Voorhies, B., Walsh, M. K., & Culleton, B. (2010). Pre-pottery farmers on the pacific coast of Southern Mexico. *Journal of Archaeological Science*, *37*(12), 3401–3411. https://doi. org/10.1016/j.jas.2010.07.035

Kennett, D. J., Prufer, K. M., Culleton, B. J., George, R. J., Robinson, M., Trask, W. R., Buckley, G. M., Moes, E., Harper, T. K., O'Donnell, L., Ray, E. E., Hill, E. C., Alsgaard, A., Merriman, C., Meredith, C., Edgar, H. J. H., Awe, J. J., & Gutierrez, S. M. (2020). Early isotopic evidence for maize as a staple grain in the Americas. *Science Advances*, *6*(23). https://doi.org/10.1126/ sciadv.aba3245

Kennett, D. J., Thakar, H. B., VanDerwarker, A. M., Webster, D. L., Culleton, B. J., Harper, T. K., Kister, L., & Hirth, K. (2017). High-precision chronology for Central American maize diversification from El Gigante Rockshelter, Honduras. *Proceedings of the National Academy of Sciences*, *114*(34), 9026–9031. https:// doi.org/10.1073/pnas.1705052114

Kistler, L., Maezumi, S. Y., de Souza, J. G., Przelomska, N. A. S., Costa, F. M., Smith, O., Loiselle, H., Ramos-Madrigal, J., Wales, N., Ribeiro, E. R., Morrison, R. R., Grimaldo, C., Prous, A. P., Arriaza, B., Gilbert, M. T. P., de Oliveira Freitas, F., & Allaby, R. (2018). Multiproxy evidence highlights a complex evolutionary legacy of maize in South America. *Science*, *362*(6420), 1309–1313. https://doi. org/10.1126/science.aav0207

Kistler, L., Thakar, H. B., VanDerwarker, A. M., Domic, A., Bergström, A., George, R. J., Harper, T. K., Allaby, R. G., Hirth, K., & Kennett, D. J. (2020). Archaeological Central American maize genomes suggest ancient gene flow from South America. *Proceedings of the National Academy of Sciences, 117*(52), 33124–33129. https://doi.org/10.1073/pnas.2015560117

Larrañaga, N., Zonneveld, M., & Hormaza, J. (2020). Holocene land and sea-trade routes explain complex patterns of pre-Columbian crop dispersion. *New Phytologist, 229*(3), 1768–1781. https://doi.org/10.1111/nph.16936

Lesure, R. G., Sinensky, R. J., Schachner, G., Wake, T. A., & Bishop, K. (2021). Large-scale patterns in the agricultural demographic transition of Mesoamerica and southwestern North America. *American Antiquity, 86*(3), 593–612. https://doi.org/10.1017/aaq.2021.23

Liao, H., Zhou, Z., Liu, Y., Luo, Y., Zhang, C., Feng, Y., Shu, Y., & Wang, J. (2024). "The three sisters" (maize/bean/squash) polyculture promotes the direct and indirect defences of maize against herbivores. *European Journal of Agronomy, 155*, 127118. https://doi.org/10.1016/j.eja.2024.127118

Llamas-Guzmán, L., Chavero, E., Perales, H., & Casas, A. (2022). Seed exchange networks of native maize, beans, and squash in San Juan Ixtenco and San Luis Huamantla, Tlaxcala, Mexico. *Sustainability, 14*(7), 3779. https://doi.org/10.3390/su14073779

Londero, P., Kelling, C., Alves, J., Bissacotti, A., Reiniger, L., & Muniz, M. (2020). Nutritional characterization and productivity evaluation of landrace maize cultivars. *Cadernos De Ciência and Tecnologia, 37*(3), 26644. https://doi.org/10.35977/0104-1096.cct2020.v37.26644

López-Ridaura, S., Barba-Escoto, L., Reyna-Ramírez, C. A., Sum, C., Palacios-Rojas, N., & Gérard, B. (2021). Maize intercropping in the milpa system. Diversity, extent and importance for nutritional security in the western highlands of Guatemala. *Scientific Reports, 11*(1). https://doi.org/10.1038/s41598-021-82784-2

Magaña-Cerino, J., Peniche-Pavía, H., Tiessen, A., & Gurrola-Díaz, C. (2019). Pigmented maize (*Zea mays* L.) contains anthocyanins with potential therapeutic action against oxidative stress—A review. *Polish Journal of Food and Nutrition Sciences, 70*(2), 85–99. https://doi.org/10.31883/pjfns/113272

Makoi, J. (2019). Yield and yield components of local cowpea (*Vigna unguiculata* L.) landraces grown in mixed culture with maize (*Zea mays* L.)

in vertic cambisols in the northern part of Tanzania. *Forestry Research and Engineering International Journal, 3*(3), 88–94. https://medcrave.com/index. php?/articles/det/19993/Yield-and-yield-components-of-local-cowpea-Vigna%02unguiculata-L-landraces-grown-in-mixed-culture-with-maize-Zea-maysL-in%02vertic-cambisols-in-the-Northern-part-of-Tanzania

Maravillas, K., Díaz-Almeyda, E., & Gerardo, N. (2019). Bacterial growth in milpa polyculture and monoculture soils [Emory University]. *Journal of Student Research, 14*(4). https://doi.org/10.47611/jsr.vi.691

Matsuoka, Y., Vigouroux, Y., Goodman, M. G. J., Buckler, E., & Doebley, J. (2002). A single domestication for maize shown by multilocus microsatellite genotyping. *Proceedings of the National Academy of Sciences, 99*(9), 6080–6084. https://doi.org/10.1073/pnas.052125199

Merrill, W. L., Hard, R. J., Mabry, J. B., Fritz, G. J., Adams, K. R., Roney, J. R., & MacWilliams, A. (2009). The diffusion of maize to the southwestern united states and its impact. *Proceedings of the National Academy of Sciences, 106*(50), 21019–21026. https://doi.org/10.1073/pnas.0906075106

Mir, C., Zerjal, T., Combes, V., Dumas, F., Madur, D., Bedoya, C., Dreisigacker, S., Franco, J., Grudloyma, P., Hao, P. X., Hearne, S., Jampatong, C., Laloë, D., Muthamia, Z., Nguyen, T., Prasanna, B. M., Taba, S., Xie, C. X., Yunus, M., … Charcosset, A. (2013). Out of America: Tracing the genetic footprints of the global diffusion of maize. *Theoretical and Applied Genetics, 126*(11), 2671–2682. https://doi.org/10.1007/s00122-013-2164-z

Mucheru-Muna, M. W., Ada, M. A., Mugwe, J. N., Mairura, F. S., Mugi-Ngenga, E., Zingore, S., & Mutegi, J. K. (2021). Socio-economic predictors, soil fertility knowledge domains and strategies for sustainable maize intensification in Embu County, Kenya. *Heliyon, 7*(2), e06345. https://doi.org/10.1016/j.heliyon.2021. e06345

Munarini, A., & Nodari, R. (2021). Effect of sowing time and density for vegetative and reproductive traits of genotypes of maize landrace in an agroecological system. *Ciência Rural, 51*(5). https://doi.org/10.1590/0103-8478cr 20200145

Murray, M., Nakamura, S., Fuentes, M., Ogawa, M., Cassidy, L., Gakuhari, T., & Nakagome, S. (2024). Ancient genomics reveals a genetic continuum with dual structure in the Classic Copan. https://doi.org/10.1101/2024.05.17.594672

Ogwu, M. C., Onah, T. P., Kosoe, E. A., Mumuni, E., Akolgo-Azupogo, H., Obahiagbon, E. G., Izah, S. C., & Imarhiagbe, O. (2024). Factors driving the acceptance of genetically modified food crops in Ghana. *Food Safety and Health*, *2*(1), 158–168. http://doi.org/10.1002/fsh3.12031

Ogwu, M. C., Osawaru, M. E., & Chime, A. O. (2014). Comparative assessment of plant diversity and utilization patterns of tropical home gardens in Edo State, Nigeria. *Scientia Africana*, *13*(2), 146–162.

Omoigui, D. O., Osawaru, M. E., Aiwansoba, R. O., & Ogwu, M. C. (2016). Morphological and phytodermological evaluation of Oka-Uselu (Maize- *Zea mays* L.). *Applied Tropical Agriculture*, *21*(3), 96–101.

Orozco–Ramírez, Q., Ross-Ibarra, J., Santacruz-Várela, A., & Brush, S. (2016). Maize diversity associated with social origin and environmental variation in Southern Mexico. *Heredity*, *116*(5), 477–484. https://doi.org/10.1038/hdy.2016.10

Pathirana, R., & Carimi, F. (2022). Management and utilization of plant genetic resources for a sustainable agriculture. *Plants*, *11*(15), 2038. https://doi.org/10.3390/plants11152038

Pennisi, E. (2024). How the three sisters shrug off pests. *Science*, *385*(6712), 920–921. https://doi.org/10.1126/science.ads7686

Piperno, D., Ranere, A., Holst, I., Iriarte, J., & Dickau, R. (2009). Starch grain and phytolith evidence for early ninth millennium B.P. maize from the Central Balsas River Valley, Mexico. *Proceedings of the National Academy of Sciences*, *106*(13), 5019–5024. https://doi.org/10.1073/pnas.0812525106

Postma, J., & Lynch, J. (2012). Complementarity in root architecture for nutrient uptake in ancient maize/bean and maize/bean/squash polycultures. *Annals of Botany*, *110*(2), 521–534. https://doi.org/10.1093/aob/mcs082

Preston-Werner, T. (2008). 4 Breaking down binaries: Gender, art, and tools in ancient Costa Rica. *Archeological Papers of the American Anthropological Association*, *18*(1), 49–59. https://doi.org/10.1111/j.1551-8248.2008.00004.x

Radel, C., Schmook, B., & McCandless, S. (2010). Environment, transnational labor migration, and gender: Case studies from Southern Yucatán, Mexico and Vermont, USA. *Population and Environment*, *32*(2–3), 177–197. https://doi.org/10.1007/s11111-010-0124-y

Ramos-Madrigal, J., Smith, B. D., Moreno-Mayar, J. V., Gopalakrishnan, S., Ross-Ibarra, J., Gilbert, M. T. P., & Wales, N. (2016). Genome sequence of a 5,310-year-old maize cob provides insights into the early stages of maize domestication. *Current Biology, 26*(23), 3195–3201. https://doi.org/10.1016/j.cub.2016.09.036

Ranum, P., Peña-Rosas, J., & García-Casal, M. (2014). Global maize production, utilization, and consumption. *Annals of the New York Academy of Sciences, 1312*(1), 105–112. https://doi.org/10.1111/nyas.12396

Read, K., & González, J. (2002). *Mesoamerican mythology: A guide to the gods, heroes, rituals, and beliefs of Mexico and Central America.* Oxford University Press. https://doi.org/10.1093/oso/9780195149098.001.0001

Rodríguez, N., Eakin, H., & Dewes, C. (2017). Perceptions of climate trends among Mexican maize farmers. *Climate Research, 72*(3), 183–195. https://doi.org/10.3354/cr01466

Rogé, P., & Astier, M. (2015). Changes in climate, crops, and tradition: Cajete maize and the rainfed farming systems of Oaxaca, Mexico. *Human Ecology, 43*(5), 639–653. https://doi.org/10.1007/s10745-015-9780-y

Rojas-Sánchez, B., Castelán-Sánchez, H., & Santoyo, G. (2023). Inoculation with pseudomonas fluorescensum UM270 alters the maize root-associated endobiome and interacting networks in a milpa model. https://doi.org/10.1101/2023.05.15.540877

Sachs, C., Jensen, L., Castellanos, P., & Sexsmith, K. (2020). *Routledge handbook of gender and agriculture.* Routledge. https://doi.org/10.4324/9780429199752

Salazar, C., Zizumbo-Villarreal, D., Colunga-GarcíaMarín, P., & Brush, S. (2016). Contemporary Maya food system in the lowlands of Northern Yucatan. In R. Lira, A. Casas, & J. Blancas (Eds.), *Ethnobotany of Mexico* (pp. 133–150). Springer. https://doi.org/10.1007/978-1-4614-6669-7_6

Serrano-Gamboa, J., Rojas-Herrera, R., González-Burgos, A., Folch-Mallol, J., Jiménez, D., & Sánchez-González, M. (2019). Degradation profile of nixtamalized maize pericarp by the action of the microbial consortium PM-06. *Amb Express, 9*(1), 85. https://doi.org/10.1186/s13568-019-0812-7

Sharma, L., Prasanna, B., & Ramesh, B. (2010). Analysis of phenotypic and microsatellite-based diversity of maize landraces in India, especially from the

North East Himalayan Region. *Genetica*, *138*(6), 619–631. https://doi.org/10.1007/s10709-010-9436-1

Smith, B. (2005). Reassessing coxcatlan cave and the early history of domesticated plants in Mesoamerica. *Proceedings of the National Academy of Sciences*, *102*(27), 9438–9445. https://doi.org/10.1073/pnas.0502847102

Stitzer, M., & Ross-Ibarra, J. (2018). Maize domestication and gene interaction. *New Phytologist*, *220*(2), 395–408. https://doi.org/10.1111/nph.15350

Therrell, M., Stahle, D., Díaz, J., Oviedo, E., & Cleaveland, M. (2006). Tree-ring reconstructed maize yield in Central Mexico: 1474–2001. *Climatic Change*, *74*(4), 493–504. https://doi.org/10.1007/s10584-006-6865-z

Ulanowska, A., & Siennicka, M. (2019). Transmission of practice, transmission of knowledge: Dynamics of textile production in the Bronze Age Aegean. In *MNHMH/MNEME. Past and memory in the Aegean Bronze Age* (pp. 753–757). Peeters Publishers. https://doi.org/10.2307/j.ctv1q26q48.94

VanDerwarker, A., & Kruger, R. (2012). Regional variation in the importance and uses of maize in the early and middle formative Olmec heartland: New archaeobotanical data from the San Carlos Homestead, Southern Veracruz. *Latin American Antiquity*, *23*(4), 509–532. https://doi.org/10.7183/1045-6635.23.4.509

van Heerwaarden, J., Doebley, J., Briggs, W. H., Glaubitz, J. C., Goodman, M. M., de Jesu Sanchez González, & Ross-Ibarra, J. (2010). Genetic signals of origin, spread, and introgression in a large sample of maize landraces. *Proceedings of the National Academy of Sciences*, *108*(3), 1088–1092. https://doi.org/10.1073/pnas.1013011108

Vwioko, D. E., Okoekhian, I., & Ogwu, M. C. (2018). Stress analysis of *Amaranthus hybridus* L. and *Lycopersicon esculentum* Mill. exposed to sulphur and nitrogen dioxide. *Pertanika Journal of Tropical Agricultural Sciences*, *41*(3), 1169–1191.

Walker, D., Reese-Taylor, K., Šprajc, I., Dunning, N., & Acuña, M. (2024). Pre-mamom pottery producers in the heart of the Yucatan Peninsula. *Estudios De Cultura Maya*, *63*, 11–40. https://doi.org/10.19130/iifl.ecm.63.2024/00171s0xw31

Worthington, M., Soleri, D., Aragón-Cuevas, F., & Gepts, P. (2012). Genetic composition and spatial distribution of farmer-managed phaseolus bean plantings: An example from a village in Oaxaca, Mexico. *Crop Science*, *52*(4), 1721–1735. https://doi.org/10.2135/cropsci2011.09.0518

Yassitepe, J., Silva, V., Hernandes-Lopes, J., Dante, R., Gerhardt, I., Fernandes, F. R., da Silva, P. A., Vieira, L. R., & Arruda, P. (2021). Maize transformation: From plant material to the release of genetically modified and edited varieties. *Frontiers in Plant Science, 12.* https://doi.org/10.3389/fpls.2021.766702

Zarrillo, S., Pearsall, D., Raymond, J., Tisdale, M., & Quon, D. (2008). Directly dated starch residues document early formative maize (*Zea mays* L.) in tropical Ecuador. *Proceedings of the National Academy of Sciences, 105*(13), 5006–5011. https://doi.org/10.1073/pnas.0800894105

Zimmerer, K. (2013). The compatibility of agricultural intensification in a global hotspot of smallholder agrobiodiversity (Bolivia). *Proceedings of the National Academy of Sciences, 110*(8), 2769–2774. https://doi.org/10.1073/pnas.1216294110

CHAPTER 10

Sorghum and Pearl Millet in Sahelian Societies

Abstract

Sorghum (*Sorghum bicolor*) and pearl millet (*Pennisetum glaucum*) are not only foundational grains in Sahelian agroecosystems but are also deeply woven into the sociocultural, ecological, and spiritual lives of the region's communities. This chapter examines the sociobiological relevance of these cereals, tracing their domestication, adaptive traits, and coevolutionary relationship with Indigenous knowledge systems in arid and semiarid environments. It explores their remarkable genetic plasticity and drought resilience, as well as the endogenous landraces maintained through farmer-led seed selection, collective memory, and gendered knowledge transmission. Traditional cultivation systems—often embedded in communal land tenure and ritual cycles—reveal a nuanced interplay between ecological stewardship and cultural continuity. Through an interdisciplinary lens that merges ethnobotany, agroecology, and anthropology, the chapter highlights how millet and sorghum farming serves as a bulwark against climate variability, food insecurity, and the erosion of cultural identity. In particular, the role of women in managing seed diversity, preparing ceremonial foods, and preserving ecological memory underscores their agency as custodians of both biological and cultural heritage. The chapter also interrogates emerging threats to these systems—such as agricultural industrialization, genetic erosion, and policy marginalization—while advocating for integrative strategies that uphold food sovereignty and resilience. Ultimately, this analysis positions sorghum and millet as more than subsistence crops: They are keystone species in a complex web of environmental, spiritual, and sociopolitical relationships essential to the future of sustainable development in the Sahel.

Keywords: Sahel, agrobiodiversity, food sovereignty, Indigenous knowledge, drought resilience, gendered practices

10.1 Introduction

Sorghum (*Sorghum bicolor*) and pearl millet (*Pennisetum glaucum*) are essential staple cereals in the food systems, economies, and identities of Sahelian societies across West and Central Africa (Onoabhagbe et al., 2024). These cereals are particularly suited for arid and semiarid environments and exhibit a remarkable capacity for adaptation to harsh climatic conditions, making them indispensable to the livelihoods of millions. Their resilience to drought, poor soil fertility, and high temperatures has enabled them to thrive in regions where other crops may fail, thus significantly contributing to food security in these areas, as evidenced by numerous studies on their agronomic characteristics and dietary importance (Chalachew et al., 2025; Morris et al., 2012). Historically, sorghum and pearl millet have been cultivated in the Sahel for thousands of years, likely originating from semiarid regions of Africa, rather than the Fertile Crescent. Indigenous farmers have selectively bred these plants for desirable traits such as drought resistance and high grain yield, which aligns with human migration patterns and agricultural practices favoring cultivation of plants best suited for local climates (Satyavathi et al., 2021; Traoré et al., 2022). Present-day agricultural practices have seen advancements in breeding techniques, enhancing the adaptability of these crops to changing climatic conditions (Sehgal et al., 2015; Srivastava et al., 2020).

As keystone species in the food systems of Sahelian societies, sorghum and pearl millet offer substantial nutritional benefits, providing vital macronutrients and micronutrients essential for combating malnutrition and supporting human health. Sorghum grains are rich in antioxidants, including polyphenols and flavonoids, which contribute to lower risks of chronic diseases (Lee et al., 2022; Mawouma et al., 2022). Pearl millet is recognized for its high protein content and beneficial vitamins and minerals and is increasingly acknowledged as a climate-resilient "nutricereal" that can help mitigate nutritional deficiencies in vulnerable communities (Mihiretu et al., 2023; Satyavathi et al., 2021). These benefits position sorghum and pearl millet not only as vital food staples but also as integral components of cultural identity and economic sustainability for many rural populations in the Sahel (Garba, 2019; Ikhajiagbe et al., 2023; Osawaru & Ogwu, 2020). The geographical distribution of sorghum and pearl millet cultivation reflects diverse environmental conditions throughout the Sahel, characterized by erratic and limited rainfall, which is typically insufficient for conventional cereal crops such as maize or wheat. The ability of these crops to withstand drought and thrive in poor soil conditions makes them the crops of choice in these regions

(Sermé et al., 2018). Agroecological practices such as intercropping and crop rotation have historically optimized land use and enhanced soil health, which are critical for sustainable agricultural development, especially as climate change poses challenges to food security (Morris et al., 2012; Sermé et al., 2018).

With advances in agronomic research, cultivation techniques for sorghum and pearl millet are evolving. Recent studies indicate significant potential for improved management practices, including better crop rotation schedules and water conservation strategies, which can enhance yield while maintaining soil health (Mihiretu et al., 2023; Sermé et al., 2018). Furthermore, genetic research aimed at identifying drought resistance traits is paving the way for developing improved varieties that can adapt to adverse growing environments (Negarestani et al., 2019; Sehgal et al., 2015). As climate change continues to affect agricultural productivity in the Sahel, the strategic development and utilization of resilient crop varieties will be crucial for ensuring food security in the region (Satyavathi et al., 2021; Traoré et al., 2022). The cultural and economic significance of sorghum and pearl millet extends beyond mere sustenance. Traditionally, these cereals have been central to community life, featuring prominently in local cuisine and agricultural festivals. Their cultivation supports food security and rural livelihoods, with many communities depending on the sale of these grains in local and regional markets (Garba, 2019; Mihiretu et al., 2023). Additionally, sorghum and pearl millet are often integral components of traditional brewing practices, further embedding these grains within the cultural fabric of society (Embashu & Nantanga, 2019). As Sahelian societies navigate the complexities of globalization, the resurgence of interest in traditional grains like sorghum and pearl millet underscores the importance of protecting Indigenous agricultural knowledge and supporting smallholder farmers. This focus aligns with ongoing international efforts to promote food sovereignty and climate resilience among vulnerable populations (Satyavathi et al., 2021; Sultan et al., 2019). Strengthening local food systems that incorporate these staple cereals is essential for fostering secure, self-sufficient communities capable of withstanding economic and environmental shocks (Mihiretu et al., 2023; Sermé et al., 2018).

This chapter aims to explore the sociobiological importance of sorghum (*Sorghum bicolor*) and pearl millet (*Pennisetum glaucum*) within the Sahelian region, highlighting their role as culturally embedded, climate-resilient staples. It investigates how these cereals have coevolved with human societies through traditional farming systems, ritual life, gendered knowledge, and ecological adaptation. Specifically, the chapter examines the genetic plasticity and drought tolerance of local landraces, traditional agronomic practices, and the role of Indigenous seed networks in maintaining agrobiodiversity. Furthermore, it delves into the culinary and spiritual significance of sorghum and millet, emphasizing how food preparation, ceremonies, and seasonal rites are tightly woven into the socioecological fabric of Sahelian communities. The chapter contributes to the growing scholarship on biocultural diversity by offering a critical lens on the intersection of crop ecology, traditional knowledge systems, and community resilience. It positions sorghum and pearl millet not merely as food sources but as cultural keystones and ecological buffers in a region increasingly vulnerable to climate change, desertification, and agricultural homogenization. By drawing on interdisciplinary perspectives from agronomy, anthropology, and food studies, this chapter provides a rich, situated account of how sustaining these cereals is central to the future of food sovereignty, heritage conservation, and adaptive livelihoods in Africa's Sahel.

10.2 Genetic Diversity and Drought Adaptation of Sorghum and Pearl Millet in Sahelian Societies

Sorghum and pearl millet have emerged as critical crops for ensuring food security in the Sahel, particularly due to their adaptability to marginal conditions characterized by erratic rainfall and low soil fertility. The morphological and genetic diversity inherent in these crops plays a significant role in their resilience to drought and other climatic stresses. For instance, pearl millet displays a wide range of phenotypic traits associated with flowering time, which can directly influence its adaptability to varying photoperiods and climatic conditions. Specific landraces, such as the early-flowering Souna and late-flowering Sanio, are particularly distinguished for their varying degrees of drought tolerance and photoperiod sensitivity, thus adding to the genetic reservoir crucial for future breeding efforts aimed at enhancing stress resilience in this crop (Diack et al., 2020; Faye et al., 2019).

On the morphological front, pearl millet exhibits notable genetic traits that contribute to its cultivation in the harsh agricultural environments of the Sahel. It is characterized by wide adaptability to low phosphorus availability in sandy soils, which favors specific traits for phosphorus uptake efficiency (Gemenet et al., 2015; Gunguniya et al., 2023). The genetic studies on pearl millet reveal a richness of natural diversity, which can be harnessed for improving not only grain yield but also nutritional quality (Gunguniya et al., 2023). The exploration of traits such as stay-green characteristics, which prolong leaf greenness during drought, as well as factors influencing panicle morphology, demonstrates a multifaceted approach to understanding how these crops have evolved in response to climatic challenges (Faye et al., 2019; Pucher et al., 2015).

Similarly, sorghum boasts a spectrum of genetic diversity that enhances its morphological adaptations under drought. A significant proportion of sorghum landraces, particularly those from Senegal, exhibit genomic signatures reflective of geographical and climatic adaptations to Sahelian and Soudanian climates. The identification of selective sweeps within these landraces has further illuminated the genetic loci associated with drought resilience and photoperiod sensitivity (Faye et al., 2019; Serba & Yadav, 2016). Furthermore, agricultural practices such as community-based seed exchanges and farmer-led selection initiatives are paramount in preserving this genetic diversity, ensuring that local adaptations are maintained through continual cultivation (Faye et al., 2019; Singh & Nara, 2022).

Farmer-led selection practices combined with robust local seed exchange networks represent vital strategies for sustaining genetic diversity within both sorghum and pearl millet. Traditional practices in these societies lead to the multiplication and preservation of valuable landraces through direct farmer involvement. Local farmers often select for traits that are crucial for their specific environmental contexts, thus contributing to the evolution of crop varieties that are more suited to their climatic realities. This localized knowledge is often integrated into broader agricultural research to enhance varietal characteristics that correspond to resilience against climate variability and food system shocks (Oberline et al., 2020; Singh & Nara, 2022). The adaptive capacity of pearl millet and sorghum is further evidenced through their robust mechanisms to cope with climate variability. For example, the morphological and physiological responses of pearl millet during drought stress have been well-documented, showcasing adaptations such as deeper root systems, which enable these plants to access moisture from deeper soil layers (Sithole et al., 2025; Palit et al., 2013). In addition to physiological traits, morphological characteristics such as stem thickness and leaf size can also influence a plant's overall drought resilience. Understanding these adaptations offers valuable insights for breeding programs aimed at enhancing the drought tolerance of these staple crops (Sithole et al., 2025).

From an ecological perspective, sorghum is increasingly recognized for its role in fostering food security within diverse agricultural systems across the Sahel. The integration of sorghum into local diets significantly contributes to nutritional resilience in the face of food system shocks. This has been highlighted in studies focusing on the incorporation of sorghum into staple diets, promoting not only agricultural sustainability but also enhancing local food systems socioeconomically (Pereira & Hawkes, 2022; Xiong et al., 2019). Farmer engagement with sorghum as a staple has also rekindled cultural ties to this crop, reinforcing its importance in food security strategies both locally and regionally (Pereira, 2021). Moreover, the resilience of pearl millet and sorghum is critically linked to their genetic variability, allowing these crops to thrive in low-nutrient soils and under fluctuating climatic conditions. Studies indicate that maintaining and enhancing this genetic diversity is key to ensuring the long-term viability of pearl millet and sorghum in the Sahel. Conservation efforts aligned with traditional farming practices reflect an understanding of the importance of local germplasm preservation amid changing environmental challenges (Singh et al., 2023; Stich et al., 2010) (Figures 10.1 and 10.2).

10.3 Agronomic Practices and Land Tenure of Sorghum and Pearl Millet in Sahelian Societies

The agronomic practices and land tenure systems surrounding sorghum and pearl millet cultivation in Sahelian societies reflect a complex interplay between traditional farming systems, seasonal agricultural rituals, and customary land ownership arrangements (Table 10.1). The socioeconomic dynamics of these practices not only define agricultural productivity but also shape the community's social fabric (Obahiagbon & Ogwu, 2023, 2024). Traditional farming systems in the Sahel often include shifting cultivation, agroforestry, and intercropping, which are integral to maintaining soil fertility and managing water resources under arid conditions. Shifting cultivation allows for the rotation of fields to prevent soil depletion, while intercropping can enhance yields by promoting complementary resource use among crops such as sorghum and millet, both staple foods in the region (Dabara et al., 2019; Traoré et al., 2021). Agroforestry practices further incorporate the cultivation of trees alongside crops, promoting biodiversity and offering additional resources, such as wood and fruits, which can be crucial for household sustenance (Maisharou et al., 2015).

Seasonal calendars and agronomic rituals hold significant importance in the agricultural timetable of Sahelian societies, often aligning with the planting and harvesting seasons. These rituals reflect the community's deep-rooted cultural practices. Seasonal rains are particularly crucial, as they directly impact crop yields because of the reliance on rain-fed agriculture in the region. Rituals may also help synchronize community efforts, strengthening social bonds and facilitating collective action, which is essential for tasks like planting and harvesting

Figure 10.1: Sorghum Plant Showing the Leaves and Flowers

Source: Pilia et al. (2025)

(Sissoko et al., 2010). The effects of climate change, including changes in rainfall patterns, threaten to disrupt these established calendars, underscoring the need for adaptive agricultural strategies to ensure food security (Traoré et al., 2021; Turner et al., 2011).

Figure 10.2: Pearl Millet Plant

Source: Srivastava et al. (2020)

Customary land tenure and inheritance systems are essential for determining access to land resources in the Sahel. Many communities practice collective land ownership, typically managed by lineage or clan structures that establish usage rights and responsibilities (Dabara et al., 2019). This customary system often contrasts sharply with statutory land policies, which may not adequately reflect local needs and practices (Effossou et al., 2022). The common practice involves transmitting land rights through familial ties, reinforcing social structures and hierarchies within the community. As land tenure systems vary widely across regions, customary tenure remains dominant in many areas, often representing a significant proportion of land rights according to various studies (Dabara et al., 2019; Zouré et al., 2017).

Table 10.1: Agronomic Practices for Sorghum and Pearl Millet in Sahelian Societies

Agronomic Practice	Sorghum (Sorghum bicolor)	Pearl Millet (Pennisetum glaucum)
Land Preparation	Manual plowing using hoes or animal-drawn plows; zai pits to retain moisture	Ridge and furrow systems; demi-lunes (half-moons) for water harvesting
Sowing Time	Early rainy season (May–June), synchronized with first rains	Early sowing just before or at the onset of rains (late May to June)
Seed Selection	Based on drought tolerance, panicle size, and resistance to birds/pests	Selection for early maturity and resilience to sandy, low-fertility soils
Sowing Methods	Broadcasting or row planting with 2–3 seeds per hole	Hill planting with wide spacing to optimize water and nutrient use
Weeding Practices	Early and manual weeding, often twice during the season	Frequent hand weeding; sometimes combined with intercropping
Fertilization	Application of manure, compost, or microdoses of mineral fertilizers	Similar use of organic matter; rock phosphate used in some areas
Water Conservation	Stone bunds, zai pits, and contour bunds to retain water	Sand barriers, mulch, and half-moon structures to reduce runoff
Pest and Disease Management	Use of ash, Neem leaves, or smoke to deter pests; resistant varieties preferred	Traditional repellents; community monitoring of migratory pests (e.g., locusts)
Harvesting Techniques	Cut with sickles or knives when panicles are mature and dry	Hand harvesting of panicles; sometimes staged to reduce postharvest losses
Postharvest Handling	Drying on mats or roofs; storage in granaries or underground pits	Sun-drying; stored in granaries raised off the ground to avoid pests/moisture

Community norms and lineage systems significantly influence cereal production, where social obligations and expectations often guide agricultural practices. For example, sharing agricultural inputs or collectively marketing products reinforces economic resilience against market fluctuations (Ayantunde et al., 2014). Moreover, lineage determines not only who works on the land but also who benefits from its produce, creating frameworks for social equity and resource distribution (Adam, 2023; Biles et al., 2024). As population density increases, traditional land tenure systems face pressures that may compel communities to rethink land allocation and utilization. This new reality may prompt individuals to pursue alternative livelihoods, often diverting attention from traditional farming practices toward more lucrative non-farm activities (Goldstein & Udry, 2008). Such shifts can adversely impact food security due to reduced investment in local agriculture and a growing reliance on outside food sources, raising sustainability concerns (Sissoko et al., 2010).

Furthermore, the interdependence between pastoralist and agricultural communities has become increasingly intricate in the Sahel, necessitating collaboration to manage shared resources effectively (Ayantunde et al., 2014; Turner et al., 2011). Increasing competition over scarce land and water resources can lead to conflicts, complicating the relationship between farmers and herders, as agricultural expansion often encroaches upon traditional grazing grounds. This competition exacerbates challenges related to climate adaptation and food security (BenYishay et al., 2024). Innovative approaches are being explored to enhance resilience within these systems, including integrated agricultural practices that align with environmental conditions. For instance, the introduction of climate-smart agricultural technologies, such as improved irrigation methods and drought-resistant crop varieties, aims to bolster productivity while mitigating the adverse effects of climate variability (Swaffield et al., 2019; Traoré et al., 2021). The pivot toward sustainable practices underscores the need for local and regional policies that prioritize resource efficiency and community engagement in agricultural management (Maisharou et al., 2015). The characterization of cereal consumption patterns in this context is also noteworthy, as the predominant reliance on staples like sorghum and millet reflects cultural preferences tied to traditional food systems. The pressures of climate change and urbanization are gradually reshaping consumption habits; however, staple grains remain essential for nutritional security among rural populations (Moutawakilou et al., 2023). Additionally, the relevance of fermented cereals emphasizes the cultural significance of local crops, reinforcing both dietary practices and community identity (Phiri et al., 2019).

10.4 Culinary Ethnoscience and Nutritional Contributions of Sorghum and Pearl Millet in Sahelian Societies

Sorghum and pearl millet play pivotal roles in the culinary heritage, nutritional practices, and health sustainability of Sahelian societies. Characterized by arid and semiarid climates, the Sahel has turned to these grains not only for their adaptability but also for the traditional dishes they help prepare, such as tô, couscous, and porridge. Beyond traditional diets, the fermentation practices associated with these grains significantly enhance flavor profiles and nutritional benefits in local cuisines. In the Sahel, tô (a starchy staple made from sorghum or pearl millet) and couscous (often derived from millet) exemplify the cultural importance of these grains. These dishes are typically prepared through processes involving soaking, grinding, and cooking, which not only preserve traditional culinary techniques but also improve the nutritional properties of the grains (Gwekwe et al., 2024). Fermentation, in particular, plays a critical role in enhancing the digestibility and nutritional value of these foods. For instance, the spontaneous fermentation of pearl millet has been shown to significantly lower levels of antinutritional factors like phytic acid, which inhibits mineral absorption, thus increasing the bioavailability of essential nutrients (Cheung et al., 2025; Krishnan & Meera, 2018).

Nutritionally, both pearl millet and sorghum are valued for their robust profiles, as they provide essential macro and micronutrients necessary for dietary security. They are particularly important for smallholder farmers, as these grains serve as both a source of food and income (Hassan et al., 2021). A review highlights pearl millet as containing vital nutrients for health, including proteins ranging from 5.8% to 20.9%, significant dietary fibers, and essential minerals like iron and zinc, though bioavailability can often be limited by factors such as the presence of phytic acid (Abah et al., 2020; Mahalakshmi et al., 2024). The natural ability of pearl millet to thrive in harsh conditions means it is often one of the few grains available that is crucial for food security in these regions (Dube et al., 2021). Moreover, the functional properties of these grains extend to their health-promoting attributes, which have been recognized in various studies. Epidemiological studies have indicated that the flavonoids in millet can have preventive effects on several chronic diseases, including heart disease and diabetes, while their antioxidant properties contribute to overall health maintenance (Nambiar et al., 2012). Specifically, pearl millet has been reported to aid in glycemic control, which is vital for managing diabetes, a condition increasingly prevalent in these regions (Raju et al., 2024).

The traditional knowledge surrounding household preparation plays an integral role in the culinary ethnoscience of Sahelian societies. In these communities, the elderly, typically women, serve as custodians of knowledge, passing down food preparation techniques such as cooking, fermentation, and mixing different grains through generations (Humblot & Guyot, 2009). This transmission of culinary knowledge is essential for maintaining the sociocultural identity of Sahelian societies, which, despite facing modernization and changes in dietary habits, continue to uphold traditional practices related to millet and sorghum (Himashree & Mahendran, 2025). Role differentiation in household food preparation highlights the practical aspects of millet and sorghum usage. Women in Sahelian societies perform most of the food processing tasks, often using locally sourced ingredients along with these grains to create meals that fit climatic conditions and food security needs (Biradar et al., 2024). This dynamic not only empowers women but also ensures the nutritional health of households, as the preparation methods employed are often optimized for nutritional efficiency (Cheung et al., 2025).

Furthermore, nutrient retention and enhancement in processed forms contribute to the upholding of traditions as well as dependencies on local flora. Fermented foods made from sorghum and millet not only retain their original nutrients but can also introduce beneficial bacteria that improve gut health, encapsulating the dual roles of food as sustenance and a medium for health benefits (Humblot & Guyot, 2009). The methodologies in fermentation and food processing are often influenced by both environmental resources and traditional practices, highlighting an interplay between environmental adaptation and cultural ethos in food preparation (Traoré et al., 2022). In essence, sorghum and pearl millet are integral to the culinary ethnoscience within Sahelian societies due to their adaptability, nutritional properties, and the cultural significance they carry. The way these grains are prepared and consumed encapsulates a narrative of resilience against environmental adversity while simultaneously promoting health and wellness within the communities. As such, the nexus between these grains and the local culinary practices represents a critical aspect of understanding nutrition and food culture in the Sahel region.

10.5 Gendered Knowledge and Labor Systems of Sorghum and Pearl Millet in Sahelian Societies

The intersection of gender, agriculture, and knowledge systems in Sahelian societies, particularly concerning sorghum and pearl millet, involves a nuanced exploration of labor division, postharvest processing, and intergenerational knowledge transfer. In the context of agricultural activities, the division of labor often reveals distinct patterns based on gender roles, with men predominantly engaging in the cultivation phase while women take precedence in postharvest activities, such as processing and storage (Yengoh, 2012). In a broader agricultural framework, this division illustrates how socioeconomic structures and cultural norms shape labor dynamics. For instance, men may focus on the physically demanding aspects of farming, reflecting traditional gender ideologies (Lavee & Katz, 2002), while women perform essential postharvest roles that ensure food security and household sustenance (Bird & Codding, 2015).

In the realm of crop processing, women engage in critical practices such as seed selection and culinary innovations, which directly impact nutritional outcomes and food diversity within their communities. Women's involvement in seed selection honors traditional knowledge passed down through generations and enhances crop resilience and adaptability by utilizing local selection criteria (Girard, 2009). This participatory approach to seed management is vital, especially in light of climate variability and the growing importance of sustainable agricultural practices in the region (Diallo et al., 2012). The transmission of agricultural knowledge and practices across generations strengthens the stewardship of gendered knowledge systems within Sahelian societies. Knowledge transfer is closely linked to the roles that both men and women play in agriculture. Women, often the custodians of knowledge related to crop processing and culinary methods, serve not just as educators for their children but also as innovators who adapt cooking techniques to enhance nutrition and preserve cultural identities (Tornello et al., 2015). This intergenerational transmission of knowledge is crucial in areas where formal education may be limited, as women impart essential survival skills necessary for resilience in changing environmental conditions (Duff et al., 2017). Furthermore, the emphasis on collective responsibility in food preparation and processing reveals a community-centric approach to sustainable agricultural practices, connecting gender roles and participatory stewardship (Yodanis, 2005).

Examining the complexities of labor division in these societies, it is evident that environmental, socioeconomic, and cultural factors play pivotal roles. The historical environmental degradation in the Sahel exacerbated by climate change underscores how agricultural practices must evolve amid decreasing natural resources (Miller, 2008; Yengoh, 2012). Agricultural sustainability requires ingenuity and cooperation, where women's traditional roles in postharvest processing and culinary development are increasingly recognized as drivers of innovation and sustainability. This perspective aligns with calls for enhancing institutional supports that integrate gender perspectives into agricultural planning, recognizing the unique contributions women make toward food systems and resilience (Duff et al., 2017).

Moreover, the significance of women's roles in these agricultural systems extends beyond mere participation; it embodies a broader framework of empowerment and recognition within the socioeconomic landscape. In Sahelian societies, the cultivation and processing of sorghum and pearl millet provide not only sustenance but also symbolize cultural heritage and identity. As women engage in postharvest roles, they cultivate not just food but also social capital, fostering community ties and resilience frameworks capable of addressing social and environmental challenges (Girard, 2009). Thus, enhancing women's involvement in decision-making related to agricultural practices can contribute to both improved household food security and greater community resilience. Furthermore, the aspects of labor division illustrate systemic issues of gender equality and empowerment within agricultural contexts. The persistence of traditional norms often constrains women's full participation in agriculture, limiting access to resources, knowledge, and decision-making power (Albanesi & Olivetti, 2009). Such constraints can be enforced by economic dependencies within family structures, where unequal distribution of labor can hinder both women's agency and family dynamics (Greenstein, 2000). To dismantle these barriers, there is a growing recognition of the need for gender-sensitive policies that not only empower women but also promote equitable sharing of agricultural responsibilities to facilitate broader societal shifts toward gender equality (Bird & Codding, 2015).

As such, agricultural initiatives aimed at improving practices in sorghum and pearl millet cultivation ought to consider gendered perspectives critically. Embedding gender-focused interventions, including education and technical support for women in postharvest roles, can significantly enhance resilience against climate variability while fostering a more equitable division of labor (Duff et al., 2017; Minnotte et al., 2010). The historical, sociopolitical, and economic contexts underpinning gendered agricultural practices must also be acknowledged in designing sustainable strategies that capitalize on women's traditional knowledge and innate capabilities (Baker & Jacobsen, 2007).

10.6 Ritual, Spirituality, and Cultural Significance of Sorghum and Pearl Millet in Sahelian Societies

Sorghum and pearl millet play a pivotal role in the cultural, spiritual, and cere-monial life of Sahelian societies, serving as more than mere staples in the diet of millions. These crops are deeply intertwined with various aspects of social practices, encompassing ceremonies, ancestral rites, and seasonal festivals that celebrate agrarian success and reinforce communal identities. The harvest of sorghum and millet often marks significant seasonal changes and is accompanied by a variety of rituals that express gratitude to deities and ancestors for their bounty (Mason et al., 2015; Sermé et al., 2018). These ceremonies can include communal feasts where the first harvest is shared, symbolizing abundance and social solidarity among community members (Zouré et al., 2019).

During pivotal life events like birth and marriage, sorghum and millet are imbued with sacred symbolism, serving as ingredients for traditional dishes that signify fertility and prosperity. In some communities, millet is used to prepare "tô," a staple dish that is essential at weddings, symbolizing the bride's contribution to her new household (Sissoko et al., 2010). Furthermore, during childbirth, millet may be presented to the mother in a ceremonial setting, signifying the return to nourishment and the strengthening of future generations (Tagesson et al., 2014). These rituals emphasize the interconnectedness between agricultural success and individual lifecycle events, demonstrating the cultural significance of these grains beyond their nutritional value.

In terms of cosmological meanings, the successful harvest of sorghum and millet is understood not merely as an economic outcome but as a reflection of divine favor and balance within the ecosystem. Communities often align agri-cultural cycles with celestial events, interpreting favorable rains and successful yields as blessings from ancestral spirits or deities (Doto et al., 2015; Egbebiyi et al., 2019). For example, the timing of sowing and harvesting is often aligned with lunar phases or astronomical markers, particularly within traditional farming communities (Badji et al., 2022). This spiritual interpretation governs agricultural practices and unites the community in shared beliefs regarding the interplay between human efforts and divine guidance. The Sahelian ecosystem is characterized by its vulnerability to climatic fluctuations and variability, which adds layers of complexity to the cultural narratives surrounding sorghum and millet cultivation. The reliance on rain-fed agriculture highlights the community's resilience and adaptive strategies to cope with erratic weather patterns, revealing a profound understanding of local environmental conditions (Salack et al., 2025; Zouré et al., 2019). Consequently, traditional knowledge systems play an essential role in shaping farming practices that respect and honor the spiritual dimensions of agriculture, including diverse rituals responsive to seasonal changes (Ndiaye et al., 2008).

Ceremony plays a vital role in invoking favorable conditions for rain and a bountiful harvest. For example, communal prayers and offerings are frequently made to the sky, asking for adequate rainfall, which underscores the simultaneous recognition of environmental dependency and spiritual belief (Sawadogo et al., 2023). The interdependence between agriculture and spirituality is reinforced through annual festivals that celebrate the harvest, resulting in public performances and communal gatherings that foster a sense of unity and shared cultural identity among community members (Diallo et al., 2012). Moreover, the cultural significance of sorghum and millet in rites of passage and communal festivals can be seen in the shared narratives and oral histories that pertain to these crops. Elders often recount tales of past harvests, weaving lessons of perseverance and gratitude into the fabric of cultural identity (Sarr et al., 2018). These stories not only serve to educate younger generations about agricultural practices but also imbue the crops with historical significance that fosters a sense of belonging within the community (Mason et al., 2015; Tagesson et al., 2014).

In times of adversity, such as drought or poor yields, the spiritual dimensions associated with sorghum and millet take on heightened importance as communities seek to navigate the complexities of shifting climate patterns while remaining connected to their agricultural roots (Badji et al., 2022; Sissoko et al., 2010). Efforts to adapt to these challenges often include reviving traditional practices and rituals that reaffirm cultural ties to the land and its produce, demonstrating the resilience of cultural identity through the agricultural lifeways of sorghum and millet cultivation (Doto et al., 2015; Egbebiyi et al., 2019). As these practices continue to evolve, integrating contemporary agricultural knowledge with traditional rituals creates an opportunity for communities to sustain their cultural significance while adapting to environmental changes and climatic challenges that threaten their agricultural productivity (Ndiaye et al., 2008; Salack et al., 2025). Each grain of sorghum and millet embodies layers of meaning that reflect a community's relationship with nature, ancestors, and the cosmos.

10.7 Challenges and Transformations of Sorghum and Pearl Millet in Sahelian Societies

The transformation and challenges faced by sorghum and pearl millet in Sahelian societies are multifaceted, reflecting a complex interplay of ecological, socioeconomic, and cultural factors. Among the primary challenges are land degradation, desertification, and rainfall variability, which hinder agricultural productivity and exacerbate food insecurity. The climatic conditions of the Sahel make it an ecologically sensitive zone, significantly influencing the growth and yield of staple crops like sorghum and pearl millet (Sermé et al., 2018). Land degradation, driven by unsustainable agricultural practices, has greatly affected soil fertility, leading to a decline in crop yields. Traditional farming methods, although effective under previous conditions, are increasingly insufficient to meet the demands of a growing population and changing climate conditions. The integration of traditional practices with modern techniques has shown promise in enhancing productivity while mitigating the impacts of degradation (Faye, 2020). Additionally, erratic rainfall patterns complicate farmers' efforts, as they struggle to plan planting and harvesting accordingly, threatening the livelihoods of rural communities dependent on these crops (Keugmeni et al., 2025).

The pressures of commercialization threaten the integrity of traditional seed systems. With the push for market-oriented approaches, local farmers are often encouraged or compelled to replace Indigenous seeds with hybrid varieties that promise higher yields. Such shifts can erode local germplasm diversity, diminishing the resilience of farming systems in the face of environmental variability (Faye, 2020). Studies indicate that traditional varieties of sorghum and pearl millet are vital for food security, yet smaller growers may gravitate toward commercially available seeds that often do not perform well under local conditions (Černý et al., 2023). The advent of agricultural modernization has not only impacted the crops themselves but has also influenced social structures. As farming becomes increasingly commercialized, younger generations may migrate to urban areas for better economic opportunities, resulting in a significant knowledge gap in agricultural practices and a decline in the cultural practices linked to traditional farming (Ogwu, 2019, 2023; O'Regan & Thompson, 2017). This disengagement can adversely affect community cohesion and the cultural heritage embodied in these agricultural traditions.

Market integration has transformed the dynamics of sorghum and pearl millet production in the Sahel. Understanding market demands and access to agricultural markets have become critical for farmers aiming to ensure their economic viability (Keugmeni et al., 2025). However, this often requires adopting new technologies and practices that may not align with traditional knowledge. The challenges and opportunities arising from agricultural commercialization highlight the broader crises of identity, adaptation, and survival faced by these communities in the contemporary Sahelian context. Youth migration complicates knowledge transfer regarding sorghum and pearl millet cultivation, as many young people abandon agricultural practices for urban employment. Many Sahelian communities rely on traditional knowledge, including cultivation techniques and seed selection practices, which are often passed down orally through generations (Shine & Dunford, 2016). The reduced involvement of youth in agricultural settings can result in the loss of critical agricultural wisdom and cultural practices essential for the community's resilience against climatic and economic shocks.

In a broader context of agricultural sustainability, the pressures to modernize crop production must be balanced with the need to maintain genetic diversity and resilience within local ecosystems. Effective policy frameworks are essential for supporting farmers in integrating traditional practices with modern advancements, thereby ensuring the persistence of both crops and the Indigenous knowledge systems that cultivate them. Various studies suggest that synergistic approaches could yield fruitful results by incorporating the strengths of both traditional and modern agricultural practices (Faye, 2020; Keugmeni et al., 2025).

Enhancing the role of youth in agriculture through targeted education and capacity-building initiatives could create new pathways for knowledge retention. By involving the younger generation in the production, processing, and marketing chains of sorghum and pearl millet, communities could nurture a new generation that values and employs Indigenous agricultural practices while also embracing innovation. Such measures could revitalize the agricultural landscape in the Sahel, ensuring the continuity of vital practices tied to sorghum and pearl millet cultivation (Faye, 2020). Addressing the socioeconomic conditions contributing to land degradation, desertification, and rainfall variability necessitates a comprehensive response that includes investment in infrastructure, education, and fertility enhancement practices. By combining traditional agricultural wisdom with modern technological interventions, stakeholders can develop more resilient agricultural systems that withstand the ongoing impacts of climate change while supporting the fabric of Sahelian societies (Eduvie & Garba, 2021; Keugmeni et al., 2025).

10.8 Sociobiological Reflection of Sorghum and Pearl Millet: Toward Resilient Futures

The sociobiological reflection of sorghum and pearl millet as fundamental crops for enhancing resilience in future agricultural systems is enriched by various dimensions, including Indigenous agricultural knowledge, policy recommendations, potential contributions to climate-resilient agriculture, and the significance of collaborative research for biocultural conservation. These dimensions highlight how these cereals are not just agricultural products but are also cultural staples with deep historical roots and an evolving role in human nutrition and sustainability. Revitalizing Indigenous agricultural knowledge is crucial for adapting to modern agricultural challenges posed by climate change and food insecurity. Indigenous communities possess a wealth of knowledge regarding the cultivation, preservation, and utilization of sorghum and pearl millet, which can inform contemporary practices. For instance, traditional cropping systems, seed selection, and post-harvest techniques derived from Indigenous practices have been shown to enhance soil health and biodiversity, both vital for fostering resilience in agriculture (Kumar et al., 2018; Obongodot & Ogwu, 2023; Satyavathi et al., 2021). Such knowledge underscores sustainable practices that have been effectively used through generations but are often overlooked in modern agronomy (Saleh et al., 2013). Additionally, efforts to integrate this Indigenous wisdom into contemporary agricultural education can establish robust frameworks for the sustainable cultivation of these crops, ensuring that valuable traditional techniques do not fade into obscurity.

Policy recommendations aimed at promoting local landraces and agroecology highlight the necessity for governments to support the conservation of genetic resources of sorghum and pearl millet. Local landraces, which demonstrate resilience to specific environmental stresses, are crucial for maintaining agricultural diversity and enhancing food security (Dias-Martins et al., 2018; Upadhyaya et al., 2012). Supportive policies could encourage the creation of community seed banks designed to conserve these varieties while also fostering their cultivation through financial incentives and improved market access for local farmers (Adebo, 2020). By promoting agroecological practices, policymakers can help develop systems that are both environmentally sustainable and financially viable for local populations. Creating markets for locally grown, diverse varieties can significantly strengthen rural economies while enhancing food security and preserving genetic diversity in these vital crops (Taylor et al., 2014). In the realm of climate-resilient agriculture, both sorghum and pearl millet offer distinctive attributes that positively contribute to food sovereignty. These grains are acknowledged for their ability to thrive in low-nutrient soils and extreme weather conditions, providing a dependable source of nutrition even when other crops fail (Kumar et al., 2018; Satyavathi et al., 2021). Pearl millet, in particular, is recognized for its high nutrient density and its potential to provide essential vitamins and minerals (Dias-Martins et al., 2018). Furthermore, the adaptability of sorghum to various environmental conditions reinforces its role as a staple in drought-prone areas, highlighting its importance in food sovereignty movements. Promoting these crops within sustainable dietary patterns can be woven into national food security strategies, offering vital support for vulnerable populations worldwide (Kudita et al., 2023).

The potential for collaborative research initiatives aimed at biocultural conservation cannot be overstated. By combining scientific research with local knowledge systems, farmers and researchers can co-create sustainable agricultural practices that enhance both crop yields and environmental health. Collaborative frameworks could leverage genomic studies to improve the resilience traits of sorghum and pearl millet while protecting Indigenous knowledge related to these crops' cultivation (Newmaster et al., 2013). Furthermore, engaging local communities in research empowers them and ensures that the outcomes are relevant to their needs and conditions. This synergy is crucial for developing resilient agricultural systems that increase food production while preserving traditional knowledge and practices (Nani & Krishnaswamy, 2021). The importance of enhancing the functional properties and health benefits of sorghum and millet through innovative processing techniques also supports their agricultural viability. Recent studies have demonstrated the advantages of various processing methods, such as dehulling and fermentation, which can elevate the bioavailability of nutrients and reduce antinutritional factors (Gwekwe et al., 2024). By improving the nutritional quality of these grains, they can be more effectively integrated into modern diets, promoting greater acceptance and consumption among wider populations (Kumar et al., 2018; Saleh et al., 2013). This advancement addresses food security while promoting consumer health through the enhanced dietary contributions of these grains.

Furthermore, addressing the economic factors related to the cultivation of sorghum and pearl millet is crucial in positioning these crops as viable alternatives to more commonly grown cereals like maize and wheat. By effectively communicating their nutritional benefits and adaptability, targeted marketing strategies can boost consumer awareness, leading to increased demand (Dias-Martins et al., 2018). Policymakers can also incentivize farmers to diversify their cropping systems to incorporate these resilient cereals, thus reducing reliance on a limited number of crops and enhancing the overall resilience of agricultural systems (Alavi et al., 2019; Kumar et al., 2018).

10.9 Conclusion

Sorghum and pearl millet serve as resilient cornerstones of Sahelian agroecology and sociocultural life. This chapter has traced their journey from ancient domestication to their enduring significance amid modern climatic and developmental pressures. Their genetic plasticity, cultivated over generations through farmer innovation and community-based seed systems, affirms their value not only as hardy cereals but also as carriers of cultural identity, ecological knowledge, and food sovereignty. Central to this sociobiological relevance is the role of gendered labor, oral traditions, and ritual practices in sustaining biodiversity and agroecological balance. The practices embedded in the cultivation, preparation, and ceremonial use of sorghum and millet reveal a sophisticated interplay between people and the environment—an interplay that has historically ensured survival under harsh ecological constraints. However, growing threats such as industrial agriculture, climate change, land tenure fragmentation, and genetic homogenization challenge the sustainability of these systems. The future of these keystone cereals, therefore, hinges on integrating Indigenous knowledge into policy frameworks, promoting community-led seed conservation, and supporting research that aligns with local priorities and practices. Revalorizing sorghum and millet as pillars of both cultural resilience and ecological adaptation offers a transformative pathway toward sustainable food systems in the Sahel. Future research must deepen our understanding of these dynamics while policy must embrace their holistic significance to ensure continuity in a region where food, identity, and landscape are inseparably linked.

Chapter Reflection

1. In what ways does the genetic diversity of sorghum and pearl millet reflect the coevolutionary relationship between humans and their environment in the Sahel?
2. How do gendered labor divisions influence both the preservation and the transformation of agricultural and culinary knowledge surrounding these grains?
3. Discuss how traditional spiritual beliefs and agricultural rituals involving sorghum and millet contribute to both cultural resilience and ecological sustainability.
4. What are the key socioeconomic and ecological risks posed by commercialization and modernization of sorghum and millet production systems?
5. How can policy frameworks better support the integration of Indigenous knowledge with modern agricultural innovations for sustainable food security?
6. Evaluate the role of youth migration in the disruption of traditional knowledge systems and propose potential strategies for reversing this trend.
7. How might collaborative research that includes both local farmers and scientists help to sustain the sociobiological importance of sorghum and millet in Sahelian societies?

REFERENCES

Abah, C., Ishiwu, C., Obiegbuna, J., & Oladejo, A. (2020). Nutritional composition, functional properties and food applications of millet grains. *Asian Food Science Journal*, *14*(2), 9–19. https://doi.org/10.9734/afsj/2020/v14i230124

Adam A. G. (2023). Systematic review of the changing land to people relationship and co-evolution of land administration. *Heliyon, 9*(10), e20637. https://doi.org/10.1016/j.heliyon.2023.e20637

Adebo, O. (2020). African sorghum-based fermented foods: Past, current and future prospects. *Nutrients*, *12*(4), 1111. https://doi.org/10.3390/nu12041111

Alavi, S., Mazumdar, S., & Taylor, J. (2019). Modern convenient sorghum and millet food, beverage and animal feed products, and their technologies. In J. R. N. Taylor & K. G. Duodu (Eds.), *Sorghum and millets* (pp. 293–329).

Woodhead Publishing and AACC International Press. https://doi.org/10.1016/ b978-0-12-811527-5.00010-1

Albanesi, S., & Olivetti, C. (2009). Home production, market production and the gender wage gap: Incentives and expectations. *Review of Economic Dynamics*, *12*(1), 80–107. https://doi.org/10.1016/j.red.2008.08.001

Ayantunde, A., Assé, R., Said, M., & Fall, A. (2014). Transhumant pastoralism, sustainable management of natural resources and endemic ruminant livestock in the sub-humid zone of West Africa. *Environment Development and Sustainability*, *16*(5), 1097–1117. https://doi.org/10.1007/s10668-014-9515-z

Badji, A., Mohíno, E., Diakhaté, M., Mignot, E., & Gaye, A. (2022). Decadal variability of rainfall in Senegal: Beyond the total seasonal amount. *Journal of Climate*, *35*(16), 5339–5358. https://doi.org/10.1175/jcli-d-21-0699.1

Baker, M., & Jacobsen, J. (2007). Marriage, specialization, and the gender division of labor. *Journal of Labor Economics*, *25*(4), 763–793. https://doi.org/10.1086/522907

BenYishay, A., Sayers, R., Singh, K., Goodman, S., Walker, M., Traoré, S., Rauschenbach, M., & Noltze, M. (2024). Irrigation strengthens climate resilience: Long-term evidence from Mali using satellites and surveys. *PNAS Nexus*, *3*(2). https://doi.org/10.1093/pnasnexus/pgae022

Biles, B. J., Serova, N., Stanbrook, G., Brady, B., Kingsley, J., Topp, S. M., & Yashadhana, A. (2024). What is Indigenous cultural health and wellbeing? A narrative review. *Lancet Regional Health: Western Pacific, 52*, 101220. https://doi.org/10.1016/j.lanwpc.2024.101220

Biradar, V. M., Kumar, P., Yallappa, M., Ramya, C. S., Nayak, P., Sekhar, M., Bhushan, S., & Babu, S. (2024). A sustainable nutricereal: A review on nutrient and bio-active composition and its potential health benefits of pearl millet. *International Journal of Advanced Biochemistry Research*, *8*(3S), 30–35. https://doi.org/10.33545/26174693.2024.v8.i3sa.684

Bird, R. B., & Codding, B. F. (2015). The sexual division of labor. In *Emerging trends in the social and behavioral sciences: An interdisciplinary, searchable, and linkable resource* (pp. 1–16). Wiley. https://doi.org/10.1002/9781118900772.etrds0300

Černý, V., Priehodová, E., & Fortes-Lima, C. (2023). A population genetic perspective on subsistence systems in the Sahel/Savannah Belt of Africa and the historical role of pastoralism. *Genes*, *14*(3), 758. https://doi.org/10.3390/genes14030758

Chalachew, E., Tsegaye, D., Biljon, A., Herselman, L., & Labuschagne, M. (2025). Kernel composition in sorghum landraces revealed via analyses of genotype-by-environment interactions. *PLOS One*, *20*(4), e0320513. https://doi .org/10.1371/journal.pone.0320513

Cheung, M., Miller, L., Deutsch, J., Sherman, R., Katz, S., & Wise, P. (2025). Sensory properties and acceptability of fermented pearl millet, a climate-resistant and nutritious grain, among consumers in the United States—A pilot study. *Foods*, *14*(5), 871. https://doi.org/10.3390/foods14050871

Dabara, D., Lawal, O., Chiwuzie, A., Omotehinshe, O., & Soladoye, J. (2019). Land tenure systems and agricultural productivity in Gombe Nigeria. *Madridge Journal of Agriculture and Environmental Sciences*, *2*(1), 51–59. https://doi. org/10.18689/mjaes-1000110

Diack, O., Kanfany, G., Guèye, M., Sy, O., Fofana, A., Tall, H., Serba, D. D., Zekraoui, L., Bertholy-Salazar, C., Vigouroux, Y., Diouf, D., & Kane, N. (2020). Gwas unveils features between early- and late-flowering pearl millets. *BMC Genomics*, *21*(1), 777. https://doi.org/10.1186/s12864-020-07198-2

Diallo, I., Sylla, M., Giorgi, F., Gaye, A., & Camara, M. (2012). Multimodel GCM-RCM ensemble-based projections of temperature and precipitation over West Africa for the early 21st century. *International Journal of Geophysics*, *2012*, 1–19. https://doi.org/10.1155/2012/972896

Dias-Martins, A. M., Pessanha, K., Pacheco, S., Rodrigues, J., & Carvalho, C. (2018). Potential use of pearl millet (*Pennisetum glaucum* (L.) R. Br.) in Brazil: Food security, processing, health benefits and nutritional products. *Food Research International*, *109*, 175–186. https://doi.org/10.1016/j.foodres.2018.04.023

Doto, V., Yacouba, H., Niang, D., & Lahmar, R. (2015). An alternative strategy for mitigating the effect of rainfall variability in Burkinabe Sahel. *Journal of Water Resource and Protection*, *7*(16), 1318–1330. https://doi.org/10.4236/jwarp.2015.716107

Dube, M., Nyoni, N., Bhebhe, S., Maphosa, M., & Bombom, A. (2021). Pearl millet as a sustainable alternative cereal for novel value-added products in sub-Saharan Africa: A review. *Agricultural Reviews*, *42*(2), 240–244. https://doi.org/10.18805/ag.r-174

Duff, A., Zedler, P., Barzen, J., & Knuteson, D. (2017). The capacity-building stewardship model: Assessment of an agricultural network as a mechanism for improving regional agroecosystem sustainability. *Ecology and Society*, *22*(1). https://doi.org/10.5751/es-09146-220145

Eduvie, M., & Garba, M. (2021). Appraisal of groundwater potential of Fadama areas within Northern Nigeria: A review. *Journal of Geoscience and Environment Protection*, *9*(3), 44–57. https://doi.org/10.4236/gep.2021.93004

Effossou, K., Cho, M., & Ramoelo, A. (2022). Impacts of conflicting land tenure systems on land acquisition by agribusiness developers in côte d'ivoire. *Journal of Agribusiness and Rural Development*, *63*(1), 25–39. https://doi.org/10.17306/j.jard.2022.01489

Egbebiyi, T., Lennard, C., Crespo, O., Mukwenha, P., Lawal, S., & Quagraine, K. (2019). Assessing future spatio-temporal changes in crop suitability and planting season over West Africa: Using the concept of crop-climate departure. *Climate*, *7*(9), 102. https://doi.org/10.3390/cli7090102

Embashu, W., & Nantanga, K. (2019). Malts: Quality and phenolic content of pearl millet and sorghum varieties for brewing nonalcoholic beverages and opaque beers. *Cereal Chemistry*, *96*(4), 765–774. https://doi.org/10.1002/cche.10178

Faye, J. (2020). Indigenous farming transitions, sociocultural hybridity and sustainability in rural senegal. NJAS—Wageningen *Journal of Life Sciences*, *92*(1), 1–8. https://doi.org/10.1016/j.njas.2020.100338

Faye, J., Maina, F., Hu, Z., Foncéka, D., Cissé, N., & Morris, G. (2019). Genomic signatures of adaptation to Sahelian and Soudanian climates in sorghum landraces of senegal. *Ecology and Evolution*, *9*(10), 6038–6051. https://doi.org/10.1002/ece3.5187

Garba, M. (2019). Mycotoxins consumption and burden of aflatoxin-induced hepatocellular carcinoma in people subsisting on sorghum based products in the derived savannah zone of Nigeria. *Asian Food Science Journal*, *8*(4), 1–10. https://doi.org/10.9734/afsj/2019/v8i429996

Gemenet, D. C., Hash, C. T., Sanogo, M. D., Sy, O., Zangre, R. G., Leiser, W. L., & Haussmann, B. I. G. (2015). Phosphorus uptake and utilization efficiency in West African pearl millet inbred lines. *Field Crops Research*, *171*, 54–66. https://doi.org/10.1016/j.fcr.2014.11.001

Girard, H. (2009). Wégoubri, the Sahelian Bocage: An integrate approach for environment preservation and social development in Sahelian agriculture (Burkina Faso). *Field Actions Science Reports*, *2*(1), 33–39. https://doi.org/10.5194/facts-2-33-2009

Goldstein, M., & Udry, C. (2008). The profits of power: Land rights and agricultural investment in Ghana. *Journal of Political Economy*, *116*(6), 981–1022. https://doi.org/10.1086/595561

Greenstein, T. (2000). Economic dependence, gender, and the division of labor in the home: A replication and extension. *Journal of Marriage and Family*, *62*(2), 322–335. https://doi.org/10.1111/j.1741-3737.2000.00322.x

Gunguniya, D. F., Kumar, S., Patel, M. P., Sakure, A. A., Patel, R., Kumar, D., & Khandelwal, V. (2023). Morpho-biochemical characterization and molecular marker based genetic diversity of pearl millet (*Pennisetum glaucum* (L.) R. Br.). *PeerJ*, *11*, e15403. https://doi.org/10.7717/peerj.15403

Gwekwe, B. N., Chopera, P., Matsungo, T. M., Chidewe, C., Mukanganyama, S., Nyakudya, E., Mtambanengwe, F., Mapfumo, P., & Nyanga, L. K. K. (2024). Effect of dehulling, fermentation, and roasting on the nutrient and anti-nutrient content of sorghum and pearl millet flour. *International Journal on Food Agriculture and Natural Resources*, *5*(1), 1–7. https://doi.org/10.46676/ij-fanres.v5i1.221

Hassan, Z., Sebola, N., & Mabelebele, M. (2021). The nutritional use of millet grain for food and feed: A review. *Agriculture and Food Security*, *10*(1), 16. https://doi.org/10.1186/s40066-020-00282-6

Himashree, P., & Mahendran, R. (2025). Effect of high-pressure soaking on the physicochemical, nutritional, and techno-functional properties of pearl millets. *Sustainable Food Technology*, *3*(3), 714–724. https://doi.org/10.1039/d5fb00042d

Humblot, C., & Guyot, J. (2009). Pyrosequencing of tagged 16s rRNA gene amplicons for rapid deciphering of the microbiomes of fermented foods such as pearl millet slurries. *Applied and Environmental Microbiology*, *75*(13), 4354–4361. https://doi.org/10.1128/aem.00451-09

Ikhajiagbe, B., Ogwu, M. C., & Omage, Z. E. (2023). Seed phenotypic variations in cowpea, *Vigna unguiculata*, from selected open markets in Edo State, Nigeria. *Cell Biology and Development*, *7*(2), 89–101. http://doi.org/10.13057/cellbioldev/v070206

Keugmeni, G. A. A., Keïta, A., Yonaba, R., Sawadogo, B., & Kengni, L. (2025). Towards sustainable food security in the Sahel: Integrating traditional conservation practices and controlled irrigation to overcome water scarcity during the dry season for onion and jute production. *Sustainability*, *17*(6), 2345. https://doi.org/10.3390/su17062345

Krishnan, R., & Meera, M. S. (2018). Pearl millet minerals: Effect of processing on bioaccessibility. *Journal of Food Science and Technology*, *55*(9), 3362–3372. https://doi.org/10.1007/s13197-018-3305-9

Kudita, S., Schoustra, S., Mubaiwa, J., Smid, E. J., & Linnemann, A. R. (2023). Substitution of maize with sorghum and millets in traditional processing of Mahewu, a non-alcoholic fermented cereal beverage. *International Journal of Food Science and Technology*, *59*(3), 1421–1431. https://doi.org/10.1111/ijfs.16887

Kumar, A., Tomer, V., Kaur, A., Kumar, V., & Gupta, K. (2018). Millets: A solution to agrarian and nutritional challenges. *Agriculture and Food Security*, *7*(1), 31. https://doi.org/10.1186/s40066-018-0183-3

Lavee, Y., & Katz, R. (2002). Division of labor, perceived fairness, and marital quality: The effect of gender ideology. *Journal of Marriage and Family*, *64*(1), 27–39. https://doi.org/10.1111/j.1741-3737.2002.00027.x

Lee, H.-S., Santana, Á. L., Peterson, J., Yücel, U., Perumal, R., De Leon, J., Lee, S.-H., & Smolensky, D. (2022). Anti-adipogenic activity of high-phenolic sorghum brans in pre-adipocytes. *Nutrients, 14*(7), 1493. https://doi.org/10.3390/nu14071493

Mahalakshmi, S., Malavika, S., Sindhuja, T., & Priya, K. (2024). Pearl millet: Potential nutraceutical properties of pearl millet and its utilization in various food products: A review. *Bhartiya Krishi Anusandhan Patrika*, *39*(1), 46–50. https://doi.org/10.18805/bkap432

Maisharou, A., Chirwa, P., Larwanou, M., Babalola, F., & Ofoegbu, C. (2015). Sustainable land management practices in the Sahel: Review of practices, techniques and technologies for land restoration and strategy for up-scaling. *The International Forestry Review*, *17*(3), 1–19. https://doi.org/10.1505/146554815816006974

Mason, S. C., Maman, N., & Pale, S. (2015). Pearl millet production practices in semi-arid West Africa: A review. *Experimental Agriculture*, *51*(4), 501–521. https://doi.org/10.1017/s0014479714000441

Mawouma, S., Condurache, N. N., Turturică, M., Constantin, O. E., Croitoru, C., & Râpeanu, G. (2022). Chemical composition and antioxidant profile of sorghum (*Sorghumbicolor* (L.) Moench) and pearl millet (*Pennisetumglaucum* (L.) R.Br.) grains cultivated in the far-north region of Cameroon. *Foods*, *11*(14), 2026. https://doi.org/10.3390/foods11142026

Mihiretu, A., Assefa, N., & Wubet, A. (2023). Pearl millet, the hope of food security in marginal arid tropics: Implications for diversifying limited cropping systems. *Journal of Agribusiness and Rural Development*, *67*(1), 93–102. https://doi.org/10.17306/j.jard.2023.01670

Miller, A. (2008). 6 changing responsibilities and collective action: Examining early North African pastoralism. *Archeological Papers of the American Anthropological Association*, *18*(1), 76–86. https://doi.org/10.1111/j.1551-8248.2008.00006.x

Minnotte, K. L., Minnotte, M. C., Pedersen, D. E., Mannon, S. E., & Kiger, G. (2010). His and her perspectives: Gender ideology, work-to-family conflict, and marital satisfaction. *Sex Roles*, *63*(5–6), 425–438. https://doi.org/10.1007/s11199-010-9818-y

Morris, G. P., Ramu, P., Deshpande, S. P., Hash, C. T., Shah, T., Upadhyaya, H. D., Riera-Lizarazu, O., Brown, P. J., Acharya, C. B., Mitchell, S. E., Harriman, J., Glaubitz, J. C., Buckler, E. S., & Kresovich, S. (2012). Population genomic and genome-wide association studies of agroclimatic traits in sorghum. *Proceedings of the National Academy of Sciences*, *110*(2), 453–458. https://doi.org/10.1073/pnas.1215985110

Moutawakilou, M. N. E. A., Chabi-Sika, J. K., Noumavo, A. D. P., Sina, H., Dossou, J., Baba-Moussa, L., & Baba-Moussa, F. (2023). Survey of cereal consumption habits in the community of Djougou, Benin. *Food and Nutrition Sciences*, *14*(9), 843–863. https://doi.org/10.4236/fns.2023.149054

Nambiar, V., Sareen, N., Daniel, M., & Gallego, E. (2012). Flavonoids and phenolic acids from pearl millet (*Pennisetum glaucum*) based foods and their functional implications. *Functional Foods in Health and Disease*, *2*(7), 251. https://doi.org/10.31989/ffhd.v2i7.85

Nani, M., & Krishnaswamy, K. (2021). Physical and functional properties of ancient grains and flours and their potential contribution to sustainable food processing. *International Journal of Food Properties*, *24*(1), 1529–1547. https://doi.org/10.1080/10942912.2021.1975740

Ndiaye, O., Goddard, L., & Ward, M. (2008). Using regional wind fields to improve general circulation model forecasts of July–September Sahel rainfall. *International Journal of Climatology*, *29*(9), 1262–1275. https://doi.org/10.1002/joc.1767

Negarestani, M., Tohidi-Nejad, E., Khajoei-Nejad, G., Nakhoda, B., & Mohammadi-Nejad, G. (2019). Comparison of different multivariate statistical methods for screening the drought tolerant genotypes of pearl millet (*Pennisetum americanum* L.) and sorghum (*Sorghum bicolor* L.). *Agronomy, 9*(10), 645. https://doi.org/10.3390/agronomy9100645

Newmaster, S., Ragupathy, S., Dhivya, S., Jijo, C., Ramalingam, S., & Patel, K. (2013). Genomic valorization of the fine scale classification of small millet landraces in Southern India. *Genome, 56*(2), 123–127. https://doi.org/10.1139/gen-2012-0183

Obahiagbon, E. G., & Ogwu, M. C. (2023). Consumer perception and demand for sustainable herbal medicine products and market. In S. C. Izah, M. C. Ogwu, & M. Akram (Eds.), *Herbal medicine phytochemistry.* Reference Series in Phytochemistry (pp. 1–34). Springer. https://doi.org/10.1007/978-3-031-21973-3_65-1

Obahiagbon, E. G., & Ogwu, M. C. (2024). Organic food preservatives: The shift towards natural alternatives and sustainability in the Global South's Markets. In M. C. Ogwu, S. C. Izah, & N. R. Ntuli (Eds.), *Food safety and quality in the Global South* (pp. 299–329). Springer. https://doi.org/10.1007/978-981-97-2428-4_10

Oberline, F. Y., Abdoul-Aziz, S., Gabriel, K., Orphé, B., Audebert, A., Bassirou, S., Daniel, F., & Hélène, J. (2020). Two contrasting patterns of crop seasonal adaptation revealed by a common garden experiment on flood recession sorghum in the Sahel. *Australian Journal of Crop Science, 14*(5), 871–879. https://doi.org/10.21475/ajcs.20.14.05.p2442

Obongodot, N. U., & Ogwu, M. C. (2023). Plant food for human health: Case study of Indigenous vegetables in Akwa Ibom State, Nigeria. In S. C. Izah, M. C. Ogwu, & M. Akram (Eds.), *Herbal medicine phytochemistry.* Reference Series in Phytochemistry (pp. 1–38). Springer. https://doi.org/10.1007/978-3-031-21973-3_2-1

Ogwu, M. C. (2019). Towards sustainable development in Africa: The challenge of urbanization and climate change adaptation. In P. B. Cobbinah & M. Addaney (Eds.), *The geography of climate change adaptation in Urban Africa* (pp. 29–55). Springer Nature. http://doi.org/10.1007/978-3-030-04873-0_2

Ogwu, M. C. (2023). Plants as monitors and managers of pollution. In A. L. Srivastav, A. S. Grewal, M. Tiwari, & T. D. Pham (Eds.), *Role of green chemistry in ecosystem restoration to achieve environmental sustainability* (pp. 51–60). Elsevier. https://doi.org/10.1016/B978-0-443-15291-7.00022-5

Onoabhagbe, O., Ogwu, M. C., & Ikhajiagbe, B. (2024). Germination characteristics of *Sorghum bicolor* (L.) Moench. under different pH regimes after chemo-priming. *VEGETOS—International Journal of Plant Research and Bio-technology, 37,* 1876–1886. http://doi.org/10.1007/s42535-024-00909-0

O'Regan, A., & Thompson, G. (2017). Indicators of young women's modern contraceptive use in Burkina Faso and Mali from demographic and health survey data. *Contraception and Reproductive Medicine, 2*(1). https://doi.org/10.1186/s40834-017-0053-6

Osawaru, M. E., & Ogwu, M. C. (2020). Survey of plant and plant products in local markets within Benin City and environs. In L. W. Filho, N. Ogugu, D. Ayal, L. Adelake, & I. da Silva (Eds.), *African handbook of climate change adaptation* (pp. 1–24). Springer. http://doi.org/10.1007/978-3-030-42091-8_159-1

Palit, P., Mathur, P., & Sharma, K. K. (2013). Pearl millet. In A. Pratap & J. Kumar (Eds.), *Alien gene transfer in crop plants* (pp. 75–83). Springer. https://doi.org/10.1007/978-1-4614-9572-7_4

Pereira, L. (2021). Follow the 'Ting: Sorghum in South Africa. *Food Culture and Society, 26*(1), 116–144. https://doi.org/10.1080/15528014.2021.1984631

Pereira, L. M., & Hawkes, C. (2022). Leveraging the potential of sorghum as a healthy food and resilient crop in the South African food system. *Frontiers in Sustainable Food Systems, 6.* https://doi.org/10.3389/fsufs.2022.786151

Phiri, S., Schoustra, S., Heuvel, J., Smid, E., Shindano, J., & Linnemann, A. (2019). Fermented cereal-based Munkoyo beverage: Processing practices, microbial diversity and aroma compounds. *PLOS One, 14*(10), e0223501. https://doi.org/10.1371/journal.pone.0223501

Pilia, S., Fontanelli, G., Santurri, L., Palchetti, E., Ramat, G., Baroni, F., Santi, E., Lapini, A., Pettinato, S., & Paloscia, S. (2025). Integration of optical and microwave satellite data for monitoring vegetation status in sorghum fields. *Remote Sensing, 17*(9), 1591. https://doi.org/10.3390/rs17091591

Pucher, A., Sy, O., Angarawai, I. I., Gondah, J., Zangre, R., Ouédraogo, M., Sanogo, M. D., Boureima, S., Hash, C. T., & Haussmann, B. I. G. (2015). Agro-morphological characterization of West and Central African pearl millet accessions. *Crop Science, 55*(2), 737–748. https://doi.org/10.2135/cropsci2014.06.0450

Raju, B., Rani, K. S., Gawande, K. N., Pandey, P. K., Nengparmoi, T., Reddy, K. B., Avinash, G., & Sharma, Y. (2024). A systematic review on the prospective significance and recommendations of pearl millet (*Pennisetum glaucum*) for diabetes mellitus. *Journal of Scientific Research and Reports*, *30*(3), 52–60. https://doi.org/10.9734/jsrr/2024/v30i31857

Salack, S., Sangaré, S. A. K. S. B., Daku, E. K., Hien, K., Sawadogo, A. M., Sanfo, S., & Ogunjobi, K. O. (2025). Crop-livestock-climate nexus: Intensification pathways under different climate realizations in the Sahel, West Africa. *Environmental Research Communications*, *7*(4). https://doi.org/10.1088/2515-7620/adc54a

Saleh, A., Zhang, Q., Chen, J., & Shen, Q. (2013). Millet grains: Nutritional quality, processing, and potential health benefits. *Comprehensive Reviews in Food Science and Food Safety*, *12*(3), 281–295. https://doi.org/10.1111/1541-4337.12012

Sarr, A. B., Camara, M., & Diba, I. (2018). Multi-model analysis of the West African monsoon: Seasonal evolution and the monsoon onset. *Journal of Scientific Research and Reports*, *20*(2), 1–17. https://doi.org/10.9734/jsrr/2018/43311

Satyavathi, C. T., Ambawat, S., Khandelwal, V., & Srivastava, R. K. (2021). Pearl millet: A climate-resilient nutricereal for mitigating hidden hunger and provide nutritional security. *Frontiers in Plant Science*, *12*. https://doi.org/10.3389/fpls.2021.659938

Sawadogo, W., Neya, T., Semdé, D., Korahiré, J. A., Combasséré, A. C. A., Do Etienne, T., Ouedraogo, P., Diasso, U. J., Abiodun, B. J., Bliefernicht, J., & Kunstmann, H. (2023). Potential impacts of climate change on the Sudan-Sahel Region in West Africa—Insights from Burkina Faso. https://doi.org/10.22541/essoar.169462048.86647744/v1

Sehgal, D., Skøt, L., Singh, R., Srivastava, R. K., Das, S. P., Taunk, J., Sharma, P. C., Pal, R., Raj, B., Hash, C. T., & Yadav, R. S. (2015). Exploring potential of pearl millet germplasm association panel for association mapping of drought tolerance traits. *PLOS One*, *10*(5), e0122165. https://doi.org/10.1371/journal.pone.0122165

Serba, D., & Yadav, R. (2016). Genomic tools in pearl millet breeding for drought tolerance: Status and prospects. *Frontiers in Plant Science*, *7*. https://doi.org/10.3389/fpls.2016.01724

Sermé, I., Ouattara, K., Bandaogo, A., & Wortmann, C. (2018). Pearl millet and sorghum yield response to fertilizer in the Sahel of Burkina Faso. *Journal of Agricultural Studies*, *6*(1), 176. https://doi.org/10.5296/jas.v6i1.12384

Shine, T., & Dunford, B. (2016). What value for pastoral livelihoods? An economic valuation of development alternatives for ephemeral wetlands in Eastern Mauritania. *Pastoralism Research Policy and Practice, 6*(1). https://doi.org/10.1186/s13570-016-0057-x

Singh, M., & Nara, U. (2022). Genetic insights in pearl millet breeding in the genomic era: Challenges and prospects. *Plant Biotechnology Reports, 17*(1), 15–37. https://doi.org/10.1007/s11816-022-00767-9

Singh, N., Bhardwaj, R., & Sohu, R. (2023). Studies on genetic variability and correlation in pearl millet inbred lines. *Agricultural Research Journal, 60*(6), 822–826. https://doi.org/10.5958/2395-146x.2023.00118.7

Sissoko, K., Keulen, H., Verhagen, A., Tekken, V., & Battaglini, A. (2010). Agriculture, livelihoods and climate change in the West African Sahel. *Regional Environmental Change, 11*(S1), 119–125. https://doi.org/10.1007/s10113-010-0164-y

Sithole, C., Sinthumule, R., Gaorongwe, J., Ruzvidzo, O., & Dikobe, T. (2025). Unraveling the complexities: Morpho-physiological and proteomic responses of pearl millet (*Pennisetum glaucum*) to dual drought and salt stress. *Frontiers in Plant Science, 16*. https://doi.org/10.3389/fpls.2025.1495562

Srivastava, R., Bollam, S., Pujarula, V., Pusuluri, M., Singh, R. B., Potupureddi, G., & Gupta, R. (2020). Exploitation of heterosis in pearl millet: A review. *Plants, 9*(7), 807. https://doi.org/10.3390/plants9070807

Srivastava, R. K., Singh, R. B., Pujarula, V. L., Bollam, S., Pusuluri, M., Chellapilla, T. S., Yadav, R. S., & Gupta, R. (2020). Genome-wide association studies and genomic selection in pearl millet: Advances and prospects. *Frontiers in Genetics, 10*. https://doi.org/10.3389/fgene.2019.01389

Stich, B., Haussmann, B. I. G., Pasam, R., Bhosale, S., Hash, C. T., Melchinger, A. E., & Parzies, H. K. (2010). Patterns of molecular and phenotypic diversity in pearl millet [*Pennisetum glaucum* (L.) R. Br.] from West and Central Africa and their relation to geographical and environmental parameters. *BMC Plant Biology, 10*(1), 216. https://doi.org/10.1186/1471-2229-10-216

Sultan, B., Defrance, D., & Iizumi, T. (2019). Evidence of crop production losses in West Africa due to historical global warming in two crop models. *Scientific Reports, 9*(1), 12834. https://doi.org/10.1038/s41598-019-49167-0

Swaffield, S., Corry, R., Opdam, P., McWilliam, W., & Primdahl, J. (2019). Connecting business with the agricultural landscape: Business strategies for sustainable rural development. *Business Strategy and the Environment*, *28*(7), 1357–1369. https://doi.org/10.1002/bse.2320

Tagesson, T., Fensholt, R., Guiro, I., Rasmussen, M., Huber, S., Mbow, C., Garcia, M., Horion, S., Sandholt, I., Holm-Rasmussen, B., Göttsche, F. M., Ridler, M.-E., Olén, N., Olsen, J. L., Ehammer, A., Madsen, M., Olesen, F. S., & Ardö, J. (2014). Ecosystem properties of semiarid savanna grassland in West Africa and its relationship with environmental variability. *Global Change Biology*, *21*(1), 250–264. https://doi.org/10.1111/gcb.12734

Taylor, J., Belton, P., Beta, T., & Duodu, K. (2014). Increasing the utilisation of sorghum, millets and pseudocereals: Developments in the science of their phenolic phytochemicals, biofortification and protein functionality. *Journal of Cereal Science*, *59*(3), 257–275. https://doi.org/10.1016/j.jcs.2013.10.009

Tornello, S., Sonnenberg, B., & Patterson, C. (2015). Division of labor among gay fathers: Associations with parent, couple, and child adjustment. *Psychology of Sexual Orientation and Gender Diversity*, *2*(4), 365–375. https://doi.org/10.1037/sgd0000109

Traoré, B., Moussa, A., Traore, A., Nassirou, Y., Ba, M., & Tabo, R. (2022). Pearl millet (*Pennisetum glaucum*) seedlings transplanting as climate adaptation option for smallholder farmers in Niger. *Atmosphere*, *13*(7), 997. https://doi.org/10.3390/atmos13070997

Traoré, B., Zemadim, B., Sangaré, S., Gumma, M., Tabo, R., & Whitbread, A. (2021). Contribution of climate-smart agriculture technologies to food self-sufficiency of smallholder households in Mali. *Sustainability*, *13*(14), 7757. https://doi.org/10.3390/su13147757

Turner, M. D., Ayantunde, A. A., Patterson, K. P., & Patterson, E. D. (2011). Livelihood transitions and the changing nature of farmer–herder conflict in Sahelian West Africa. *The Journal of Development Studies*, *47*(2), 183–206. https://doi.org/10.1080/00220381003599352

Upadhyaya, H. D., Reddy, K. N., Ahmed, M. I., & Gowda, C. L. L. (2012). Identification of gaps in pearl millet germplasm from East and Southern Africa conserved at the ICRISAT Genebank. *Plant Genetic Resources*, *10*(3), 202–213. https://doi.org/10.1017/s1479262112000275

Xiong, Y., Zhang, P., Warner, R., & Fang, Z. (2019). Sorghum grain: From genotype, nutrition, and phenolic profile to its health benefits and food applications. *Comprehensive Reviews in Food Science and Food Safety*, *18*(6), 2025–2046. https://doi.org/10.1111/1541-4337.12506

Yengoh, G. T. (2012). Climate and food production: Understanding vulnerability from past trends in Africa's Sudan-Sahel. *Sustainability*, *5*(1), 52–71. https://doi.org/10.3390/su5010052

Yodanis, C. (2005). Divorce culture and marital gender equality. *Gender and Society*, *19*(5), 644–659. https://doi.org/10.1177/0891243205278166

Zouré, C. O., Koïta, M., Niang, D., Baba, I. I., Yonaba, O. R., Dara, A. M., Fowé, T., Queloz, P., & Karambiri, H. (2017). Relationship between soil water content and crop yield under Sahelian climate conditions: Case study of Tougou experimental site in Burkina Faso. *Journal of Advances in Physics*, *13*(10), 5177–5184. https://rajpub.com/index.php/jap/article/view/6681

Zouré, C., Queloz, P., Koïta, M., Niang, D., Fowé, T., Yonaba, R., Consuegra, D., Yacouba, H., & Karambiri, H. (2019). Modelling the water balance on farming practices at plot scale: Case study of Tougou watershed in Northern Burkina Faso. *Catena*, *173*, 59–70. https://doi.org/10.1016/j.catena.2018.10.002

CHAPTER 11

Wild and Semidomesticated Fruits in the Amazon

Abstract

This chapter explores the sociobiological significance of wild and semidomesticated fruits in the Amazon, focusing on emblematic species such as açaí (*Euterpe oleracea*), Brazil nut (*Bertholletia excelsa*), and camu camu (*Myrciaria dubia*). These fruits are not just nutritional resources but are deeply embedded in intricate ecological, cultural, and spiritual systems supporting Indigenous Amazonian livelihoods. Drawing from ecological science, ethnobotany, and political ecology, the chapter highlights how these species are keystone components of multispecies agroforestry systems, contributing to biodiversity conservation, forest regeneration, and carbon sequestration. It examines the symbiotic relationships between humans, animals, and plants, demonstrating how traditional harvesting and processing practices maintain ecological balance while reinforcing intergenerational knowledge transmission. Central to this discussion are the spiritual and symbolic roles of fruits in ritual practices, oral traditions, and forest cosmologies, which frame the forest not as wilderness but as a culturally inhabited and managed landscape. The chapter also interrogates the sociopolitical tensions surrounding the commodification of non-timber forest products, including concerns about intellectual property, cultural appropriation, and the erosion of local food sovereignty. Case studies, such as the açaí export boom and Brazil nut trade, illustrate market integration opportunities and risks. Ultimately, the chapter advocates for decolonial policy approaches that foreground Indigenous knowledge systems and reinforce the ecological and cultural agency of Amazonian communities. These fruits and the knowledge systems sustaining them are critical to envisioning sustainable and just futures in the Amazon.

Keywords: wild fruits, agroforestry, Indigenous knowledge, non-timber forest products, food sovereignty, decolonial ecology, cultural resilience

11.1 Introduction

The Amazon rainforest represents a biodiversity hotspot with a significant wealth of wild and semidomesticated fruits, providing crucial nutritional, ecological, and economic resources for local communities and beyond. Among these, fruits such as açaí (Euterpe oleracea), Brazil nut (Bertholletia excelsa), and camu camu (*Myrciaria dubia*) hold historic and cultural importance, influencing food systems and traditional practices. The sociobiological lens provides insight into the relationships among food, forest vitality, and cultural systems, emphasizing how humans interact with and manage these natural resources. Açaí, often celebrated for its antioxidant properties, has surged globally, particularly within health-conscious markets (Izah et al., 2023; Yang, 2009). Its cultivation offers significant economic opportunities for local communities within the Amazon, as its production and commercialization have evolved due to increasing demand (Yang, 2009). The historical significance of açaí is deeply rooted in the dietary practices of Indigenous populations, reflecting a blend of traditional knowledge and modern economic models (Ribeiro et al., 2014). Combined with the ecological understanding of its growth in floodplain environments, açaí cultivation exemplifies the interplay between natural resources and human sustenance (Faruk et al., 2023).

The Brazil nut, recognized as one of the primary non-timber forest products in the Amazon, carries significant socioeconomic implications (Kluczkovski et al., 2023). It is not merely a food source but is also a vital component of local and national economies (Yang, 2009). Studies indicate that Brazil nuts thrive in specific forest types, particularly as parts of anthropogenic landscapes managed by Indigenous peoples, who have historically cultivated these trees to enhance biodiversity and maintain ecological balance (Ribeiro et al., 2014). Higher yields are often linked to practices that respect traditional ecological knowledge, which emphasizes sustainability in harvesting practices, ensuring the long-term viability of both the species and the associated ecosystems (Yang, 2009). These nuts are rich in selenium, which contributes to their health benefits and dietary significance (Carvalho et al., 2015). This makes them integral to local and global nutrition (Silva et al., 2017). Camu camu is another prominent fruit native to the Amazon, renowned for its extraordinarily high vitamin C content (Kluczkovski et al., 2023). This fruit has garnered interest for its nutritional value and potential in dietary supplements and functional foods (Faruk et al., 2023). The sociocultural context surrounding camu camu highlights its significance in local diets and traditional medicine (Faruk et al., 2023). Indigenous practices surrounding the harvesting and utilization of camu camu exemplify a harmonious relationship between human communities and their natural environment, wherein sustainability is central to their usage patterns (Ribeiro et al., 2014). Furthermore, the cultivation and marketing of camu camu can create economic incentives for conservation, promoting the preservation of biodiversity within the Amazon.

When viewed through a sociobiological lens, the interplay between wild and semidomesticated fruits such as açaí, Brazil nut, and camu camu reveals dynamic relationships within ecosystems and human cultures. These Indigenous fruits contribute to nutritional security and cultural identity, fostering community and continuity among Indigenous peoples (Silva et al., 2017). As modern pressures like deforestation and climate change threaten these ecosystems, integrating traditional ecological knowledge and sustainable practices becomes increasingly vital for conserving and managing these resources (Ribeiro et al., 2014). Additionally, the global market for these fruits underscores the implicit necessity for balanced economic development that respects both cultural heritage and ecological integrity (Yang, 2009). In considering the impacts on ecological health and food systems, it is crucial to recognize the multifaceted roles these fruits play. They do not exist in isolation but are deeply intertwined with the forest ecosystems from which they derive. Sustainability in harvesting and cultivation, informed by Indigenous practices, can help ensure that these vital resources are preserved for future generations. Understanding the ecological processes that underpin the growth and reproduction of these fruits offers insights into effective conservation strategies that are culturally sensitive and ecologically sound (Carvalho et al., 2015). Through these frameworks, we can appreciate the broader implications of wild and semidomesticated fruits in the Amazon. Their roles extend beyond mere food sources to become symbols of Indigenous knowledge, ecological diversity, and economic potential in the face of pressing environmental challenges. The ongoing relationship between these fruits and the communities that rely on them provides a unique perspective on the broader narratives of biodiversity conservation, cultural resilience, and sustainable livelihoods (Silva et al., 2017).

This chapter aims to shed light on the sociobiological significance of wild and semidomesticated fruits in Amazonian Indigenous societies, focusing on ecologically vital and culturally revered species like açaí, Brazil nut, and camu camu. By integrating ecological science, ethnobotany, and cultural anthropology, the chapter emphasizes how these fruits are intertwined in multispecies relationships, food cosmologies, and traditional forest-based livelihoods. It examines how Indigenous agroforestry practices—often overlooked in dominant conservation paradigms—reflect complex ecological knowledge systems that encourage biodiversity, resilience, and long-term sustainability. The chapter also highlights the symbolic and spiritual dimensions of fruit cultivation and consumption, stressing their roles in ritual practices, oral traditions, and community cohesion. Additionally, it critically assesses the sociopolitical dynamics surrounding the commercialization of non-timber forest products (NTFPs), pointing out tensions between global market demand, cultural integrity, and forest stewardship. This chapter contributes to broader discourses on food sovereignty, ecological justice, and sustainable development in the Amazon by centering Indigenous perspectives and knowledge. It advocates for policy frameworks that acknowledge these fruits' economic value and safeguard the rights, cosmologies, and ecological agency of the communities that have long nurtured them.

11.2 Ecological Adaptations and Wild Multispecies Interactions in the Amazon

The Amazon rainforest is renowned for its unmatched biodiversity and ecological intricacy; it offers a plethora of fruit species that are essential both ecologically and economically. It is estimated that over 14,000 species inhabit the lowland Amazon rainforests, accounting for between 3.8% and 4.7% of all known seed plant species worldwide (Cardoso et al., 2017). Among these, various fruit species such as *Bactris gasipaes* (peach palm), *Bertholletia excelsa* (Brazil nut), and *Colossoma macropomum* (tambaqui) stand out for their ecological interactions and economic value. These fruits nourish local communities and play critical roles in intricate multispecies interactions involving animals, pollinators, and other organisms within their ecosystems. Tropical fruit species like the peach palm have been recognized for producing highly nutritious fruits with significant market potential and culinary uses, often enjoyed in the region (Costa et al., 2022). This palm and its fruits are not merely a dietary staple but also bolster local economies by being cultivated in agroforestry systems (Spacki et al., 2022). Nevertheless, the commercial viability of these species largely depends on sustaining the diverse array of pollinating species crucial for their reproductive success (Sales et al., 2020).

Pollination is a critical ecological service within the Amazon ecosystem, and studies have shown that fruit species often rely on specific pollinators that exhibit a degree of specialization (Cavalcante et al., 2012). For instance, Brazil nut trees primarily depend on specific bee species that visit these flowers for nectar, facilitating pollination (Paz et al., 2021). Research highlights a vibrant and diverse community of pollinators, with bees and beetles as the primary agents responsible for fruit set, thus influencing the reproductive outcomes of these native fruit species (Barros et al., 2024). Indeed, fruit yields would plummet without these pollinators, illustrating the interconnectedness of floral reproduction and animal behavior in this biodiverse habitat. Seed dispersal is another vital function in the Amazon rainforest, actively involving frugivorous animals such as fish and birds that frequently interact with fruiting plants throughout the floodplain ecosystems (Lucas, 2008). The tambaqui fish, for example, demonstrates significant seed dispersal capabilities as it consumes various fruits, thereby aiding in the regeneration of forested areas and maintaining ecological balance (Lucas, 2008). These dynamic interactions emphasize the complexity of the Amazon's ecological web, where each species plays a critical role in sustaining the habitat and promoting biodiversity.

The interplay between human communities and these fruit species further exemplifies symbiosis within the Amazon ecosystem. Local populations utilize over 220 species of edible fruits that contribute not only to nutrition but also to cultural practices and traditional medicine (Pires et al., 2019). Medicinal plants from families such as Fabaceae remain fundamental for health management in local communities, demonstrating how human interaction with fruit species influences biodiversity preservation and utilization (Júnior & Souza, 2021). Additionally, agricultural practices have evolved to integrate sustainable management of fruit-bearing species, evidencing a robust understanding of ecological practices among Indigenous and local farmers (de Souza, 2023).

As climate conditions continue to change, the resilience of these fruit species and their respective ecosystems is often put to the test. Research indicates that ongoing shifts in environmental conditions can threaten the unique multispecies interactions critical to forest health and regeneration (Sales et al., 2020). The adaptive capacities of various species to cope with these changes are vital, particularly as the frequency of extreme weather events increases. Understanding how specific fruit species, such as those in the Moraceae family, respond to climate stressors is crucial for developing conservation strategies in the face of biodiversity loss (García-Cox et al., 2023). Given the rich diversity of flora found within the Amazon, higher conservation priority should be assigned to those regions that exhibit both high species richness and important ecosystem functions related to fruit production (Catenacci & Simon, 2017). Effective management practices that protect pollinator habitats alongside sustainable fruit species harvesting can significantly enhance biodiversity conservation and community livelihoods, strengthening ecological and economic resilience (Borges et al., 2020).

11.3 Indigenous Agroforestry Systems and Foraging Practices in Amazonian Societies

Indigenous agroforestry systems in the Amazon represent a remarkable integration of biodiversity and cultural practices that have evolved over millennia. These systems can be categorized into various typologies, notably including home gardens and chagras, which exhibit significant agrobiodiversity and complex management strategies. The chagras, as defined by local Indigenous communities, represent a polycultural mode of agriculture that provides food and sustains cultural identity through the careful management of time and space (Marentes et al., 2021). This is consistent with findings from the Ecuadorian and Colombian Amazon, where agroforestry systems are recognized for their role in producing essential food items and contributing to household incomes (Bucheli et al., 2021). The home gardens of Amazonian societies serve as another example of Indigenous agroforestry. These systems integrate a variety of plant species, often reaching high levels of biodiversity; for instance, a study recorded 205 species in just 40 home gardens (Lins et al., 2015). These multifunctional gardens contribute to food security and maintain traditional ecological knowledge (TEK) that fosters sustainable practices (Jung, 2025). Indigenous practices in home gardens embody a deep understanding of local ecosystems and contribute to resilience against climate fluctuations (Jung, 2025).

The intentional cultivation and preservation of wild and semidomesticated species characterize management practices within these agroforestry systems. For instance, species such as guarana, which are central to the cultural and economic lives of various Indigenous tribes, are interspersed with other crops in biodiverse agroforestry setups (Vignoli et al., 2022). This approach has been shown to enhance food security and contribute to the sustainability of natural resources and biodiversity conservation (Díaz-Ricaurte et al., 2020). The cooperative management of these systems helps ensure that essential ecosystem services—including pollination, pest control, and nutrient cycling—are upheld (Smith et al., 2012). The dynamics of foraging in Amazonian societies exemplify the integration of seasonal cycles, mobility, and spatial knowledge. Seasonally driven resource availability compels Indigenous communities to develop intricate knowledge concerning the timing and location of foraging activities (Bucheli & Bokelmann, 2017). This knowledge is crucial for the sustainable management of both cultivated and wild resources, demonstrating a nuanced interaction with the surrounding environment. Such foraging practices underscore the importance of preserving ecological integrity while facilitating a diverse procurement strategy, which contributes to dietary diversity and resilience (Smith et al., 2012).

Moreover, the contributions of these Indigenous agroforestry systems to biodiversity conservation and carbon sequestration are significant. Therefore, by maintaining diverse plant and animal species within traditional agroforestry landscapes, Indigenous practices promote biodiversity and provide habitats that support various life forms (Doddabasawa et al., 2018). The integrated approach of agroforestry facilitates effective carbon sequestration, as dense, biodiverse plantings can absorb substantial amounts of carbon from the atmosphere, aiding in climate change mitigation (Smith et al., 2012). Studies indicate that traditional systems, when properly managed, can outperform conventional agricultural methods in terms of carbon uptake and ecosystem service provision (Villani et al., 2017). The intergenerational transfer of TEK is crucial in ensuring that these systems continue to thrive amid modernization and environmental change (Marentes et al., 2021). For instance, in the Sibundoy Valley, the integration of biodiversity into local livelihoods has shown that agroforestry systems are vital for food security and essential for sustaining cultural practices (Bucheli & Bokelmann, 2017). This supports the argument for acknowledging Indigenous peoples as stewards of the forests, fostering greater collaboration between state and Indigenous entities in conservation efforts (Albuquerque, 2020; Osawaru et al., 2013). Thus, the Amazon presents a compelling case study for the efficacy of Indigenous agroforestry systems in fostering ecological balance, community resilience, and cultural preservation. Future research must prioritize integrating Indigenous knowledge and practices within contemporary environmental management frameworks to enhance the sustainability of these crucial systems (Doddabasawa et al., 2018; Ogwu, 2023). Incorporating such practices could offer a pathway toward addressing the dual challenges of biodiversity loss and climate change, ensuring that the invaluable ecological legacies of Indigenous peoples are preserved for future generations.

11.4 Knowledge Systems Associated with Wild and Semidomesticated Fruits in the Amazon

Various sociocultural factors profoundly shape the intricate relationship between knowledge systems associated with wild and semidomesticated fruits in the Amazon, including oral traditions, kinship, and apprenticeships. These dimensions are crucial for transmitting ecological knowledge, particularly among Indigenous populations who rely heavily on biodiversity and natural resources for their subsistence. Heckenberger and Neves (2009) highlight how Indigenous knowledge in the Brazilian Amazon fosters a deep ecological understanding and cultural identity, asserting that preserving these knowledge systems is vital to resource management and biodiversity conservation. This encompasses oral traditions where stories and practices are passed down through generations, providing a framework for understanding and appreciating ecological relationships with fruit-bearing plants.

Moreover, kinship ties play a poignant role in sharing this ecological knowledge, ensuring that practices around the harvesting and processing of fruits are imbued with cultural significance, which is often gender-specific. Among the Maijuna communities, specific roles related to the harvesting of *Mauritia flexuosa* are gendered, with men broadly engaging in hunting and women in gathering fruits, thus not only delineating a clear division of labor but also reinforcing community bonds and cultural narratives surrounding these practices (Gilmore et al., 2013). Apprenticeship, particularly in the context of the Indigenous socio-ecological systems, serves as a critical mechanism through which younger generations acquire knowledge on the ecological functions of these fruits, enabling the continuation of both practical and theoretical aspects of their traditional ecological understanding (Ogwu, 2019, 2020; Schmidt et al., 2021).

In terms of gender dynamics, the roles of women and men in harvesting and processing wild fruits differ significantly, reflecting broader social structures within Amazonian Indigenous societies. Women often serve as the primary gatherers and processors of fruits, placing them in a critical role within their economic and nutritional systems (Pilnik et al., 2023). Their extensive knowledge of fruit species is practical and intertwined with storytelling traditions that convey cultural history and ecological understanding, exemplifying how traditional storytelling serves as a vessel for complex environmental wisdom. Rosa et al. (2021) have noted the profound emotional and cultural connections that different gender groups maintain with their environmental practices, emphasizing the need for research to account for these sociocultural differences. The integration of fruits into local pharmacopeias and nutritional ontologies is another pivotal aspect influenced by Indigenous ecological knowledge. Many wild fruits are recognized for their medicinal properties and nutritional benefits, embedded within the local practices and beliefs of Indigenous peoples across the Amazon (Tenea et al., 2021). For instance, the camu camu fruit, known for its high vitamin C content and presence of various phytochemicals, is essential not just as a food source but also as a traditional remedy, illustrating the crucial intersection between biodiversity, nutrition, and health within local knowledge systems (Castro et al., 2018; Osawaru & Ogwu, 2020). These practices are sustained through forms of traditional ecological knowledge that emphasize the importance of diverse fruit species in local diets and their contributions to health and wellness in the context of food sovereignty.

Furthermore, the socioeconomic potential of these wild fruit species is increasingly recognized, leading to initiatives that promote the sustainable harvesting of non-timber forest products (NTFPs) in conjunction with conservation strategies. Pilnik et al. (2023) argue that traditional botanical knowledge facilitates sustainable management practices that align local economies with conservation efforts, thereby providing an economic incentive to preserve the rich biodiversity associated with these fruits. This kind of integration is vital for developing agroecological strategies that support both the livelihoods of local communities and the health of the ecosystems they depend on (Lagneaux et al., 2021; Imarhiagbe & Ogwu, 2022; Imarhiagbe et al., 2022). The relationship between Indigenous people and the semidomesticated fruit species is equally noteworthy. These fruits often occupy a transitional space that reflects the historical interactions between Indigenous agricultural practices and natural landscapes. Levis et al. (2018) noted that the domestication of Amazonian forests involved both the preservation of wild species and their integration into managed landscapes, showcasing a dynamic interplay between human agency and ecological systems. Moreover, wild fruits often exhibit greater genetic diversity than domesticated varieties, which can be crucial for future breeding programs to enhance food security, particularly in climate change (Adin et al., 2004).

This reflection on ecological knowledge associated with wild and semidomesticated fruits in the Amazon ultimately underscores a critical narrative of cultural resilience. The passing of knowledge through generations serves as a bulwark against the pressures of globalization and environmental degradation, reinforcing cultural identity and sustainability through economic and subsistence-oriented practices. Freitas et al. highlighted that the transfer of traditional ecological knowledge is an essential process for the resilience of both communities and ecosystems within the Amazon, emphasizing the relevance of cultural heritage in managing ecological landscapes (Freitas et al., 2015). As the future unfolds regarding environmental change and socioeconomic challenges, the role of traditional ecological knowledge in conserving biodiversity and fostering sustainable practices will remain increasingly vital. Collective efforts to recognize and integrate these knowledge systems in broader conservation policies can enable a more holistic approach to environmental management that honors and elevates the inherent wisdom of Indigenous populations (Miller & Nair, 2006; Moegenburg & Levey, 2002). The continued focus on the sociocultural dimensions surrounding wild and semidomesticated fruits can illuminate pathways toward sustainable development that also respect and affirm the rights of Indigenous peoples within their ecological contexts (Neves & Heckenberger, 2019).

11.5 Symbolism, Cosmologies, and Ritual Practices Associated with Wild and Semidomesticated Fruits in the Amazon

The intricate relationship between wild and semidomesticated fruits in the Amazon is deeply embedded in the cultural, spiritual, and ecological fabrics of the region. These fruits play a crucial role in spiritual practices and forest cosmologies. Many Amazonian tribes perceive fruits as offerings to forest spirits and sacred trees, reflecting a belief system that sees the natural world as interconnected with the spiritual realm. Such practices often involve rituals where fruits are presented to infer divine favor or to maintain harmony between humans and nature. For instance, in the northwestern Amazon, entities of the forest, land, water, and air are referred to in familial terms, such as grandparents or brothers, demonstrating a profound kinship with these beings (Rezende, 2024). Rituals conducted before harvesting fruits are commonly performed to seek permission from these spirits, ensuring that the gathering is respectful and in balance with nature (Rezende, 2024).

Furthermore, sacred trees, which are often associated with particular fruits, symbolize life, continuity, and ancestral heritage. The offerings made at these trees highlight the importance of fruit-bearing plants' importance as food sources and as integral aspects of the community's cosmology. In this context, fruits embody a life force, linking the material and spiritual worlds. As shown by the Baniwa people, spiritual narratives around the pandemic invoked a return to traditional respect for the spirits that inhabit the environment, illustrating how fruit and plant-based rituals can surface in public health narratives and cultural meanings (Wright, 2022).

The ethnographic insights into fruit-related myths, taboos, and festivals provide a vivid narrative of how Indigenous cultures in the Amazon experience and celebrate these natural offerings. Festivals often coincide with fruit harvesting periods, transforming the act of gathering into a collective celebration of abundance. Such gatherings not only reinforce social ties but also reaffirm cultural heritage by cherishing traditional knowledge passed down through generations. Rituals during these festivals often include songs, dances, and food sharing, where specific fruits symbolize community bonding and social cohesion. The intergenerational transfer of knowledge about fruit harvesting, preparation, and the significance of various species enriches the community dynamic, allowing for interconnectedness among individuals.

Additionally, the social meaning of fruit exchanges, feasts, and kin relations is profound. In many Indigenous cultures, sharing fruits can signify goodwill between families or tribes, fostering alliances and reinforcing kinship bonds. The ritual aspects associated with fruit sharing often transcend mere sustenance; they symbolize ecological relationships and cultural continuity. Moreover, the management of fruit resources, as examined by Moegenburg (2002), shows that human interaction with the Amazon's fruiting plants has significant implications for other species that rely on these fruits, leading to broader ecological ramifications. Understanding these relationships is essential for sustainable practices and reflects the delicate balance that Indigenous populations maintain with their environment. The fruit harvest is an activity that activates spirits and ancient narratives that speak of fertility, renewal, and sustenance. These narratives underscore environmental stewardship, tying the survival of fruit plants directly to moral obligations held by the community toward the ecosystem. There is a recognition that disrespect toward these spirits or improper harvesting practices can lead to diminished fruit yields, affecting both spiritual and physical well-being (Wright, 2022). This reciprocity emphasizes the dual role of fruits as both a resource and a medium of spiritual engagement, intertwining ecological knowledge with ethical retribution.

Moreover, it is noteworthy how the introduction of external influences, such as contemporary agricultural practices and religious shifts, can disturb traditional fruit-related rituals. The conversion of Indigenous peoples to Christianity has often led to a decline in rituals associated with fruit offerings, which were previously designed to maintain active relationships with nature spirits (Stanley, 2020). Such transformations challenge the sustainability of traditional practices and the ecological balance that these Indigenous narratives have historically upheld. The current context necessitates reevaluating these relationships and visualizing how sacred connections to the land can be integrated with modern conservation efforts. The importance of understanding the relationship between fruits and ecological practices is bolstered by the knowledge that many fruits are crucial not only for human consumption but also for the broader ecosystem, enhancing biodiversity and sustaining wildlife that depend on these plants

(Paz et al., 2021). These interactions shape a unique ethnobotanical landscape, wherein fruits are at the center of both economic exchange and cultural identity. The acknowledgment of fruits as "guardians" of traditional knowledge links cultural practices with ecological sustainability, anchoring these practices deeply within the community's identity and legacy (Rezende, 2024).

Communities that embrace these practices often demonstrate resilience in the face of environmental degradation, relying on their rich tradition of viewing the natural world as a complex interplay of interdependent relationships. The ongoing management and harvesting of fruits signify an enduring connection to ancient practices that ensure survival and foster a cultural ethos centered on respect and reciprocity towards nature. Ultimately, preserving fruit-related rituals and their associated cosmologies is as essential as conserving the Amazon's rich biodiversity.

11.6 Forest Economies and the Politics of Non-Timber Forest Products

The increasing prominence of non-timber forest products (NTFPs) in both re-gional and global markets reflects a significant shift in the perception of forests from mere timber sources to multipurpose resource systems capable of providing various ecological, economic, and social benefits. Non-timber forest products encompass a broad range of goods such as fruits, nuts, resins, medicinal plants, and fibers, which are essential for local livelihoods and are becoming significant players in international trade networks. This trend is supported by various studies exploring NTFPs' diverse applications and their growing market relevance, which highlights their potential for generating income for rural communities (Meinhold & Darr, 2019; Thapa & Singh, 2023). The economic benefits derived from NTFPs can be instrumental in alleviating poverty, supporting local development, and promoting environmental conservation when managed sustainably (Belcher & Schreckenberg, 2007; Mukul et al., 2015).

However, the burgeoning commercialization of NTFPs does not occur in a vacuum; it is fraught with tensions arising from competing interests in conser-vation, local livelihoods, and Indigenous rights. The dual pressure of market demands and conservation objectives often creates conflicts that must be navigated judiciously. Successful integration of NTFPs into the local economy requires consideration of the social and ecological contexts in which they are produced (He et al., 2014; Ramadhan et al., 2022). Efforts to promote NTFP commercialization must ensure that the income generated supports local harvesters' livelihoods and broader conservation goals, fostering a symbiotic relationship between resource use and ecological sustainability (Asamoah et al., 2024). Case studies, such as the açaí boom in Brazil and the Brazil nut trade, provide practical insights into the complexities surrounding NTFPs. The açaí berry, once a modest product, is now a global health-food phenomenon, illustrating the transformative potential of NTFPs in local economies (Olawuyi et al., 2021). However, the commercialization of açaí raises critical questions about environmental sustainability and social equity, particularly regarding the rights of Indigenous communities who depend on these resources for their livelihoods. Similarly, the Brazil nut trade highlights the delicate balance between exploitation and conservation; unsustainable harvesting practices threaten the ecosystems that support these valuable resources (Aluko et al., 2020; Belcher & Schreckenberg, 2007). Local communities often confront challenges related to equitable benefit-sharing and overexploitation of natural resources.

The legal landscape surrounding NTFPs is complicated by intellectual property and biopiracy issues, necessitating a deep examination of benefit-sharing frameworks. Indigenous communities frequently face challenges in asserting their rights to the genetic resources they have historically managed, particularly when external entities seek to commercialize these resources without equitable compensation (Magry et al., 2023; Rambey et al., 2024). Developing policies that protect Indigenous knowledge and ensure fair distribution of economic benefits is critical for fostering trust and cooperation among stakeholders. Such frameworks must address the intersection of conservation, economic viability, and social justice to accommodate the diverse interests involved in NTFP commercialization. In reviewing the factors shaping NTFP markets, it becomes evident that successful management relies on understanding local ecological knowledge and community involvement in decision-making processes. Research indicates that regions where local communities actively participate in managing their forest resources tend to experience more sustainable outcomes (Akpan et al., 2023; Shobika et al., 2021). Addressing the socioeconomic contexts in which NTFPs are embedded is essential for fostering resilience among rural households and ensuring that economic activities align with conservation efforts (Gelan, 2023).

Given the complexities outlined above, the multifaceted nature of NTFPs requires a range of adaptive strategies to enhance sustainable use while promoting local livelihoods. Researchers, policymakers, and practitioners must collaborate to develop innovative NTFP management approaches that prioritize ecological health and community well-being (Ahenkan & Boon, 2011; Miina et al., 2020). These strategies should integrate traditional practices with modern market dynamics, thereby creating pathways for sustainable commercialization that benefits both local communities and global consumers (Nguyễn et al., 2020; Sundriyal, 2021). Education and capacity-building initiatives are crucial for equipping local actors with the skills and knowledge required to navigate the commercial landscape effectively. Empowerment, particularly of marginalized groups such as women, should be a focal point of intervention strategies, as studies indicate the substantial role NTFPs can play in fostering women's economic independence and community development (Mortimer et al., 2012; Olawuyi et al., 2021).

11.7 Challenges and Opportunities in a Changing Amazon

The Amazon rainforest represents one of the most biodiverse ecosystems on the planet, yet it faces multifaceted challenges and opportunities stemming from land-use change, climate change, and varying governance strategies.

11.7.1 Land-Use Change and Deforestation Pressures

Deforestation remains one of the most pressing threats to the Amazon rainforest. Indigenous territories, which comprise over 20% of Brazil's Amazon, have notably provided a buffer against large-scale deforestation. Studies indicate that these territories maintain higher biodiversity levels compared to non-Indigenous areas, which are often subjected to extractive practices like agriculture, mining, and logging (Tourneau, 2015; Walker et al., 2020). While Indigenous stewardship has shown to mitigate some of the adverse effects of deforestation, these territories are simultaneously experiencing pressures from economic development and social changes that threaten their integrity (Tourneau, 2015). Furthermore, recent projections suggest that, under current land-use scenarios, approximately 36% of Amazonian species could be classified as globally threatened by 2050 due to the compounded impacts of deforestation and climate change (Feeley & Silman, 2016). Protecting the remaining forests is critical not only for conserving species but also for sustaining the ecological services provided by the Amazon, such as carbon sequestration, which is essential in combating climate change. Agricultural expansion is mainly responsible for deforestation, with policies often favoring short-term economic gain over long-term sustainability. This scenario is exacerbated by global commodity markets that drive deforestation as land is cleared for soybean cultivation and cattle ranching (Feeley & Silman, 2016; Tourneau, 2015). Indigenous communities possess traditional ecological knowledge and are crucial in implementing sustainable land-use practices. However, with the increasing encroachment of capitalist interests, balancing economic aspirations with the urgent need for environmental conservation is the challenge.

11.7.2 Climate Change and Its Impacts on Biodiversity

Climate change is reshaping the ecological landscape of the Amazon, with far-reaching implications for biodiversity. Altered precipitation patterns and rising temperatures are contributing to a phenomenon known as "savannization," where tropical forests convert to degraded savannah-like ecosystems. This shift poses a significant risk to high-diversity species, particularly mammals that rely on intact forest habitats (Rocha & Sollmann, 2023). The climatic alterations have been documented to disrupt established ecological relationships, threatening food webs and species persistence (Braz-Mota & Val, 2024; Röpke et al., 2017). Additionally, the future of prominent fruit species in the Amazon is at risk due to climate change. For example, studies have demonstrated that the iconic Brazil nut tree is projected to face serious challenges, potentially leading to localized extinction by the end of the century unless climate-related mitigation strategies are implemented (Anjos et al., 2024). The cascading effects of losing keystone species like the Brazil nut can disrupt not only the ecological balance but also the socioeconomic practices of Indigenous communities that rely on these resources.

11.7.3 Pathways for Sustainability Through Indigenous Stewardship and Adaptive Governance

The future of the Amazon hinges on adaptive governance strategies that integrate Indigenous knowledge and practices alongside contemporary scientific approaches to conservation. As evidenced in various studies, Indigenous territories in the Amazon are not only physical spaces but also represent dynamic cultural landscapes where traditional ecological knowledge is pivotal in fostering resilience against climate change (Tourneau, 2015; Walker et al., 2020). Such stewardship demonstrates remarkable efficacy in preventing deforestation and preserving intricate ecological frameworks. Indigenous practices often include rotational agriculture, sustainable harvesting, and habitat preservation, which collectively contribute to ecosystem integrity. The implementation of adaptive governance frameworks that recognize and empower Indigenous land management strategies presents an opportunity for holistic management of Amazonian ecosystems. Governance models must adapt to the current realities of rapid climate changes and socioeconomic pressures. More proactive measures, such as establishing flexible protected areas that adjust over time in response to ecological dynamics, are essential for ensuring biodiversity conservation (Frederico et al., 2021; Ogden, 2019).

Furthermore, international cooperation and policy frameworks need to emphasize integrating economic development with ecological stewardship. Collaborative efforts at local, regional, and global levels are critical in promoting sustainable practices that not only protect the environment but also support the livelihood of local communities (Walker et al., 2020). Efforts to promote agroecological farming can be fostered, leveraging both traditional knowledge and scientific innovation to create resilience against climate change impacts while sustaining local economies. Adopting a participatory approach that involves all stakeholders, including Indigenous communities, governmental bodies, and civil society, will enhance decision-making processes and lead to more equitable and effective conservation outcomes. Such strategies can contribute significantly to achieving not only ecological restoration but also the cultural revitalization of Indigenous nations across the Amazon.

11.8 Sociobiological Reflection on Wild and Semidomesticated Fruits in the Amazon

The Amazon rainforest, renowned for its biodiversity, houses a rich array of wild and semidomesticated fruits that play integral roles in the sociobiological and nutritional frameworks of local communities. This exploration focuses on the implications of these fruits from biological, ecological, and sociocultural perspectives, particularly regarding their domestication, cultivation, and the influence of local practices and perceptions. The unique constellation of environmental conditions in the Amazon has fostered a remarkable variety of fruit species, many of which are essential not only for local diets but also for the sustainable management of ecosystems.

The genetic diversity of Amazonian fruits is noteworthy, as many species undergo gradual processes of domestication influenced by historical and ongoing human activity. For instance, the peach palm (*Bactris gasipaes*) demonstrates substantial physical and nutritional characteristics that enhance its value as a functional food, reflecting traditional agricultural practices aimed at ensuring sustainable yields through intercropping systems within agroforestry environments (Clément et al., 2015; Costa et al., 2022). Furthermore, research on the domestication of fruit species such as Inga edulis highlights that, despite their cultivated status, many still retain significant genetic variation from their wild relatives, which is critical for the resilience and adaptability of these crops to environmental changes (Clément et al., 2010; Rollo et al., 2020). Wild fruits like bacuri (*Platonia insignis*) and camu camu (*Myrciaria dubia*) exemplify the underutilization of Amazonian flora in broader food systems. The bacuri tree, though valued for its delicious fruit, faces challenges such as slow germination and genetic self-incompatibility in propagation (Alves et al., 2025; Marinho et al., 2022). Camu camu, conversely, stands out for its health-promoting phytochemicals, including high vitamin C content, which confer various health benefits and underline the nutritional importance of Indigenous species in combating modern health issues (Castro et al., 2018). The juxtaposition of these fruits illustrates not only their cultural significance but also the imperative need for conservation efforts that recognize both wild genetic stocks and cultivated varieties in local diets (Duarte et al., 2024; Sánchez-Capa et al., 2023).

Ecologically, the relationship between human management and the hybridization of fruit species is evident, as seen with the domesticated tree gourd (*Crescentia cujete*) and its wild counterpart (*C. amazonica*). This relationship underscores the important role that local communities play in shaping fruit characteristics through selective harvesting and cultivation methods (Moreira et al., 2017a, 2017b). The preservation of these practices is crucial, as they foster genetic diversity and contribute to the socioeconomic well-being of local populations. Moreover, the perception of wild and cultivated biodiversity in the Amazon influences dietary choices, with community preferences often dictating harvesting and conservation strategies (Machado et al., 2021). The introduction of processed foods into traditional diets marks a significant nutritional transition affecting fruit consumption patterns. As rural communities increasingly rely on ultra-processed foods, the role of Indigenous fruits is threatened, warranting urgent attention to sustain the health benefits associated with these traditional dietary components (Machado et al., 2021). The complexities of food habits in the Amazon necessitate a culturally nuanced approach to nutritional science, acknowledging both the historical significance and contemporary challenges faced by these fruit species.

Moreover, bioactive compounds found in wild Amazonian fruits have garnered interest for their potential in addressing metabolic syndrome and other health-related issues. The examination of fruits such as Arazá (*Eugenia stipitata*) and Amazon grape (*Pourouma cecropiifolia*) illustrates their potential applications in functional food development, serving both traditional and modern nutritional needs (Duarte et al., 2024; Sánchez-Capa et al., 2023). These fruits harbor a variety of phytochemicals that can bolster health and mitigate chronic diseases, further emphasizing the importance of preserving Indigenous knowledge regarding their uses. The evolution of fruit cultivation practices in the Amazon also reflects broader ecological dynamics, where the interplay of climatic variability and anthropogenic factors influences fruit yield and quality. For example, studies on the impact of harvest timing on the chemical composition of Patauá (*Oenocarpus bataua*) fruits reveal how environmental factors, such as climate and soil type, intricately dictate the nutrient profile of these fruits (Souza et al., 2012). Such findings reinforce the necessity for adaptive management strategies that can ensure sustainable fruit production amid changing ecological conditions.

Additionally, the Amazon region has historically served as a major center for plant domestication, exemplifying how traditional agricultural practices have coevolved with local ecosystems. Investigations into fruits like cacao (Theobroma cacao) provide insight into the socio-ecological systems that sustain these crops, showcasing their cultivation in dark earths—regions modified by ancient agricultural societies (Clément et al., 2010, 2015). This rich agricultural history suggests that local communities have a profound understanding of their environment and its productive capacities, which is critical for continued biodiversity conservation. Sociologically, the cultivation and consumption of these fruits intertwine with cultural identities and social practices among Indigenous peoples. The significance of fruits such as the Jocote (*Spondias purpurea*) extends beyond nutrition, encompassing social symbolism and cultural heritage, thus reflecting how food practices are deeply embedded in the fabric of community life (Dawson et al., 2007; Neves & Heckenberger, 2019). This cultural dimension underscores the importance of integrating traditional practices into modern conservation strategies, allowing for a holistic approach to biodiversity management. Research findings illustrating the health benefits and nutritional potential of various Amazonian fruits emphasize the urgent need for conservation and sustainable use practices. Many of these fruits not only provide vital nutrients but are also linked to traditional medicine, highlighting their multifaceted roles in local health systems. Strategically promoting these fruits as functional foods can serve as a pathway to enhance food security and promote sustainable economic development (Tenea et al., 2021).

Furthermore, the increased interest in the genetic characteristics of both wild and cultivated species propels the need for innovative conservation methods that reflect the genetic structure and biodiversity dynamics within fruit populations. This could involve initiatives to document and study the genetic variability observed in key species to ensure their resilient future in the face of environmental challenges (Filho et al., 2023; Rollo et al., 2020). Such strategies would not only preserve genetic diversity but also enrich the nutritional offerings of the Amazon, aligning health outcomes with ecological sustainability.

11.9 Conclusion

Wild and semidomesticated fruits such as açaí, Brazil nut, and camu camu are far more than forest commodities; they are dynamic nodes within Amazonian sociobiological systems that interweave ecological resilience, cultural continuity, and spiritual meaning. Through generations of careful stewardship, Indigenous communities have cultivated agroforestry mosaics that preserve biodiversity, mitigate climate change, and sustain intricate knowledge systems rooted in lived experience and ancestral cosmologies. This chapter has demonstrated how these fruits serve as both sustenance and symbols—markers of ecological relationships, spiritual reverence, and sociopolitical identity. From ritual offerings to kin-based seed exchanges, the cultural vitality and ecological functionality of these species reveal the deeply reciprocal relationship between Amazonian people and their forests. Simultaneously, the encroachment of extractive economies and global market integration poses serious challenges to the sustainability of these systems, threatening to sever the ties between traditional knowledge, biodiversity, and place-based sovereignty. Moving forward, research and policy must recognize Amazonian fruits not only for their economic value but also for their position as critical vehicles of cultural resilience, climate adaptation, and decolonial environmental governance. Strengthening Indigenous rights, securing territorial claims, and supporting endogenous food systems are essential for ensuring the future viability of forest ecologies. Ultimately, reclaiming forest food systems is not only an ecological imperative but also a profound act of cultural and epistemic justice.

Chapter Reflection

1. Name three key wild or semidomesticated fruits discussed in this chapter.
2. Why are pollinators important for Brazil nut production?
3. What is a chagra in Amazonian agroforestry?
4. How does traditional ecological knowledge (TEK) support sustainable harvesting?
5. Give one example of a spiritual practice related to fruit harvesting.

6. What global market trend has affected açaí production?
7. How does gender influence fruit gathering in Amazonian communities?
8. What is one threat posed by climate change to Amazonian fruit species?
9. What does NTFP stand for?
10. Why is protecting Indigenous knowledge important for conservation?

REFERENCES

Adin, A., Weber, J., Montes, C., Vidaurre, H., Vosman, B., & Smulders, M. (2004). Genetic differentiation and trade among populations of peach palm (*Bactris gasipaes* kunth) in the Peruvian Amazon—implications for genetic resource management. *Theoretical and Applied Genetics*, *108*(8), 1564–1573. https://doi.org/10.1007/s00122-003-1581-9

Ahenkan, A., & Boon, E. (2011). Non-timber forest products farming and empowerment of rural women in Ghana. *Environment Development and Sustainability*, *13*(5), 863–878. https://doi.org/10.1007/s10668-011-9295-7

Akpan, J., Umunnakwe, O., & Omadewu, L. (2023). Effects of lime and poultry manure on soil properties and cucumber (*Cucumis sativus* L) performance in the humid tropics of southern Nigeria. *Global Journal of Agricultural Sciences*, *22*(1), 101–107. https://doi.org/10.4314/gjass.v22i1.11

Albuquerque, M. (2020). Linking agrobiodiversity and culture through adopting agroforestry practices: The agroforestry Indigenous agents. *Parks Stewardship Forum*, *36*(1). https://doi.org/10.5070/p536146395

Aluko, O., Adejumo, A., & Bobadoye, A. (2020). Adaptive strategies to deforestation among non-timber forest products (NTFPs) collectors across the gender line in Oluwa Forest Reserve Area of Ondo State, Nigeria. *Agro-Science*, *19*(2), 48–52. https://doi.org/10.4314/as.v19i2.8

Alves, K., Lima, A., Rivas, P., Albuquerque, I., Pinheiro, J., Catunda, P., & Ferraz, T. (2025). Platonia insignis: A systematic synthesis of scientific studies on its biology, ecology, and potential applications. *Plants*, *14*(6), 884. https://doi.org/10.3390/plants14060884

Anjos, L., Gonçalves, G., Dutra, V., Rosa, A., Santos, L., Barros, M., & Toledo, P. (2024). Brazil nut journey under future climate change in the Amazon. *PLOS One*, *19*(11), e0312308. https://doi.org/10.1371/journal.pone.0312308

Asamoah, O., Danquah, J., Bamwesigye, D., Boakye, E., Appiah, M., & Pappinen, A. (2024). Perception of locals on multiple contributions of NTFPs to the livelihoods of forest fringe communities in Ghana. *Forests*, *15*(5), 861. https://doi.org/10.3390/f15050861

Barros, P., Costa, A., Gomes, M., Castellani, D., Kato, O., & Vasconcelos, S. (2024). Floristic composition and temporal dynamics of palm oil agroforests in the eastern Amazon. https://doi.org/10.21203/rs.3.rs-4425875/v1

Belcher, B., & Schreckenberg, K. (2007). Commercialisation of non-timber forest products: A reality check. *Development Policy Review*, *25*(3), 355–377. https://doi.org/10.1111/j.1467-7679.2007.00374.x

Borges, R., Brito, R., Imperatriz-Fonseca, V., & Giannini, T. (2020). The value of crop production and pollination services in the eastern Amazon. *Neotropical Entomology*, *49*(4), 545–556. https://doi.org/10.1007/s13744-020-00791-w

Braz-Mota, S., & Val, A. (2024). Fish mortality in the Amazonian drought of 2023: The role of experimental biology in our response to climate change. *Journal of Experimental Biology*, *227*(17). https://doi.org/10.1242/jeb.247255

Bucheli, V., & Bokelmann, W. (2017). Agroforestry systems for biodiversity and ecosystem services: The case of the Sibundoy Valley in the Colombian province of Putumayo. *International Journal of Biodiversity Science Ecosystems Services and Management*, *13*(1), 380–397. https://doi.org/10.1080/21513732.2017.1391879

Bucheli, V., Mallen, R., Álvarez-Macías, A., Coral, C., & Bokelmann, W. (2021). Indigenous family labor in agroforestry systems in the context of global transformations: The case of the Inga and Camëntsá communities in Putumayo, Colombia. *Forests*, *12*(11), 1503. https://doi.org/10.3390/f12111503

Cardoso, D., Särkinen, T., Alexander, S., Amorim, A., Bittrich, V., Celis, M., & Forzza, R. (2017). Amazon plant diversity revealed by a taxonomically verified species list. *Proceedings of the National Academy of Sciences*, *114*(40), 10695–10700. https://doi.org/10.1073/pnas.1706756114

Carvalho, R., Huguenin, G., Luiz, R., Moreira, A., Oliveira, G., & Rosa, G. (2015). Intake of partially defatted Brazil nut flour reduces serum cholesterol in hypercholesterolemic patients- a randomized controlled trial. *Nutrition Journal*, *14*(1). https://doi.org/10.1186/s12937-015-0036-x

Castro, J., Maddox, J., Cobos, M., & Imán, S. (2018). *Myrciaria dubia* "camu camu" fruit: Health-promoting phytochemicals and functional genomic characteristics.

In J. R. Soneji & M. Nageswara-Rao (Eds.), *Breeding and health benefits of fruit and nut crops*. IntechOpen. https://doi.org/10.5772/intechopen.73213

Catenacci, F., & Simon, M. (2017). A checklist of lecythidaceae in the upper Madeira River, Rondônia, Brazil, with comments on diversity and conservation. *Brittonia*, *69*(4), 447–456. https://doi.org/10.1007/s12228-017-9482-4

Cavalcante, M., Oliveira, F., Maués, M., & Freitas, B. (2012). Pollination requirements and the foraging behavior of potential pollinators of cultivated Brazil nut (*Bertholletia excelsa* Bonpl.) trees in the central Amazon rainforest. *Psyche: A Journal of Entomology*, *2012*, 1–9. https://doi.org/10.1155/2012/978019

Clément, C., Cristo-Araújo, M., D'Eeckenbrugge, G., Alves-Pereira, A., & Rodrigues, D. (2010). Origin and domestication of native Amazonian crops. *Diversity*, *2*(1), 72–106. https://doi.org/10.3390/d2010072

Clément, C., Denevan, W., Heckenberger, M., Junqueira, A., Neves, E., Teixeira, W., & Woods, W. (2015). The domestication of Amazonia before European conquest. *Proceedings of the Royal Society B Biological Sciences*, *282*(1812), 20150813. https://doi.org/10.1098/rspb.2015.0813

Costa, R., Rodrigues, A., & Silva, L. (2022). The fruit of peach palm (*Bactris gasipaes*) and its technological potential: An overview. *Food Science and Technology*, *42*. https://doi.org/10.1590/fst.82721

Dawson, I. K., Hollingsworth, P. M., Doyle, J. J., Kresovich, S., Weber, J. C., Montes, C. S., Pennington, T. D., & Pennington, R. T. (2007). Origins and genetic conservation of tropical trees in agroforestry systems: A case study from the Peruvian Amazon. *Conservation Genetics*, *9*(2), 361–372. https://doi.org/10.1007/s10592-007-9348-5

de Souza, L. A. G. (2023). Biodiversity of Fabaceae in the Brazilian Amazon and its timber potential for the future. https://doi.org/10.5772/intechopen.110374

Díaz-Ricaurte, J., Villegas, N., Coronado, J., Garzón, G., & Fiorillo, B. (2020). Effects of agricultural systems on the anuran diversity in the Colombian Amazon. *Studies on Neotropical Fauna and Environment*, *57*(1), 18–28. https://doi.org/10.1080/01650521.2020.1809334

Doddabasawa, B., Chittapur, B. M., Pampangouda, & Gurumurthy, H. (2018). Microbial density in Azadirachta indica A. Juss and Tectona grandis Linn. based agroforestry systems. *Legume Research*, *41*(6), 856–861.

Duarte, R., González-Jaramillo, N., Bailón-Moscoso, N., Rojas-Le-Fort, M., & Romero-Benavides, J. (2024). Five underutilized Ecuadorian fruits and their bioactive potential as functional foods and in metabolic syndrome: A review. *Molecules*, *29*(12), 2904. https://doi.org/10.3390/molecules29122904

Faruk, A., Nersesyan, A., Papikyan, A., Galstyan, S., Hakobyan, E., Barblish-vili, T., Mikatadze-Pantsulaia, T., Darchidze, T., Kuchukhidze, M. Kereselidze, N., Kikodze, D., Willey, T., Ryan, P., & Breman, E. (2023). Multigenerational differences in harvesting and use of wild edible fruits and nuts in the South Caucasus. *Plants People Planet*, *6*(1), 238–248. https://doi.org/10.1002/ppp3.10434

Feeley, K., & Silman, M. (2016). Disappearing climates will limit the efficacy of Amazonian protected areas. *Diversity and Distributions*, *22*(11), 1081–1084. https://doi.org/10.1111/ddi.12475

Filho, G., Souza, C., Lúcia, C., Sant'ana, H., & Santos, R. (2023). Nutrients and bioactive compounds in wild fruits from the Brazilian Amazon rainforest. *Food Science and Technology*, *43*. https://doi.org/10.5327/fst.17823

Frederico, R., Dias, M., Jézéquel, C., Tedesco, P., Hugueny, B., Zuanon, J., & Oberdorff, T. (2021). The representativeness of protected areas for Amazonian fish diversity under climate change. *Aquatic Conservation Marine and Freshwater Ecosystems*, *31*(5), 1158–1166. https://doi.org/10.1002/aqc.3528

Freitas, C., Shepard, G., & Piedade, M. (2015). The floating forest: Traditional knowledge and use of Matupá Vegetation Islands by riverine peoples of the Central Amazon. *PLOS One*, *10*(4), e0122542. https://doi.org/10.1371/journal.pone.0122542

García-Cox, W., López-Tobar, R., Herrera-Feijoo, R. J., Tapia, A., Heredia-R, M., Toulkeridis, T., & Torres, B. (2023). Floristic composition, structure, and aboveground biomass of the Moraceae family in an evergreen Andean Amazon forest, Ecuador. *Forests*, *14*(7), 1406. https://doi.org/10.3390/f14071406

Gelan, A. (2023). Socio-economic factors influencing household dependency on non-timber forest products from Chilimo Forest, Ethiopia. *Indonesian Journal of Social and Environmental Issues (IJSEI)*, *4*(1), 81–88. https://doi.org/10.47540/ijsei.v4i1.722

Gilmore, M., Endress, B., & Horn, C. (2013). The socio-cultural importance of *Mauritia flexuosa* palm swamps (aguajales) and implications for multi-use management in two Maijuna communities of the Peruvian Amazon. *Journal of Ethnobiology and Ethnomedicine*, *9*(1). https://doi.org/10.1186/1746-4269-9-29

He, J., Dong, M., & Stark, M. (2014). Small mushrooms for big business? Gaps in the sustainable management of non-timber forest products in southwest China. *Sustainability*, *6*(10), 6847–6861. https://doi.org/10.3390/su6106847

Heckenberger, M., & Neves, E. (2009). Amazonian archaeology. *Annual Review of Anthropology*, *38*(1), 251–266. https://doi.org/10.1146/annurev-anthro-091908-164310

Imarhiagbe, O., & Ogwu, M. C. (2022). Sacred groves in the Global South: A panacea for sustainable biodiversity conservation. In S. C. Izah (Ed.), *Biodiversity in Africa: Potentials, threats and conservation* (Vol. 29, pp. 525–546). Sustainable Development and Biodiversity. http://doi.org/10.1007/978-981-19-3326-4_20

Imarhiagbe, O. Onyeukwu, I. I., Egboduku, W., Mukah, F. E., & Ogwu, M. C. (2022). Forest conservation strategies in Africa: Historical perspectives, status and sustainable avenues for progress. In S. C. Izah (Ed.), *Biodiversity in Africa: Potentials, threats and conservation* (Vol. 29, pp. 547–572). Sustainable Development and Biodiversity. http://doi.org/10.1007/978-981-19-3326-4_21

Izah, S. C., Richard, G., Stanley, H. O., Sawyer, W. E., Ogwu, M. C., & Uwaeme, O. R. (2023). Integrating the one health approach and statistical analysis for sustainable aquatic ecosystem management and trace contamination mitigation. *ES Food and Agroforestry*, *14*, 1012. https://www.espublisher.com/uploads/article_pdf/esfaf1012.pdf

Jung, D. (2025). Agroforestry backyards for Indigenous resilience: A participatory study in Brazil. https://doi.org/10.21203/rs.3.rs-6550288/v1

Júnior, J., & Souza, I. (2021). Medicinal plants used in the Amazon region: A systematic review. *Research Society and Development*, *10*(14), e163101419965. https://doi.org/10.33448/rsd-v10i14.19965

Kluczkovski, A., Barros, H., Barroncas, J., Viana, C., & Lima, É. (2023). Aflatoxins in raw brazil nut (bertholletia excelsa H.B.K.). *Journal of Agricultural Studies*, *11*(2), 14. https://doi.org/10.5296/jas.v11i2.20741

Lagneaux, E., Jansen, M., Quaedvlieg, J., Zuidema, P., Anten, N., Roca, M., & Kettle, C. (2021). Diversity bears fruit: Evaluating the economic potential of undervalued fruits for an agroecological restoration approach in the Peruvian Amazon. *Sustainability*, *13*(8), 4582. https://doi.org/10.3390/su13084582

Levis, C., Flores, B., Moreira, P., Luize, B., Alves, R., Franco-Moraes, J., & Clément, C. (2018). How people domesticated Amazonian forests. *Frontiers in Ecology and Evolution*, *5*. https://doi.org/10.3389/fevo.2017.00171

Lins, J., Lima, H., Baccaro, F., Kinupp, V., Shepard, G., & Clément, C. (2015). Pre-Columbian floristic legacies in modern homegardens of Central Amazonia. *PLOS One*, *10*(6), e0127067. https://doi.org/10.1371/journal.pone.0127067

Lucas, C. (2008). Within flood season variation in fruit consumption and seed dispersal by two characin fishes of the Amazon. *Biotropica*, *40*(5), 581–589. https://doi.org/10.1111/j.1744-7429.2008.00415.x

Machado, C., Prata, E., & Kinupp, V. (2021). Human food dynamics in highly seasonal ecosystems: A case study of plant-eating in riverine communities in central Amazon. *Journal of Ethnobiology*, *41*(2), 247–262. https://doi.org/10.2993/0278-0771-41.2.247

Magry, M., Cahill, D., Rookes, J., & Narula, S. (2023). Do socio-economic factors impact non-timber forest products-based incomes: An analysis employing structural equation modelling (SEM); A case of India. https://doi.org/10.21203/rs.3.rs-3278110/v1

Marentes, M., Venturi, M., Scaramuzzi, S., Focacci, M., & Santoro, A. (2021). Traditional forest-related knowledge and agrobiodiversity preservation: The case of the chagras in the Indigenous reserve of Monochoa (Colombia). *Biodiversity and Conservation*, *31*(10), 2243–2258. https://doi.org/10.1007/s10531-021-02263-y

Marinho, T., Corrêa, T., Vieira, K., Albuquerque, I., Alves, G., Pinheiro, M., & Ferraz, T. (2022). Genetic variability during in vitro establishment of bacurizeiro (*Platonia insignis* Mart.): An Amazon species. *Australian Journal of Crop Science*, *16*(6), 819–825. https://doi.org/10.21475/ajcs.22.16.06.p3575

Meinhold, K., & Darr, D. (2019). The processing of non-timber forest products through small and medium enterprises—A review of enabling and constraining factors. *Forests*, *10*(11), 1026. https://doi.org/10.3390/f10111026

Miina, J., Kurttila, M., Calama, R., de-Miguel, S., & Pukkala, T. (2020). Modelling non-timber forest products for forest management planning in Europe. *Current Forestry Reports*, *6*(4), 309–322. https://doi.org/10.1007/s40725-020-00130-7

Miller, R., & Nair, P. (2006). Indigenous agroforestry systems in Amazonia: From prehistory to today. *Agroforestry Systems*, *66*(2), 151–164. https://doi.org/10.1007/s10457-005-6074-1

Moegenburg, S. (2002). *Harvesting and managing forest fruits by humans: Implications for fruit-frugivore interactions* (pp. 479–494). CABI Digital Library. https://doi.org/10.1079/9780851995250.0479

Moegenburg, S., & Levey, D. (2002). Prospects for conserving biodiversity in Amazonian extractive reserves. *Ecology Letters, 5*(3), 320–324. https://doi.org/10.1046/j.1461-0248.2002.00323.x

Moreira, P. A., Aguirre-Dugua, X., Mariac, C., Zekraouï, L., Couderc, M., Rodrigues, D. P., Casas, A., Clement, C. R., & Vigouroux, Y. (2017a). Diversity of treegourd (*Crescentia cujete*) suggests introduction and prehistoric dispersal routes into Amazonia. *Frontiers in Ecology and Evolution, 5.* https://doi.org/10.3389/fevo.2017.00150

Moreira, P. A., Mariac, C., Zekraouï, L., Couderc, M., Rodrigues, D. P., Clément, C. R., & Vigouroux, Y. (2017b). Human management and hybridization shape treegourd fruits in the Brazilian Amazon Basin. *Evolutionary Applications, 10*(6), 577–589. https://doi.org/10.1111/eva.12474

Mortimer, P., Karunarathna, S., Qiao-Hong, L., Gui, H., Yang, X., Yang, X., & Hyde, K. (2012). Prized edible Asian mushrooms: Ecology, conservation and sustainability. *Fungal Diversity, 56*(1), 31–47. https://doi.org/10.1007/s13225-012-0196-3

Mukul, S., Rashid, A., Uddin, M., & Khan, N. (2015). Role of non-timber forest products in sustaining forest-based livelihoods and rural households' resilience capacity in and around protected area: A Bangladesh study†. *Journal of Environmental Planning and Management, 59*(4), 628–642. https://doi.org/10.1080/09640568.2015.1035774

Neves, E., & Heckenberger, M. (2019). The call of the wild: Rethinking food production in ancient Amazonia. *Annual Review of Anthropology, 48*(1), 371–388. https://doi.org/10.1146/annurev-anthro-102218-011057

Nguyễn, T., Lv, J., Vũ, T., & Zhang, B. (2020). Determinants of non-timber forest product planting, development, and trading: Case study in central Vietnam. *Forests, 11*(1), 116. https://doi.org/10.3390/f11010116

Ogden, L. (2019). Climate change outpaces species shifts in the Amazon. *Bioscience, 69*(3), 232–232. https://doi.org/10.1093/biosci/biy162

Ogwu, M. C. (2019). Towards sustainable development in Africa: The challenge of urbanization and climate change adaptation. In P. B. Cobbinah & M. Addaney (Eds.), *The geography of climate change adaptation in Urban Africa* (pp. 29–55). Springer Nature. http://doi.org/10.1007/978-3-030-04873-0_2

Ogwu, M. C. (2020). Value of *Amaranthus* [L.] species in Nigeria. In V. Wais-undara (Ed.), *Nutritional value of Amaranth* (pp. 1–21). IntechOpen. http://doi.org/10.5772/intechopen.86990

Ogwu, M. C. (2023). Local food crops in Africa: Sustainable utilization, threats, and traditional storage strategies. In S. C. Izah & M. C. Ogwu (Eds.), *Sustainable utilization and conservation of Africa's biological resources and environment.* Sustainable Development and Biodiversity (Vol. 888, pp. 353–374). Springer. https://doi.org/10.1007/978-981-19-6974-4_13

Olawuyi, E., Odeyale, O., Ugege, B., & Adenuga, D. (2021). Socio-economic analysis of non-timber forest products: A case of wrapping leaves in Oluwa Forest Reserve, Ondo State, Nigeria. *Agro-Science*, *20*(2), 9–13. https://doi.org/10.4314/as.v20i2.2

Osawaru, M. E., & Ogwu, M. C. (2020). Survey of plant and plant products in local markets within Benin City and environs. In L. W. Filho, N. Ogugu, D. Ayal, L. Adelake, & I. da Silva, (Eds.), *African handbook of climate change adaptation* (pp. 1–24.). Springer Nature. http://doi.org/10.1007/978-3-030-42091-8_159-1

Osawaru, M. E., Ogwu, M. C., Ogbeifun, N. S., & Chime, A. O. (2013). Microflora diversity of the phylloplane of wild Okra (*Corchorus olitorius* L. Jute). *Bayero Journal of Pure and Applied Sciences*, *6*(2), 136–142.

Paz, F., Pinto, C., Brito, R., Imperatriz-Fonseca, V., & Giannini, T. (2021). Edible fruit plant species in the Amazon forest rely mostly on bees and beetles as pollinators. *Journal of Economic Entomology*, *114*(2), 710–722. https://doi.org/10.1093/jee/toaa284

Pilnik, M., Argentim, T., Kinupp, V., Haverroth, M., & Ming, L. (2023). Traditional botanical knowledge: Food plants from the Huni Kuĩ Indigenous people, Acre, Western Brazilian Amazon. *Rodriguésia*, *74*. https://doi.org/10.1590/2175-7860202374016

Pires, M., Amante, E., Lopes, A., Rodrigues, A., & Silva, L. (2019). Peach palm flour (*Bactris gasipae* Kunth): Potential application in the food industry. *Food Science and Technology*, *39*(3), 613–619. https://doi.org/10.1590/fst.34617

Ramadhan, R., Tosepu, R., Phuwapraisirisan, P., Amirta, R., Phontree, K., Firdaus, Y. F. H., Abdulgani, N., Muttaqin, M. Z., & Saparwadi, S. (2022). Evaluation of non-timber forest products used as medicinal plants from East

Kalimantan (Indonesia) to inhibit ?-glucosidase and free radicals. *Biodiversitas Journal of Biological Diversity*, *23*(11). https://doi.org/10.13057/biodiv/d231102

Rambey, R., Lubis, R. A., Kembaren, Y., Rahmawaty, R., Delvian, D., Saraan, M., Rauf, A., Suratman, M. N., Gandaseca, S., & Abdullah, M. F. (2024). The sustainable economic value of agroforestry products in community forest areas in Suka Tendel Village, Tiga Nderket District, Karo Regency, North Sumatra Province. *IOP Conference Series Earth and Environmental Science*, *1413*(1), 012110. https://doi.org/10.1088/1755-1315/1413/1/012110

Rezende, J. (2024). Tõkowiseri: Kumuánica, bayaroánica and yaiwánica cosmovivences. *Estudos Avançados*, *38*(112), 95–112. https://doi.org/10.1590/s0103-4014.202438112.006-en

Ribeiro, M. B. N., Jerozolimski, A., de Robert, P., Salles, N. V., Kayapó, B., Pimentel, T. P., & Magnusson, W. E. (2014). Anthropogenic landscape in Southeastern Amazonia: Contemporary impacts of low-intensity harvesting and dispersal of brazil nuts by the Kayapó Indigenous people. *PLOS One*, *9*(7), e102187. https://doi.org/10.1371/journal.pone.0102187

Rocha, D., & Sollmann, R. (2023). Habitat use patterns suggest that climate-driven vegetation changes will negatively impact mammal communities in the Amazon. *Animal Conservation*, *26*(5), 663–674. https://doi.org/10.1111/acv.12853

Rollo, A., Ribeiro, M., Costa, R., Santos, C. P. Z., Mandák, B., & Lojka, B. (2020). Genetic structure and pod morphology of Inga edulis cultivated vs. wild populations from the Peruvian Amazon. *Forests*, *11*(6), 655. https://doi.org/10.3390/f11060655

Röpke, C., Amadio, S., Zuanon, J., Ferreira, E., Deus, C., Pires, T., & Winemiller, K. (2017). Simultaneous abrupt shifts in hydrology and fish assemblage structure in a floodplain lake in the central Amazon. *Scientific Reports*, *7*(1). https://doi.org/10.1038/srep40170

Rosa, D., Higuchi, M., & Roazzi, A. (2021). Attachment to the Amazon rainforest: Constitutive aspects and their predictors. *Paidéia (Ribeirão Preto)*, *31*. https://doi.org/10.1590/1982-4327e3128

Sales, L., Rodrigues, L., & Masiero, R. (2020). Climate change drives spatial mismatch and threatens the biotic interactions of the Brazil nut. *Global Ecology and Biogeography*, *30*(1), 117–127. https://doi.org/10.1111/geb.13200

Sánchez-Capa, M., Corell, M., & Mestanza-Ramón, C. (2023). Edible fruits from the Ecuadorian Amazon: Ethnobotany, physicochemical characteristics, and bioactive components. *Plants, 12*(20), 3635. https://doi.org/10.3390/plants12203635

Schmidt, M., Ikpeng, Y., Kayabi, T., Sanches, R., Ono, K., & Adams, C. (2021). Indigenous knowledge and forest succession management in the Brazilian Amazon: Contributions to reforestation of degraded areas. *Frontiers in Forests and Global Change, 4.* https://doi.org/10.3389/ffgc.2021.605925

Shobika, R., Selvanayaki, S., Deepa, N., & Vasanthi, R. (2021). An empirical analysis of timber trade in India. *Asian Journal of Agricultural Extension Economics and Sociology, 39*(11), 1–10. https://doi.org/10.9734/ajaees/2021/v39i1130712

Silva, E., Wadt, L., Silva, K., Lima, R., Batista, K., Guedes, M., & Guilherme, L. (2017). Natural variation of selenium in brazil nuts and soils from the Amazon region. *Chemosphere, 188*, 650–658. https://doi.org/10.1016/j.chemosphere.2017.08.158

Smith, J., Pearce, B., & Wolfe, M. (2012). Reconciling productivity with protection of the environment: Is temperate agroforestry the answer?. *Renewable Agriculture and Food Systems, 28*(1), 80–92. https://doi.org/10.1017/s1742170511000585

Souza, R., Andrade, J., & Costa, S. (2012). Effect of the harvest date on the chemical composition of patauá (*Oenocarpus bataua* Mart.) fruits from a forest reserve in the Brazilian Amazon. *International Journal of Agronomy, 2012*, 1–6. https://doi.org/10.1155/2012/524075

Spacki, K., Corrêa, R., Uber, T., Barros, L., Ferreira, I., Peralta, R., & Peralta, R. (2022). Full exploitation of peach palm (*Bactris gasipaes* Kunth): State of the art and perspectives. *Plants, 11*(22), 3175. https://doi.org/10.3390/plants11223175

Stanley, E. (2020). Religious conversion and the decline of environmental ritual narratives. *Journal for the Study of Religion, Nature and Culture, 13*(3), 266–285. https://doi.org/10.1558/jsrnc.36277

Sundriyal, M. (2021). Development of the NTFP sector for income generation and environmental conservation. *Journal of Graphic Era University, 9*(1), 83–104 https://doi.org/10.13052/jgeu0975-1416.915

Tenea, G., Jarrín-V, P., & Yépez, L. (2021). Microbiota of wild fruits from the Amazon region of Ecuador: Linking diversity and functional potential of lactic acid bacteria with their origin. In H. J. Mikkola (Ed.), *Ecosystem and biodiversity of Amazonia.* IntechOpen. https://doi.org/10.5772/intechopen.94179

Thapa, A., & Singh, K. (2023). Environmental policy influencing the contribution of non-timber forest products to rural livelihoods: Case studies from Himachal Pradesh, India. *IOP Conference Series Earth and Environmental Science, 1279*(1), 012027. https://doi.org/10.1088/1755-1315/1279/1/012027

Tourneau, F. (2015). The sustainability challenges of Indigenous territories in Brazil's Amazonia. *Current Opinion in Environmental Sustainability, 14*, 213–220. https://doi.org/10.1016/j.cosust.2015.07.017

Vignoli, C., Leeuwen, J., Miller, R., Ticona-Benavente, C., Silva, B., Striffler, B., & Alfaia, S. (2022). Soil management in Indigenous agroforestry systems of guarana (*Paullinia cupana* Kunth) of the Sateré-Mawé ethnic group, in the lower Amazon River region. *Sustainability, 14*(22), 15464. https://doi.org/10.3390/su142215464

Villani, F. T., Ribeiro, G. A. A., de Albuquerque Villani, E. M., Teixeira, W. G., de Souza Moreira, F. M., Miller, R., & Alfaia, S. S. (2017). Microbial carbon, mineral-n and soil nutrients in Indigenous agroforestry systems and other land use in the upper solimand#245;es region, western Amazonas state, Brazil. *Agricultural Sciences, 8*(7), 657–674. https://doi.org/10.4236/as.2017.87050

Walker, W., Gorelik, S., Baccini, A., Aragón-Osejo, J., Josse, C., Meyer, C., & Schwartzman, S. (2020). The role of forest conversion, degradation, and disturbance in the carbon dynamics of Amazon Indigenous territories and protected areas. *Proceedings of the National Academy of Sciences, 117*(6), 3015–3025. https://doi.org/10.1073/pnas.1913321117

Wright, R. (2022). The "sparks of Kuwai." *Journal for the Study of Religion Nature and Culture, 16*(1), 50–76. https://doi.org/10.1558/jsrnc.20769

Yang, J. (2009). Brazil nuts and associated health benefits: A review. *LWT-Food Science and Technology, 42*(10), 1573–1580. https://doi.org/10.1016/j.lwt.2009.05.019

CHAPTER 12

Sociobiology of Fermented Indigenous Foods and Probiotic Cultures

Abstract

This chapter explores the sociobiological dimensions of fermented Indigenous foods and probiotic cultures, focusing on the evolutionary, cultural, microbial, and health-related significance of these practices. Fermentation, a widespread and ancient biotechnological process, is examined as a co-adaptive strategy between human communities and beneficial microbes, shaped by environmental conditions, social structures, and cultural knowledge systems. Drawing from diverse Indigenous traditions across Africa, Asia, and the Americas, the chapter highlights how fermented foods like ogi (Nigeria), injera (Ethiopia), chicha (Andean region), Gundruk (Nepal), and natto (Japan) serve not only as essential nutritional resources but also as embodiments of cultural identity, social memory, spiritual practice, and ecological adaptation. These foods, often prepared through low-tech and sustainable methods, reflect complex microbial ecologies that are locally maintained and passed down through generations. The chapter emphasizes the role of women and elders in safeguarding fermentation knowledge and the symbolic meanings attached to these food traditions. It further investigates the probiotic potential of fermented Indigenous foods in promoting gut health, nutrient bioavailability, and immune function, thereby offering valuable insights for modern health interventions. At the same time, the chapter critiques the threats posed by globalization, industrialization, and bioprospecting to Indigenous microbial heritage. Through an integrated sociobiological and biocultural lens, it calls for the recognition of Indigenous fermentation systems in food policy, education, and biodiversity conservation frameworks.

Keywords: fermentation, probiotic cultures, Indigenous knowledge, microbial ecology, biocultural heritage, traditional food systems, sociobiology, gut health, fermented foods, cultural transmission

12.1 Introduction

Fermentation can be defined as a biochemical process that converts organic compounds, often carbohydrates, into alcohol or acids through the action of microorganisms such as bacteria and yeast. This process is crucial not only for the production of various food products but also for enhancing their safety, nutritional value, and preservation (Adesulu-Dahunsi et al., 2020; Elechi et al., 2022). Fermented foods often contain probiotic cultures, which refer to live microorganisms that, when consumed in adequate amounts, confer health benefits to the host by improving intestinal microbiota and overall gut health (Gupta & Sharma, 2017; Sindhu & Khetarpaul, 2002). Though the term "probiotic" is largely associated with certain strains of lactic acid bacteria (LAB) such as *Lactobacillus* and *Bifidobacterium*, the functionality and survival of these cultures during fermentation are pivotal to their efficacy in promoting human health (Adesulu-Dahunsi et al., 2022; Rezac et al., 2018). The relevance of fermented foods is particularly pronounced within Indigenous food systems, where these foods hold significant cultural, social, and nutritional importance. Indigenous fermented foods have been integrated into the dietary customs of many communities worldwide, serving as vital sources of essential nutrients and means of sustainable food production (Jernigan et al., 2021; Lima et al., 2022). Furthermore, fermentation processes contribute significantly to food security and the availability of safe food options for marginalized populations, thereby reinforcing the nutritional frameworks within these communities (Irakoze et al., 2021; Jernigan et al., 2021).

A sociobiological framing of fermented foods encompasses several core concepts: adaptive significance, cultural transmission, and microbial symbiosis. Adaptive significance refers to the evolutionary advantages that fermentation provides, such as increased food preservation and the detoxification of antinutritional factors inherent in raw ingredients (Adesulu-Dahunsi et al., 2022; Patra et al., 2016). Cultural transmission highlights how knowledge of fermentation techniques and the identification of robust probiotic strains is passed down through generations, fostering resilience and community identity (Lima et al., 2022; Sousa et al., 2023). Microbial symbiosis emphasizes the collaborative interactions between Indigenous cultures and the microbial communities driving fermentation, illustrating a model of coevolution where both humans and microorganisms adapt to optimize the fermentation process (Irakoze, 2024; Tamang et al., 2016). The various roles that fermented foods serve in Indigenous settings are underscored by the preservation of traditional practices that have evolved over centuries. As noted in studies, traditional fermented products not only serve functional roles in nutrition but also facilitate social interactions and cultural practices among community members (Cano & Suárez, 2020; Lima et al., 2022). The myriad flavors and health benefits attributed to these foods are rooted in both microbiological and sociocultural contexts (Majumdar et al., 2015).

Furthermore, the nutritional benefits of fermented foods are notable; they can reduce antinutritional factors and improve digestibility, thereby enhancing the bioavailability of essential nutrients (Irakoze et al., 2021; Sindhu & Khetarpaul, 2002). For example, lactic fermentation has been shown to detoxify Indigenous vegetables, thus improving their safety and nutrient absorption while extending shelf life (Irakoze, 2024; Patra et al., 2016). The interaction of fermentation with local ecosystems exemplifies the wisdom of Indigenous food systems, which utilize local biodiversity and traditional knowledge to produce nutrient-dense foods (Lamba et al., 2019; Lima et al., 2022). In the wider context of health and nutrition, the effect of probiotic cultures in fermented foods is significant. These microorganisms can have positive effects on digestive health, boost immune responses, and even help manage some chronic diseases (Gupta & Sharma, 2017; Taale et al., 2024). Recent studies have highlighted specific strains of LAB that can function as probiotic agents, competing with harmful microorganisms and producing beneficial metabolites that support host health (Adesulu-Dahunsi et al., 2017, 2020). The presence of such probiotics in established dietary contexts showcases the potential for microbial health to be integrated into local and traditional food practices.

Cultural narratives surrounding fermented foods further cement their positions as cornerstones of identity among Indigenous populations. It is essential to recognize that these understandings and practices not only reflect nutritional strategies but are also intrinsically linked to communal identities and local ecological knowledge (Cano & Suárez, 2020; Jernigan et al., 2021). The sharing of fermented foods during communal gatherings serves not only as sustenance but also as a medium for cultural expression and continuity. Such practices resonate with the principles governing food sovereignty, emphasizing the right of Indigenous peoples to define their food systems and maintain autonomy over their food sources and production methods (Cano & Suárez, 2020; Jernigan et al., 2021). Moreover, the technical know-how attributed to fermentation has implications for contemporary food sovereignty movements, as communities seek to regain control over food production through the revival of traditional microbial practices (Sousa et al., 2023; Suwannasom et al., 2025). This renewed focus on local fermentation processes can contribute to resilient food systems that withstand the pressures of industrialization and globalization while fostering community resilience (Jernigan et al., 2021; Lima et al., 2022). By aligning fermentation practices with ecological sustainability, Indigenous peoples can create pathways toward addressing food security challenges in an ever-changing environmental landscape.

The multifaceted advantages of fermented foods extend into realms beyond nutrition and health; they also engage with the concept of biodiversity (Adesulu-Dahunsi et al., 2022; Patra et al., 2016). Distinct cultural practices of fermentation are shaped by local climates, available resources, and traditional knowledge systems, which together influence the uniqueness of regional food patterns (Motlhanka et al., 2018; Sousa et al., 2023). This complexity of fermentation emphasizes the potential for science to collaborate with traditional practices, enhancing appreciation for both cultural and biological diversity in Indigenous food systems (Adesulu-Dahunsi et al., 2020; Cano & Suárez, 2020). Furthermore, the genetic diversity of LAB strains used in fermentation contributes to the specific flavors and health-promoting properties of Indigenous foods (Adesulu-Dahunsi et al., 2017; Gupta & Sharma, 2017). Diversity among microbial populations ensures a robust fermentation process, enhancing the probiotic properties of the final product. The strategic selection and cultivation of local LAB strains can serve as a springboard for improving local food products and optimizing them for both culinary and health benefits (Adesulu-Dahunsi et al., 2017, 2020; Lamba et al., 2019). These aspects can position Indigenous fermented foods not only as cultural artifacts but also as innovative solutions to modern dietary deficiencies and chronic health issues.

This chapter examines the evolutionary and adaptive significance of fermentation in Indigenous societies through a sociobiological lens, exploring how this ancient practice serves both biological and cultural functions. It investigates the symbolic, gendered, and social roles of fermented foods within traditional knowledge systems while highlighting the microbial ecologies that support Indigenous fermentation techniques across diverse regions. The chapter further assesses the nutritional and medicinal values of fermented foods, particularly their probiotic properties, and considers how intergenerational knowledge transmission sustains these practices. It also addresses contemporary challenges, such as globalization, commodification, and bioprospecting, which threaten Indigenous microbial heritage. It demonstrates how these foods function not only as health-promoting resources but also as expressions of biocultural identity, resilience, and ecological adaptation.

12.2 Evolutionary Context of Fermentation in Human Societies

The historical emergence of fermentation practices represents a pivotal chapter in human ecology and biology, informing not only dietary habits but also broader cultural and technological advancements. The earliest evidence of directed fermentation by humans can be traced back to around 4300 B.C., with some research suggesting that such practices may have originated as far back as 12,500 years ago, during the Paleolithic period. These early endeavors involved the spontaneous fermentation of fruits and grains, while modern advances include microbiome engineering, with both emphasizing the vital role of microbes in transforming raw materials into consumable forms (Figure 12.1; Amato et al., 2021; McGovern et al., 2004). Figure 12.1 illustrates the progressive stages of food fermentation practices, from ancient spontaneous processes to advanced microbiome engineering, each building on previous knowledge to improve consistency, safety, and functional benefits. Central to this evolution is the growing body of scientific knowledge and tools that aim to guide and optimize fermentation for stability, standardization, and enhanced health outcomes (Mannaa et al., 2021). This practice exemplifies humanity's long-standing relationship with fermentation, rooted in the ecological and nutritional contexts of early human societies.

Fermentation is as much an art as it is a science, representing a multifaceted survival and preservation strategy that dates back to when humans transitioned from nomadic hunter-gatherer lifestyles to settled agricultural communities. The preservation of food through fermentation not only enhances the safety and nutritional quality of food but also ultimately contributes to the overall resilience of early human settlements in the face of seasonal variations and resource scarcity (Dunn et al., 2021; Selhub et al., 2014). Fermented foods were crucial in enriching diets, providing essential nutrients, and extending the shelf life of perishable items, thus allowing early agricultural societies to thrive despite environmental uncertainties (Raghuvanshi et al., 2019). The use of fermentation as a food preservation strategy illustrates the adaptive skills of humans in effectively utilizing available resources, serving as a testament to our evolutionary ingenuity (Dunn et al., 2021).

The coevolution of humans and microbial consortia highlights the profound impact that fermentation practices have had throughout human evolution. Humans have not only influenced microbial ecosystems through dietary choices, but they have also experienced significant genetic adaptations in response to

Figure 12.1: Evolution of Fermentation Practices
and Technological Innovations

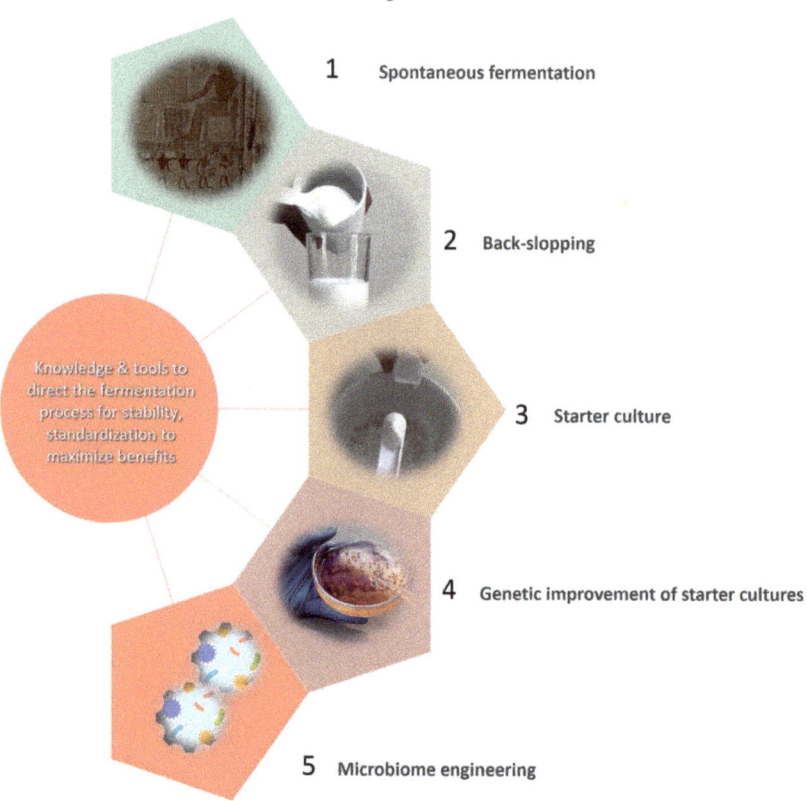

Source: Mannaa et al. (2021)

the consumption of fermented products. Genetic evidence suggests that modern humans may have developed enhanced capabilities for metabolizing ethanol, which was prevalent in naturally occurring sources like fermenting fruits (Janiak et al., 2020). These evolutionary changes hint at complex interactions between the human microbiome and dietary habits, emphasizing a coevolutionary dynamic where both parties influence each other's development. For instance, the domestication of LAB began with agricultural practices and the cultivation of fermentation processes, establishing a symbiotic relationship that has persisted through the ages as a fundamental component of human diets across diverse cultures (Makarova et al., 2006). Fermentation practices have continually diversified across human societies, transcending agriculture to include methodologies employed by pastoralists and hunter-gatherers alike (Dunn et al., 2021). For example, in regions where traditional agriculture may have faltered, the ability to ferment local fruits or grains provided a means to derive both sustenance and social cohesion among communities. The social implications of fermentation encompass rituals, celebrations, and communal gatherings, further embedding these practices into the cultural fabric of human societies. Historical records from ancient China indicate that fermented beverages played significant roles in marking societal and cultural milestones, underscoring their importance beyond mere sustenance (McGovern et al., 2004).

Moreover, the technological advancements associated with fermentation practices evolved alongside changes in human societies, leading to increasingly refined methods for beverage production and food preservation. From the ancient techniques used in winemaking to the more complex fermentation processes developed in the modern era, each cultural context adapted fermentation technology to local resources and needs, creating unique products that often reflected societal values and preferences (Cosme et al., 2024). Furthermore, the significance of fermented foods in promoting gut health and overall human well-being has garnered scientific interest, linking traditional practices with contemporary nutritional science and public health understanding (Raghuvanshi et al., 2019; Selhub et al., 2014). The implications of such research emphasize the untapped biodiverse potential of fermentation in contemporary diets. Thus, the story of fermentation within human evolution encapsulates an intricate web of historical, ecological, and cultural elements that illustrate humanity's adaptive strategies in resource utilization. The genetics of ethanol metabolism, the role of fermentation in food preservation, and its social dimensions, all demonstrate how such a seemingly straightforward process has evolved into a cornerstone of human survival and societal development. As our understanding of the microbiome deepens, it is likely that further insights will emerge, revealing more about the influence of fermented foods on physical and mental health, as well as their grounding in the evolutionary trajectory of our species (Selhub et al., 2014). The intersection of behavioral practices and evolutionary biology continues to unveil why fermentation remains a universal human practice, integral to our culture and survival, underscoring its historical significance in human development.

The importance of fermentation as a survival strategy cannot be overstated; it facilitated human adaptability to diverse environments and resource availability. As groups transitioned to sedentary lifestyles, fermentation became crucial for community sustainability, allowing societies to thrive amid fluctuations in climate, food supply, and health challenges (Dunn et al., 2021). Understanding these evolutionary contexts reveals how adaptive strategies, like controlled fermentation, were practiced not just out of necessity but also as expressions of cultural identity, reflecting a deep connection between humans and the microbial world that facilitates such processes. Fermentation encapsulates a rich tapestry of interactions between humans and their environment, punctuating the course of human history with lessons on resilience, adaptability, and the indelible marks left by microbial consortia on human societies. This historical overview not only elucidates the significance of fermentation in past human societies but also invites ongoing inquiry into its implications for future food security, health, and social practices as we confront the pressing challenges of contemporary times.

12.3 Cultural and Ritual Significance of Fermented Indigenous Foods and Probiotic Cultures

The cultural and ritual significance of fermented Indigenous foods, including their various societal roles, is paramount in historical and contemporary contexts. Beyond their nutritional value, fermented foods encompass deep symbolic and spiritual meanings within numerous cultures (Table 12.1). Beyond their nutritional and probiotic benefits, these foods are deeply woven into the fabric of community life, ritual practice, and ecological wisdom. From ancestral offerings and rites of passage to harvest festivals and daily sustenance, fermented products like *chicha*, *ogi*, *natto*, and *injera* embody Indigenous knowledge systems that bridge health, heritage, and cosmology. Also, in the Faroese culture, fermented fish, notably "Ræstur fiskur," emerges as a central component of their identity, symbolizing local heritage and resulting from traditional practices that date back generations (Figure 12.2; Svanberg, 2015). Such foods often play pivotal roles during rituals and celebrations, reinforcing community bonds while reflecting historical continuity. Fermented foods can also be fundamental in healing traditions, where certain items are believed to possess medicinal properties aiding in recovery and well-being, emphasizing the interplay between food, culture, and health (Ojeda-Linares et al., 2020). The transmission of fermentation knowledge across generations further illustrates the centrality of these foods in cultural practices. Many Indigenous communities maintain this wisdom, which is often passed down through oral traditions or practical demonstrations, which reinforce social structures and roles within family and community settings. In some cultures, such as the Shui communities of Southwest China, fermentation practices intertwine with rituals and medicinal knowledge (Hong et al., 2015). The preservation of

Table 12.1: Cultural and Ritual Significance of Fermented Indigenous Foods

Region/ Culture	Fermented Food	Cultural/Ritual Function	Associated Beliefs/ Practices
West Africa (Nigeria)	Ogi, Palm wine, Dawadawa, etc.	Used in naming ceremonies, funerals, and as offerings to deities	Believed to connect the living with ancestors; fermentation seen as spiritual transformation
Ethiopia	Injera (fermented teff)	Staple in communal meals; shared during coffee ceremonies and Orthodox Christian fasting	Represents hospitality and sacred community bonding
South Asia	Idli, Dosa (fermented rice/lentils)	Integral to daily meals and temple offerings	Symbolizes purity and renewal in Ayurvedic and Hindu traditions
Japan	Natto, Miso, Sake	Consumed during Shinto rituals and seasonal festivals	Fermentation aligns with the Shinto concept of purification and harmony with nature
Andean region (Peru, Bolivia)	Chicha (fermented maize beverage)	Consumed in harvest festivals and ancestor worship	Used to invoke Pachamama (Earth Mother) and foster community reciprocity
Mongolia	Airag (fermented mare's milk)	Central to hospitality; offered to guests and in wedding rituals	Viewed as a symbol of strength, health, and ancestral continuity
Southeast Asia (Indonesia)	Tempeh, Tape	Featured in birth, marriage, and agricultural festivals	Believed to cleanse the body and honor ancestral spirits

Region/ Culture	Fermented Food	Cultural/Ritual Function	Associated Beliefs/ Practices
Eastern Europe	Kefir, Sauerkraut	Prepared during seasonal transitions and preserved for winter	Seen as protective food, preserving health during hardship and scarcity
Indigenous North America	Fermented corn and fish products	Used in healing ceremonies and seasonal feasts	Represents transformation and the cyclical nature of life
Pacific Islands	Poi (fermented taro)	Central in communal feasts and sacred gatherings	Symbolizes ancestral nourishment and land stewardship

Figure 12.2: Illustrated Sketch of Ræstur Fiskur: Traditional Faroese Fermented Fishz

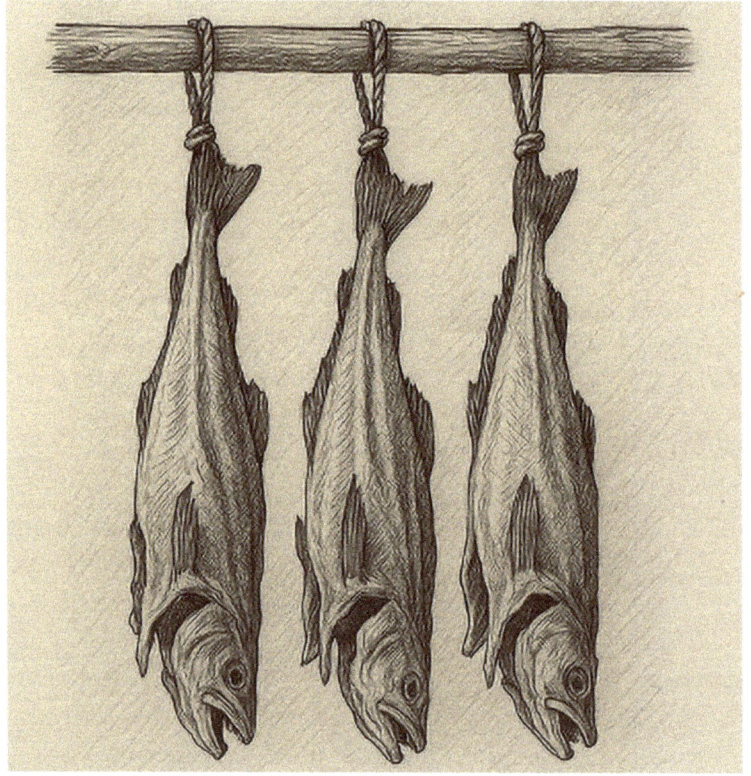

fermentation technologies not only aids in maintaining nutritional health but also enhances cultural identity and resilience against external influences.

Gendered knowledge and roles are critical when examining fermentation processes. Typically, these tasks are viewed as women's work in many cultures, with women often serving as the primary custodians and practitioners of fermentation techniques. This traditional role reinforces their status within community and family systems (Alhareky & Nazir, 2021). Female practitioners frequently engage in the fermentation of vegetables and fruits while also educating younger generations about these practices (Parkouda et al., 2009). Such practices symbolize the transmission of cultural values and social structures, intertwining personal identity with roles in food production and preservation. Research indicates that traditional fermented foods are dynamic systems shaped by the cultural, environmental, and historical contexts of their production (Tamang & Samuel, 2010). This perspective emphasizes that agricultural practices associated with fermentation are as vital as the fermentation process itself, highlighting the symbiotic relationship between culture and food production. In particular, studies focusing on Indigenous fermented beverages have shown how these foods evolve alongside societal changes while retaining traditional techniques that define local food cultures (Ojeda-Linares et al., 2021). The fermentation of these Indigenous products leads to unique flavors and health benefits, which are central to cultural identity and continuity.

Moreover, the observed variations in microbial communities across different fermented foods highlight the ecological dimensions of these practices: Specific fermentation processes not only aid in food preservation but also enhance the organoleptic qualities of food (Devirgiliis et al., 2013). For example, dairy-based fermented products in Africa illustrate a scenario where local microbial flora contribute distinct flavors that are pivotal to local diets (Jans et al., 2017). Since the fermentation process relies heavily on the specific contexts in which it occurs, regional fermentation practices foster a unique dialogue between food, microbial communities, and cultural narratives. In a broader context, fermented foods are often employed as tools in cultural diplomacy and economic development. The global appreciation of fermented foods like kimchi shows how local culinary practices can gain international acclaim and promote community pride (Young & Kiefer, 2014). Fermented foods frequently act as ambassadors of culture, representing local traditions on a global stage and creating pathways for sustainable tourism and economic opportunities. The globalization of fermentation practices emphasizes the importance of safeguarding traditions that more homogenized food products can easily overshadow.

The intergenerational transmission of fermentation knowledge illustrates the complexities of cultural evolution. This knowledge, as it transforms and adapts over time, demonstrates resilience in preserving a community's heritage while navigating the pressures of modernization. At the same time, as various cultural practices spread and intermingle, there is space for innovation within traditional frameworks, allowing for new dimensions of cultural expression through food (Tamang et al., 2020). The creative adaptations observed in fermentation processes emphasize the dynamism of cultural practices as they respond to changing environmental and societal landscapes.

The exploration of fermented foods thus offers insights into understanding the unique interplay of culture, health, and identity on a global scale. The richness of fermentation narratives, tied intimately with symbolic, spiritual, and practical dimensions, reveals the extensive impacts these culinary traditions have on social structures, human health, and cultural continuity (Song et al., 2016). Maintaining these practices supports local communities in their efforts to retain identity and autonomy amid a rapidly changing global food landscape, emphasizing the need for recognition and support of traditional knowledge systems. The practice of fermentation exemplifies a cultural narrative that encapsulates the act of not just preserving food but preserving identity itself. Each ferment conveys a story interwoven with the lives of those who create it, embodying a confluence of history, environment, and social dynamics. Thus, understanding the cultural significance of these foods extends beyond nutrition into the societal frameworks that sustain communities and preserve their heritage.

12.4 Microbial Ecology of Indigenous Fermented Foods

The microbial ecology of Indigenous fermented foods is shaped by centuries of localized knowledge, environmental factors, and specific food substrates. These ecosystems are rich in naturally occurring microorganisms that initiate and sustain fermentation, often without commercial starter cultures. Across the globe, several genera of bacteria and fungi consistently dominate traditional fermentation systems, notably *Lactobacillus, Saccharomyces, Leuconostoc, Pediococcus,* and *Bacillus* (Evivie, Ogwu, Abdelazez, et al., 2020; Evivie, Ogwu, Ebabhamiegbebho, et al. 2020; Evivie et al., 2021; Masumuzzaman et al., 2021). These microbial consortia not only contribute to desirable organoleptic properties such as taste, aroma, and texture but also enhance food safety, nutritional quality, and shelf life through acidification, enzymatic activity, and microbial antagonism

(Marco et al., 2017; Tamang et al., 2020). *Lactobacillus*, a genus of LAB, is particularly prevalent in cereal- and dairy-based fermentations due to its ability to thrive in anaerobic, carbohydrate-rich environments and produce lactic acid, which suppresses spoilage organisms. *Saccharomyces*, especially *S. cerevisiae*, plays a central role in alcoholic fermentations and leavening processes, while *Leuconostoc* and *Pediococcus* contribute to heterofermentative processes that produce both lactic acid and carbon dioxide, crucial in products like pickled vegetables and dough-based foods (Swain et al., 2014; Wolfe & Dutton, 2015).

Indigenous fermentation methods often rely on spontaneous fermentation, where environmental and utensil-associated microbes initiate the process. The biodiversity within these methods is immense, varying not only between regions but also from household to household. This variability leads to rich microbial landscapes that reflect local environments, practices, and food traditions (Ogwu, 2023; Tamang et al., 2016).

Several case studies illustrate the distinct microbial ecologies in Indigenous fermented foods, revealing the diversity of raw materials, local techniques, and microbial communities adapted to specific cultural and environmental contexts:

- **Ogi (Nigeria):** Ogi is a staple fermented cereal gruel commonly consumed as a weaning food or breakfast in many West African communities (Figure 12.3). It is typically prepared from maize, sorghum, or millet by soaking the grains for 24 to 72 hours, followed by wet milling, sieving, and fermentation of the starch-rich filtrate. The fermentation process is spontaneous mainly and driven by naturally occurring LAB, notably *Lactobacillus plantarum*, *L. fermentum*, and *Leuconostoc mesenteroides*. These LAB species play crucial roles in acidifying the medium, suppressing pathogenic organisms, and contributing to the characteristic sour flavor and thickened texture of ogi. In addition to LAB, yeasts such as *Saccharomyces cerevisiae* are frequently detected, enhancing the flavor profile and increasing bioavailability of micronutrients such as iron and zinc through phytate breakdown. The microbial profile and fermentation dynamics of ogi may vary with cereal type, water quality, ambient temperature, and household-specific practices, reflecting the intimate link between local ecology and microbial succession (Akinrele, 1970; Oyedeji et al., 2013).
- **Injera (Ethiopia):** Injera, a sour, spongy flatbread made from *teff* (*Eragrostis tef*), is central to Ethiopian cuisine and cultural life (Figure 12.4). Its preparation involves mixing teff flour with water and allowing it to ferment for approximately two to three days. The resulting batter is poured

Figure 12.3: Ogi Prepared from Fermented Maize (*Zea mays*)

onto a hot griddle to create a large, pancake-like bread with characteristic holes on the surface, formed by gas bubbles released during fermentation. The microbial community responsible for injera fermentation includes yeasts such as *Candida milleri* and *Saccharomyces exiguus*, along with LAB like *Lactobacillus sanfranciscensis*. These organisms function synergistically: Yeasts produce carbon dioxide that contributes to the bread's spongy texture, while LAB generate lactic acid that imparts sourness and inhibits spoilage organisms. The fermentation also improves protein digestibility and reduces antinutritional factors. Regional variations in microbial composition and fermentation time reflect local traditions and microecologies, reinforcing the notion that injera is both a biochemical product and a cultural artifact (Yetneberk et al., 2005).

Figure 12.4: Injera Made from *teff* (*Eragrostis tef*)

- **Chicha (Andean Region):** Chicha is a traditional fermented beverage made primarily from maize and widely consumed in parts of the Andes, particularly in Peru, Bolivia, and Ecuador (Figure 12.5). The process of making chicha can differ significantly by region but often involves chewing the maize kernels to initiate enzymatic starch breakdown (in enzymatic chicha) or malting and boiling the grain (in germinated chicha). The fermentation is spontaneous, occurring in earthenware vessels or open containers. It is driven by a mixture of yeasts—especially *Saccharomyces cerevisiae*—and LAB such as *Lactobacillus fermentum*. The dual fermentation leads to mild alcoholic content, acidity, and microbial stability. Chicha holds deep cultural significance, often consumed during communal rituals, celebrations, and agricultural festivals. Its preparation and fermentation reflect coevolved human–microbial relationships shaped by social customs and environmental conditions (Puerari et al., 2015).
- **Gundruk (Nepal):** Gundruk is a fermented and dried leafy vegetable product, commonly produced from mustard greens (*Brassica juncea*), radish leaves (*Raphanus sativus*), or cauliflower leaves (Figure 12.6). It is an important source of micronutrients and fiber during lean agricultural seasons, especially in the hilly and mountainous regions of Nepal. The

Figure 12.5: Peruvian Chichia

Source: Kuoda Travel (2025)

Figure 12.6: Gundruk Made from Mustard, Cauliflower, or Radish

traditional process involves crushing the leaves, packing them tightly in earthen or plastic containers, and allowing them to ferment anaerobically for seven to ten days (Evivie, Ogwu, Abdelazez, et al. 2020; Ogwu & Osawaru, 2022). The dominant fermentative organisms include LAB such as *Lactobacillus plantarum* and *Pediococcus pentosaceus*, which lower the pH through lactic acid production, inhibiting spoilage and pathogenic bacteria. The degree of fermentation and microbial diversity can vary depending on altitude, ambient temperature, seasonal conditions, and the type of leaf used. Gundruk exemplifies how Indigenous fermentation practices are deeply embedded in ecological adaptation, resource preservation, and cultural identity (Tamang et al., 2005).

- **Natto (Japan):** Natto is a unique Japanese fermented soybean product distinguished by its sticky texture and pungent aroma (Figure 12.7). Unlike other fermented foods dominated by LAB or yeasts, natto relies

Figure 12.7: Japanese Natto Made from Fermented Soybean

Source: Adapted from JNTO (2022)

on *Bacillus subtilis natto*, a spore-forming bacterium that thrives under aerobic conditions. The fermentation process typically involves steaming soybeans and inoculating them with *B. subtilis*, followed by incubation at warm temperatures for 18 to 24 hours. This bacterium produces enzymes such as nattokinase, which has fibrinolytic and cardiovascular health benefits, and polyglutamic acid, responsible for the characteristic mucilaginous texture. Natto's fermentation also enhances protein digestibility and vitamin K content. It is commonly consumed as a breakfast food and is valued both for its probiotic qualities and as a symbol of Japanese culinary heritage. The microbial ecology of natto reflects precise human management of environmental conditions and strain selection, illustrating the sophistication of traditional microbial biotechnology (Yonetani, 2015).

Collectively, these examples reveal the microbial richness and ecological specificity of Indigenous fermentation practices. They underscore how communities have, through empirical observation and cultural transmission, curated microbial landscapes that support nutrition, health, and cultural continuity.

12.5 Nutritional and Health Implications of Fermented Indigenous Foods

Fermented Indigenous foods play a crucial role in enhancing nutrient bioavailability, representing cultural practices, and promoting overall health. Fermentation not only preserves food but also transforms its biochemical composition, significantly increasing the bioavailability of nutrients, making them more digestible and accessible to human health (Table 12.2). Several studies have documented the effects of fermentation on various foodstuffs and how they relate to these nutritional advantages. Through fermentation, the growth of LAB, particularly members of the *Lactobacillus* genus, enhances the nutritional profile of food. Fermented products exhibit increased levels of essential nutrients such as proteins, vitamins, and minerals, while simultaneously reducing antinutritional factors like phytic acid that inhibit mineral absorption. This transformation is particularly relevant in traditional fermented foods derived from grains and legumes, which are staples in many Indigenous communities across Africa and Asia. For example, studies indicate that fermentation can significantly enhance the bioavailability of essential micronutrients, such as iron and calcium, in plant-based foods (Samtiya et al., 2021; Suvarna et al., 2018). This process effectively positions fermented

Table 12.2: Cross-Cultural Overview of the Nutritional and Health Implications of Fermented Indigenous Foods

Fermented Indigenous Food	Region of Origin	Key Nutrients	Probiotics/ Microorganisms	Health Benefits
Ogi/ Akamu/ Pap	West Africa (e.g., Nigeria)	Carbohydrates, B-vitamins, folate, etc.	*Lactobacillus plantarum,* and *Saccharomyces cerevisiae*	Enhances gut health, improves digestibility, and boosts immune function
Chicha de Jora	Andes (Peru, Bolivia, Ecuador)	Complex carbs, vitamin B1, potassium, etc.	*Saccharomyces* spp., and wild yeasts	Mild diuretic, supports cardiovascular health, and provides natural energy
Natto	Japan	Protein, vitamin K2, nattokinase enzyme, isoflavones, etc.	*Bacillus subtilis* natto	Supports bone health, reduces blood clot risk, improves gut flora, and lowers cholesterol
Injera	Ethiopia and Eritrea	Iron, fiber, protein (from teff), folate, etc.	*Lactobacillus fermentum,* and *Candida milleri*	Enhances iron bioavailability, regulates blood sugar, and aids in digestion
Kenkey	Ghana	Carbohydrates, B-complex vitamins, etc.	Lactic acid bacteria (*Lactobacillus* spp.)	Improves gut health, and lowers pH of the gut to deter pathogens
Kimchi	Korea	Vitamins A, B, and C, dietary fiber, etc.	*Lactobacillus kimchii,* and *Leuconostoc* spp.	Antioxidant properties, anti-obesity effects, and improves immune and digestive health

Fermented Indigenous Food	Region of Origin	Key Nutrients	Probiotics/ Microorganisms	Health Benefits
Idli/Dosa Batter	South India	Carbs, protein, B-vitamins	*Leuconostoc mesenteroides, Lactobacillus delbrueckii*	Easier digestion of rice and lentils, promotes gut health, enhances nutrient bioavailability
Kefir	Caucasus region	Protein, calcium, B12, riboflavin	Kefir grains (*Lactobacillus, Streptococcus,* yeasts)	Strong probiotic activity, anti-inflammatory, supports bone and digestive health
Uji (Fermented Porridge)	East Africa (e.g., Kenya)	Carbs, iron, zinc, vitamin C	*Lactobacillus* spp., *Enterococcus* spp.	Enhances mineral absorption, improves digestion, especially in infants and elderly
Pito	West Africa (Ghana, Nigeria)	Carbohydrates, trace minerals	*Saccharomyces cerevisiae,* lactic acid bacteria	Stimulates appetite, contains antioxidants, may aid in mild antimicrobial activity

foods as optimal dietary components to combat nutrient deficiencies prevalent in many populations.

The relationship between gut microbiota and fermented foods is another pivotal aspect of their health implications. The probiotic properties associated with fermented foods contribute positively to gut microbiota diversity, which is integral for gut health, immune function, and disease prevention. Fermented products are rich in probiotics, organisms that confer health benefits, including improving digestive health and potentially modulating immune responses. The presence of LAB during fermentation can lead to beneficial interactions within the gastrointestinal microbiome, supporting nutrient absorption and enhancing the body's immune defenses by affecting gut permeability and promoting an anti-inflammatory environment (Knez et al., 2023; Leeuwendaal et al., 2022). This perspective underscores the importance of incorporating fermented foods into diets to bolster immune function and reduce the risk of infections and diseases.

Moreover, traditional fermented dairy products like yogurt and kefir have shown to alleviate symptoms for many individuals suffering from lactose intolerance. The fermentation process decreases lactose content, as LAB metabolize lactose into lactic acid, thus making these foods generally more tolerable for individuals with lactose intolerance (Fischer et al., 2025). This phenomenon highlights a significant benefit of consuming fermented dairy products, expanding dietary options for individuals who otherwise might avoid dairy due to discomfort. Biocultural evidence pertaining to the health-promoting effects of fermented foods incorporates the integration of traditional practices with empirical findings. Historically, many cultures have emphasized the value of fermentation not only as a method for food preservation but also as a means of enhancing nutrient absorption and health benefits. For instance, traditional fermented fish products in North-Eastern India have been ingrained in local diets for centuries, praised for their diverse nutritional benefits, which include high levels of omega-3 fatty acids and essential micronutrients (Sasirekhamani et al., 2025). Such ethnographic observations align with scientific inquiries confirming the health benefits of these foods, thereby supporting the concept that fermentation intertwines cultural heritage and nutritional science.

Additionally, the potential economic implications of promoting fermented Indigenous foods cannot be overlooked. These traditional practices contribute to local economies while enhancing food security. The enhanced marketability of fermented products—due to their nutritional benefits and prolonged shelf life—could lead to increased incomes for local producers while ensuring that communities maintain access to nutritious foods (Irakoze, 2024). This combination of health benefits with economic viability emphasizes the multifaceted importance of fermented foods in Indigenous cultures. However, it is essential to acknowledge potential risks associated with fermented food production, particularly if hygiene practices are compromised, leading to possible contamination with harmful pathogens. Studies have indicated that while fermentation generally improves food safety, poor practices can allow for the growth of undesirable microorganisms that pose health risks (Dey et al., 2023). Hence, it is vital for communities to maintain traditional knowledge of safe fermentation practices to effectively capitalize on the benefits while minimizing health threats.

12.6 Social Learning and Knowledge Systems Associated with of Fermented Indigenous Foods

The preservation and transmission of knowledge associated with fermentation techniques for Indigenous foods are deeply rooted in systems of social learning within various communities. Apprenticeship and intergenerational teaching form a crucial backbone of this knowledge transfer, ensuring that techniques are not only retained but also adapted over time to meet changing environmental and social contexts. In Indigenous communities, these practices frequently encompass methods of fermentation that have been fine-tuned over generations, highlighting a relationship with the land and local resources that transcends mere survival to encompass cultural significance and identity (Domingo et al., 2021; Ghosh-Jerath et al., 2018). For instance, Ghosh-Jerath et al. (2018) emphasize the importance of traditional diets among Indigenous populations, showing how local food systems are integral not only for nutrition but also for sustaining cultural practices associated with food gathering and preparation. These practices can be viewed as a form of cultural heritage, reflecting a community's identity while also contributing to resilience against environmental changes. The narrative around fermented foods often includes storytelling, where each dish tells a story about the community, the ingredients drawn from the local ecosystem, and historical practices of preparation and consumption (Swiderska et al., 2022).

Moreover, sensory cues play a pivotal role in the learning process. Many Indigenous learners utilize sensory modalities such as taste, smell, and sight to understand fermentation. Swiderska et al. argue that sensory experiences are a critical component of Indigenous knowledge systems, underscoring how fermentation techniques are often learned through direct interaction with food and through communal gatherings where food is prepared and shared (Swiderska et al., 2022). This sensory learning fosters a deeper connection with the foods being prepared, promoting not only skill development but also a sense of belonging and community. Local innovations often emerge alongside traditional practices as communities navigate contemporary challenges. For example, Richmond et al. highlight how Indigenous peoples have increasingly adopted methods and technologies that enhance traditional practices without undermining their cultural significance (Richmond et al., 2020). As these communities face the implications of climate change and biodiversity loss, innovative approaches to fermentation—such as the use of locally sourced fermenting agents—create pathways for both survival and cultural expression. The integration of traditional knowledge with scientific insights can enhance the sustainability of these foods, promoting ecological health alongside cultural vitality (Nikolaus et al., 2022).

Additionally, the role of local ecological knowledge in these practices cannot be overstated. As communities adapt their fermentation techniques to new environmental conditions, they demonstrate resilience and ingenuity. The adaptation of fermentation processes to utilize seasonal ingredients helps preserve nutritional diversity and contributes to food security, as noted by Sidiq et al. (2022a, 2022b). This adaptive knowledge system ensures that fermentation techniques remain relevant, fostering both food sovereignty and community health. The intersection of fermentation practices with broader community dynamics further illustrates the social complexities of Indigenous food systems. As Neufeld et al. (2017) suggest, continuous community-engaged research is essential for understanding and navigating the local food systems that manifest through traditional practices. This research underscores the interconnectedness of social, cultural, and ecological dimensions that shape fermentation practices within Indigenous contexts. Furthermore, the revival of interest in traditional fermented foods provides an opportunity to engage younger generations, promoting cultural pride and identity (Jernigan et al., 2023; Materia et al., 2021). Community-based educational initiatives have been paramount in facilitating this engagement, combining formal and informal educational strategies to teach not only the techniques but also the significance behind them. Such programs often employ participatory action research methods, enabling Indigenous communities to assert their values and priorities in food governance and education toward revitalization of their food sovereignty (Michnik et al., 2021).

12.7 Gender, Labor, and the Political Economy of Fermented Indigenous Foods

The intersection of gender, labor, and the political economy of fermented Indigenous foods is an essential area of study that highlights complex social dynamics and traditional practices. Fermentation, being both an economic and cultural activity, illustrates how gender roles and labor divisions are deeply embedded in knowledge and practice within various communities. The division of labor in fermentation practices historically exhibits significant gender disparities, where women often assume the role of custodians of probiotic knowledge. This is particularly evident in many Indigenous populations, where women's expertise in fermentation not only contributes to food security but also reinforces their status within the community. Women's roles in fermentation practices reflect broader trends in the sexual division of labor, particularly in agricultural economies where they are frequently responsible for tasks centered around food production, including fermentation. Historically, women have been instrumental in developing and maintaining these practices, passing down their knowledge through generations. Studies have shown that women's participation in fermentation and related agricultural activities is crucial for maintaining food security in communities (Christine & Emmanuel, 2023), and this knowledge is critical for the health and nutritional benefits derived from fermented foods (Guerra et al., 2022). These roles often emerge from a context where women's unpaid labor and knowledge are undervalued, despite their critical contributions to family and community sustenance.

In examining the commodification of fermented foods, a dual effect can be observed: While commercialization opens new market opportunities for women, it also risks undermining traditional practices and knowledge. The move toward formal markets can create a tension between preserving Indigenous methods of production and adapting to consumer demands in the global marketplace (Ogwu, 2019; Osawaru & Ogwu, 2020; Phommavong et al., 2020; Pyle & Ward, 2003). The case of coffee commercialization in the Bolaven Plateau exemplifies the complexities of this interaction, where traditional production methods meet modern economic frameworks, often resulting in a nuanced division of labor impacted by gender (Phommavong et al., 2020). Such developments emphasize the need for a framework that does not merely exploit women's labor but recognizes and values their knowledge and contributions.

Furthermore, the commodification of fermented foods can lead to inequalities within alternative food networks, where gender disparities in labor distribution may persist. Alternative food networks often aim to provide equitable access to food and promote sustainable practices. However, studies indicate that these networks can still perpetuate gender inequalities by failing to adequately recognize women's labor roles and experiences (Boillat et al., 2023; Castellano, 2014). The question of who benefits from the marketization of Indigenous fermentation practices thus becomes paramount. Social networks and community support also play a vital role in these dynamics. As women engage in fermentation, they often create social spaces that foster community solidarity, enhancing their overall well-being and identity (Yunindyawati et al., 2025). The crafting and sharing of traditional fermented foods serve not only as a means for economic sustenance but also as a cultural connection, empowering women through shared experiences and socialization. The emotional and psychological benefits derived from these communal activities help mitigate stresses associated with their dual roles in the domestic and public spheres (Yunindyawati et al., 2025). The interplay between traditional knowledge and modern economic pressures exemplifies the complexity of gender roles in the production of fermented foods. As communities navigate these changes, there is a recognition of the need to advocate for equitable resource ownership and decision-making opportunities for women, which is crucial for enforcing food security in the long-term (Christine & Emmanuel, 2023; Njeri et al., 2024). Enabling women's engagement in these realms aligns with broader goals of social equity and sustainable development, as it encourages diversified and resilient local food systems.

Moreover, the political economy of food must consider the broader implications of gendered labor roles in the production of fermented foods. Policies aimed at supporting smallholder farmers have increasingly acknowledged the role women play in labor and production processes, although further efforts are required to embed gender equity in agricultural policies and practices (Belesky, 2019). Recognizing women's contributions as not only crucial for production but also integral to sociocultural identity can lead to more profound transformations in food systems and community health overall. The cultural significance of fermented foods further complicates their commodification. In many Indigenous cultures, these foods carry historical and ceremonial importance that transcends mere economic value. This aspect of fermentation can conflict with global market dynamics where the focus is primarily on generating profit, often leading to the marginalization of traditional practices. Preserving the cultural heritage of these foods is essential for maintaining community identity and resilience against global economic pressures (García-Díez et al., 2021; Guerra et al., 2022).

As the landscape of labor, gender, and fermented foods continues to evolve with globalization, it becomes imperative to reflect critically on the social, economic, and political dimensions of this intersection. The recognition of women as active agents in the preservation and innovation of fermentation practices signifies a necessary shift toward a more inclusive understanding of labor in the food economy. Therefore, addressing the inequities faced by women in fermented food production and championing their roles in these processes remain vital. Ultimately, further research in this area should focus on developing frameworks that capture the richness of women's experiences and contributions in the world of fermented Indigenous foods, as well as how these can be supported and expanded in contemporary settings. This exploration is crucial not only for ensuring food security but also for promoting gender equity in agriculture and beyond. By acknowledging the dynamic interplay between gender roles, cultural practices, and economic transformations in fermented food production, we can better understand the pathways to fostering inclusive food systems that honor both tradition and innovation.

12.8 Sustainability of Fermented Indigenous Foods and Probiotic Cultures

The sustainability of fermented Indigenous foods and probiotic cultures is a significant topic in discussions around food security, ecological footprints, and community resilience. Fermentation, recognized as a low-resource technology, presents an ecofriendly approach that aids in food preservation and enhances nutritional profiles. Indigenous fermented foods illustrate this well, especially in areas with limited access to advanced preservation techniques. By utilizing local microbial biodiversity, such as LAB, fermentation successfully extends the shelf life of food products while promoting the health benefits associated with probiotics (Ibrahim et al., 2023; Irakoze, 2024; Irakoze et al., 2021). Indigenous vegetables, particularly those high in phytonutrients, undergo transformations through fermentation. This process helps mitigate the effects of antinutrients, making essential nutrients more bioavailable, which supports maternal and child nutrition in vulnerable communities (Irakoze et al., 2021). Furthermore, lactic fermentation has been identified as a cost-effective preservation method that enhances both food safety and nutritional value. Such fermented products often command higher prices in both local and international markets due to their perceived health benefits and cultural significance. However, market dynamics can limit the full economic potential of these food systems, as consumer price sensitivity may hinder increased adoption (Irakoze, 2024).

Additionally, fermentation acts as a barrier against food scarcity and spoilage, which is crucial for combating food insecurity. The microbial ecosystems actively participate in maintaining food quality by inhibiting spoilage organisms and pathogens through the production of antimicrobial compounds (Dahiya & Nigam, 2022; Ibrahim et al., 2023; Mokoena et al., 2016). The resilience offered by fermented foods is particularly important in regions prone to climate variability or disruptions within supply chains. This resilience is increasingly crucial in light of globalization and the standardization of food systems, reinforcing the importance of reviving Indigenous fermentation practices that employ local raw materials (Das et al., 2016; Johansen et al., 2019; Tamang et al., 2016). Community networks that support fermentation also enhance resilience by reflecting social and cultural dynamics that promote knowledge sharing, resource pooling, and innovative food processing methods (dos Santos et al., 2024; Nuraida, 2015). These collaborative efforts can strengthen food sovereignty by empowering local producers and consumers to take control of their food systems. Participative approaches to fermentation allow communities to preserve their traditions while adapting to contemporary challenges, thus supporting biodiversity in agricultural practices and the microbial strains essential for fermentation (Johansen et al., 2019; Wolfe & Dutton, 2015).

Another aspect of the sustainability of fermented foods is their comparatively low ecological footprint relative to industrial food systems. Fermentation typically does not require the extensive resource inputs associated with modern agriculture, such as chemical fertilizers and preservatives. Moreover, using locally sourced grains and vegetables minimizes transport emissions and bolsters local economies (Das et al., 2016; Irakoze, 2024; Irakoze et al., 2021). As consumer concerns about sustainability grow, fermented products marketed as traditional or artisanal can attract consumer interest and stimulate market demand (Küçükgöz & Trząskowska, 2022; Marco et al., 2021). The health advantages attributed to probiotics in fermented foods also play a significant role in discussions about sustainability. Probiotics are known to modulate gut microbiota, thereby promoting overall health and nutrition (Dahiya & Nigam, 2022; Ibrahim et al., 2023). The production of probiotic-rich fermented foods can alleviate pressure on healthcare systems, especially in areas where access to medical interventions is limited. Engaging with fermentation enables communities to establish food systems that support health and wellness while improving nutritional profiles (Dahiya & Nigam, 2022; Ibrahim et al., 2023).

The microbial diversity found in Indigenous fermented foods, particularly in regions rich in cultural heritage, represents a valuable resource for improving food security and health. The functional capabilities of LAB and yeasts, combined with traditional knowledge systems, form a strong basis for innovative and sustainable food practices (Johansen et al., 2019; Mokoena et al., 2016). The importance of spontaneous fermentation, which allows Indigenous microbiota to thrive through local processing methods, underscores the need for further research into these complex microbial ecosystems that have successfully functioned over centuries (Johansen et al., 2019; Tamang et al., 2016; Wolfe & Dutton, 2015).

Despite these advantages, challenges remain, particularly concerning the scientific validation of health claims related to Indigenous fermented foods. As global interest in probiotics rises, it is essential to address issues of safety and efficacy to build public trust in these products (Capozzi et al., 2017; Dahiya & Nigam, 2022; Ibrahim et al., 2023). Balancing traditional fermentation techniques with modern food safety standards will enhance the sustainability of Indigenous foods, optimize their benefits while minimize associated risks (Capozzi et al., 2017).

12.9 Threats and Transformations of Fermented Indigenous Foods and Probiotic Cultures

The threats and transformations of fermented Indigenous foods and probiotic cultures are multifaceted, originating from both historical and contemporary influences. The preservation of traditional knowledge associated with the fermentation of Indigenous foods is jeopardized by modernization and shifting socioeconomic contexts. As these processes evolve, there is an observable decline in the intergenerational transmission of practices fundamental to the utilization of local microbial strains and traditional fermentation methods. This decline has significant implications, with increased reliance on commercially produced and processed foods leading to a diminished intake of traditional, nutrient-dense options possessing health benefits inherent from their local cultivation and processing (Heaney et al., 2024; Luppens & Power, 2018). The loss of traditional knowledge is compounded by the impact of climate change, which alters the availability of Indigenous food sources, challenging existing food sovereignty and security among Indigenous populations (Hansell, 2025). As traditional landscapes undergo transformation, foundational practices that once ensured dietary diversity and cultural identity are threatened, risking the overall health and well-being of Indigenous communities (Poirier et al., 2024). Preserving these foodways is not only a matter of maintaining cultural identity but is also necessary for achieving health equity (Jernigan et al., 2023). Therefore, systemic efforts to secure food sovereignty are essential in supporting Indigenous peoples in maintaining their traditional practices while counteracting the pressures of globalization and modernization (Nguyen et al., 2023).

Another serious concern affecting the landscape of Indigenous fermented foods is biopiracy and the appropriation of microbial strains. Recent research illustrates the phenomenon where valuable Indigenous microbial cultures are extracted, commoditized, or exploited by external entities without adequate recognition or compensation for Indigenous communities (Ibnouf, 2020). This practice undermines the traditional knowledge associated with these microbial strains and poses ethical challenges regarding ownership and rights to the biological resources utilized for food fermentation (Hodge et al., 2011; Lips, 2021). Biopiracy can lead to a scenario where Indigenous communities are effectively alienated from their own culinary heritage, creating further vulnerabilities amid already precarious health situations caused by diet-related diseases (Nguyen et al., 2023; Timler et al., 2019). The commodification of these strains often necessitates expensive industrial processes that prioritize economic gains over cultural significance, thus fostering competition that detracts from the intrinsic value of Indigenous fermentation practices (Yao et al., 2022). This act of appropriation diminishes the potential health benefits that could arise from the diverse offerings of traditional foods (Chen & Liu, 2020; Takaidza, 2024).

In addition to these underlying threats, there are challenges resulting from health standardization pressures and competition in the industrial food landscape. As commercial food producers cater to regulatory frameworks demanding uniformity and safety, the unique attributes of fermented Indigenous foods often encounter barriers related to these standards. Traditional fermentation processes frequently conflict with industrial health regulations, limiting the participation of small-scale producers within formal markets (Jernigan et al., 2023; Mailer, 2013). This regulatory environment marginalizes traditional food producers and diminishes the availability of probiotic-rich foods that have historically supported community health (Jernigan et al., 2023). Such dysregulation of Indigenous food practices promotes a reliance on processed foods, further exacerbating chronic health conditions within Indigenous populations (Timler et al., 2019; Willows et al., 2024). As prevailing public health narratives lean toward standardization, there is a risk of ignoring the diverse and culturally pertinent food systems that exist within Indigenous communities (Ray et al., 2019; Takaidza, 2024). As Indigenous peoples reclaim their food sovereignty, recognizing traditional foods as a counterforce to chronic illness becomes vital, establishing a relationship between health, culture, and culinary practice (Harper et al., 2019; Jernigan et al., 2023).

To mitigate the various threats to Indigenous fermented foods, it is essential that Indigenous communities engage in assertive advocacy for the protection of their intellectual property rights regarding food traditions and microbial knowledge (Chicmana-Zapata et al., 2023; Ibnouf, 2020). Creating an environment where Indigenous voices are prioritized in discussions about health, nutrition, and food systems fosters a more inclusive and equitable discourse about the necessity of preserving these traditional practices (Andani, 2020; Liao & Kuo, 2024). Alongside such advocacy efforts, there is a need for interdisciplinary research that can frame food sovereignty initiatives within broader socioeconomic and environmental contexts (Asryan, 2024; Timler et al., 2019). By bridging traditional Indigenous knowledge with contemporary scientific understanding, there exists significant potential for revitalizing fermented Indigenous foods—not only as staples of nutrition but also as cultural cornerstones that embody identity and resilience against external pressures (Dennis & Robin, 2020). Implementing community-based initiatives can further facilitate the valorization of traditional fermentation practices, enhancing public health benefits while fostering a renewed appreciation for the cultural significance of these food systems (Jernigan et al., 2023; Poirier et al., 2024). Addressing these threats and transformations demands a holistic approach encompassing policy reform, community engagement, and recognition of Indigenous cultural practices as fundamental to health and well-being. Increasing awareness and education about the significance of Indigenous fermented foods will play a crucial role in reversing trends of loss and appropriation (Willows et al., 2024). Supporting economic models valuing traditional knowledge while ensuring equitable access to local food systems strengthens the fabric of food sovereignty and cultural continuity (Chicmana-Zapata et al., 2023; Harper et al., 2019). Communities empowered with the tools to navigate these challenges will emerge resilient, with vibrant health outcomes and sustainable food practices that honor their rich heritage. Through thoughtful integration of traditional and scientific efforts, Indigenous fermented foods can continue to thrive in both local and global contexts, ensuring that the cultural wealth they represent remains an integral part of contemporary nutrition and health frameworks.

12.10 Future Directions in the Sociobiology of Fermented Indigenous Foods

As global interest in gut health, probiotics, and the human microbiome continues to rise, there is increasing awareness of the value of Indigenous fermented foods as complex, coevolved systems of microbial and cultural knowledge. These foods, often produced through nonindustrial, low-energy, and ecologically embedded methods, represent biocultural reservoirs of microbial diversity. However, they remain significantly underrepresented in both microbiome science and public health policy. Integrating microbiome science with Indigenous knowledge offers a vital pathway for advancing equitable, sustainable, and scientifically rich understandings of probiotic potential rooted in tradition (Marco et al., 2017; Tamang et al., 2016). A truly integrative approach must center on respectful coproduction of knowledge, where Indigenous epistemologies and empirical microbiology are placed in dialogue rather than hierarchy. This involves moving beyond extractive sampling and classification practices to methodologies that are participatory, collaborative, and context aware. Community-based microbiome mapping, ethnomicrobiology, and participatory strain characterization are emerging strategies that allow for scientific validation while respecting cultural protocols and custodianship (Tamang et al., 2020; Wolfe & Dutton, 2015). Such frameworks elevate the expertise of Indigenous communities and foster reciprocal learning that enhances both scientific insight and cultural continuity.

Policy recognition of probiotic heritage is another critical frontier. While international conventions like the Convention on Biological Diversity and the Nagoya Protocol address benefit-sharing for genetic resources, traditional fermented foods and their associated microbial strains often fall outside formal recognition structures. There is an urgent need to develop policy instruments that safeguard the microbial commons of Indigenous peoples such as intellectual property rights for microbial strains, geographical indications for culturally specific fermented products, and their inclusion in food-based dietary guidelines and national biodiversity strategies (Marco et al., 2017; Tamang et al., 2016). These mechanisms would not only protect against biopiracy and commodification but also enhance the visibility and value of Indigenous foodways in national and global arenas.

Looking ahead, future research should prioritize the identification, documentation, and microbiological analysis of underexplored fermented foods, particularly those from marginalized or understudied communities. Thousands of Indigenous fermented products—from leafy vegetables and wild fruits to tubers and root beverages remain undocumented in global food databases. These foods may harbor unique strains of LAB, yeasts, and bacilli with probiotic, enzymatic, or antimicrobial properties yet to be characterized. Their study could lead to new insights in food security, functional nutrition, and microbial ecology (Puerari et al., 2015; Swain et al., 2014). To facilitate this, the creation of open-access microbial biobanks, linked with metadata on cultural origin, ecological context, and community ownership, is essential. These biobanks should operate with transparency, ethical oversight, and adherence to fair benefit-sharing principles, ensuring that microbial knowledge and resources are not divorced from their cultural and geographic roots.

12.11 Conclusion

Fermented Indigenous foods and their associated probiotic cultures exemplify the intricate interplay between biology, culture, and ecology within traditional food systems. This chapter has demonstrated that fermentation is far more than a biochemical process—it is a sociobiological phenomenon rooted in evolutionary adaptation, community resilience, and intergenerational knowledge transmission. Through diverse practices across the globe, Indigenous peoples have cultivated microbial partnerships that enhance food preservation, nutrient availability, and health, while also reinforcing cultural identity, social cohesion, and environmental sustainability. The microbial ecologies of these foods represent living legacies, maintained through localized expertise, ritual practice, and gendered labor. However, the survival of these traditions is increasingly threatened by industrial homogenization, intellectual property conflicts, and the undervaluing of Indigenous epistemologies in global food and health discourse. To address these challenges, this chapter calls for greater recognition of the cultural and biological value of fermented Indigenous foods, the protection of microbial heritage, and the inclusion of Indigenous knowledge holders in food sovereignty and policy dialogues. Ultimately, understanding the sociobiology of fermentation enriches our appreciation of the microbial commons and reaffirms the need to preserve biocultural diversity as a foundation for sustainable and just food futures.

Chapter Reflection

1. How does fermentation serve as both a survival strategy and cultural expression in Indigenous societies?
2. In what ways do human societies and microbial communities coevolve through fermentation practices?
3. Discuss the role of gender, particularly women's labor, in preserving and transmitting fermentation knowledge.
4. How does fermentation improve the nutritional value and bioavailability of nutrients in Indigenous foods?
5. What are some examples of the cultural and ritual significance of fermented foods across different Indigenous communities?
6. How do local environmental factors influence the microbial ecology of Indigenous fermented foods?

7. What are the major threats posed by globalization, industrialization, and biopiracy to Indigenous fermentation systems?
8. Why is the protection of Indigenous microbial heritage important for both cultural identity and food sovereignty?
9. How can participatory, community-based research support the sustainable future of Indigenous fermentation practices?
10. What policy measures could help safeguard Indigenous fermentation knowledge and microbial resources?

REFERENCES

Adesulu-Dahunsi, A., Dahunsi, S., & Ajayeoba, T. (2022). Co-occurrence of *Lactobacillus* species during fermentation of African Indigenous foods: Impact on food safety and shelf-life extension. *Frontiers in Microbiology*, *13*. https://doi.org/10.3389/fmicb.2022.684730

Adesulu-Dahunsi, A., Dahunsi, S., & Olayanju, T. (2020). Synergistic microbial interactions between lactic acid bacteria and yeasts during production of Nigerian Indigenous fermented foods and beverages. *Food Control*, *110*, 106963. https://doi.org/10.1016/j.foodcont.2019.106963

Adesulu-Dahunsi, A., Sanni, A., Jeyaram, K., & Banwo, K. (2017). Genetic diversity of *Lactobacillus* plantarum strains from some Indigenous fermented foods in Nigeria. *LWT-Food Science and Technology*, *82*, 199–206. https://doi.org/10.1016/j.lwt.2017.04.055

Akinrele, I. A. (1970). Fermentation studies on maize during the preparation of traditional African starch-cake food. *Journal of the Science of Food and Agriculture*, *21*(12), 619–625.

Alhareky, M., & Nazir, M. (2021). Dental visits and predictors of regular attendance among female schoolchildren in Dammam, Saudi Arabia. *Clinical Cosmetic and Investigational Dentistry*, *13*, 97–104. https://doi.org/10.2147/ccide.s300108

Amato, K. R., Chaves, Ó. M., Mallott, E. K., Eppley, T. M., Abreu, F., Baden, A. L., Barnett, A. A., Bicca-Marques, J. C., Boyle, S. A., Campbell, C. J., Chapman, C. A., De la Fuente, M. F., Fan, P., Fashing, P. J., Felton, A., Fruth, B., Fortes, V. B., Grueter, C. C., Hohmann, G., … Zeng, Y. (2021). Fermented food consumption in wild nonhuman primates and its ecological drivers. *American Journal of Physical Anthropology, 175*(3), 513–530. https://doi.org/10.1002/ajpa.24257

Andani, A. (2020). Welfare effects of the production of Indigenous food crops in farming communities in the northern region of Ghana. *South Asian Journal of Social Studies and Economics, 8*(2), 35–45. https://doi.org/10.9734/sajsse/2020/v8i230208

Asryan, V. (2024). Fundamentals of bakery biotechnology. In *State and development prospects of agribusiness* (pp. 113–114). EDP Sciences. https://doi.org/10.23947/interagro.2024.113-114

Belesky, P. (2019). *Rice, politics and power: The political economy of food insecurity in East Asia* [Thesis, University of Queensland]. https://doi.org/10.31237/osf.io/hn264

Boillat, S., Bottazzi, P., & Sabaly, I. (2023). The division of work in Senegalese conventional and alternative food networks: A contributive justice perspective. *Frontiers in Sustainable Food Systems, 7*. https://doi.org/10.3389/fsufs.2023.1127593

Cano, A., & Suárez, M. (2020). Ethnobiology of algarroba beer, the ancestral fermented beverage of the Wichí people of the Gran Chaco I: A detailed recipe and a thorough analysis of the process. *Journal of Ethnic Foods, 7*(1), 4. https://doi.org/10.1186/s42779-019-0028-0

Capozzi, V., Fragasso, M., Romaniello, R., Berbegal, C., Russo, P., & Spano, G. (2017). Spontaneous food fermentations and potential risks for human health. *Fermentation, 3*(4), 49. https://doi.org/10.3390/fermentation3040049

Castellano, R. (2014). Alternative food networks and food provisioning as a gendered act. *Agriculture and Human Values, 32*(3), 461–474. https://doi.org/10.1007/s10460-014-9562-y

Chen, G., & Liu, X. (2020). On the future fermentation. *Microbial Biotechnology, 14*(1), 18–21. https://doi.org/10.1111/1751-7915.13674

Chicmana-Zapata, V., Arotoma-Rojas, I., Anza-Ramirez, C., Ford, J., Galappaththi, E. K., Pickering, K., Sacks, E., Togarepi, C., Perera, C. D., van Bavel, B., Hyams, K., Akugre, F. A., Nkalubo, J., Dharmasiri, I., Nakwafila, O., Mensah, A., Miranda, J. J., & Zavaleta-Cortijo, C. (2023). Justice implications of health and food security policies for Indigenous peoples facing Covid-19: A qualitative study and policy analysis in Peru. *Health Policy and Planning, 38*(Supplement_2), ii36–ii50. https://doi.org/10.1093/heapol/czad051

Christine, A., & Emmanuel, K. (2023). Refocusing on gender roles in agriculture and their impact on household food security: An in-depth analysis of chosen wards within Kisarawe District, Tanzania. *IDOSR Journal of Humanities and Social Sciences*, *8*(2), 109–124. https://doi.org/10.59298/idosrjhss/2023/12.1.5702

Cosme, F., Nunes, F. M., & Filipe-Ribeiro, L. (2024). Winemaking: Advanced technology and flavor research. *Foods*, *13*(12), 1937. https://doi.org/10.3390/foods13121937

Dahiya, D., & Nigam, P. (2022). Nutrition and health through the use of probiotic strains in fermentation to produce non-dairy functional beverage products supporting gut microbiota. *Foods*, *11*(18), 2760. https://doi.org/10.3390/foods11182760

Das, G., Patra, J. P., Singdevsachan, S. K., Gouda, S., & Shin, H. (2016). Diversity of traditional and fermented foods of the Seven Sister States of India and their nutritional and nutraceutical potential: A review. *Frontiers in Life Science*, *9*(4), 292–312. https://doi.org/10.1080/21553769.2016.1249032

Dennis, M., & Robin, T. (2020). Healthy on our own terms. *Journal of Critical Dietetics*, *5*(1), 4–11. https://doi.org/10.32920/cd.v5i1.1333

Devirgiliis, C., Zinno, P., & Perozzi, G. (2013). Update on antibiotic resistance in foodborne *Lactobacillus* and *Lactococcus* species. *Frontiers in Microbiology*, *4*. https://doi.org/10.3389/fmicb.2013.00301

Dey, T. K., Lindahl, J. F., Sanjukta, R., Milton, A. A. P., Das, S., Kannan, P., Lundkvist, Å., Sen, A., & Ghatak, S. (2023). Characterization of lactic acid bacteria and pathogens isolated from traditionally fermented foods, in relation to food safety and antimicrobial resistance in tribal hill areas of Northeast India. *Journal of Food Quality*, *2023*, 1–12. https://doi.org/10.1155/2023/6687015

Domingo, A., Charles, K., Jacobs, M., Brooker, D., & Hanning, R. (2021). Indigenous community perspectives of food security, sustainable food systems and strategies to enhance access to local and traditional healthy food for partnering Williams Treaties First Nations (Ontario, Canada). *International Journal of Environmental Research and Public Health*, *18*(9), 4404. https://doi.org/10.3390/ijerph18094404

dos Santos, V. M., Tan, Y., Zhu, Y., Wijffels, R., Zhang, H., Scott, W., & Xu, Y. (2024). Controlling metabolic stability of food microbiome for stable Indigenous liquor fermentation. https://doi.org/10.21203/rs.3.rs-3745207/v1

Dunn, R. R., Wilson, J., Nichols, L. M., & Gavin, M. C. (2021). Toward a global ecology of fermented foods. *Current Anthropology, 62*(S24), S220–S232. https://doi.org/10.1086/716014

Elechi, J., Nwiyi, I., & Oboh, E. (2022). Fermentation and diet diversity: Biochemical and functional properties of fermented mango (mangifera indica l) pulp flour. *European Food Science and Engineering, 3*(2), 44–51. https://doi.org/10.55147/efse.1181022

Evivie, S. E., Ogwu, M. C., Abdelazez, A., Bian, X., Liu, F., Li, B., & Huo, G. (2020). Suppressive effects of *Streptococcus thermophilus* KLDS 3.1003 on some foodborne pathogens revealed through *in vitro, in vivo* and genomic insights. *Food and Function, 11*, 6573–6587. http://doi.org/10.1039/D0FO01218A

Evivie, S. E., Ogwu, M. C., Abdelazez, A., Bian, X., Liu, F., Li, B., & Huo, G. (2021). Correction: Suppressive effects of *streptococcus thermophilus* KLDS 3.1003 on some foodborne pathogens revealed through *in vitro, in vivo* and genomic insights. *Food and Function, 12*(7), 3280. http://doi.org/10.1039/D1FO90021H

Evivie, S. E., Ogwu, M. C., Ebabhamiegbebho, P. A., Abel, E. S., Imaren, J. O., & Igene, J. O. (2020). Packaging and the Nigerian food industry: Challenges and opportunities. In C. A. Ogunlade, K. M. Adeleke, & M. T. Oladejo (Eds.), *Food technology and culture in Africa* (pp. 28–99). Reamsworth Publishing.

Fischer, J., Zutphen, K., & Freymond, M. (2025). Exploring traditional fermentation: An introduction. https://doi.org/10.52439/iqpu9681

García-Díez, J., Gonçalves, C., Grispoldi, L., Cenci-Goga, B., & Saraiva, C. (2021). Determining food stability to achieve food security. *Sustainability, 13*(13), 7222. https://doi.org/10.3390/su13137222

Ghosh-Jerath, S., Singh, A., Lyngdoh, T., Magsumbol, M., Kamboj, P., & Goldberg, G. (2018). Estimates of Indigenous food consumption and their contribution to nutrient intake in Oraon tribal women of Jharkhand, India. *Food and Nutrition Bulletin, 39*(4), 581–594. https://doi.org/10.1177/0379572118805652

Guerra, L., Cevallos-Cevallos, J., Weckx, S., & Ruales, J. (2022). Traditional fermented foods from Ecuador: A review with a focus on microbial diversity. *Foods, 11*(13), 1854. https://doi.org/10.3390/foods11131854

Gupta, A., & Sharma, N. (2017). In vitro characterization of lactic acid bacteria isolated from lasoda bari—A rare fermented food of Himachal Pradesh-India for potential probiotic attributes. *Journal of Microbiology Biotechnology and Food Sciences*, *6*(6), 1323–1328. https://doi.org/10.15414/jmbfs.2017.6.6.1323-1328

Hansell, R. (2025). Indigenous food sovereignty: Literature review. *Fourth World Journal*, *24*(2), 147–184. https://doi.org/10.63428/7zjhvq86

Harper, S. L., Berrang-Ford, L., Cárcamo, C., Cunsolo, A., Edge, V. L., Ford, J. D., Llanos, A., Lwasa, S., & Namanya, D. B. (2019). The Indigenous climate–food–health nexus: Indigenous voices, stories, and lived experiences in Canada, Uganda, and Peru. In L. R. Mason & J. Rigg (Eds.), *People and climate change: Vulnerability, adaptation, and social justice* (pp. 184–207). Oxford University Press. https://doi.org/10.1093/oso/9780190886455.003.0010

Heaney, D., Padilla-Zakour, O., & Chen, C. (2024). Processing and preservation technologies to enhance Indigenous food sovereignty, nutrition security and health equity in North America. *Frontiers in Nutrition*, *11*. https://doi.org/10.3389/fnut.2024.1395962

Hodge, A., Cunningham, J., Maple-Brown, L., Dunbar, T., & O'Dea, K. (2011). Plasma carotenoids are associated with socioeconomic status in an urban Indigenous population: An observational study. *BMC Public Health*, *11*(1), 76. https://doi.org/10.1186/1471-2458-11-76

Hong, L., Zhuo, J., Lei, Q., Jiang-ju, Z., Ahmed, S., Wang, C., Long, Y., Li, F., & Long, C. (2015). Ethnobotany of wild plants used for starting fermented beverages in Shui communities of Southwest China. *Journal of Ethnobiology and Ethnomedicine*, *11*(1). https://doi.org/10.1186/s13002-015-0028-0

Ibnouf, F. O. (2020). Can Indigenous foods play role as 'the food of survival'?. *European Journal of Agriculture and Food Sciences*, *2*(4). https://doi.org/10.24018/ejfood.2020.2.4.26

Ibrahim, S. A., Yeboah, P. J., Ayivi, R. D., Eddin, A. S., Wijemanna, N. D., Paidari, S., & Bakhshayesh, R. (2023). A review and comparative perspective on health benefits of probiotic and fermented foods. *International Journal of Food Science and Technology*, *58*(10), 4948–4964. https://doi.org/10.1111/ijfs.16619

Irakoze, M. (2024). The role of lactic fermentation in ensuring the safety and extending the shelf life of African Indigenous vegetables and its economic potential. *Applied Research*, *4*(1). https://doi.org/10.1002/appl.202400131

Irakoze, M. L., Wafula, E. N., & Owaga, E. (2021). Potential role of African fermented Indigenous vegetables in maternal and child nutrition in sub-Saharan Africa. *International Journal of Food Science, 2021*(1), 1–11. https://doi.org/10.1155/2021/3400329

Janiak, M., Pinto, S., Duytschaever, G., Carrigan, M., & Melin, A. (2020). Genetic evidence of widespread variation in ethanol metabolism among mammals: Revisiting the "myth" of natural intoxication. *Biology Letters, 16*(4), 20200070. https://doi.org/10.1098/rsbl.2020.0070

Jans, C., Meile, L., Kaindi, D. W. M., Kogi-Makau, W., Lamuka, P., Renault, P., Kreikemeyer, B., Lacroix, C., Hattendorf, J., Zinsstag, J., Schelling, E., Fokou, B., & Bonfoh, B. (2017). African fermented dairy products—Overview of predominant technologically important microorganisms focusing on African streptococcus infantarius variants and potential future applications for enhanced food safety and security. *International Journal of Food Microbiology, 250*, 27–36. https://doi.org/10.1016/j.ijfoodmicro.2017.03.012

Jernigan, V. B. B., Maudrie, T. L., Nikolaus, C. J., Benally, T., Johnson, S., Teague, T., Mayes, M., Jacob, T., & Taniguchi, T. (2021). Food sovereignty indicators for Indigenous community capacity building and health. *Frontiers in Sustainable Food Systems, 5*. https://doi.org/10.3389/fsufs.2021.704750

Jernigan, V. B. B., Nguyen, C. J., Maudrie, T. L., Demientieff, L. X., Black, J. C., Mortenson, R., Wilbur, R. E., Clyma, K. R., Lewis, M., & Lopez, S. V. (2023). Food sovereignty and health: A conceptual framework to advance research and practice. *Health Promotion Practice, 24*(6), 1070–1074. https://doi.org/10.1177/15248399231190367

Jernigan, V. B. B., Taniguchi, T., Nguyen, C. J., London, S. M., Henderson, A., Maudrie, T. L., Blair, S., Clyma, K. R., Lopez, S. V., & Jacob, T. (2023). Food systems, food sovereignty, and health: Conference shares linkages to support Indigenous community health. *Health Promotion Practice, 24*(6), 1109–1116. https://doi.org/10.1177/15248399231190360

JNTO [Japan National Tourism Organization]. (2022). *Learn about Japan's unique superfood, natto* [natto bowl]. Japan Travel. https://www.japan.travel/en/sg/story/learn-about-japans-unique-superfood-natto/

Johansen, P., Owusu-Kwarteng, J., Parkouda, C., Padonou, S., & Jespersen, L. (2019). Occurrence and importance of yeasts in Indigenous fermented food and beverages produced in sub-Saharan Africa. *Frontiers in Microbiology, 10*. https://doi.org/10.3389/fmicb.2019.01789

Knez, E., Kadac-Czapska, K., & Grembecka, M. (2023). Effect of fermentation on the nutritional quality of the selected vegetables and legumes and their health effects. *Life*, *13*(3), 655. https://doi.org/10.3390/life13030655

Küçükgöz, K., & Trząskowska, M. (2022). Nondairy probiotic products: Functional foods that require more attention. *Nutrients*, *14*(4), 753. https://doi.org/10.3390/nu14040753

Kuoda Travel. (2025, April 9). *Chicha: The ancient Andean beverage that connects Peru's past and present* [Chichia]. Kuoda Travel. https://www.kuoda-travel.com/blog/chicha-andean-beverage/

Lamba, J., Goomer, S., & Nain, L. (2019). Exploring Indigenous fermented foods of India for the presence of lactic acid bacteria. *Nutrition and Food Science*, *49*(5), 942–954. https://doi.org/10.1108/nfs-08-2018-0228

Leeuwendaal, N. K., Stanton, C., O'Toole, P. W., & Beresford, T. P. (2022). Fermented foods, health and the gut microbiome. *Nutrients*, *14*(7), 1527. https://doi.org/10.3390/nu14071527

Liao, Z., & Kuo, F. (2024). Building Indigenous health: Insights from Indigenous adults on knowledge integration and adaptive strategies. *Innovation in Aging*, *8*(Suppl. 1), 558–559. https://doi.org/10.1093/geroni/igae098.1826

Lima, T., Hosken, B., Venturim, B., Lopes, I., & Martin, J. (2022). Traditional Brazilian fermented foods: Cultural and technological aspects. *Journal of Ethnic Foods*, *9*(1), 35. https://doi.org/10.1186/s42779-022-00153-4

Lips, D. (2021). Fuelling the future of sustainable sugar fermentation across generations. *Engineering Biology*, *6*(1), 3–16. https://doi.org/10.1049/enb2.12017

Luppens, L., & Power, E. (2018). "Aboriginal isn't just about what was before, it's what's happening now": Perspectives of Indigenous peoples on the foods in their contemporary diets. *Canadian Food Studies/La Revue Canadienne Des Études Sur L Alimentation*, *5*(2), 142–161. https://doi.org/10.15353/cfs-rcea.v5i2.219

Mailer, G. (2013). Decolonizing the diet: Synthesizing Native-American history, immunology, and nutritional science. *Journal of Evolution and Health an Ancestral Health Society Publication*, *1*(1). https://doi.org/10.15310/2334-3591.1014

Majumdar, R., Roy, D., Bejjanki, S., & Bhaskar, N. (2015). Chemical and microbial properties of *shidal*, a traditional fermented fish of Northeast India.

Journal of Food Science and Technology, 53(1), 401–410. https://doi.org/10.1007/s13197-015-1944-7

Makarova, K., Slesarev, A., Wolf, Y., Sorokin, A., Mirkin, B., Koonin, E., Pavlov, A., Pavlova, N., Karamychev, V., Polouchine, N., Shakhova, V., Grigoriev, I., Lou, Y., Rohksar, D., Lucas, S., Huang, K., Goodstein, D. M., Hawkins, T., Plengvidhya, V., … Mills, D. (2006). Comparative genomics of the lactic acid bacteria. *Proceedings of the National Academy of Sciences, 103*(42), 15611–15616. https://doi.org/10.1073/pnas.0607117103

Manna, M. S., Tamer, Y. T., Gaszek, I., Poulides, N., Ahmed, A., Wang, X., Toprak, F. C., Woodard, D. R., Koh, A. Y., Williams, N. S., Borek, D., Atilgan, A. R., Hulleman, J. D., Atilgan, C., Tambar, U., & Toprak, E. (2021). A trimethoprim derivative impedes antibiotic resistance evolution. *Nature Communications, 12*(1), 2949. https://doi.org/10.1038/s41467-021-23191-z

Marco, M. L., Heeney, D., Binda, S., Cifelli, C. J., Cotter, P. D., Foligné, B., Gänzle, M., Kort, R., Pasin, G., Pihlanto, A., Smid, E. J., & Hutkins, R. (2017). Health benefits of fermented foods: Microbiota and beyond. *Current Opinion in Biotechnology, 44*, 94–102.

Marco, M. L., Sanders, M. E., Gänzle, M., Arrieta, M. C., Cotter, P. D., De Vuyst, L., Hill, C., Holzapfel, W., Lebeer, S., Merenstein, D., Reid, G., Wolfe, B. E., & Hutkins, R. (2021). The international scientific association for probiotics and prebiotics (ISAPP) consensus statement on fermented foods. *Nature Reviews Gastroenterology and Hepatology, 18*(3), 196–208. https://doi.org/10.1038/s41575-020-00390-5

Masumuzzaman, Md., Evivie, S. E., Ogwu, M. C., Bailiang, L., Jin-Cheng, D., Wan, L., Guicheng, H., Fei, L., & Song, W. (2021). Genomic and *in vitro* properties of the dairy *Streptococcus thermophilus* SMQ-301 strain against selected pathogens. *Food and Function, 12*(8), 7017–7028. http://doi.org/10.1039/D0FO02951C

Materia, V., Linnemann, A., Smid, E., & Schoustra, S. (2021). Contribution of traditional fermented foods to food systems transformation: Value addition and inclusive entrepreneurship. *Food Security, 13*(5), 1163–1177. https://doi.org/10.1007/s12571-021-01185-5

McGovern, P. E., Zhang, J., Tang, J., Zhang, Z., Hall, G. R., Moreau, R. A., Nuñez, A., Butrym, E. D., Richards, M. P., Wang, C.-S., Cheng, G., Zhao, Z., & Wang, C. (2004). Fermented beverages of pre- and proto-historic China.

Proceedings of the National Academy of Sciences, *101*(51), 17593–17598. https://doi.org/10.1073/pnas.0407921102

Michnik, K., Thompson, S., & Beardy, B. (2021). Moving your body, soul, and heart to share and harvest food. *Canadian Food Studies/La Revue Canadienne Des Études Sur L Alimentation*, *8*(2). https://doi.org/10.15353/cfs-rcea.v8i2.446

Mokoena, M., Mutanda, T., & Olaniran, A. (2016). Perspectives on the probiotic potential of lactic acid bacteria from African traditional fermented foods and beverages. *Food and Nutrition Research*, *60*(1), 29630. https://doi.org/10.3402/fnr.v60.29630

Motlhanka, K., Zhou, N., & Lebani, K. (2018). Microbial and chemical diversity of traditional non-cereal based alcoholic beverages of sub-Saharan Africa. *Beverages*, *4*(2), 36. https://doi.org/10.3390/beverages4020036

Neufeld, H., Richmond, C., & Centre, S. (2017). Impacts of place and social spaces on traditional food systems in Southwestern Ontario. *International Journal of Indigenous Health*, *12*(1), 93–115. https://doi.org/10.18357/ijih112201716903

Nguyen, C. J., Wilbur, R. E., Henderson, A., Sowerwine, J., Mucioki, M., Sarna-Wojcicki, D., Ferguson, G. L., Maudrie, T. L., Moore-Wilson, H., Wark, K., & Jernigan, V. B. B. (2023). Framing an Indigenous food sovereignty research agenda. *Health Promotion Practice*, *24*(6), 1117–1123. https://doi.org/10.1177/15248399231190362

Nikolaus, C. J., Johnson, S., Benally, T., Maudrie, T., Henderson, A., Nelson, K., Lane, T., Segrest, V., Ferguson, G. L., Buchwald, D., Jernigan, V. B. B., & Sinclair, K. (2022). Food insecurity among American Indian and Alaska native people: A scoping review to inform future research and policy needs. *Advances in Nutrition*, *13*(5), 1566–1583. https://doi.org/10.1093/advances/nmac008

Njeri, R., Mainah, M., Okemwa, P., & Guettou, N. (2024). Gender roles and labor dynamics in household composting practices in Ndeiya, Kiambu County, Kenya. *International Journal of Research and Scientific Innovation*, *XI*(XII), 778–782. https://doi.org/10.51244/ijrsi.2024.11120069

Nuraida, L. (2015). A review: Health promoting lactic acid bacteria in traditional Indonesian fermented foods. *Food Science and Human Wellness*, *4*(2), 47–55. https://doi.org/10.1016/j.fshw.2015.06.001

Ogwu, M. C. (2019). Towards sustainable development in Africa: The challenge of urbanization and climate change adaptation. In P. B. Cobbinah & M. Addaney (Eds.), *The geography of climate change adaptation in Urban Africa* (pp. 29–55). Springer Nature. http://doi.org/10.1007/978-3-030-04873-0_2

Ogwu, M. C. (2023). Local food crops in Africa: Sustainable utilization, threats, and traditional storage strategies. In S. C. Izah & M. C. Ogwu (Eds.), *Sustainable utilization and conservation of Africa's biological resources and environment.* Sustainable Development and Biodiversity (Vol. 888, pp. 353–374). Springer. https://doi.org/10.1007/978-981-19-6974-4_13

Ogwu, M. C., & Osawaru, M. E. (2022). Traditional methods of plant conservation for sustainable utilization and development. In S. C. Izah (Ed.), *Biodiversity in Africa: Potentials, threats and conservation.* Sustainable Development and Biodiversity (Vol. 29, pp. 451–472). Springer. http://doi.org/10.1007/978-981-19-3326-4_17

Ojeda-Linares, C., Álvarez-Ríos, G. D., Figueredo-Urbina, C. J., Islas, L. A., Lappe-Oliveras, P., Nabhan, G. P., Torres-García, I., Vallejo, M., & Casas, A. (2021). Traditional fermented beverages of Mexico: A biocultural unseen foodscape. *Foods*, *10*(10), 2390. https://doi.org/10.3390/foods10102390

Ojeda-Linares, C., Vallejo, M., Lappe-Oliveras, P., & Casas, A. (2020). Traditional management of microorganisms in fermented beverages from cactus fruits in Mexico: An ethnobiological approach. *Journal of Ethnobiology and Ethnomedicine, 16*(1). https://doi.org/10.1186/s13002-019-0351-y

Osawaru, M. E., & Ogwu, M. C. (2020). Survey of plant and plant products in local markets within Benin City and environs. In L. W. Filho, N. Ogugu, D. Ayal, L. Adelake, & I. da Silva (Eds.), *African handbook of climate change adaptation* (pp. 1–24). Springer Nature. http://doi.org/10.1007/978-3-030-42091-8_159-1

Oyedeji, O., Ayeni, F. A., & Adeboye, A. S. (2013). Comparative microbial analysis of spontaneous fermented ogi prepared from different cereals. *Journal of Food Research, 2*(5), 13–21.

Parkouda, C., Nielsen, D. S., Azokpota, P., Ouoba, L. I. I., Amoa-Awua, W. K., Thorsen, L., Hounhouigan, J. D., Jensen, J. S., Tano-Debrah, K., Diawara, B., & Jakobsen, M. (2009). The microbiology of alkaline-fermentation of Indigenous seeds used as food condiments in Africa and Asia. *Critical Reviews in Microbiology, 35*(2), 139–156. https://doi.org/10.1080/10408410902793056

Patra, J., Das, G., Paramithiotis, S., & Shin, H. (2016). Kimchi and other widely consumed traditional fermented foods of Korea: A review. *Frontiers in Microbiology*, *7*. https://doi.org/10.3389/fmicb.2016.01493

Phommavong, S., Douangphachanh, M., & Svengsucksa, K. (2020). Coffee commercialization in the Bolaven Plateau in the Southern of Lao PDR. In D. T. Castanheira (Ed.), *Coffee - production and research*. IntechOpen. https://doi.org/10.5772/intechopen.90105

Poirier, B., Soares, G., Neufeld, H., Hedges, J., Sethi, S., & Jamieson, L. (2024). Conceptualising the relationships between food sovereignty, food security and oral health among global Indigenous communities: A scoping review. *Public Health Nutrition*, *27*(1). https://doi.org/10.1017/s1368980024001198

Puerari, C., Magalhães-Guedes, K. T., & Schwan, R. F. (2015). Diversity of lactic acid bacteria during fermentation of *chicha*, a traditional beverage of Brazilian Amerindians. *World Journal of Microbiology and Biotechnology*, *31*, 1767–1775.

Pyle, J. L., & Ward, K. B. (2003). Recasting our understanding of gender and work during global restructuring. *International Sociology*, *18*(3), 461–489. https://doi.org/10.1177/02685809030183002

Raghuvanshi, R., Grayson, A., Schena, I., Amanze, O., Suwintono, K., & Quinn, R. (2019). Microbial transformations of organically fermented foods. *Metabolites*, *9*(8), 165. https://doi.org/10.3390/metabo9080165

Ray, L., Burnett, K., Cameron, A., Joseph, S., LeBlanc, J., Parker, B., Recollet, A., & Sergerie, C. (2019). Examining Indigenous food sovereignty as a conceptual framework for health in two urban communities in Northern Ontario, Canada. *Global Health Promotion*, *26*(3_suppl), 54–63. https://doi.org/10.1177/1757975919831639

Rezac, S., Kok, C., Heermann, M., & Hutkins, R. (2018). Fermented foods as a dietary source of live organisms. *Frontiers in Microbiology*, *9*. https://doi.org/10.3389/fmicb.2018.01785

Richmond, C., Steckley, M., Neufeld, H., Kerr, R. B., Wilson, K., & Dokis, B. (2020). First nations food environments: Exploring the role of place, income, and social connection. *Current Developments in Nutrition*, *4*(8), nzaa108. https://doi.org/10.1093/cdn/nzaa108

Samtiya, M., Aluko, R. E., Puniya, A. K., & Dhewa, T. (2021). Enhancing micronutrients bioavailability through fermentation of plant-based foods: A concise review. *Fermentation, 7*(2), 63. https://doi.org/10.3390/fermentation7020063

Sasirekhamani, M., Selvi, A., Mathuravalli, S., & Bhanumathi, K. (2025). Nutritional and cultural significance of traditional fermented fish products in North-East India—A review. *International Journal of Scientific Research in Science and Technology, 12*(2), 1321–1328. https://doi.org/10.32628/ijsrst251222692

Selhub, E., Logan, A., & Bested, A. (2014). Fermented foods, microbiota, and mental health: Ancient practice meets nutritional psychiatry. *Journal of Physiological Anthropology, 33*(1). https://doi.org/10.1186/1880-6805-33-2

Sidiq, F. F., Coles, D., Hubbard, C., Clark, B., & Frewer, L. J. (2022a). Factors influencing consumption of traditional diets: Stakeholder views regarding sago consumption among the Indigenous peoples of West Papua. *Agriculture and Food Security, 11*(1). https://doi.org/10.1186/s40066-022-00390-5

Sidiq, F. F., Coles, D., Hubbard, C., Clark, B., & Frewer, L. J. (2022b). The role of traditional diets in promoting food security for Indigenous peoples in low- and middle-income countries: A systematic review. *IOP Conference Series Earth and Environmental Science, 978*(1), 012001. https://doi.org/10.1088/1755-1315/978/1/012001

Sindhu, S., & Khetarpaul, N. (2002). Effect of probiotic fermentation on antinutrients and in vitro protein and starch digestibilities of Indigenously developed RWGT food mixture. *Nutrition and Health, 16*(3), 173–181. https://doi.org/10.1177/026010600201600303

Song, Y., Sun, Z., Guo, C., Wu, Y., Liu, W., Yu, J., Menghe, B., Yang, R., & Zhang, H. (2016). Genetic diversity and population structure of *Lactobacillus delbrueckii* subspecies *bulgaricus* isolated from naturally fermented dairy foods. *Scientific Reports, 6*(1), 22704. https://doi.org/10.1038/srep22704

Sousa, N. S. O., Souza, É. S., Canto, E. S. M., Silva, J. P. A., Carneiro, L. M., Franco-de-Sá, J. F. O., & Souza, J. V. B. (2023). Amazonian fermentations: An analysis of industrial and social technology as tools for the development of bioeconomy in the region. *Brazilian Journal of Biology, 83*. https://doi.org/10.1590/1519-6984.276493

Suvarna, V., Nivetha, N., Shraddha, A., & Abhishek, R. (2018). Enhancement of bioavailable iron and calcium contents in fermented linseed (*Linum usitatissimum* L.) beverages. *Asian Journal of Dairy and Food Research, 37*(4), 331–334. https://doi.org/10.18805/ajdfr.dr-1397

Suwannasom, N., Siriphap, A., Japa, O., Thephinlap, C., Thepmalee, C., & Khoothiam, K. (2025). Lactic acid bacteria from Northern Thai (Lanna) fermented foods: A promising source of probiotics with applications in symbiotic formulation. *Foods*, *14*(2), 244. https://doi.org/10.3390/foods14020244

Svanberg, I. (2015). Ræstur fiskur: Air-dried fermented fish the Faroese way. *Journal of Ethnobiology and Ethnomedicine*, *11*(1), 76. https://doi.org/10.1186/s13002-015-0064-9

Swain, M. R., Anandharaj, M., Ray, R. C., & Rani, R. P. (2014). Fermented fruits and vegetables of Asia: A potential source of probiotics. *Biotechnology Research International*, *2014*, 1–19.

Swiderska, K., Argumedo, A., Wekesa, C., Ndalilo, L., Song, Y., Rastogi, A., & Ryan, P. (2022). Indigenous peoples' food systems and biocultural heritage: Addressing Indigenous priorities using decolonial and interdisciplinary research approaches. *Sustainability*, *14*(18), 11311. https://doi.org/10.3390/su141811311

Taale, E., Gambogou, B., Sawadogo, A., Cissé, H., Souho, T., Amouzou, S. K., & Tchabi, A. (2024). Involved microorganisms in the production of Indigenous fermented food from West Africa: Technological characteristics and probiotic power. In M. C. Hueda (Ed.), *The science of fermentation*. IntechOpen. https://doi.org/10.5772/intechopen.114893

Takaidza, S. (2024). Indigenous South African food: Nutrition and health benefits. In M. Soto-Hernández, E. Aguirre-Hernández & M. Palma-Tenango (Eds.), *Phytochemicals in agriculture and food*. IntechOpen. https://doi.org/10.5772/intechopen.110732

Tamang, J. P., Cotter, P. D., Endo, A., Han, N. S., Kort, R., Liu, S. Q., Mayo, B., Westerik, N., & Hutkins, R. (2020). Fermented foods in a global age: East meets west. *Comprehensive Reviews in Food Science and Food Safety*, *19*(1), 184–217. https://doi.org/10.1111/1541-4337.12520

Tamang, J. P., & Samuel, D. (2010). Dietary cultures and antiquity of fermented foods and beverages. In J. P. Tamang & K. Kailasapathy (Eds.), *Fermented foods and beverages of the world* (pp. 1–40). CRC Press.

Tamang, J. P., Tamang, B., Schillinger, U., Franz, C. M. A. P., Gores, M., & Holzapfel, W. H. (2005). Identification of predominant lactic acid bacteria isolated from traditionally fermented vegetable products of the Eastern Himalayas. *International Journal of Food Microbiology*, *105*(3), 347–356.

Tamang, J. P., Watanabe, K., & Holzapfel, W. (2016). Review: Diversity of microorganisms in global fermented foods and beverages. *Frontiers in Microbiology*, *7*, 377. https://doi.org/10.3389/fmicb.2016.00377

Timler, K., Varcoe, C., & Brown, H. (2019). Growing beyond nutrition. *International Journal of Indigenous Health*, *14*(2), 95–114. https://doi.org/10.32799/ijih.v14i2.31938

Willows, N., Loewen, O., Blanchet, R., Godrich, S., Veugelers, P., & Committee, A. (2024). Indigenous identity and household food insecurity are associated with poor health outcomes in Canada. *Canadian Journal of Dietetic Practice and Research*, *85*(2), 76–82. https://doi.org/10.3148/cjdpr-2023-024

Wolfe, B., & Dutton, R. (2015). Fermented foods as experimentally tractable microbial ecosystems. *Cell*, *161*(1), 49–55. https://doi.org/10.1016/j.cell.2015.02.034

Yao, S., Hao, L., Zhou, R., Jin, Y., Huang, J., & Wu, C. (2022). Multispecies biofilms in fermentation: Biofilm formation, microbial interactions, and communication. *Comprehensive Reviews in Food Science and Food Safety*, *21*(4), 3346–3375. https://doi.org/10.1111/1541-4337.12991

Yetneberk, S., De Kock, H. L., Rooney, L. W., & Taylor, J. R. N. (2005). Effects of sorghum cultivar on injera quality. *Cereal Chemistry*, *82*(3), 312–318.

Yonetani, T. (2015). Natto: A Japanese fermented soybean food. *Microbial Ecology in Health and Disease*, *26*, 27799.

Young, V., & Kiefer, A. (2014). Kimchi: Spicy science for the undergraduate microbiology laboratory. *Journal of Microbiology and Biology Education*, *15*(2), 297–298. https://doi.org/10.1128/jmbe.v15i2.695

Yunindyawati, Y., Lidya, E., & Azni, U. (2025). Weaving life: Women's livelihood strategies based on local wisdom and stress reduction through purun (*Eleocharis dulcis*) crafting among female artisans in rural peatland areas. *Society*, *13*(1), 1–17. https://doi.org/10.33019/society.v13i1.701

CHAPTER 13

Medicinal Plants as Nutritional Food Sources in Indigenous Societies: Sociobiological Perspectives

Abstract

Medicinal plants serve as food and therapeutic agents and represent a critical yet underexplored intersection in human ecology with profound implications for nutrition, health, and sustainability. This chapter reviews the sociobiological dimensions of such plants in Indigenous societies, drawing on ethnobotanical research, phytochemical data, and ecological insights. It explores how traditional knowledge systems shape the dual use of species like *Moringa oleifera* (drumstick tree), *Curcuma longa* (turmeric), *Bidens pilosa* (blackjack), and *Gongronema latifolium*, highlighting coevolved human–plant relationships where cultural practices are deeply intertwined with empirical health benefits. The review examines key sociobiological drivers like ecological availability, disease burden, and modes of cultural transmission that influence the selection, preparation, and persistence of these multifunctional plants. It also addresses pressing threats to these systems, including the erosion of biocultural knowledge and unsustainable commercialization. Framing these plants within a sociobiological paradigm reveals their central role in sustaining resilient food systems, promoting health equity, and advancing integrative approaches in planetary health and One Health initiatives. The recognition and preservation of Indigenous food–medicine systems such as the use of *M. oleifera* leaves in West Africa, *C. longa* in South Asia, *B. pilosa* in the Americas, and *G. latifolium* in West African cuisines is essential for global food security, biodiversity conservation, and the development of culturally grounded, functional nutrition interventions.

Keywords: Indigenous knowledge, ethnobotany, functional foods, biocultural diversity, food–medicine continuum, one health, sustainable diets

13.1 Introduction

In the context of Indigenous societies, a profound interconnection exists between medicinal plants and nutritional food sources, a relationship that can be conceptualized as a dual-purpose agent within their food systems. This is particularly relevant in communities where plants are not simply viewed through rigid classifications as either food or medicine; rather, they occupy a continuum that acknowledges their simultaneous roles in nutrition and healthcare (Azam et al., 2014; de Medeiros et al., 2021). The dual-purpose nature of these plants is underscored by the historical roots of the food–medicine continuum observed across various traditional societies. In many cultures, what is defined as food often overlaps with medicinal properties, reflecting a long-standing understanding that health can be maintained through nutrition and that certain dietary choices can also have therapeutic effects (Duguma, 2020; García & Price, 2012). Historical records indicate that early human societies recognized the significance of plants as both sustenance and remedies. This historical amalgamation of food and medicine is critical for understanding how Indigenous societies use local biodiversity to meet both their nutritional needs and healthcare demands, particularly in light of resource scarcity (Azam et al., 2014; Gallegos et al., 2023). Studies show that during times of famine or nourishment shortages, Indigenous populations often resort to plants that fulfill both nutritional and medicinal requirements (Mishra et al., 2021). Such utilization emphasizes the sociobiological dynamics of human–plant interactions where the ecological context greatly informs cultural practices (Borelli et al., 2020; Yang et al., 2014).

Sociobiological analysis is instrumental for elucidating these human–plant interactions, as it clarifies the cultural values, social structures, and environmental contexts that influence how plants are perceived and utilized within Indigenous communities. This multifaceted approach allows scholars to understand the ecological and nutritional roles of medicinal plants and their social importance and cultural meanings within different Indigenous traditions (de Medeiros et al., 2021; Jumare et al., 2022). For example, among the Naxi community in Yunnan (China), the cultivation of both food and medicinal plants reflects deep-rooted communal practices and knowledge transferred through generations (Yang et al., 2014), further revealing that these plants are embedded within the sociocultural fabric of societies (Alonso-Castro et al., 2012). Understanding the integration of medicinal plants in Indigenous nutrition hence requires an examination of the specific phytochemical properties these plants exhibit, which are crucial for both dietary sustenance and healthcare (Matenanga et al., 2024). The richness of phytochemicals found in many Indigenous plants, such as polyphenols, flavonoids, and carotenoids, demonstrates their capacity to provide essential nutrients while simultaneously offering protective health benefits against a variety of ailments (Ali-Shtayeh et al., 2008; Matenanga et al., 2024). Furthermore, the extraction of these compounds can lead to enhanced recognition of Indigenous food resources that would otherwise be overlooked, reaffirming their role not merely as food but in fact as vital elements in the health and well-being of those communities (Gupta et al., 2022; Kaur & Roy, 2021). The historical context of plant use echoes the necessity for continued research into the medicinal properties of wild edible plants. Ethnobotanical studies have underscored that many plants regarded as nutritional also harbor medicinal qualities, providing a robust base for health interventions within the Indigenous framework (de Medeiros et al., 2021; García & Price, 2012). This phenomenon is particularly crucial in settings where access to conventional pharmaceutical medicines is limited, facilitating a reliance on local biodiversity to alleviate health issues (Abuka & Feyissa, 2024).

Moreover, as the dietary needs of communities evolve due to changing environmental conditions, maintaining knowledge of these food–medicine plants becomes increasingly essential to support adaptive strategies (Borelli et al., 2020; Yang et al., 2014). Evaluating traditional food systems' roles in ensuring nutritional security and health resilience within Indigenous populations cannot be overstated. Incorporating such knowledge into contemporary dietary practices may offer insights into sustainable food systems that prioritize local flora while addressing modern nutritional challenges (Anand et al., 2019; Duguma, 2020). In essence, the dual-purpose nature of medicinal plants in Indigenous societies embodies a complex, interwoven relationship that transcends mere classification as either food or medicine. These plants are pivotal in meeting immediate nutritional needs and providing avenues for traditional health practices to flourish in the face of globalization and environmental degradation (Azam et al., 2014; Borelli et al., 2020). Understanding this relationship through a sociobiological lens can inform policies and practices that honor cultural heritage and prioritize the sustainability of traditional knowledge surrounding medicinal plant use.

This chapter explores the dual function of medicinal plants as both nutritional resources and therapeutic agents, offering a sociobiological perspective on their integral role in Indigenous food systems. Examining the coevolutionary relationships between humans and plants illuminates how cultural practices, ecological knowledge, and health imperatives have shaped the selection and use of specific species as food medicines. The chapter integrates biochemical, ethnobotanical, and environmental insights to demonstrate how these multifunctional plants contribute to dietary diversity, disease prevention, and overall community well-being (Abajue et al., 2023). Drawing on case studies and traditional knowledge systems, it deepens the understanding of the sociobiology of Indigenous foods, revealing how these plants sustain physical health, reinforce cultural identity, and enhance resilience in the face of ecological and nutritional stressors.

13.2 The Food-Medicine Continuum

Indigenous medicinal food represents a valuable intersection of nutrition and healing, offering powerful contributions to immune health through naturally occurring vitamins, minerals, and bioactive compounds. These traditional food medicines have long been embedded in cultural knowledge systems and serve as both sustenance and primary disease prevention and treatment tools. Figure 13.1 illustrates the critical roles that nutrients such as vitamins A, C, D, and E, zinc, iron, folic acid, and selenium from local foods play in regulating and supporting the human immune system. Many of these nutrients are abundantly found in Indigenous plant-based foods and remedies. For instance, leafy greens from amaranths; roots from yams, cassava, and potato; seeds from locust beans;

Figure 13.1: Immunonutritional Roles of Indigenous Medicinal Foods: Dual Functions as Medicine and Diet

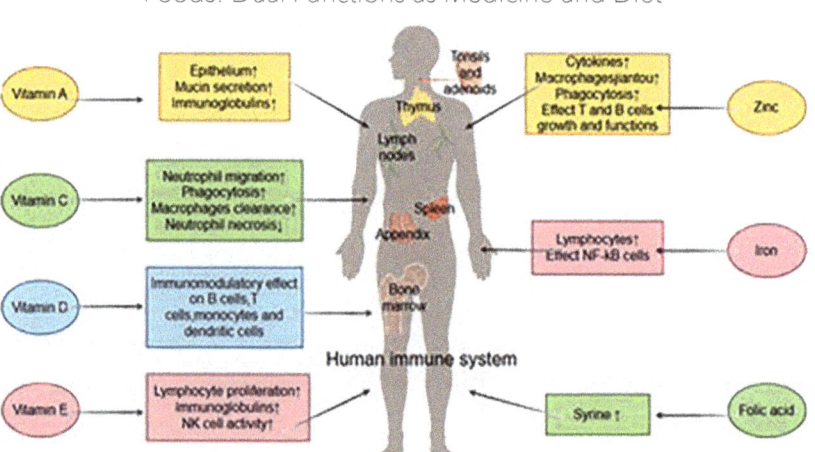

Source: Adapted from Zhou et al. (2024)

and diverse domesticated and wild fruits used in traditional diets often provide synergistic benefits by strengthening epithelial barriers, modulating lymphocyte function, and enhancing phagocytic activity. Indigenous medicinal foods should be recognized as functional foods with immense potential for promoting immune resilience and addressing modern health challenges (Obongodot & Ogwu, 2023; Osawaru & Ogwu, 2023). Bridging ancestral wisdom with contemporary immu-nonutrition science reinforces the relevance of these food systems in developing holistic, sustainable healthcare strategies.

The concept of the food–medicine continuum increasingly engages scholars and practitioners in nutrition, medicine, and public health. An emerging area of interest revolves around defining nutraceuticals and functional foods and exploring their cultural interpretations across various societies. Nutraceuticals are foods or products that offer numerous health benefits, including the prevention and treatment of diseases; Stephen De Felice popularized the term in 1989 (Golla, 2018). In contrast, functional foods provide additional health benefits beyond basic nutrition, influencing physiological functions and potentially reducing disease risk (Maroyi, 2019). The discourse surrounding nutraceuticals often incorporates their classification under various terms, such as dietary supplements, medical foods, or functional foods, depending on the context and the associated claims (Santini et al., 2018). This overlap is crucial as it highlights the fluid nature of these categories, which can change based on cultural understandings and regu-latory frameworks in different geographical regions. For instance, in countries like China, nutraceuticals are protected under specific laws, allowing them to be utilized as complementary treatment options. In contrast, the United States and the European Union generally regulate them as food supplements (Dima et al., 2020). This regulatory disparity underscores the diverse cultural interpretations of what constitutes a health-promoting food or supplement.

Furthermore, the cultural context shapes the recognition and acceptance of foods as medicine. Ethnographic studies provide rich examples of plants serving dual roles as nutritional staples and therapeutic agents (Izah et al., 2023; Obongodot & Ogwu, 2023). Various societies utilize native plants for their caloric content and medicinal properties, often embedded within traditional practices and beliefs about healthcare and diet. For instance, plants like turmeric (Curcuma longa), known for its active compound curcumin, have been integral to traditional Indian medicine as a spice and its anti-inflammatory and antioxidant properties (Kotha & Luthria, 2019). Such ethnobotanical knowledge reflects a deep understanding of the therapeutic potential of food, rooted in centuries of human interaction with the environment. Human selection, coevolution, and the development of adaptive traits in plants also play significant roles in this food–medicine continuum (Ogwu et al., 2016, 2017, 2018; Osawaru & Ogwu, 2014). The dynamic relationship between humans and plants has led to the selection of species based on their edible qualities and health-promoting attributes. The process of domestication has favored plants that possess enhanced bioactive compounds, thus making them effective as nutraceuticals over generations (Kriššáková et al., 2022). Coevolution signifies a mutual adaptation where humans and plants have evolved to meet each other's needs; as humans cultivate certain plants for nutritional benefits, those plants simultaneously develop traits that enhance their medicinal value.

Moreover, sociobiological mechanisms highlight how dietary choices shape and reflect cultural identities. In societies where herbal medicine has thrived, community knowledge and practices related to plants have played a pivotal role in establishing what is considered a healthy diet (Aronson, 2016). This sociocultural dimension also influences policymaking in healthcare and agriculture, asserting that the understanding of food extends beyond mere sustenance to encompass healthcare strategies. A significant aspect of nutraceutical intervention is its scientific verification and bioavailability in the human body, which indicates how effectively these compounds can provide health benefits once consumed. Research shows that the food matrix and processing conditions can critically impact the bioavailability of nutraceuticals, underscoring the importance of integrating traditional dietary practices with modern nutritional science (Dima et al., 2020). The potential of nutraceuticals to alleviate various health concerns, ranging from cardiovascular diseases to inflammation, is increasingly supported by clinical studies that demonstrate their efficacy and safety profiles as preventive measures (Kim & Jung, 2020; Maliar et al., 2010).

Nutraceuticals hold promise for individual health and present significant economic implications in reducing healthcare costs associated with chronic diseases. The 1994 Dietary Supplement Health and Education Act catalyzed the growth of the nutraceuticals market in the United States, leading to increased public interest and consumption, which has resulted in considerable pharmaceutical savings perceived through reduced disease burden (Finley, 2016). This trend reflects an ongoing shift in health paradigms, emphasizing preventive care and prioritizing diet and lifestyle choices over traditional pharmaceutical interventions. The interplay of food medicine within the interactive realms of culture, science, and economics provides a comprehensive framework for understanding nutraceuticals and functional foods. It reveals the intricate balance between traditional knowledge systems and contemporary scientific advancements, articulating a future where food is a source of energy and a catalyst for health. Nutraceuticals, in their multifaceted role, illustrate the unity of dietary practices and medicinal use, embedding them deeply within the cultural tapestry of global society.

13.3 Ethnobotanical Knowledge of Medicinal Plants and Their Social Value and Transmission

Ethnobotanical knowledge surrounding medicinal plants is deeply embedded in the cultural and social fabric of Indigenous communities. These plants serve not only as remedies for physical ailments but also hold symbolic, culinary, and spiritual significance. The transmission of knowledge regarding their use often occurs through oral traditions, rituals, apprenticeships, and hands-on practice, ensuring continuity across generations (Table 13.1). This body of knowledge is predominantly passed down through generations, wherein family heritage and community practices play vital roles (Table 13.1). The transfer of this knowledge typically happens through informal avenues such as storytelling, hands-on training from elders, and participation in the gathering and use of plants for medicinal purposes (Abdoulaye et al., 2022; Megenas et al., 2019). The oral tradition serves as a means of preserving knowledge while embedding cultural narratives and spiritual significance associated with these plants, reflecting a holistic view of health that is deeply tied to identity and community cohesion (Megenas et al., 2019; Rinto et al., 2023).

The specific pathways of knowledge transfer can vary significantly by demographics, particularly gender and age. Research indicates that older informants generally possess richer ethnobotanical knowledge than younger generations because of accumulated experiences and cultural practices (Aparicio et al., 2021; Osman et al., 2020). Women often serve as essential custodians of this knowledge, primarily responsible for plant gathering and preparation within many societies, sustaining both the practical and cultural dimensions of ethnobotanical practices (Bhattarai et al., 2010; Qamariah et al., 2020). In various Indigenous communities, gender is significant, indicating social hierarchies and cultural roles these women embody concerning plant medicine (Kindie & Tamiru, 2021; Supiandi et al., 2019). Additionally, studies have shown that the intergenerational transfer of knowledge encompasses a wide range of practices, from recognizing beneficial species to applying them in various therapeutic contexts (Akpojosevbe et al., 2023; Mutie et al., 2020).

The roles of custodians in preserving plant wisdom are critical, with healers, elders, and women being identified as primary transmitters of ethnobotanical

Table 13.1: Examples of Ethnobotanical Knowledge on Medicinal Plants: Social Value and Modes of Transmission

Plant Species	Local/Traditional Use	Social Value	Knowledge Transmission
Moringa oleifera (drumstick tree)	Used to treat malnutrition, inflammation, and fatigue	Symbol of vitality; used in maternal and child health	Oral tradition and elder-to-youth instruction
Curcuma longa (turmeric)	Anti-inflammatory, antiseptic, and digestive aid	Ritual significance; used in weddings and purification ceremonies	Intergenerational, through culinary and ritual use
Bidens pilosa (blackjack)	Treats infections, wounds, and gastrointestinal discomfort	Accessible medicine for the poor, and wild edible	Community herbalists, observation, trial, and error
Gongronema latifolium (utazi)	Manages diabetes and is used as a soup vegetable	Integral to local cuisine and disease prevention	Domestic knowledge, and particularly among women
Ocimum gratissimum (African basil)	Treats colds, flu, and skin infections	Sacred plant, grown near shrines, and is used in spiritual cleansing	Religious rituals, and apprenticeships with healers
Zanthoxylum zanthoxyloides (artar root)	Treats dental pain and malaria	Valued for spiritual protection and medicinal efficacy	Healer mentorship and forest foraging knowledge
Vernonia amygdalina (bitter leaf)	Antiparasitic, used for gastrointestinal health	Central in postpartum care and bitter tonics	Passed down through maternal lines

knowledge. Healers often operate within a framework that navigates traditional practices and modern medicinal paradigms, which may emphasize empirical validation over traditional wisdom (Ahmad et al., 2014; Bhattarai et al., 2010). Elders offer historical contexts that enrich the understanding of plant use while weaving in cultural narratives that maintain the significance of these medicinal practices (Alebie et al., 2017; Mutie et al., 2020). Furthermore, women's active involvement in these traditions emphasizes a gendered distribution of knowledge that preserves medicinal practices and conduits for cultural identity (Bayih & Usman, 2018; Yelianti et al., 2023). The convergence of these roles illustrates the fragility and resilience of traditional knowledge systems, which are often at risk from external factors, such as environmental changes and modernization (Mutie et al., 2020).

Intracultural variation in ethnobotanical knowledge among different age groups and educational backgrounds reflects broader educational influences on traditional knowledge retention (Aparicio et al., 2021; Luijk et al., 2021). The loss of this knowledge is often correlated with declining intergenerational engagement, compounded by shifting cultural values favoring modern medicines (Tedila & Dida, 2019). A significant concern arises from the commodification of traditional medicines, where medicinal plant knowledge becomes vulnerable to appropriation without respecting the Indigenous communities possessing this knowledge (Megenas et al., 2019). The role of educational institutions is increasingly recognized as a potential avenue for rekindling interest in traditional medicine among younger generations, emphasizing curricula that balance scientific methods and Indigenous practices (Rinto et al., 2023). Practices surrounding the preparation and application of medicinal plants are not static; they evolve with the changing realities faced by communities. Studies have shown how local environmental changes, cultural transformations, and shifts in societal roles shape the ongoing use of ethnobotanical knowledge for human and livestock ailments (Akpojosevbe et al., 2023; Kindie & Tamiru, 2021). The understanding of effective plant-based therapy can change, necessitating adaptability within practices that rely on traditional knowledge (Akpojosevbe et al., 2023; Girmay et al., 2012). As observed in various Indigenous contexts, cultural beliefs surrounding plants often inform their medicinal use and perceived spiritual significance (Hanazaki et al., 2013; Zemede et al., 2024).

As globalization continues to influence traditional practices, the urgency to document and sustain ethnobotanical knowledge grows. Engaging younger community members in these practices offers opportunities for revitalization, facilitating the integration of Indigenous knowledge into modern healthcare conversations (Akpojosevbe et al., 2023; Megenas et al., 2019; Rinto et al., 2023). The decreasing interaction with nature among younger generations may lead to a lack of familiarity and a subsequent decline in the use of local plants, posing a threat to the knowledge cultivated over centuries within these communities

(Tedila & Dida, 2019; Yelianti et al., 2023). Thus, preserving ethnobotanical knowledge on medicinal plants is critical for cultural and environmental sustainability. Incorporating traditional practices into modern educational frameworks can validate and enhance this knowledge, facilitating its transmission and evolution (Luijk et al., 2021; Rinto et al., 2023). Given the pressing challenges of modernity and environmental degradation, fostering intergenerational links that honor ecological knowledge and community traditions is imperative for ensuring the continuity of practices that have historically provided health solutions rooted in local biodiversity (Klooster et al., 2019; Mutie et al., 2020).

13.4 Biochemical and Nutritional Synergy on Medicinal Plants in Indigenous Cultures

The interconnection between the biochemical and nutritional profiles of medicinal plants in Indigenous cultures reveals a complex tapestry of health benefits provided by these plants through their rich array of bioactive compounds. Many medicinal plants play an essential role in nutrition, offering crucial macronutrients and micronutrients necessary for maintaining health within Indigenous populations. Research indicates that the synergistic effects of these bioactive compounds and vital nutrients can enhance overall human health and improve disease resistance.

Studies highlight key bioactive compounds found in Indigenous medicinal plants, which contribute to health improvements and serve as vital dietary supplements. For instance, phenolic compounds and flavonoids, abundant in many plant species, demonstrate significant antioxidant properties that can mitigate oxidative stress linked to chronic diseases (Samtiya et al., 2021; Zamora-Ros et al., 2018). These phytochemicals work synergistically with vitamins and minerals to enhance physiological functions. The bioactive compounds derived from the Cucurbitaceae family, such as those in squashes and pumpkins, exemplify this synergy well, as they are rich in micronutrients like vitamins A, C, and E, and also contain peptide-based bioactive compounds with functional properties (Yadav et al., 2010). Investigations into the agronomic profiles of various Indigenous plants show that many provide a robust synthesis of both macro- and micro-nutrients. For example, a comprehensive examination of Pereskia species reveals promising nutritional potential (Maciel et al., 2019). The dietary contributions of these plants are crucial for sustaining health, especially in areas where conventional agricultural practices may not be as prevalent. The Chenchu tribe's nutritional practices emphasize the importance of local, edible plants rich in macro- and micro-nutrients, thereby supporting nutritional self-sufficiency through traditional knowledge and practices.

Furthermore, studies emphasize the importance of ethnobotanical knowledge passed down through generations, guiding Indigenous communities in selecting plants with both medicinal and nutritional attributes. This accumulated knowledge is fundamental to food security among these populations, where traditional food systems can directly influence health outcomes. Documentation from ethnomedicinal studies reveals that plants such as *Evolvulus alsinoides* offer medicinal benefits and nutritional value (Gomathi et al., 2013). Exploring the phytochemical profiles of specific plants uncovers a wealth of potential health benefits derived from their complex biochemical compositions. The interactions among different phytochemicals can produce synergistic effects that enhance their health benefits. For instance, a diet rich in polyphenols, lignans, quercetin, and resveratrol, found abundantly in various fruits, has been shown to reduce the risk of gastrointestinal cancers, demonstrating a clear link between nutrient-rich plant foods and enhanced health outcomes (Lin et al., 2014). Additionally, the anti-inflammatory and immunomodulatory properties of these compounds contribute to an overall improved resistance to various diseases, including cancer and metabolic disorders (Samtiya et al., 2021).

Indigenous diets, which heavily rely on plant-based foods, sustain nutritional needs while enhancing health through various phytochemicals that play crucial roles in disease prevention. Integrating diverse medicinal plants into daily meals contributes to the overall health of these communities, fostering resilience against chronic illnesses. These plants often contain potent antimicrobial and antioxidant compounds that protect against pathogens and oxidative damage (Ortega-Ramírez et al., 2014). The health benefits arise from a diet enriched with essential vitamins and minerals and fortified with bioactive compounds that work synergistically to enhance human health. Research emphasizes the vital role of these bioactive compounds in promoting human resilience against common ailments. Incorporating bioactive food components into conventional diets offers a practical approach to combating lifestyle-related diseases (Mondal et al., 2021). This supports the argument that traditional diets rich in Indigenous plants can be effective interventions for contemporary health crises, especially in populations facing high malnutrition rates amid dietary deficiencies prevalent in modern times (Fungo et al., 2015). Continued exploration of the synergy between macronutrients and phytochemicals is crucial for understanding the comprehensive health benefits provided by Indigenous plants. For instance, organic compounds such as carotenoids and glucosinolates found in various plants strengthen the immune system's response to inflammation through antioxidative properties that counteract free radical damage occurring during metabolic processes (Samtiya et al., 2021). As the world embraces a more integrative approach to nutrition and health, acknowledging and respecting the wisdom embedded in Indigenous practices becomes essential. The preservation of Indigenous knowledge regarding these medicinal plants is vital for cultural heritage and a scientific asset that could revolutionize modern nutrition and health paradigms (Lin et al., 2013; Perbawasari et al., 2023).

13.5 Case Studies of Some Medicinal Plants Across Cultures

The selection of *Moringa oleifera, Curcuma longa, Bidens pilosa, Gongronema latifolium, Artemisia annua,* and *Phytolacca americana* for this chapter's case studies was intentional and rooted in their widespread use across diverse Indigenous cultures as both food and medicine. These plants exemplify the food–medicine continuum, offering rich insights into how traditional knowledge systems integrate nutritional and therapeutic values. Each species represents a different geographic and cultural context, ranging from Africa and South Asia to the Americas, and allows for a comparative analysis of their ethnobotanical significance. They were chosen not only for their known bioactive and nutritional profiles but also because of their enduring presence in Indigenous diets, their contributions to community health, and their symbolic or spiritual value within local traditions. These plants serve as illustrative examples of how Indigenous communities have long maintained holistic health through localized ecological knowledge, showcasing the resilience and adaptability of traditional food systems. By highlighting both widely known species (like turmeric and moringa) and lesser-studied yet culturally important ones (like pokeweed and blackjack), this synthesis aims to deepen appreciation for the sociobiological richness embedded in Indigenous food-medicine systems.

Moringa oleifera, commonly known as the drumstick tree, exemplifies the intersection of nutrition and traditional medicine. This plant is revered in many African and South Asian communities for its nutrient-rich leaves, which contain high concentrations of vitamins A, C, and E, as well as essential amino acids and minerals that are pivotal in combating malnutrition in these regions (Berhan et al., 2006; Porto et al., 2023). Ethnobotanical research reveals that Moringa is not only consumed as a vegetable but is also recognized for its role in traditional medicine, where it is used to enhance immunity and provide energy (Mbelebele et al., 2024; Mufungizi et al., 2025). The dual-use of *Moringa* as both a food source and a medicinal plant exemplifies the food–medicine continuum noted in ethnobiological studies, where species utilized for food often possess considerable health benefits that transcend basic nutritional needs (Esakkimuthu et al., 2018) (Figure 13.2).

Figure 13.2: *Moringa oleifera*: Tree, Leaves, Seeds, and Flowers

Source: Adapted from Anzano et al. (2021)

Curcuma longa, or turmeric, holds a similar stature in South Asia, where it is deeply entrenched in culinary traditions and medicinal practices. Historically utilized for its anti-inflammatory properties, turmeric contains curcumin, a compound with therapeutic outcomes against various ailments, including some cancers and neurodegenerative diseases (Choo & Shaikh, 2021; Tamam et al., 2010). Turmeric is not only a staple in many South Asian diets but also serves as a vital ingredient in traditional medicine systems, highlighting the sociobiological relevance of culinary plants as multifaceted resources crucial to Indigenous health practices (Mark et al., 2017). Moreover, the potency of curcumin has spurred interest in its application beyond traditional medicine, leading to contemporary investigations into its role in disease management, further reinforcing the blend of nutritional and therapeutic utilization of this plant (Njugi, 2018) (Figure 13.3).

Figure 13.3: *Curcuma longa* (Turmeric) Plant and Rhizome

Source: Adapted from Kępińska-Pacelik and Biel (2023)

Bidens pilosa, known as blackjack in various regions of the Americas, serves as an important example of a medicinal and dietary plant with renowned antimicrobial properties. Indigenous communities have historically used *Bidens pilosa* for treating infections and enhancing digestive health, showcasing its multifaceted contributions to health and nutrition (Pérez-Ochoa et al., 2019; Sam et al., 2022). This plant's use reflects a traditional knowledge system that emphasizes the significance of wild edible plants in ensuring dietary diversity and health, which is particularly vital in areas where access to conventional pharmaceuticals may be limited (Calzada et al., 2010). Current studies highlight the critical role of such plants in traditional societies, underscoring the importance of preserving Indigenous knowledge associated with them to ensure sustainable practices (Bodeker & Kariippanon, 2020; Mbelebele et al., 2024) (Figure 13.4).

Figure 13.4: Watercolors of *B. pilosa*. Creator: Frances Worth Horne. Taken from: Archives of The New York Botanical Garden

Source: The New York Botanical Garden (2023)

Gongronema latifolium, known for its hypoglycemic properties, holds a significant culinary and medicinal place in West African cuisine. This plant is used for flavor enhancement and as a dietary measure to manage blood sugar levels among diabetic populations (Jeambey et al., 2009; Mark et al., 2017). In many communities, *Gongronema latifolium* is combined with other staple foods, reflecting the integrative approach of Indigenous diets where culinary practices heighten health-promoting effects. This use further illustrates the intertwined relationship between food, nutrition, and medical practices prevalent in various cultures (Esakkimuthu et al., 2018; Oliveros-Garay & González, 2025). Attention to such plants is fundamental to understanding the dietary practices of Indigenous peoples who continue to rely heavily on local flora for health and nutrition (Figure 13.5).

Figure 13.5: Watercolors of *Gongronema latifolium* Benth.
Leaf and Flower

To further enrich the discourse on medicinal plants in Indigenous diets, *Artemisia annua*, known for its antimalarial properties, offers insights into how traditional knowledge informs modern pharmacology (Berhan et al., 2006; Lulekal et al., 2008). Indigenous populations in different regions have used Artemisia annua as a medicinal herb and a food additive in traditional dishes, showcasing a culture of utilizing plants that serve dual functions as food and medicine (Poudel, 2022; Rankoana, 2022). This dual application underscores the importance of such plants in dietary diversity while providing potential avenues for disease amelioration, reinforcing the notion of plants as integral components of holistic health systems within these communities (Esakkimuthu et al., 2018) (Figure 13.6).

Phytolacca americana, or pokeweed, provides another compelling case study within food and medicine. Historically, this plant has been utilized for its purgative properties and in treating various ailments. Additionally, its young shoots may be eaten as food, though caution is warranted due to its toxic components when improperly prepared (Berhan et al., 2006; Tamam et al., 2010). This exemplifies the critical intersection of knowledge and practice in Indigenous communities where the nuances of plant-based dietary practices demand a strong understanding of their health implications (Calzada et al., 2010; Pérez-Ochoa et al., 2019). Sustaining and understanding these Indigenous practices is critical in addressing contemporary health challenges, promoting dietary diversity, and preserving cultural heritage (Figure 13.7).

Figure 13.6: Watercolors of *Artemisia annua* Plant

Figure 13.7: *Phytolacca americana* (Pokeweed): Plant, Fruit, and Processed Powder; **A** Mature *Phytolacca americana* plant showing clusters of ripe blackberries; **B** Close-up of harvested pokeweed berries; and **C** Dried and ground powder derived from pokeweed fruits, used in traditional preparations

Source: Adapted from Veleşcu et al. (2025)

13.6 Sociobiological Determinants of Medicinal Plant Use in Indigenous Cultures

Various ecological, cultural, and adaptive factors shape the sociobiological determinants of medicinal plant use among Indigenous cultures, illustrating a complex interplay between the environment and traditional health practices. One significant aspect of this relationship is the ecological distribution, seasonal availability, and cultural valuation of medicinal plants. Indigenous populations often rely on local flora intimately connected to their environmental context and cultural beliefs. Studies, such as those by Porto et al. (2023), highlight that adherence to herbal medication is often driven by the accessibility of these plants and cultural practices surrounding their use, indicating how local biodiversity influences traditional herbal medicine. The use of medicinal plants in Indigenous cultures is shaped by a complex interplay of social, cultural, and biological factors (Table 13.2). These sociobiological determinants influence how plants are selected, valued, and transmitted across generations. From ecological availability and disease prevalence to symbolic meanings and gender roles, each factor is vital in shaping plant-based healing practices.

Table 13.2: Sociobiological Determinants of Medicinal Plant Use

Determinant	Defining Features	Examples
Ecological Availability	Use of plants that are locally abundant and seasonally accessible	*Moringa oleifera* in sub-Saharan Africa; *Bidens pilosa* in Central America
Cultural Beliefs and Symbolism	Ritualistic, spiritual, or symbolic meanings assigned to plants	*Curcuma longa* in South Asian purification ceremonies; *Ocimum* sp. in shrines
Traditional Knowledge Systems	Transmission of ethnobotanical knowledge across generations	Maternal teaching of *Gongronema latifolium* preparation in West Africa
Perceived Efficacy	Continued use based on observed therapeutic outcomes	*Artemisia annua* for malaria; *Phytolacca americana* for purgative effects

Determinant	Defining Features	Examples
Health and Disease Patterns	Selection influenced by local disease burden and nutritional needs	Use of *Vernonia amygdalina* for digestive health; hypoglycemics for diabetes
Gender and Social Roles	Gendered roles in plant identification, preparation, and healing practices	Women as custodians of postpartum herbs; male shamans as spiritual healers
Accessibility and Affordability	Preference for wild or home-grown plants over commercial pharmaceuticals	Foraging of *Bidens pilosa* and *Vernonia amygdalina* in low-income regions
Adaptability and Versatility	Plants used for multiple purposes: food, medicine, and cultural rituals	*Moringa oleifera* as a food, tonic, and health enhancer

Moreover, the seasonal availability of medicinal plants plays a crucial role in their utilization. The timing of harvesting plants, influenced by ecological cycles, affects the pharmacological efficacy and cultural practices surrounding their use (Kiasi et al., 2020). For example, specific plants may only be harvested at certain times of the year, making their availability vital to cultural rituals and healthcare practices (Ralte & Singh, 2024). This relationship is further mediated by ethnoclassification systems, where Indigenous cultures classify plants based on their medicinal properties, cultural significance, and ecological presence (Leonti et al., 2001). The work of Lulekal et al. (2008) reinforces the idea that numerous medicinal plant species within a given region foster a rich tapestry of ethnomedicinal knowledge essential for the survival of these cultural systems. On a different level, food taboos and ritual significance amplify the importance of certain plants within Indigenous health systems. Particular species may be imbued with spiritual significance, illustrating how cultural belief systems intersect with practical health considerations (Chaachouay & Zidane, 2021). For instance, some plants may be used in healing rituals or as offerings, reinforcing their role in the broader cultural narrative of health and well-being. The cultural valuation of these plants goes hand in hand with their ecological roles, as species can only thrive in specific ecological niches, thus linking cultural practices directly to environmental stewardship and local biodiversity (Akunna et al., 2023).

The adaptive responses of Indigenous cultures to local disease burdens and nutrient scarcities reveal their deep connection with their surrounding ecosystems. Many Indigenous communities have developed extensive knowledge systems that allow them to utilize local plants effectively to cope with prevalent health issues, such as malaria or other endemic diseases (Maema et al., 2017). Ethnobotanical studies in districts like Jabitehnan in Ethiopia demonstrate that local knowledge about medicinal plants is often tailored to address specific health concerns linked to the environment (Berhan et al., 2006). This responsiveness showcases an intricate understanding of local biodiversity and a strategic approach to public health crises these populations face. Specific examples of medicinal plants Indigenous societies use to highlight this adaptive utilization of flora. In Mizoram, India, Ralte and Singh (2024) document a diverse array of 302 medicinal plants employed by various ethnic tribes for their primary healthcare. The diversity of uses—ranging from treating gastrointestinal issues to more serious ailments—illustrates how traditional knowledge is reflected in the practical, everyday health-seeking behaviors of these communities. Similarly, studies conducted by Jindal and Seth demonstrate how traditional practices inform contemporary pharmacological developments, establishing a continuity of knowledge from ancient practices to current medical innovations (Jindal & Seth, 2022).

Furthermore, the impacts of globalization and modern healthcare systems introduce challenges to preserving and using traditional medicinal practices. Rapid urbanization, climate change, and the rise of conventional medicine have begun to alter the landscape of medicinal plant utilization among Indigenous populations (Luitel et al., 2014). As Luitel et al. suggest, there is a significant risk of losing traditional plant knowledge and cultural identity due to these changes, leading to decreased usage of local medicinal plants (Luitel et al., 2014). This knowledge erosion, compounded by environmental stressors like climate alterations and habitat destruction, threatens both the biodiversity of medicinal plants and the cultural heritage attached to their use (Shibambu & Maluleke, 2023). Indigenous knowledge systems are, therefore, essential for maintaining the viability of medicinal plant use and its associated health practices. They embody not just the means of addressing physical ailments but are also repositories of cultural history and social structures. For example, the differentiated knowledge of plants held by elders compared to younger community members, as shown in the study by Lulekal et al. for the Amhara region, indicates the transmission of crucial traditional knowledge that could be jeopardized as modernization progresses (Lulekal et al., 2014). Such knowledge encapsulates the collective memory of Indigenous communities and informs their strategies for healing and survival (Dahlberg & Trygger, 2009).

Moreover, the cooperative engagement between Indigenous cultures and the scientific community can lead to a fruitful exploration of pharmacologically active compounds from these plants. For instance, studies on the ethnopharmacological potential of plants utilized in traditional practices have been associated with modern drug discovery, illustrating the importance of maintaining Indigenous plant knowledge as a starting point for future medicinal developments (Nugraha et al., 2020). This relationship emphasizes the relevance of traditional medicinal practices and their significant contribution to global health and pharmacology, thus underscoring the need for prioritizing Indigenous voices in health dialogues (Semenya et al., 2012). The persistence of local medicinal plant knowledge among Indigenous populations points to their resilience and profound relationship with nature. They navigate complex social and environmental landscapes through a rich understanding of local medicinal resources, demonstrating a model of sustainable practice that blends cultural beliefs with ecological realities. Sustainability efforts must involve recognizing the value of this Indigenous knowledge to prevent further degradation of both the plants and the cultural heritage that relies upon them. As the world increasingly acknowledges the potential of traditional medicine, it becomes imperative to integrate these Indigenous perspectives into contemporary healthcare frameworks, ensuring a more holistic and inclusive approach to public health. As existing systems undergo transitions due to external pressures, there remains a pressing need to protect and promote these Indigenous traditions, ensuring both the sustainability of biodiversity and the cultural integrity of the communities that uphold these practices.

13.7 Challenges and Threats to Traditional Knowledge Systems on Medicinal Plants

Preserving and promoting traditional knowledge systems surrounding medicinal plants face significant challenges due to many factors, including globalization, urbanization, unsustainable harvesting, habitat degradation, and policy neglect (Ogwu et al., 2014; Ogwu, 2023). The intricate relationship between these factors threatens the essence of traditional ethnobotanical knowledge and practices that have persisted for generations among Indigenous communities. One of the most pressing concerns is the loss of biocultural heritage attributed to globalization and urbanization. Traditional knowledge systems that focus on the medicinal use of plants are increasingly at risk as economic and scientific advancements tend to overshadow Indigenous practices. For example, Cheng et al. (2022) highlight how the rapid economic development in regions like northwest Yunnan, China, leads to a diminishing interest in traditional medicines among younger populations, driven by the allure of modern medical practices. This shift reflects a broader global trend where the commodification of medicinal plants prioritizes economic gains over Indigenous cultural practices, thus eroding the traditional knowledge passed down through generations (Kala et al., 2006). As urbanization expands, it transforms social structures and relationships, thereby marginalizing communities that hold valuable traditional wisdom regarding medicinal plants (Wanjohi et al., 2020).

The impact of unsustainable harvesting practices and habitat degradation cannot be understated. These practices are typically driven by local and global market demands, leading communities to exploit medicinal plants at unsustainable rates. Bayih and Usman (2018) delineate this crisis by mentioning how traditional healers, as knowledge bearers, often lack the understanding of best practices for harvesting, leading to a decline in medicinal plant populations and associated knowledge. This unsustainable exploitation ultimately threatens biodiversity and the ecosystem services these plants provide, risking the very existence of vital species that contribute to traditional healing systems (Ugwu et al., 2023). Moreover, habitat degradation, fueled by agriculture and urban sprawl, further compounds the problems faced by medicinal plants. The complexities of climate change in altering the natural habitats of these plants pose additional threats, as reported by Roy and Roy, emphasizing the vulnerabilities of medicinal plant species native to Bangladesh (Roy & Roy, 2016).

Furthermore, the commodification of medicinal plants tends to distance local users from their medicinal heritage. The increasing global demand for herbal products prioritizes marketability over sustainable practices, undermining local traditions (Obahiagbon & Ogwu, 2023a, 2023b, 2024a, 2024b). Lodha posits that while traditional medicine has historical relevance, the rising commercial interest in herbal remedies frequently erodes the Indigenous methods that govern their use (Lodha, 2016). The reduction of traditional knowledge frameworks also emerges from changes in lifestyle and educational priorities as younger generations increasingly embrace modern medicine without understanding their ancestors' practices (Gautam, 2023). In some regions, there is a palpable disconnect where traditional healers become uncertain about the identity and efficacy of the medicinal plants they once relied on, complicating the transmission of this knowledge (Bayih & Usman, 2018). Policy neglect also plays a crucial role in the marginalization of Indigenous health and food systems. Despite recognizing the significance of medicinal plants in various cultures, governmental frameworks often fail to support these traditional systems adequately. Ugwu et al. (2023) stress the necessity for regulatory frameworks to safeguard traditional medicine practices while integrating sound research and conservation strategies. The lack of policy support leads to a continual obfuscation of conventional knowledge, resulting in a diminished role for Indigenous communities in preserving and promoting their practices.

The need for documentation and functional integration of traditional knowledge with contemporary scientific approaches is critical. Calibrating empirical knowledge with Indigenous practices allows for an enriching collaboration that can lead to the resilience of both systems. For instance, integrating traditional ecological knowledge with modern pharmacological methods has been hailed as a promising avenue for enriching the understanding and cultivation of medicinal plants, thereby preserving both ecological and cultural heritage (Pratama et al., 2024). The collaboration between scientists and Indigenous communities is vital in creating sustainable approaches to conservation that respect both the ecosystem and the Indigenous lifestyles centered around medicinal plants (Ugwu et al., 2023). A comprehensive approach that incorporates educational initiatives focusing on traditional knowledge can aid in countering the loss of biocultural heritage. Pratama et al. (2024) propose that educational tourism can create pathways for sharing traditional ethnobotanical knowledge while fostering appreciation and understanding of the cultural significance of medicinal plants. This could facilitate a renewed interest among younger generations in learning about and utilizing these plants for health benefits, thus bridging the gap between traditional and modern medicinal practices. The imperative remains not just to preserve these invaluable systems but also to adapt them meaningfully to engage future generations, ensuring that traditional wisdom concerning medicinal plants continues to thrive within a contemporary setting.

13.8 Contemporary Significance and Future Directions of the Sociobiology of Medicinal Plants

The intersection of Indigenous plant utilization in contemporary food systems and public health discourse is critical for addressing global challenges such as malnutrition, chronic diseases, climate change, and sustainable development frameworks. Indigenous medicinal plants serve as a vital nutritional resource within their communities, offering culinary benefits and therapeutic properties that foster health resilience (Ambu et al., 2020; Gallegos et al., 2023; Oduor et al., 2023). Integrating these plants into global food systems allows for diversifying dietary sources, which is integral for combating nutritional deficiencies prevalent in many Indigenous populations. The recognition of such plants as both food and medicine highlights their dual significance in promoting health and well-being (Jarapala et al., 2021; Ju et al., 2013).

Indigenous societies have long relied on a diverse array of plants for nourishment and medicinal purposes. Research has illustrated that these societies utilize a plethora of local flora, significantly contributing to their overall dietary intake. The Chenchu tribes in India, for example, gather a variety of green leafy vegetables and fruits that not only provide sustenance but also contain higher micronutrient levels than many conventional food sources (Jarapala et al., 2021). This dietary practice aligns with findings that associate increased consumption of Indigenous plants with lower incidences of chronic diseases, such as cardiovascular diseases and diabetes (Oduor et al., 2023; Xu et al., 2020). Such knowledge further emphasizes the role of traditional diets in maintaining health and preventing malnutrition, which is critical in the face of growing global health challenges (Kumar, 2021).

Furthermore, the significance of Indigenous foods is amplified within the context of climate resilience and sustainable agricultural practices. Many Indigenous plants demonstrate superior adaptability to local environmental conditions, making them crucial for food security in an era marked by climate variability (Matenanga et al., 2024; Rankoana, 2021). For instance, the nutrition and phytochemical properties of Indigenous plants like *Sclerocarya birrea* and *Adansonia digitata* (baobab) underscore their potential as sources of essential nutrients and their roles in combating food insecurity in regions like sub-Saharan Africa (Baiyeri & Olajide, 2023; Matenanga et al., 2024). The sustainable cultivation and consumption of these plants not only bolster local economies but also enhance ecosystem stability, fostering a more resilient agricultural framework (Rankoana, 2016; Takaidza, 2024).

The holistic approach inherent in Indigenous knowledge systems, which acknowledges the interconnectedness of health, environment, and community well-being, is increasingly recognized in contemporary discussions on One Health and planetary health frameworks (Mbelebele et al., 2024). By fostering a greater appreciation for the role of traditional ecological knowledge in biodiversity management, we can promote sustainable practices that align modern agricultural strategies with Indigenous wisdom (Ghosh-Jerath et al., 2019). This culturally sensitive framework can facilitate the development of sustainable dietary interventions that address malnutrition and enhance agrobiodiversity conservation. Incorporating Indigenous practices into modern food systems offers a myriad of possibilities for addressing pressing global health issues (Ogwu & Osawaru, 2022). The participation of Indigenous communities in such initiatives is paramount to their success. Engaging these groups in co-creating nutritional guidelines ensures that cultural values and traditional knowledge are upheld while addressing contemporary health concerns (Oliveira & Sou, 2020; Ssenku et al., 2022). Moreover, educating local communities about the nutritional value of Indigenous plants can bridge generational gaps in knowledge, as seen in studies on the culinary and medicinal heritage of Ecuadorian communities (Gallegos et al., 2023). The economic implications of integrating Indigenous plants into global food systems extend beyond health benefits. By promoting the use of wild edible plants and encouraging their sustainable harvesting, we can create new income streams for Indigenous communities, enhancing their economic stability while fostering biodiversity conservation (Duguma, 2020; Kohila & Kensa, 2019). For example, wild edible plants in places like Turkana County in Kenya have been noted to significantly contribute to local food security and nutrition, demonstrating the value of these resources beyond mere survival (Oduor et al., 2023).

Furthermore, the infusion of traditional ecological practices into contemporary agriculture supports sustainable management of natural resources, contributing to broader environmental goals. Practices such as in situ conservation of Indigenous plant varieties promote ecological resilience and align with global sustainable development goals (Kumar, 2021; Mbelebele et al., 2024). Strategies that transcend conventional agricultural models by integrating native foods can mitigate the adverse effects of climate change and enhance food sovereignty among Indigenous populations (Matenanga et al., 2024; Rankoana, 2016). The continuous exploration and valuation of Indigenous plants as functional food sources play a pivotal role in addressing nutritional deficiencies, chronic diseases, and the challenges posed by climate change. As we advance toward an integrated approach that encompasses One Health and sustainable development, it is essential to recognize and respect the knowledge and practices of Indigenous peoples. This will promote dietary diversity and foster ecological preservation and community empowerment, ensuring that their rich traditions continue to thrive in a rapidly changing world (Gajurel & Doni, 2020; Rankoana, 2021; Xu et al., 2020).

13.9 Conclusion

Medicinal plants serving nutritional and therapeutic purposes form a critical pillar of Indigenous food systems, reflecting the deep interdependence between biological adaptation and cultural innovation. These multifunctional plants exemplify a coevolutionary relationship between humans and their environments shaped by generations of observation, experimentation, and ecological stewardship. This chapter has demonstrated how species such as *M. oleifera*, *C. longa*, *B. pilosa*, and *G. latifolium* are embedded within sociobiological systems that integrate food, health, and cultural identity. Examining their phytochemical profiles, nutritional contributions, and cultural contexts, also reveal complex sophisticated knowledge systems that underpin their use and highlight their significance in maintaining community well-being. However, these food-medicine systems are increasingly vulnerable to erosion due to globalization, environmental degradation, and the marginalization of Indigenous knowledge in formal health and food policy frameworks. Unsustainable harvesting practices and commercial exploitation further threaten both plant biodiversity and the cultural fabric that sustains their use. Recognizing and preserving these systems is not merely an act of cultural conservation—it is a strategic imperative for advancing global food security, public health, and ecological resilience. A sociobiological perspective invites a reimagining of food and medicine not as separate domains but as interconnected pathways toward sustainable and equitable futures. Embracing the wisdom embedded in Indigenous food-medicine systems offers valuable insights for addressing nutritional deficiencies, combating noncommunicable diseases, and promoting holistic, culturally grounded approaches to planetary health.

Chapter Reflection

1. How does the concept of the food–medicine continuum challenge modern distinctions between nutrition and healthcare in Indigenous societies?
2. In what ways do sociobiological factors like gender roles, ecological availability, and cultural beliefs influence the use of medicinal plants in Indigenous food systems?
3. Why is the preservation of ethnobotanical knowledge critical for both cultural identity and global food and health security?

4. How do bioactive compounds in Indigenous medicinal plants provide synergistic nutritional and therapeutic benefits?
5. What role do women and elders play in sustaining the intergenerational transmission of knowledge about medicinal plants?
6. Discuss how globalization and commercialization threaten the integrity of traditional knowledge systems on food-medicinal plants.
7. How do case studies like Moringa oleifera and Curcuma longa illustrate the sociobiological richness of Indigenous plant use?
8. What policy or educational measures could help bridge Indigenous knowledge with modern healthcare and nutrition science?
9. In what ways can the integration of Indigenous food-medicine systems into One Health or Planetary Health frameworks contribute to sustainable development goals?
10. How can the recognition of medicinal plants as functional foods reshape dietary interventions for addressing malnutrition and chronic diseases in Indigenous and global contexts?

REFERENCES

Abajue, M., Sawyer, W., Izah, S. C., & Ogwu, M. C. (2023). Diversity of medicinal plants used in the treatment and management of viral diseases transmitted by mosquitoes in the tropics. In S. C. Izah, M. C. Ogwu, & M. Akram (Eds.), *Herbal medicine phytochemistry.* Reference Series in Phytochemistry (pp. 1–35). Springer. https://doi.org/10.1007/978-3-031-21973-3_60-1

Abdoulaye, A., Noumavo, A., Glodjinon, N., Adjanohoun, A., Baba-Moussa, L., & Baba-Moussa, F. (2022). Characteristics of medicinal plants used in traditional medicine for oral diseases treatment in southern Benin. *European Journal of Medicinal Plants*, *33*(9), 22–36. https://doi.org/10.9734/ejmp/2022/v33i930488

Abuka, M., & Feyissa, G. (2024). Unlocking the prospective of neglected and underutilized wild plants for human food, nutrition and ethno-medicine. https://doi.org/10.21203/rs.3.rs-3996737/v1

Ahmad, M., Sultana, S., Fazl-i-Hadi, S., ben Hadda, T., Rashid, S., Zafar, M., Ajab Khan, M., Zada Khan, M. P., & Yaseen, G. (2014). An ethnobotanical study of medicinal plants in high mountainous region of Chail Valley (District

Swat-Pakistan). *Journal of Ethnobiology and Ethnomedicine, 10*(1), 36. https://doi.org/10.1186/1746-4269-10-36

Akpojosevbe, E. J., Ishaku, L. E., Akogwu, E. I., Tondo, B. K., Hong, J., Muhammad, Z., Oyebade, K. F., Makoshi, M. S., Shok, B. Z., Okpalaeke, E. E., Usman, J. G., Gotep, J. G., Kwaja, E. Z., Barde, I. J., Bitrus, Y., & Muhammad, M. (2023). An inventory of medicinal and poisonous plants of the National Veterinary Research Institute VOM, Plateau State, Nigeria. https://doi.org/10.21203/rs.3.rs-3220915/v1

Akunna, G., Lucyann, C., & Saalu, L. (2023). Rooted in tradition, thriving in the present: The future and sustainability of herbal medicine in Nigeria's healthcare landscape. *Journal of Innovations in Medical Research, 2*(11), 28–40. https://doi.org/10.56397/jimr/2023.11.05

Alebie, G., Urga, B., & Worku, A. (2017). Systematic review on traditional medicinal plants used for the treatment of malaria in Ethiopia: Trends and perspectives. *Malaria Journal, 16*(1). https://doi.org/10.1186/s12936-017-1953-2

Ali-Shtayeh, M. S., Jamous, R. M., Al-Shafie', J. H., Elgharabah, W. A., Kherfan, F. A., Qarariah, K. H., Khdair, I. S., Soos, I. M., Musleh, A. A., Isa, B. A., Herzallah, H. M., Khlaif, R. B., Aiash, S. M., Swaiti, G. M., Abuzahra, M. A., Haj-Ali, M. M., Saifi, N. A., Azem, H. K., & Nasrallah, H. (2008). Traditional knowledge of wild edible plants used in Palestine (Northern West Bank): A comparative study. *Journal of Ethnobiology and Ethnomedicine, 4*(1), 13. https://doi.org/10.1186/1746-4269-4-13

Alonso-Castro, Á. J., Maldonado-Miranda, J. J., Zarate-Martinez, A., del Rosario Jacobo-Salcedo, M., Fernández-Galicia, C., Figueroa-Zúñiga, L. A., Rios-Reyes, N. A., de León-Rubio, M. A., Medellín-Castillo, N. A., Reyes-Munguia, A., Méndez-Martínez, R., & Carranza-Álvarez, C. (2012). Medicinal plants used in the huasteca potosina, Mexico. *Journal of Ethnopharmacology, 143*(1), 292–298. https://doi.org/10.1016/j.jep.2012.06.035

Ambu, G., Chaudhary, R. P., Mariotti, M., & Cornara, L. (2020). Traditional uses of medicinal plants by ethnic people in the Kavrepalanchok District, Central Nepal. *Plants, 9*(6), 759. https://doi.org/10.3390/plants9060759

Anand, U., Jacobo-Herrera, N., Altemimi, A., & Lakhssassi, N. (2019). A comprehensive review on medicinal plants as antimicrobial therapeutics: Potential avenues of biocompatible drug discovery. *Metabolites, 9*(11), 258. https://doi.org/10.3390/metabo9110258

Anzano, A., Ammar, M., Papaianni, M., Grauso, L., Sabbah, M., Cappar-elli, R., & Lanzotti, V. (2021). *Moringa oleifera* Lam.: A phytochemical and pharmacological overview. *Horticulturae*, *7*(10), 409. https://doi.org/10.3390/horticulturae7100409

Aparicio, J., Voeks, R., & Funch, L. (2021). Are Mixtec forgetting their plants? Intracultural variation of ethnobotanical knowledge in Oaxaca, Mexico. *Economic Botany*, *75*(3–4), 215–233. https://doi.org/10.1007/s12231-021-09535-2

Aronson, J. (2016). Defining 'nutraceuticals': Neither nutritious nor pharmaceutical. *British Journal of Clinical Pharmacology*, *83*(1), 8–19. https://doi.org/10.1111/bcp.12935

Azam, F., Biswas, A., Mannan, A., Afsana, N., Jahan, R., & Rahmatullah, M. (2014). Are famine food plants also ethnomedicinal plants? An ethnomedicinal appraisal of famine food plants of two districts of Bangladesh. *Evidence-Based Complementary and Alternative Medicine*, *2014*(1). https://doi.org/10.1155/2014/741712

Baiyeri, K., & Olajide, K. (2023). Lesser known African Indigenous tree and fruit plants: Recent evidence from literatures and regular cultivation culture. In M. S. Khan (Ed.), *Tropical plant species and technological interventions for improvement*. IntechOpen. https://doi.org/10.5772/intechopen.104890

Bayih, T., & Usman, A. (2018). Assessment of traditional practices of healers to treat human illness in Shashamene Town in Ethiopia. *Internal Medicine Open Access*, *8*(3). https://doi.org/10.4172/2165-8048.1000278

Berhan, A., Asfaw, Z., & Kelbessa, E. (2006). Ethnobotany of plants used as insecticides, repellents and antimalarial agents in Jabitehnan District, West Gojjam. *Sinet Ethiopian Journal of Science*, *29*(1). https://doi.org/10.4314/sinet.v29i1.18263

Bhattarai, S., Chaudhary, R., Quave, C., & Taylor, R. (2010). The use of medicinal plants in the trans-Himalayan arid zone of Mustang District, Nepal. *Journal of Ethnobiology and Ethnomedicine*, *6*(1), 14. https://doi.org/10.1186/1746-4269-6-14

Bodeker, G., & Kariippanon, K. (2020). *Traditional medicine and Indigenous health in Indigenous hands*. Oxford Research Encyclopedia of Global Public Health. https://doi.org/10.1093/acrefore/9780190632366.013.155

Borelli, T., Hunter, D., Powell, B., Ulian, T., Mattana, E., Termote, C., Pawera, L., Beltrame, D., Penafiel, D., Tan, A., Taylor, M., & Engels, J. (2020). Born to eat wild: An integrated conservation approach to secure wild food plants for food security and nutrition. *Plants*, *9*(10), 1299. https://doi.org/10.3390/plants9101299

Calzada, F., Arista, R., & Pérez, H. (2010). Effect of plants used in Mexico to treat gastrointestinal disorders on charcoal–gum acacia-induced hyperperistalsis in rats. *Journal of Ethnopharmacology*, *128*(1), 49–51. https://doi.org/10.1016/j.jep.2009.12.022

Chaachouay, N., & Zidane, L. (2021). Neurological phytotherapy by Indigenous people of Rif, Morocco. In M. Bernardo-Filho, R. Taiar, D. da Cunha de Sá-Caputo, & A. Seixas (Eds.), *Therapy approaches in neurological disorders*. IntechOpen. https://doi.org/10.5772/intechopen.97175

Cheng, Z., Hu, X., Lu, X., Fang, Q., Meng, Y., & Long, C. (2022). Medicinal plants and fungi traditionally used by Dulong people in northwest Yunnan, China. *Frontiers in Pharmacology*, *13*. https://doi.org/10.3389/fphar.2022.895129

Choo, B., & Shaikh, M. (2021). Mechanism of curcuma longa and its neuroactive components for the management of epileptic seizures: A systematic review. *Current Neuropharmacology*, *19*(9), 1496–1518. https://doi.org/10.2174/15701 59x19666210517120413

Dahlberg, A., & Trygger, S. (2009). Indigenous medicine and primary health care: The importance of lay knowledge and use of medicinal plants in rural South Africa. *Human Ecology*, *37*(1), 79–94. https://doi.org/10.1007/s10745-009-9217-6

de Medeiros, P., Figueiredo, K. F., Gonçalves, P. H. S., de Almeida Caetano, R., da Costa Santos, É. M., do Santos, G. M. C., Barbosa, D. M., de Paula, M., & Mapeli, A. M. (2021). Wild plants and the food-medicine continuum—An ethnobotanical survey in Chapada Diamantina (Northeastern Brazil). *Journal of Ethnobiology and Ethnomedicine*, *17*(1), 37. https://doi.org/10.1186/s13002-021-00463-y

Dima, C., Assadpour, E., Dima, Ş., & Jafari, S. (2020). Bioavailability of nutraceuticals: Role of the food matrix, processing conditions, the gastrointestinal tract, and nanodelivery systems. *Comprehensive Reviews in Food Science and Food Safety*, *19*(3), 954–994. https://doi.org/10.1111/1541-4337.12547

Duguma, H. (2020). Wild edible plant nutritional contribution and consumer perception in Ethiopia. *International Journal of Food Science*, *2020*, 1–16. https://doi.org/10.1155/2020/2958623

Esakkimuthu, S., Darvin, S., Mutheeswaran, S., Paulraj, M., Pandikumar, P., Ignacimuthu, S., & Al-Dhabi, N. A. (2018). A study on food-medicine continuum among the non-institutionally trained siddha practitioners of Tiruvallur district, Tamil Nadu, India. *Journal of Ethnobiology and Ethnomedicine, 14*(1), 45. https://doi.org/10.1186/s13002-018-0240-9

Finley, J. W. (2016). The nutraceutical revolution: Emerging vision or broken dream? Understanding scientific and regulatory concerns. *Clinical Research and Regulatory Affairs, 33*(1), 1–3. https://doi.org/10.3109/1060 1333.2016.1117096

Fungo, R., Muyonga, J., Kaaya, A., Okia, C., Tieguhong, J., & Baidu-Forson, J. (2015). Nutrient and bioactive compounds content of *Baillonella toxisperma*, *Trichoscypha abut*, and *Pentaclethra macrophylla* from Cameroon. *Food Science and Nutrition, 3*(4), 292–301. https://doi.org/10.1002/fsn3.217

Gajurel, P., & Doni, T. (2020). Diversity of wild edible plants traditionally used by the Galo tribe of the Indian Eastern Himalayan State of Arunachal Pradesh. *Plant Science Today, 7*(4). https://doi.org/10.14719/pst.2020.7.4.855

Gallegos, R., Aráuz, M., Saeteros-Hernández, A., Chávez, R., & Moyano, M. (2023). The Indigenous bioculture of the Pungalá Parish of Ecuador: An approach to their culinary and medicinal heritage. https://doi.org/10.21203/rs.3.rs-3161621/v1

García, G., & Price, L. (2012). Weeds as important vegetables for farmers. *Acta Societatis Botanicorum Poloniae, 81*(4), 397–403. https://doi.org/10.5586/asbp.2012.047

Gautam, G. (2023). Advancement of network pharmacology in multi-targeted therapeutic evaluation of medicinal plants. *Journal of CAM Research Progress, 1*(1). https://doi.org/10.33790/jcrp1100113

Ghosh-Jerath, S., Downs, S., Singh, A., Paramanik, S., Goldberg, G., & Fanzo, J. (2019). Innovative matrix for applying a food systems approach for developing interventions to address nutrient deficiencies in Indigenous communities in India: A study protocol. *BMC Public Health, 19*(1), 944. https://doi.org/10.1186/s12889-019-6963-2

Girmay, Z., Mohammed, Z., & Zewdie, S. (2012). An ethnobotanical study of medicinal plants in Asgede Tsimbila District, Northwestern Tigray, Northern Ethiopia. *Ethnobotany Research and Applications, 10*, 305. https://doi.org/10.17348/era.10.0.305-320

Golla, U. (2018). Emergence of nutraceuticals as the alternative medications for pharmaceuticals. *International Journal of Complementary and Alternative Medicine, 11*(3), 155–158. https://doi.org/10.15406/ijcam.2018.11.00388

Gomathi, D., Kalaiselvi, M., Ravikumar, G., Devaki, K., & Uma, C. (2013). Gc-ms analysis of bioactive compounds from the whole plant ethanolic extract of *Evolvulus alsinoides* (L.) L. *Journal of Food Science and Technology, 52*(2), 1212–1217. https://doi.org/10.1007/s13197-013-1105-9

Gupta, R., Nigam, A., & Kapila, R. (2022). Cultivation and conservation of underutilized medicinal and agricultural plants in India. *Proceedings of the National Academy of Sciences India Section B Biological Sciences, 92*(4), 741–745. https://doi.org/10.1007/s40011-022-01405-8

Hanazaki, N., Herbst, D., Marques, M., & Vandebroek, I. (2013). Evidence of the shifting baseline syndrome in ethnobotanical research. *Journal of Ethnobiology and Ethnomedicine, 9*(1), 75. https://doi.org/10.1186/1746-4269-9-75

Izah, S. C., Ogwu, M. C., & Akram, M. (2023). *Herbal medicine phytochemistry: Applications and Trends*. Springer. https://doi.org/10.1007/978-3-031-21973-3

Jarapala, S., Shivudu, G., Mangathya, K., Rathod, A., Panda, H., & Reddy, P. (2021). A new approach for identifying potentially effective Indigenous plants consumed by Chenchu tribes and their nutritional composition- India. *American Journal of Plant Sciences, 12*(8), 1180–1196. https://doi.org/10.4236/ajps.2021.128082

Jeambey, Z., Johns, T., Talhouk, S., & Batal, M. (2009). Perceived health and medicinal properties of six species of wild edible plants in north-east Lebanon. *Public Health Nutrition, 12*(10), 1902–1911. https://doi.org/10.1017/s1368980009004832

Jindal, A., & Seth, C. (2022). Medicinal plants: The rising strategy for synthesis of modern medicine. *International Journal of Plant and Environment, 8*(1), 76–80. https://doi.org/10.18811/ijpen.v8i01.09

Ju, Y., Zhuo, J., Liu, B., & Long, C. (2013). Eating from the wild: Diversity of wild edible plants used by Tibetans in Shangri-la region, Yunnan, China. *Journal of Ethnobiology and Ethnomedicine, 9*(1), 28. https://doi.org/10.1186/1746-4269-9-28

Jumare, A., Dogara, A., & Amlabu, W. (2022). Traditional medicinal plants used in the management of cutaneous leishmaniasis diseases in Sokoto State, Northern Nigeria. *Ethnobotany Research and Applications, 23*. https://doi.org/10.32859/era.23.38.1-21

Kala, C. P., Dhyani, P. P., & Sajwan, B. S. (2006). Developing the medicinal plants sector in northern India: Challenges and opportunities. *Journal of Ethnobiology and Ethnomedicine*, *2*(1), 32. https://doi.org/10.1186/1746-4269-2-32

Kaur, S., & Roy, A. (2021). A review on the nutritional aspects of wild edible plants. *Current Traditional Medicine*, *7*(4), 552–563. https://doi.org/10.2174/2 215083806999201123201150

Kępińska-Pacelik, J., & Biel, W. (2023). Turmeric and curcumin—Health-promoting properties in humans versus dogs. *International Journal of Molecular Sciences*, *24*(19), 14561. https://doi.org/10.3390/ijms241914561

Kiasi, Y., Forouzeh, M., Mirdeilami, S., & Niknahad–Gharmakher, H. (2020). Ethnobotanical study on the medicinal plants in Khosh Yeilagh Rangeland, Golestan Province, Iran. https://doi.org/10.21203/rs.3.rs-103978/v1

Kim, M., & Jung, S. (2020). Nutraceuticals for prevention of atherosclerosis: Targeting monocyte infiltration to the vascular endothelium. *Journal of Food Biochemistry*, *44*(6), e13200. https://doi.org/10.1111/jfbc.13200

Kindie, B., & Tamiru, C. (2021). Assessment of traditional medicinal plant ethnomedicinal value and its sustainable conservation status used by Indigenous people to treat different ailments in Babile District, Oromia Region, Ethiopia. *MOJ Biology and Medicine*, *6*(3), 101–106. https://doi.org/10.15406/mojbm.2021.06.00140

Klooster, C. V., Haabo, V., & Van Andel, T. (2019). Our children do not have time anymore to learn about medicinal plants: How an ethnobotanical school assignment can contribute to the conservation of saramaccan maroon traditional knowledge. *Ethnobotany Research and Applications*, *18*, 1–47. https://doi.org/10.32859/era.18.9.11.1-47

Kohila, A., & Kensa, M. (2019). Survey of wild edible plants of Dhanakarkulam Panchayath, Tirunelveli District, Tamil Nadu, India. *Kongunadu Research Journal*, *6*(2), 20–27. https://doi.org/10.26524/krj297

Kotha, R., & Luthria, D. (2019). Curcumin: Biological, pharmaceutical, nutraceutical, and analytical aspects. *Molecules*, *24*(16), 2930. https://doi.org/10.3390/molecules24162930

Kriššáková, Z., Čierniková, M., Vykouková, I., Hrabovský, A., Masarovičová, E., & Beracko, P. (2022). Functional traits of medicinal plant species under different ecological conditions. https://doi.org/10.21203/rs.3.rs-1310161/v1

Kumar, A. (2021). How can India leverage its botanic gardens for the conservation and sustainable utilization of wild food plant resources through the implementation of a global strategy for plant conservation? *Journal of Zoological and Botanical Gardens, 2*(4), 586–599. https://doi.org/10.3390/jzbg2040042

Leonti, M., Vibrans, H., Sticher, O., & Heinrich, M. (2001). Ethnopharmacology of the popoluca, Mexico: An evaluation. *Journal of Pharmacy and Pharmacology, 53*(12), 1653–1669. https://doi.org/10.1211/0022357011778052

Lin, Y., Wang, C., Chen, I., Jheng, J., Li, J., & Tung, C. (2013). Tipdb: A database of anticancer, antiplatelet, and antituberculosis phytochemicals from Indigenous plants in Taiwan. *The Scientific World Journal, 2013*(1). https://doi.org/10.1155/2013/736386

Lin, Y., Yngve, A., Lagergren, J., & Lu, Y. (2014). A dietary pattern rich in lignans, quercetin and resveratrol decreases the risk of oesophageal cancer. *British Journal of Nutrition, 112*(12), 2002–2009. https://doi.org/10.1017/s0007114514003055

Lodha, A. (2016). Traditional medicinal plant resources from Maval Taluka, District Pune, Maharashtra, India. *International Journal of Research in Ayurveda and Pharmacy, 7*(3), 87–91. https://doi.org/10.7897/2277-4343.073118

Luijk, N., Soldati, G., & Fonseca-Kruel, V. (2021). The role of schools as an opportunity for transmission of local knowledge about useful restinga plants: Experiences in southeastern Brazil. *Journal of Ethnobiology and Ethnomedicine, 17*(1). https://doi.org/10.1186/s13002-021-00461-0

Luitel, D., Rokaya, M., Timsina, B., & Münzbergová, Z. (2014). Medicinal plants used by the Tamang community in the Makawanpur District of Central Nepal. *Journal of Ethnobiology and Ethnomedicine, 10*(1). https://doi.org/10.1186/1746-4269-10-5

Lulekal, E., Asfaw, Z., Kelbessa, E., & Damme, P. (2014). Ethnoveterinary plants of Ankober District, North Shewa Zone, Amhara Region, Ethiopia. *Journal of Ethnobiology and Ethnomedicine, 10*(1). https://doi.org/10.1186/1746-4269-10-21

Lulekal, E., Kelbessa, E., Bekele, T., & Yineger, H. (2008). An ethnobotanical study of medicinal plants in Mana Angetu District, Southeastern Ethiopia. *Journal of Ethnobiology and Ethnomedicine, 4*(1), 10. https://doi.org/10.1186/1746-4269-4-10

Maciel, V., Yoshida, C., & Goycoolea, F. (2019). Agronomic cultivation, chemical composition, functional activities and applications of pereskia species—A mini review. *Current Medicinal Chemistry*, *26*(24), 4573–4584. https://doi.org/10.2174/0929867325666180926151615

Maema, L. P., Mahlo, S. M., & Potgieter, M. J. (2017). Ethnomedicinal uses of Indigenous plant species in Mogalakwena municipality of Waterberg district, Limpopo province, South Africa. *International Journal of Traditional and Complementary Medicine*, *2*, 18. https://doi.org/10.28933/maema-ijtcm-2016

Maliar, T., Drobná, J., Kraic, J., Maliarová, M., & Jurovatá, J. (2010). Proteinase inhibition and antioxidant activity of selected forage crops. *Biologia*, *66*(1), 96–103. https://doi.org/10.2478/s11756-010-0149-9

Mark, U., Ikyese, C., & Ekhuemelo, D. (2017). Ethnomedical study of plants used by Indigenous people of Nyiev and Mbawa districts, Makurdi, Benue State, Nigeria. *GSC Biological and Pharmaceutical Sciences*, *1*(3), 001–011. https://doi.org/10.30574/gscbps.2017.1.3.0023

Maroyi, A. (2019). Utilization of *Bridelia mollis* as herbal medicine, nutraceutical, and functional food in Southern Africa: A review. *Tropical Journal of Pharmaceutical Research*, *18*(1), 203. https://doi.org/10.4314/tjpr.v18i1.30

Matenanga, O. K., Boitumelo, W., Mareko, M., Mojeremane, W., Seifu, E., Pholo-Tait, M., Michel, S., Bayram, M., & Haki, G. D. (2024). Nutritional and phytochemical properties of morula (*Sclerocarya birrea*), Moretologa (*Ximenia americana*), Mowana (*Adansonia digitata* L.), and Mopane (*Bauhinia petersiana*): Indigenous plants in Botswana. *European Journal of Agriculture and Food Sciences*, *6*(3), 16–21. https://doi.org/10.24018/ejfood.2024.6.3.780

Mbelebele, Z., Mdoda, L., Ntlanga, S., Nontu, Y., & Gidi, L. (2024). Harmonizing traditional knowledge with environmental preservation: Sustainable strategies for the conservation of Indigenous medicinal plants (IMPs) and their implications for economic well-being. *Sustainability*, *16*(14), 5841. https://doi.org/10.3390/su16145841

Megenas, J., Gelaye, K., & Dara, P. (2019). Indigenous knowledge and practices on medicinal plants used by local communities of the Gambella region, south-west Ethiopia. *International Journal of Tropical Disease and Health*, *39*(2), 1–14. https://doi.org/10.9734/ijtdh/2019/v39i230203

Mishra, A., Swamy, S., Thakur, T., Bhat, R., Bijalwan, A., & Kumar, A. (2021). Use of wild edible plants: Can they meet the dietary and nutritional needs of Indigenous communities in Central India. *Foods, 10*(7), 1453. https://doi.org/10.3390/foods10071453

Mondal, S., Soumya, N., Mini, S., & Sivan, S. (2021). Bioactive compounds in functional food and their role as therapeutics. *Bioactive Compounds in Health and Disease, 4*(3), 24. https://doi.org/10.31989/bchd.v4i3.786

Mufungizi, A., Musakwa, W., & Chanza, N. (2025). Exploring perceived impacts of shifting mopane woodland on medicinal plants in Vhembe, South Africa. *Environmental Research Communications, 7*(1), 015026. https://doi.org/10.1088/2515-7620/ada8fb

Mutie, F. M., Gao, L., Kathambi, V., Rono, P., Musili, P., Ngugi, G., Hu, G.-W., & Wang, Q. (2020). An ethnobotanical survey of a dryland botanical garden and its environs in Kenya: The Mutomo Hill Plant Sanctuary. *Evidence-Based Complementary and Alternative Medicine, 2020*(1). https://doi.org/10.1155/2020/1543831

Njugi, W. (2018). A constructive approach on lethal plants for medicinal use. *Journal of Medical Toxicology and Clinical Forensic Medicine, 4*(1). https://doi.org/10.21767/2471-9641.100032

Nugraha, A., Triatmoko, B., Wangchuk, P., & Keller, P. (2020). Vascular epiphytic medicinal plants as sources of therapeutic agents: Their ethnopharmacological uses, chemical composition, and biological activities. *Biomolecules, 10*(2), 181. https://doi.org/10.3390/biom10020181

Obahiagbon, E. G., & Ogwu, M. C. (2023a). Sustainable supply chain management in the herbal medicine industry. In S. C. Izah, M. C. Ogwu, & M. Akram (Eds.), *Herbal medicine phytochemistry*. Reference Series in Phytochemistry (pp. 1–29). Springer. https://doi.org/10.1007/978-3-031-21973-3_64-1

Obahiagbon, E. G., & Ogwu, M. C. (2023b). Consumer perception and demand for sustainable herbal medicine products and market. In S. C. Izah, M. C. Ogwu, & M. Akram (Eds.), *Herbal medicine phytochemistry*. Reference Series in Phytochemistry (pp. 1–34). Springer. https://doi.org/10.1007/978-3-031-21973-3_65-1

Obahiagbon, E. G., & Ogwu, M. C. (2024a). The nexus of business, sustainability, and herbal medicine. In S. C. Izah, M. C. Ogwu, & M. Akram (Eds.), *Herbal medicine phytochemistry*. Reference Series in Phytochemistry (pp. 1–34). Springer. https://doi.org/10.1007/978-3-031-21973-3_67-1

Obahiagbon, E. G., & Ogwu, M. C. (2024b). Organic food preservatives: The shift towards natural alternatives and sustainability in the Global South's Markets. In M.

C. Ogwu, S. C. Izah, & N. R. Ntuli (Eds.), *Food safety and quality in the Global South* (pp. 299–329). Springer. https://doi.org/10.1007/978-981-97-2428-4_10

Obongodot, N. U., & Ogwu, M. C. (2023). Plant food for human health: Case study of Indigenous vegetables in Akwa Ibom State, Nigeria. In S. C. Izah, M. C. Ogwu, & M. Akram (Eds.), *Herbal medicine phytochemistry*. Reference Series in Phytochemistry (pp. 1–38). Springer. https://doi.org/10.1007/978-3-031-21973-3_2-1

Oduor, F., Kaindi, D., Abong, G., Thuita, F., & Termote, C. (2023). Diversity and utilization of Indigenous wild edible plants and their contribution to food security in Turkana County, Kenya. *Frontiers in Sustainable Food Systems*, 7. https://doi.org/10.3389/fsufs.2023.1113771

Ogwu, M. C. (2023). Local food crops in Africa: Sustainable utilization, threats, and traditional storage strategies. In S. C. Izah & M. C. Ogwu (Eds.), *Sustainable utilization and conservation of Africa's biological resources and environment*. Sustainable Development and Biodiversity (Vol. 888, pp. 353–374). Springer. https://doi.org/10.1007/978-981-19-6974-4_13

Ogwu, M. C., Chime, A. O., & Oseh, O. M. (2018). Ethnobotanical survey of tomato in some cultivated regions in Southern Nigeria. *Maldives National Research Journal*, 6(1), 19–29.

Ogwu, M. C., & Osawaru, M. E. (2022). Traditional methods of plant conservation for sustainable utilization and development. In S. C. Izah (Ed.), *Biodiversity in Africa: Potentials, threats and conservation*. Sustainable Development and Biodiversity (Vol. 29, pp. 451–472). Springer. http://doi.org/10.1007/978-981-19-3326-4_17

Ogwu, M. C., Osawaru, M. E., & Ahana, C. M. (2014). Challenges in conserving and utilizing plant genetic resources (PGR). *International Journal of Genetics and Molecular Biology*, 6(2), 16–23. https://academicjournals.org/journal/IJGMB/article-abstract/DE9B5EB47127

Ogwu, M. C., Osawaru, M. E., Aiwansoba, R. O., & Iroh, R. N. (2016). Ethnobotany and collection of West African Okra [*Abelmoschus caillei* (A. Chev.) Stevels] germplasm in some communities in Edo and Delta States, Southern Nigeria. *Borneo Journal of Resource Science and Technology*, 6(1), 25–36. https://doi.org/10.33736/bjrst.212.2016

Ogwu, M. C., Osawaru, M. E., & Obahiagbon, G. E. (2017). Ethnobotanical survey of medicinal plants used for traditional reproductive care by Usen people of Edo State, Nigeria. *Malaya Journal of Biosciences*, 4(1), 17–29.

Oliveira, P., & Sou, B. (2020). Traditional knowledge of forest medicinal plants of Munduruku Indigenous People—Ipaupixuna. *European Journal of Medicinal Plants*, *31*(13), 20–35. https://doi.org/10.9734/ejmp/2020/v31i1330309

Oliveros-Garay, O., & González, A. (2025). Detection of Colombian datura virus infecting brugmansia × candida medicinal cultivars and evaluation of sap inoculation in Solanaceae plants. https://doi.org/10.21203/rs.3.rs-5845050/v1

Ortega-Ramírez, L. A., Rodriguez-Garcia, I., Leyva, J. M., Cruz-Valenzuela, M. R., Silva-Espinoza, B. A., González-Aguilar, G. A., Siddiqui, Md. W., & Ayala-Zavala, J. F. (2014). Potential of medicinal plants as antimicrobial and antioxidant agents in food industry: A hypothesis. *Journal of Food Science*, *79*(2), R129–R137. https://doi.org/10.1111/1750-3841.12341

Osawaru, M. E., & Ogwu, M. C. (2014). Ethnobotany and germplasm collection of two genera of *cocoyam* (*Colocasia* [Schott] and *Xanthosoma* [Schott], Araceae) in Edo State, Nigeria. *Science, Technology and Arts Research Journal*, *3*(3), 23–28. http://doi.org/10.4314/star.v3i3.4

Osawaru, M. E., & Ogwu, M. C. (2023). Plants used in the management and treatment of female reproductive health issues: Case study from Southern Nigeria. In S. C. Izah, M. C. Ogwu, & M. Akram (Eds.), *Herbal medicine phytochemistry*. Reference Series in Phytochemistry (pp. 1–37). Springer. https://doi.org/10.1007/978-3-031-21973-3_5-1

Osman, A., Sbhatu, D., & Giday, M. (2020). Medicinal plants used to manage human and livestock ailments in the Raya Kobo district of the Amhara regional state, Ethiopia. *Evidence-Based Complementary and Alternative Medicine*, *2020*(1). https://doi.org/10.1155/2020/1329170

Perbawasari, S., Sjoraida, D., Anisa, R., Subekti, P., & Bakti, I. (2023). Communication and cultural inheritance through a traditional school in Dangiang Village, Garut. *Jurnal Kajian Komunikasi*, *11*(1), 18. https://doi.org/10.24198/jkk.v11i1.45121

Pérez-Ochoa, M., Chávez-Servia, J., Vera-Guzmán, A., Aquino-Bolaños, E., & Carrillo-Rodríguez, J. (2019). Medicinal plants used by Indigenous communities of Oaxaca, Mexico, to treat gastrointestinal disorders. In S. Perveen & A. Al-Taweel (Eds.), *Pharmacognosy - Medicinal plants*. IntechOpen. https://doi.org/10.5772/intechopen.82182

Porto, A. C. L., de A. Holanda, E., Fernandes, R. L., Sousa, A. B. F., da Silva, K. L. F., Nonato, D. T. T., & da S. Pantoja, P. (2023). Medicinal plants vs. conventional medicine: Treatment assessments for Indigenous populations. *Brazilian Journal of Health and Biomedical Sciences*, *22*(2), 98–105. https://doi.org/10.12957/bjhbs.2023.80040

Poudel, A. (2022). Indigenous knowledge on medicinal plants practiced in the Myagdi district, Nepal. https://doi.org/10.21203/rs.3.rs-1214404/v1

Pratama, H., Hikmat, A., & Hidayati, S. (2024). Ethnobotany and traditional medicine as educational tourism. *IOP Conference Series Earth and Environmental Science*, *1366*(1), 012053. https://doi.org/10.1088/1755-1315/1366/1/012053

Qamariah, N., Mulia, D., & Fakhrizal, D. (2020). Indigenous knowledge of medicinal plants by the Dayak community in Mandomai village, Central Kalimantan, Indonesia. *Pharmacognosy Journal*, *12*(2), 386–390. https://doi.org/10.5530/pj.2020.12.60

Ralte, L., & Singh, Y. (2024). Ethnobotanical survey of medicinal plants used by various ethnic tribes of Mizoram, India. *PLOS One*, *19*(5), e0302792. https://doi.org/10.1371/journal.pone.0302792

Rankoana, S. (2016). Sustainable use and management of Indigenous plant resources: A case of Mantheding Community in Limpopo Province, South Africa. *Sustainability*, *8*(3), 221. https://doi.org/10.3390/su8030221

Rankoana, S. (2021). Climate change impacts on Indigenous health promotion: The case study of Dikgale community in Limpopo province, South Africa. *Global Health Promotion*, *29*(1), 58–64. https://doi.org/10.1177/17579759211015183

Rankoana, S. (2022). Indigenous medicinal plants administered for the prevention and treatment of influenza. *E-Journal of Humanities Arts and Social Sciences*, *3*(12), 589–596. https://doi.org/10.38159/ehass.20223126

Rinto, R., Iswari, R., Mindyarto, B., & Saptono, S. (2023). Bridging the generational gap: Exploring youth understanding on ethnobotanical knowledge and its integration in higher education curricula. *Ethnobotany Research and Applications*, *26*, 1–16. https://doi.org/10.32859/era.26.48.1-16

Roy, S., & Roy, D. (2016). Use of medicinal plant and its vulnerability due to climate change in northern part of Bangladesh. *American Journal of Plant Sciences*, *7*(13), 1782–1793. https://doi.org/10.4236/ajps.2016.713166

Sam, J., Chan, Y., Nandong, J., Siner, A., Kansedo, J., & Panau, F. (2022). Antioxidant properties of underutilized bornean dabai fruit and its potential applications as a nutraceutical product. *Materials Science Forum, 1077*, 211–218. https://doi.org/10.4028/p-2i1sm4

Samtiya, M., Aluko, R., Dhewa, T., & Moreno-Rojas, J. (2021). Potential health benefits of plant food-derived bioactive components: An overview. *Foods, 10*(4), 839. https://doi.org/10.3390/foods10040839

Santini, A., Cammarata, S. M., Capone, G., Ianaro, A., Tenore, G., Pani, L., & Novellino, E. (2018). Nutraceuticals: Opening the debate for a regulatory framework. *British Journal of Clinical Pharmacology, 84*(4), 659–672. https://doi.org/10.1111/bcp.13496

Semenya, S., Potgieter, M., & Erasmus, L. (2012). Ethnobotanical survey of medicinal plants used by Bapedi healers to treat diabetes mellitus in the Limpopo province, South Africa. *Journal of Ethnopharmacology, 141*(1), 440–445. https://doi.org/10.1016/j.jep.2012.03.008

Shibambu, N., & Maluleke, W. (2023). A review study on alternative conservation and management methods to sustain medicinal plants in South Africa. *E-Journal of Humanities Arts and Social Sciences, 4*(8), 1006–1015. https://doi.org/10.38159/ehass.2023489

Ssenku, J. E., Okurut, S. A., Namuli, A., Kudamba, A., Tugume, P., Matovu, P., Wasige, G., Kafeero, H. M., & Walusansa, A. (2022). Medicinal plant use, conservation, and the associated traditional knowledge in rural communities in eastern Uganda. *Tropical Medicine and Health, 50*(1). https://doi.org/10.1186/s41182-022-00428-1

Supiandi, M., Mahanal, S., Zubaidah, S., Julung, H., & Ege, B. (2019). Ethnobotany of traditional medicinal plants used by the Dayak Desa community in Sintang, West Kalimantan, Indonesia. *Biodiversitas Journal of Biological Diversity, 20*(5). https://doi.org/10.13057/biodiv/d200516

Takaidza, S. (2024). Indigenous South African food: Nutrition and health benefits. In M. Soto-Hernández, E. Aguirre-Hernández & M. Palma-Tenango (Eds.), *Phytochemicals in agriculture and food*. IntechOpen. https://doi.org/10.5772/intechopen.110732

Tamam, S., Madbouly, H., & Amin, F. (2010). Antiviral activity of Curcuma longa against Newcastle disease virus (in vitro and in vivo studies). *Journal of Veterinary Medical Research, 20*(1), 290–295. https://doi.org/10.21608/jvmr.2020.77624

Tedila, H., & Dida, G. (2019). Ethno-medicinal study of plants used for treatment of human and livestock ailments by traditional healers in Goba Woreda, Bale Zone, Oromia, Ethiopia. *Advances in Life Science and Technology, 71*. https://doi.org/10.7176/alst/71-03

The New York Botanical Garden. (2023). *Original watercolor of Bidens pilosa by Frances W. Horne for Flora Borinqueña*. Retrieved April 17, 2023, from https://sweetgum.nybg.org/science/vh/multimedia-details/?irn=114024

Ugwu, O., Alum, E., Obeagu, E., Subbarayan, S., & Sankarapandiyan, V. (2023). Integrating medicinal plant diversity in post-COVID Uganda for holistic healthcare management. *International Academic Association Journals, 10*(3), 32–41. https://doi.org/10.59298/iaajb/2023/1.4.1000

Veleşcu, I. D., Crivei, I. C., Balint, A. B., Arsenoaia, V. N., Robu, A. D., Stoica, F., & Raţu, R. N. (2025). Valorization of Betalain pigments extracted from *Phytolacca americana* L. Berries as Natural Colorant in Cheese Formulation. *Agriculture, 15*(1), 86. https://doi.org/10.3390/agriculture15010086

Wanjohi, B., Sudoi, V., Njenga, E., & Kipkore, W. (2020). An ethnobotanical study of traditional knowledge and uses of medicinal wild plants among the Marakwet community in Kenya. *Evidence-Based Complementary and Alternative Medicine, 2020*(1). https://doi.org/10.1155/2020/3208634

Xu, Y., Liang, D., Wang, G., Wen, J., & Wang, R. (2020). Nutritional and functional properties of wild food-medicine plants from the coastal region of South China. *Journal of Evidence-Based Integrative Medicine, 25*, 2515690X2091326. https://doi.org/10.1177/2515690x20913267

Yadav, M., Jain, S., Tomar, R., Prasad, G. B. K. S., & Yadav, H. (2010). Medicinal and biological potential of pumpkin: An updated review. *Nutrition Research Reviews, 23*(2), 184–190. https://doi.org/10.1017/s0954422410000107

Yang, L., Ahmed, S., Stepp, J. R., Mi, K., Zhao, Y., Ma, J., Liang, C., Pei, S., Huai, H. Xu, G., Hamilton, A. C., Yang, Z.-W., & Xue, D. (2014). Comparative home garden medical ethnobotany of Naxi healers and farmers in northwestern Yunnan, China. *Journal of Ethnobiology and Ethnomedicine, 10*(1), 6. https://doi.org/10.1186/1746-4269-10-6

Yelianti, U., Muswita, M., & Aswan, D. (2023). Medicinal plant used by Indigenous people namely Suku Anak Dalam (SAD) in Nyogan Village Jambi Province. *Jurnal Penelitian Pendidikan IPA, 9*(2), 977–980. https://doi.org/10.29303/jppipa.v9i2.1008

Zamora-Ros, R., Cayssials, V., Jenab, M., Rothwell, J. A., Fedirko, V., Aleksandrova, K., Tjønneland, A., Kyrø, C., Overvad, K., Boutron-Ruault, M.-C., Carbonnel, F., Mahamat-Saleh, Y., Kaaks, R., Kühn, T., Boeing, H., Trichopoulou, A., Valanou, E., Vasilopoulou, E., Masala, G., … Scalbert, A. (2018). Dietary intake of total polyphenol and polyphenol classes and the risk of colorectal cancer in the European prospective investigation into cancer and nutrition (epic) cohort. *European Journal of Epidemiology*, *33*(11), 1063–1075. https://doi.org/10.1007/s10654-018-0408-6

Zemede, J., Mekuria, T., Ochieng, C., Onjalalaina, G., & Hu, G. (2024). Ethnobotanical study of traditional medicinal plants used by the local Gamo people in Borda Abaya District, Gamo Zone, Southern Ethiopia. *Journal of Ethnobiology and Ethnomedicine*, *20*(1), 28. https://doi.org/10.1186/s13002-024-00666-z

Zhou, F., Ma, Z., Rashwan, A. K., Khaskheli, M. B., Abdelrady, W. A., Abdelaty, N. S., Hassan Askri, S. M., Zhao, P., Chen, W., & Shamsi, I. H. (2024). Exploring the Interplay of Food Security, Safety, and Psychological Wellness in the COVID-19 Era: Managing Strategies for Resilience and Adaptation. *Foods*, *13*(11), 1610. https://doi.org/10.3390/foods13111610

CHAPTER 14

Cultural Food Taboos and Nutritional Consequences

Abstract

This chapter critically examines the cultural, ecological, and sociobiological dimensions of food taboos in Indigenous societies and their implications for health and nutrition. Food taboos are ritualized and socially enforced restrictions on certain foods. They are deeply embedded in gender norms, religious cosmologies, environmental knowledge, and systems of social order. Drawing from sociological, anthropological, evolutionary, and ecological frameworks, the chapter explores how food taboos are classified by gender, age, ritual practice, seasonal rhythms, and kinship systems. Through cross-cultural case studies from Africa, Asia, the Americas, and Oceania, it highlights the diverse rationales and functions of food taboos, including their roles in reinforcing identity, promoting ecological sustainability, and regulating health practices. While food taboos often reflect adaptive cultural logic and traditional ecological knowledge, they can simultaneously lead to nutritional vulnerabilities, particularly among pregnant women, children, and marginalized social groups. On the other hand, prohibitions against certain foods, such as pork in Islamic traditions or beef in Hindu cultures, can promote communal health by mitigating disease risks associated with those food items. The chapter analyzes how these taboos intersect with gendered power structures, maternal health, and food insecurity. It further explores the evolving landscape of taboos amid globalization, religious conversion, urbanization, and climate change, illustrating how traditional beliefs are being reinterpreted or challenged in new sociocultural contexts. The chapter calls for culturally responsive health interventions that engage respectfully with Indigenous knowledge holders, challenge harmful practices without eroding cultural heritage, and promote food sovereignty. It contributes to sociobiological scholarship by advancing an interdisciplinary and ethically grounded understanding of how food taboos mediate the dynamic relationships between culture, health, and the environment.

Keywords: Indigenous food systems, Food taboos, Nutritional consequences, Gender and diet, Cultural ecology, Public health nutrition, Sociocultural beliefs

14.1 Introduction

Food taboos are rules prohibiting the consumption of certain foods for various reasons, including health, cultural, religious, or social beliefs. These prohibitions often extend beyond mere dislike or personal preference, representing deep-seated cultural norms that can significantly shape individuals' dietary practices across diverse communities. Among Indigenous societies, these taboos are not merely personal choices but are integral to the identity and cultural heritage of these groups, serving as markers of belonging and adherence to tradition (Meyer-Rochow, 2009; Vasilevski & Carolan, 2016). They may arise from historical precedents, environmental adaptations, or societal hierarchy, reflecting intricate sociocultural dynamics (Landim et al., 2023). The significance of food taboos in Indigenous societies cannot be overstated, as they often dictate what foods are accessible or acceptable for consumption, substantially impacting nutrition and health outcomes. Numerous studies indicate that these taboos can lead to nutritional deficiencies, particularly among vulnerable demographics such as pregnant women and children (Lekey et al., 2024; Vasilevski & Carolan, 2016; Zerfu et al., 2016). For example, in particular Indigenous populations, adherence to food taboos restricts the intake of essential micronutrients during critical periods of development and growth, thus presenting substantial health challenges (Lekey et al., 2024; Tripura, 2025). Therefore, the analysis of food taboos becomes vital in understanding the nutritional landscape of distinct communities.

Notably, the emergence of food taboos often correlates with the socioeconomic context of the community. In regions where access to a diverse diet is limited, taboos can exacerbate problems of undernutrition and health disparities. This relationship underscores the importance of contextual factors, such as socioeconomic status, in understanding the persistence and implications of such taboos (Chakona & Shackleton, 2019; Tsegaye et al., 2021). Cultural norms and traditional beliefs surrounding foods can further entrench these dietary restrictions, creating a barrier to optimal health that is challenging to surmount (Akiso et al., 2024; Vasilevski & Carolan, 2016).

In examining the multifaceted nature of food taboos, it is essential to consider their origins and their implications for health behaviors and dietary choices. Research suggests that enforcing these taboos may reflect broader social norms and values, illustrating a complex interaction of cultural identity and nutrition (Meyer-Rochow, 2009; Tsegaye et al., 2021). This interaction often extends to gender dynamics, as many food taboos disproportionately affect women, particularly during pregnancy and lactation, thereby influencing maternal and child health outcomes (Ekwochi et al., 2016; Köhler et al., 2018). The exploration of food taboos within Indigenous societies offers profound insights into the myriad ways in which culture and nutrition intersect. Investigating how these practices manifest and their effects on dietary intake and health can inform public health strategies and nutritional interventions tailored to specific cultural contexts. Hence, understanding food taboos is crucial for fostering cultural appreciation and respect and addressing the pressing health challenges these communities face (Ekwochi et al., 2016; Gibson et al., 2021).

This chapter critically examines the origins, functions, and impacts of cultural food taboos within Indigenous communities, highlighting how these practices shape dietary behaviors, reinforce social structures, and influence health outcomes. By integrating sociological, anthropological, and nutritional perspectives, the chapter offers a nuanced analysis of how food taboos operate as adaptive cultural mechanisms and potential sources of nutritional vulnerability. Its contribution to sociobiological knowledge lies in uncovering the evolutionary and ecological underpinnings of food avoidance behaviors while also interrogating their gendered and symbolic dimensions. Furthermore, the chapter provides a cross-cultural synthesis that deepens our understanding of the interplay between culture, biology, and health, ultimately advancing a more holistic framework for analyzing food systems within Indigenous societies.

14.2 Theoretical Perspectives on Food Taboos

Food taboos represent a complex aspect of human culture, emerging from various sociological, anthropological, evolutionary, and ecological factors. Therefore, understanding food taboos in Indigenous societies requires a multidisciplinary framework, drawing from sociological, anthropological, evolutionary, and environmental perspectives (Figure 14.1). Sociological and anthropological theories provide foundational insights into how taboos function within social systems. Structuralism, as proposed by Claude Lévi-Strauss, views food taboos as part of the underlying structures that organize human thought and cultural classifications, particularly the dichotomies of nature versus culture or raw versus cooked (Lévi-Strauss, 1969). Functionalist theories, such as those advanced by

Figure 14.1: Theoretical Perspectives of Food Taboos

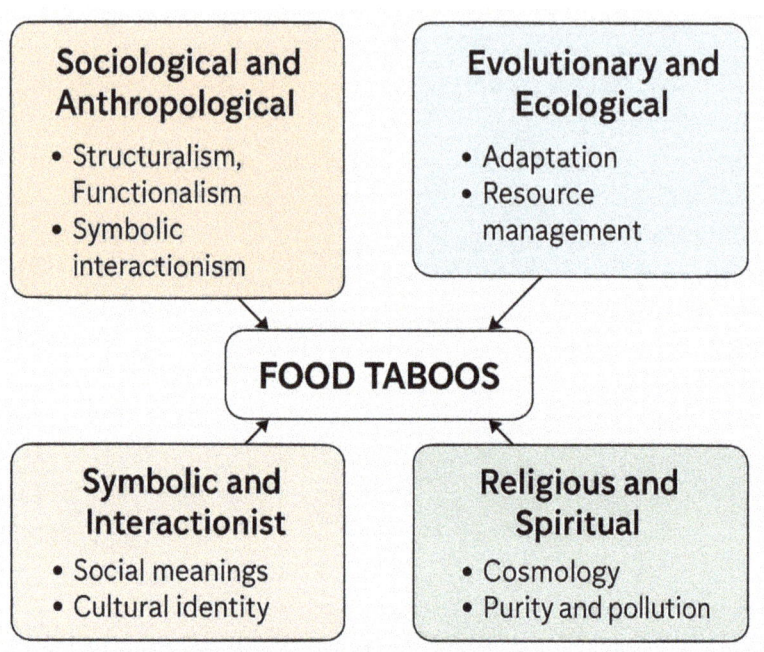

Malinowski (1944) and Radcliffe-Brown (1952), argue that food taboos serve social cohesion by regulating behavior, maintaining group identity, and reinforcing moral boundaries. From a symbolic interactionist perspective, as developed by George Herbert Mead and later Erving Goffman, food taboos are seen as dynamic symbols embedded in everyday interactions, reflecting meanings that individuals and groups construct through social engagement (Blumer, 1969).

Evolutionary and ecological explanations take a different approach, focusing on the adaptive functions of taboos in human survival and resource management. From this perspective, some food taboos may have developed as mechanisms to prevent the consumption of toxic or spoiled foods, protect scarce resources, or reduce disease transmission—thus promoting group fitness and long-term ecological balance (Fessler & Navarrete, 2003; Rozin & Fallon, 1987). For instance, dietary restrictions on undercooked pork or shellfish may have emerged due to their higher risk of harboring parasites in certain ecological contexts. From a sociological perspective, structuralism emphasizes that food taboos serve specific functions that maintain social order and cohesion. Gowder discusses how food and nutrition are influenced by cultural, economic, and social factors, which ultimately reproduce social norms and reinforce collective identity (Gowder, 2024). This perspective aligns with Bourdieu's theory that posits an individual's habitus shapes their dietary choices, indicating that food taboos reflect broader societal imperatives about what is acceptable or forbidden within a cultural context (Costa et al., 2014).

Functionalism further explores how food taboos can arise as mechanisms to ensure food safety and communal health. Prohibitions against certain foods, such as pork in Islamic traditions or beef in Hindu cultures, can promote communal health by mitigating disease risks associated with those food items (Gowder, 2024). This functional perspective underscores the role of food taboos in fostering group identity and social cohesion. Symbolic interactionism provides an additional lens through which to consider food taboos, emphasizing the significance of social interactions in constructing and perpetuating these practices. Thomas and Emond's (2017) work illustrates how societal norms and personal narratives shape the meanings attached to mealtime experiences. This approach suggests that food can embody meanings that transcend basic nutrition, asserting its role in cultural identity and resilience, especially during significant disruptions such as those experienced post-Hurricane Katrina (Harvey, 2017). Food rituals reinforce collective memories and continuity in such contexts, highlighting their importance in social interactions. The interplay between evolved behaviors and ecological contexts also informs our understanding of food taboos. Evolutionary theories suggest that specific prohibitions may stem from adaptive behaviors that maximize survival, such as avoiding spoiled or contaminated foods, which can be framed within natural selection (Cook, 2023). Furthermore, ecological perspectives note how environmental conditions shape food availability and cultural practices. The dietary restrictions among Indigenous communities must be understood within their unique ecological circumstances and historical contexts, such as colonialism and resource extraction (Henryks & Brimblecombe, 2016).

Additionally, the religious and spiritual dimensions significantly influence food taboos, often reflecting shared cosmologies and ethical frameworks. For instance, Jewish dietary laws prohibiting certain animals illustrate the interconnection between spirituality and dietary restrictions, shaping community identity (Cook, 2023). Many cultures view abstaining from specific foods as a demonstration of commitment to a higher moral order, reinforcing community bonds through shared rituals (Smith, 2023). These prohibitions are often encoded in mythologies and sacred texts, interpreted as expressions of divine will or spiritual order. Mary Douglas (1966), in her seminal work *Purity and Danger*, argued that food taboos serve to maintain symbolic boundaries between purity and pollution, reinforcing a cosmological order. In many

African, Native American, and Pacific Island cultures, food taboos are directly tied to beliefs in ancestral spirits, totemic affiliations, and ritual cycles, where violating such taboos is believed to bring spiritual harm or communal imbalance (Meyer-Rochow, 2009; Simoons, 1994). Global dietary shifts and cultural exchanges further complicate the construction of food taboos. Acculturation processes among immigrant populations highlight how traditional food practices can evolve in response to new cultural contexts (Martínez, 2013). Such transformations often prompt renegotiation of identity, as individuals balance pressures of assimilation with efforts to maintain cultural heritage through food choices.

14.3 Classification of Food Taboos

Food taboos are culturally constructed dietary restrictions that dictate which foods are permissible for consumption by individuals in varying contexts. They manifest through diverse factors, including gender, age, ceremonial contexts, seasonal and ecological considerations, and social stratification, such as caste or lineage. The classification of food taboos reveals the social structures that underpin these prohibitions and their implications on nutrition and health across different populations. Classifying food taboos by gender reveals significant disparities in dietary allocation and preferences. Many societies impose distinct dietary restrictions on men and women, reflecting, in part, the prevailing social hierarchies that elevate male dietary needs over those of females. Research indicates that in various cultural contexts, men, particularly adult males, are often prioritized in distributing nutrient-dense foods, leading to inequitable nutritional outcomes for women and children (Harris-Fry et al., 2017). For instance, a South Asian study found that adult males receive larger portions of nutrient-rich foods, symptomatic of the underlying gender dynamics that place higher value on male consumption (Harris-Fry et al., 2017). Similarly, in rural contexts, food taboos can be heavily gendered, wherein women face explicit restrictions that limit their access to essential nutrients during critical periods such as pregnancy (Gulema et al., 2023). This supports the notion that food taboos function as instruments of social control, delineating acceptable consumption patterns for each gender, thus reinforcing patriarchal norms within these cultures (Bukachi et al., 2022).

Another salient factor influencing food taboos is age, particularly in the contexts of children and older people. Certain dietary restrictions are uniquely imposed on different age groups, and special considerations often come into play during pivotal life stages, such as adolescence and pregnancy. Pregnant women confront a myriad of explicit food prohibitions that aim to safeguard both maternal health and fetal development (Wamoyi et al., 2014). Research indicates that in South Africa, pregnant women frequently encounter numerous dietary restrictions that can exacerbate nutritional deficits, subsequently affecting both the mother's and child's health throughout gestation and lactation (Wamoyi et al., 2014). The cultural beliefs surrounding age and dietary restrictions emphasize the interplay between life stages and food taboos, which can significantly impact individuals' nutritional choices and overall health.

Ceremonial and ritualistic contexts also contribute substantially to the classification of food taboos. Specific food prohibitions are often tied to significant life events such as menstruation, familial mourning, and religious ceremonies. Individuals may abstain from certain foods during these occasions to exhibit reverence, comply with religious tenets, or adhere to societal norms (Gitungwa et al., 2021). Research highlights how food practices aligned with religious rituals are not merely acts of consumption but are in fact vital components of community identity, reinforcing social cohesion and cultural heritage (Gitungwa et al., 2021). Therefore, ritualistic food taboos determine acceptable dietary behaviors and foster a sense of belonging within a community through shared cultural practices.

Seasonal and ecological considerations further shape food taboos, guiding dietary practices based on agricultural cycles and environmental sustainability. Numerous cultures have developed restrictions preventing the consumption of particular foods during specific seasons, reflecting societal norms prioritizing ecological rhythms and sustainability (Bukachi et al., 2022). For instance, dietary prohibitions may allow wildlife populations to replenish or conserve resources during critical agricultural periods. This ecological awareness reflects societal norms emphasizing environmental conservation in conjunction with culturally embedded food taboos, often aligning community practices with sustainable agrarian methods (Domínguez-Salas et al., 2016; Ogwu & Osawaru, 2022).

In more hierarchical societies, caste- or lineage-based taboos significantly inform dietary practices, intertwining food customs with social stratification. In India, the food practices of various castes demonstrate how dietary choices often reflect social status and identity. Research indicates that upper-caste groups have historically adopted vegetarianism to assert social dominance. At the same time, lower castes contest these dietary restrictions in advocacy for their rights to consume a more diverse range of foods, thereby asserting their agency within the confines of caste-based dietary taboos (Angsongna et al., 2016). Furthermore, the intersection of caste and social stratification has been shown to directly affect malnutrition rates among children, underscoring the detrimental health effects that arise from discriminatory dietary practices (Golden et al., 2016) (Table 14.1).

Table 14.1: Classification of Food Taboos

Category of Taboos	Focus	Examples	Function
Gender-Based Taboos	Restrictions based on sex or gender roles	Women are forbidden from eating eggs or meat during pregnancy (e.g., Gogo, Tanzania)	Reinforce gender roles, regulate fertility, control symbolic power
Age-Specific Taboos	Restrictions targeting specific age groups such as children or initiates	Tikopia (Polynesia) children are restricted from eating octopus or sea turtle	Protect the health, signify social maturity or transition
Ritual and Ceremonial Taboos	Applied during menstruation, pregnancy, mourning, and so on	Yoruba (Nigeria) menstruating women barred from preparing certain foods; Andean pregnancy taboos	Preserve ritual purity, protect spiritual and physical health
Seasonal and Ecological Taboos	Tied to ecological cycles or resource conservation	Kayapo (of Brazil) hunting bans during mating seasons	Promote sustainability, ensure food security
Caste/Clan/Lineage-Based Taboos	Linked to social stratification, totemic affiliations, or kinship structures	Naga tribes' (India) clan-based animal consumption taboos	Maintain social order, express group identity, respect spiritual links

14.4 Case Studies from Indigenous Communities

Indigenous communities worldwide maintain various cultural taboos that reflect their beliefs, social structures, and relationships with the environment. Food taboos are practiced within Indigenous communities across regions to reinforce social norms and control behaviors. They also embody the intricate relationships these groups maintain with their spiritual beliefs and the environment (Ogwu et al., 2023, 2025). From the delicate balance of food restrictions to the deep cultural narratives that sustain these practices, the study of Indigenous taboos offers critical insights into the resilience and adaptability of traditional societies. Examining case studies from Africa, Asia, the Americas, and Oceania provides a comprehensive understanding of these taboos and their significance within different cultural contexts.

14.4.1 Africa: Taboos Among Indigenous Groups

Many African societies exhibit strong adherence to taboos that reflect deep-rooted cultural practices and beliefs. Among the Igbo in southern Nigeria, taboos concerning sexual behavior, such as the stigma surrounding pregnancies out of wedlock, illustrate how societal expectations can dictate personal conduct (Dumbili, 2016). Breaking these taboos may result in severe social repercussions, including ostracism from the community (Ezeifeka, 2019). Furthermore, traditional Igbo beliefs attribute misfortunes to violations of taboos, intertwined with a belief in supernatural influences like the Almighty God and lesser deities, which structure moral conduct and social harmony (Ikwuka et al., 2013). Similar observations can be made among the Yoruba and Maasai. For example, the Maasai of Kenya and Tanzania have stringent taboos governing the consumption of certain animals, which are often viewed as sacred. These prohibitions serve to regulate their relationship with nature and ensure the tribe's survival through sustainable practices (Ikhajiagbe et al., 2021; Imarhiagbe & Ogwu, 2022). Additionally, among the Yoruba, spiritual and cultural taboos concerning land use and the treatment of ancestors play crucial roles in maintaining socioeconomic stability (Blythe, 2013).

14.4.2 Asia: Taboos in Tribal India and Southeast Asia

In tribal India, various taboos arise from complex social structures and local deities. For instance, certain tribes observe dietary restrictions linked to spiritual practices, which dictate what members can consume and when. This is similarly observed among the Indigenous peoples of Southeast Asia, where taboos surrounding food, particularly in sacred contexts, illustrate the intersection of spirituality and daily life (Nwankwo & Itanyi, 2019). These Indigenous populations' taboos often serve as social control mechanisms and frameworks for maintaining ecological balance. Indigenous groups in the Amazon, for example, enforce prohibitions on hunting specific animals during certain seasons, which not only aligns with their cosmology but also promotes biodiversity conservation (Ogwu, 2023).

14.4.3 The Americas: Andean and Amazonian Practices

In the Andean highlands, food restrictions are often linked to agricultural practices, aligning with seasonal changes and cultural festivals. Many Andean communities impose taboos on certain crops to promote biodiversity and sustainability within their agrarian systems. For instance, cultivating specific plants may be forbidden during religious observances, reflecting a deep respect for ancestral traditions and the environment (Ọnụkawa, 2021). Likewise, among Amazonian tribes, shamanic practices dictate prohibitions that are integral to their worldview. These taboos often surround the use of specific plants and animals in rituals, and violations are thought to invoke supernatural retribution, thus sustaining the cultural narratives that reinforce social cohesion and spiritual beliefs.

14.4.4 Oceania: Taboos Among Aboriginal Australians and Polynesians

Indigenous Australians have a rich tapestry of taboos that significantly influence their cultural identity. Food taboos, such as restrictions on hunting particular species during specific seasons, symbolize a profound relationship with the land and a commitment to sustainable practices inherent to their worldview (Yen et al., 2017). Moreover, these taboos often stem from spiritual beliefs regarding the connection between the people, the land, and ancestral spirits. In Polynesian cultures, taboos (or "kapu" in Hawaiian) govern various aspects of life, from social interactions to environmental stewardship. These prohibitions are not merely restrictions but in fact reflect a sophisticated understanding of ecological balance and community norms (Maria et al., 2023). As Polynesian societies face external pressures from modernization, these traditional taboos often transform, as seen in contemporary discussions about the reengagement with Indigenous practices (Nweke & Okpaleke, 2019).

14.5 Cultural Rationales Behind Food Taboos

Food taboos are a complex interplay of cultural beliefs and practices that shape dietary choices across different societies. These taboos can serve various functions—from health and hygiene considerations to social control mechanisms reinforcing group identities and ecological balance. Examining these dimensions helps illuminate the broad implications of food taboos on maternal health and community dynamics.

14.5.1 Health and Hygiene Justifications

Health and hygiene often serve as fundamental rationales behind food taboos. Certain dietary restrictions may evolve as protective mechanisms against foodborne illnesses, particularly in regions where food safety is compromised. For example, pregnant women are frequently advised to avoid specific foods believed to be harmful or potentially allergenic, which may contribute positively to maternal and fetal health outcomes. Tofik et al. (2025) note that food taboos can help curb the consumption of items that are prone to contamination, thus reducing the risk of foodborne illnesses during pregnancy. Additionally, studies highlight how women adhere to these taboos out of fear of delivering larger babies or otherwise risking their health and the health of their unborn children (Amare et al., 2022; Zinyemba, 2023). Cultural beliefs surrounding food consumption often involve ancestral wisdom about food safety, aligning eating practices with a broader understanding of health risks (Babu & Patnaik, 2021). Moreover, traditional food taboos have been shown to encourage healthier dietary choices in various cultural contexts. Tsegaye et al. elaborate on how adherence to food customs significantly contributes to maintaining maternal health. In their study, they found that cultural influences are critical in shaping the perceptions of food safety and hygiene among pregnant women (Tsegaye et al., 2021). Similarly, Lekey et al. (2024) indicate a corresponding pattern in Tanzania, where adherence to food taboos during pregnancy is linked with perceptions of health and safety for both mothers and infants. Hence, food taboos stem from traditional knowledge and manifest as practical responses to environmental and health-related challenges.

14.5.2 Social Control and Identity

Beyond health considerations, food taboos also perform essential roles in social cohesion and community identity formation. They often reinforce cultural identity boundaries, differentiating one group from another through unique dietary practices. For instance, specific food taboos might emerge from historical narratives that shape a community's collective identity. Golden and Comaroff (2015) assert that food taboos delineate shared history and cultural substance boundaries, solidifying community kinship ties and collective identity. This insight extends to how social norms dictate dietary restrictions, often anchored in shared values, traditions, and beliefs, contributing to group cohesion (Zinyemba, 2023). The transmission of food taboos, particularly among women, strengthens intergenerational ties, as these customs are typically passed down through maternal lines. In various studies, it has been found that older women, such as mothers-in-law, enforce adherence to taboos, which exemplifies how these practices function as not only a health strategy but also a tool for maintaining culture and identity (Tela et al., 2020; Tugume et al., 2024). Consequently, women may adhere to such taboos to align with cultural expectations, suggesting that food practices can augment social bonding while emphasizing conformance to group norms.

14.5.3 Conservation and Ecological Balance

Food taboos also reflect a collective awareness of ecological balance and conservation efforts. Many cultures adopt dietary restrictions to manage wildlife populations and preserve ecological integrity. For example, taboos against hunting certain species during their breeding seasons serve to conserve biodiversity and ensure sustainable practices within local ecosystems. Landim et al. (2023) evidence this idea, stating that cultural expressions, such as food taboos, significantly influence human interaction with wildlife species, thus promoting conservation through traditional practices. Similarly, Begossi et al. (2004) discuss how food taboos related to fishing practices among Indigenous communities in South America contribute to the sustainability of aquatic ecosystems. These restrictions on consuming specific fish species serve ecological and cultural purposes, enabling communities to avoid overexploitation of resources while maintaining traditional lifestyles (Begossi et al., 2004). Food taboos can synergize human needs and ecological sustainability by integrating environmental consciousness with cultural practices.

14.5.4 Taboos and Maternal Nutrition

Throughout various studies, the relationship between food taboos and maternal nutrition emerges as a significant area of concern. The repercussions of adhering to food taboos can sometimes adversely affect the nutritional intake of pregnant women. For instance, Zinyemba (2023) highlights how culturally driven food prohibitions can lead to inadequate maternal diets, thereby increasing vulnerability to maternal mortality and compromising fetal health outcomes. Despite the conscious intention to protect maternal and infant health, such restrictions can paradoxically exacerbate health risks if not aligned with balanced dietary practices. Conversely, other studies suggest that some food taboos are formed based on a community's understanding of how various foods can affect pregnancy and maternal health. Amare et al. (2022) indicate how traditional knowledge about nutrition can inform food taboos, positioning certain foods as detrimental during pregnancy based on culturally specific health beliefs. Consequently, while the cultural aspect of food taboos protects and promotes health, there is a complex dynamic at play where the efficacy of these taboos depends heavily on the overall dietary context.

14.6 Nutritional Consequences of Food Taboos

The nutritional consequences of food taboos during pregnancy and early childhood are profound and complex, significantly impacting micronutrient deficiency, maternal health, and child development. The restriction of certain foods, often deemed taboo due to cultural or traditional beliefs, can substantially limit the dietary diversity available to pregnant women and their children, leading to inadequate intake of crucial nutrients such as iron, calcium, protein, and vitamins. Studies have highlighted that these restricted foods typically include sources rich in essential micronutrients vital for mothers and their developing fetuses. For instance, research indicates that the avoidance of animal proteins such as fish, eggs, and dairy products standard in various cultures can contribute to maternal anemia and low birth weight, ultimately jeopardizing both maternal and child health outcomes (Arzoaquoi et al., 2015; Tofik et al., 2025).

Specific taboos surrounding dietary practices during pregnancy have been documented extensively in various cultural contexts. Arzoaquoi et al. (2015) pointed out that in rural Ghana, food prohibitions often eliminate high-energy, nutrient-rich foods critical for the caloric needs of pregnant women. Similarly, a meta-analysis by Maggiulli et al. (2022) corroborated that significant cross-cultural differences exist in food prohibitions, with direct implications for childbirth and health outcomes among agricultural populations, where nutritional needs are critical during pregnancy. This restriction of dietary sources can contribute to severe micronutrient deficiencies, affecting maternal health and leading to long-term developmental issues for the child, reinforcing cycles of undernutrition within families and communities (Gebregziabher et al., 2023).

The impact of these taboos can be seen in examples such as those within the Khmu ethnic group in Laos, where strict adherence to postpartum food taboos has been shown to hinder maternal nutritional status significantly (de et al., 2012). These cultural practices are deeply embedded in societal norms and often lead to a lack of awareness regarding the nutritional needs of women during pregnancy. Consequently, many pregnant women are left without adequate sources of dietary iron and calcium, essential for fetal development and maternal health. Additionally, avoiding certain foods can be linked to sociocultural practices to prevent perceived harm to the unborn baby (Chakona & Shackleton, 2019). Such avoidance can lead to a nutrient deficiency cycle, exacerbating poor health outcomes in maternal and child populations.

Compensatory food practices have emerged as critical strategies within these communities to counterbalance the absence of traditional food sources. Despite the potential for substituting restricted foods, evidence suggests that compensatory foods often do not fully meet the nutritional needs of pregnant women. For example, Ramulondi et al. (2021) observed that traditional substitutes frequently lack essential micronutrients found in taboo foods, which are pivotal for both pregnancy and nursing. While compensatory diets may provide some level of nutrition, they often fail to encompass the diversity and richness of the protective foods being avoided, resulting in inadequate dietary intake (Gebregziabher et al., 2023). Long-term public health implications of food taboos extend beyond individual health, impacting community health. Nutritional deficiencies stemming from these taboos can result in increased rates of maternal anemia, preterm births, and low birth weights, ultimately leading to developmental delays in children (Gebregziabher et al., 2023; Tofik et al., 2025). Moreover, the cyclical nature of poverty and unhealthy dietary practices reinforced by cultural food taboos can impose significant burdens on healthcare systems, necessitating targeted public health interventions. Addressing the cultural underpinnings of food taboos becomes increasingly important for health advisors and policymakers who aim to improve maternal and child health outcomes. Interventions must be culturally sensitive and focused on educating communities about the essential role of diverse foods in promoting health (Esimai et al., 2001).

Ultimately, the ramifications of food taboos are pervasive, affecting numerous aspects of health and nutrition, particularly among vulnerable populations. The divergence in dietary practices and subsequent health implications underscores the need for a comprehensive understanding of cultural attitudes toward food during pregnancy and early childhood. Addressing these complex issues will require collaborative efforts from health professionals, cultural leaders, and community stakeholders to foster an environment that supports healthy nutritional practices while respecting traditional beliefs. By tackling the issues of food taboos comprehensively, communities can move toward healthier practices that support both maternal and child health in a sustainable manner.

14.7 Influence of Gender and Power Dynamics in Food Taboos

To understand the influence of gender and power dynamics in food taboos, it is imperative to delve into how these taboos reflect and reinforce gender roles, their impacts on women's nutrition and reproductive health, and the dual nature of taboos as instruments of patriarchy or empowerment. First and foremost, food taboos are often deeply intertwined with cultural beliefs, which are frequently influenced by gender roles. For instance, in many societies, women are typically the primary custodians of food preparation and consumption, directly reflecting their societal roles as caretakers. Therefore, when specific foods are deemed taboo, it determines the dietary patterns of pregnant women and emphasizes their subordinate role in patriarchal structures that govern these dietary restrictions (Tsegaye et al., 2021; Vasilevski & Carolan, 2016). Much of the taboo's women adhere to during pregnancy are rooted in cultural traditions, which expect them to maintain specific diets for the perceived well-being of their families, particularly infants (Geta et al., 2022; Zerfu et al., 2016). Such practices further reinforce existing gender norms by positioning women as primarily responsible for familial health amid restrictive dietary practices.

The implications of food taboos extend beyond cultural preservation; they often lead to nutritional deficiencies for women, particularly during pregnancy. Numerous studies have shown that women who avoid certain foods due to taboos usually face inadequate dietary diversity. For instance, in Ethiopia, pregnant women who abstain from consuming poultry or dairy products as part of their food taboos are significantly more likely to exhibit poorer nutritional outcomes, reflecting suboptimal dietary diversity (Alamirew et al., 2024; Mohammed et al., 2019). This emphasizes that food taboos can severely impact women's health and nutrition, particularly during critical periods such as pregnancy, affecting maternal and fetal health (Amare et al., 2022; Mohammed et al., 2019). The disparities resulting from these taboos can contribute to conditions such as anemia and hinder proper fetal development (Amare et al., 2022; Mohammed et al., 2019).

Moreover, food taboos also function within the dynamics of power in households. Enforcing these taboos is often synonymous with power relationships, typically dominated by male figures in the family, such as husbands and mothers-in-law. The decision-making power regarding which foods are taboo and how strictly these restrictions are enforced illustrates the intersection of gender and power (McNamara & Wood, 2019; Nguyen et al., 2021). Community-level research has demonstrated that men's attitudes toward women's diets can significantly influence dietary patterns and adherence to food taboos during pregnancy (Amare et al., 2022), suggesting that pregnant women may often lack practical agency over their nutritional choices. This patriarchal enforcement of food taboos not only affects nutrition but also perpetuates gender-based inequalities within households. Food taboos can also act as instruments of empowerment or cultural identity in some contexts. In Eastern Ethiopian communities, for instance, women often express that adherence to food taboos can foster a sense of identity and community belonging, as these practices are deeply rooted in the cultural fabric (Tofik et al., 2025). By participating in these traditions, women can reinforce social bonds within their communities, albeit at a cost to their nutritional health (Ekwochi et al., 2016; Tofik et al., 2025). Yet, the perception of empowerment is also nuanced. While maintaining cultural practices can offer a sense of belonging, it simultaneously constrains women's autonomy concerning dietary choices, thus revealing the paradox inherent in culturally prescribed roles (Zerfu et al., 2016).

Further complicating this dynamic, women with higher educational levels and socioeconomic status are more likely to challenge or break such taboos, reflecting an emerging form of empowerment where knowledge and economic independence play pivotal roles in decision-making about diet during pregnancy (Placek et al., 2017; Rayna et al., 2024). Educated women often develop a critical awareness of the nutritional implications of food and may prioritize health over adherence to traditional taboos. Nevertheless, the collective force of religious and cultural norms can still apply significant pressure, often leading to the continued prevalence of restrictive dietary beliefs, even among those aware of the potential health implications (Tela et al., 2020; Tugume et al., 2024). Investigating the diverse landscape of food taboos also reveals their socioeconomic underpinnings. Research shows that women of lower socioeconomic backgrounds tend to adhere more to food taboos, correlating with limited access to information, resources, and healthcare (Getnet et al., 2018; Girma et al., 2022). Economic conditions can dictate not only the availability of food but also the extent to which a woman can challenge existing dietary restrictions, thus situating food taboos within a broader framework of economic inequality and access to healthcare resources

(Amare et al., 2022). The intricate relationships between a woman's socioeconomic status, her dietary options, health outcomes, and her adherence to taboos reflect a complex interplay that maintains gendered disparities in nutrition and health (Girma et al., 2022).

In many instances, the intersectionality of race, culture, and gender significantly influences the adherence to food taboos. Cultural beliefs surrounding food taboos are often reinforced across generations, deeply entrenched in community identities (Chakona & Shackleton, 2019; Tsegaye et al., 2021). In rural areas of South Africa and Uganda, for example, women express the belief that following food taboos is crucial for their health and the health of their children, demonstrating the powerful hold of these cultural constructs (Ramulondi et al., 2021). Yet, narratives of resistance exist, where some women strive to reformulate healthy practices while contending with it within their cultural frameworks (Ekwochi et al., 2016; Tela et al., 2020). Such complexities highlight the dynamic nature of food taboos not only as exertors of patriarchal power but also as spaces for potential agency and cultural renegotiation.

Moreover, examination of food taboos through a public health lens reveals that educational interventions can play a crucial role in mitigating the negative impacts of these restrictions. Health communication strategies focused on increasing maternal knowledge about nutrition can empower women, thus challenging the traditional power dynamics that dictate food consumption patterns (Arzoaquoi et al., 2015; Nguyen et al., 2021). Engaging community leaders, particularly men, can also enhance the effectiveness of these interventions by fostering supportive family environments that promote healthy dietary practices for pregnant women (Nguyen et al., 2021). Taboos, while often casting a shadow on women's nutrition and health, also serve to highlight the resilience and agency women can harness in the face of restrictive cultural norms. Overcoming such cultural barriers requires collective efforts that engage communities in dialogue about gender equality, health, and cultural practices. By fostering environments where women can negotiate their dietary needs against the backdrop of cultural expectations, we can better support their nutritional and health outcomes (Amare et al., 2022; Tugume et al., 2024).

14.8 Changing Landscapes of Food Taboos

The evolving landscapes of food taboos reflect a complex interplay of factors, including globalization, conversion to religions such as Islam and Christianity, and the rise of modernization, each acting in tandem to reshape dietary practices across diverse cultures. Globalization, with its rapid dissemination of ideas, cultures, and practices, has significantly altered traditional food taboos (Table 14.2). Societal norms that once dictated what foods were deemed acceptable or forbidden have been challenged by the introduction of global cuisines and culinary practices. Meyer-Rochow (2009) emphasizes that food taboos often emerge from deep-seated cultural beliefs serving specific purposes, such as health preservation, spiritual adherence, or social cohesion. Yet, the encroachment of globalization prompts a reevaluation of these beliefs, as individuals and communities navigate between traditional dietary restrictions and the allure of new, unfamiliar food styles (Vasilevski & Carolan, 2016). For instance, in Ethiopia, food taboos are often aligned with local customs and health concerns, particularly during vulnerable periods such as pregnancy. Tsegaye et al. found that adherence to pregnancy-specific food taboos stems from culturally ingrained beliefs that can lead to nutritional deficiencies if significant foods are avoided (Tsegaye et al., 2021). This situation is compounded by the influence of rising educational levels and changing socioeconomic statuses in these communities. Studies indicate that increased knowledge and shifting attitudes toward food can lead to the gradual relaxation or outright breaking of longstanding taboos, particularly if individuals no longer observe negative consequences from consuming previously tabooed foods (Mohammed et al., 2019; Rayna et al., 2024).

Table 14.2: Drivers of the Changing Landscapes of Food Taboos

Drivers of Change	Impact on Food Taboos	Examples
Globalization	Introduction of foreign foods and exposure to new dietary norms can lead to the erosion or hybridization of traditional taboos.	In Ghana, imported fast foods have reduced adherence to traditional food rules among urban youth.
Religious Conversion	Conversion to Christianity or Islam often alters or eliminates Indigenous food taboos due to new religious dietary laws or discrediting of local beliefs.	Among the Yoruba, Christian converts may no longer observe menstruation-based taboos.

Drivers of Change	Impact on Food Taboos	Examples
Urbanization and Migration	Rural-to-urban migration exposes individuals to diverse cuisines and weakens community enforcement of taboos.	Igbo migrants in Nigerian cities often relax clan-specific taboos due to mixed urban settings.
Education and Modern Healthcare	Increased nutritional awareness and medical advice can challenge food taboos previously justified by spiritual or symbolic reasoning.	Expectant mothers in Andean communities increasingly follow biomedical advice over traditional food restrictions.
Climate Change and Environmental Stress	Resource scarcity and changing ecosystems may force communities to abandon previously taboo foods for survival.	Maasai in drought-affected regions now consume previously avoided wild plant foods.
Cultural Revitalization Movements	Some Indigenous communities are reasserting traditional taboos as part of broader movements to reclaim cultural identity and heritage.	Aboriginal communities in Australia are revitalizing traditional food knowledge, including seasonal taboos.

Religious conversion, including the influence of Islam and Christianity, introduces another dimension to the transformation of food taboos. In many communities, certain foods are prohibited based on religious edicts that can override or augment existing cultural norms (Acire et al., 2023; Kahsay et al., 2017). As conversion increases, dietary practices also shift, often reflecting a blend of traditional and religious influences; this syncretism emphasizes an adaptive response to changing ideologies and lifestyles. It becomes evident that modern religious adherents may adhere to food taboos dictated by faith while simultaneously negotiating new dietary guidelines presented by globalization (Abdullah, 2022; Ramulondi et al., 2021).

Urbanized Indigenous communities display distinctly evolving dietary practices, intricately tied to the loss of traditional food systems and the increasing reliance on imported, processed foods, which are often at odds with their cultural taboos.

Urbanization prompts a disconnect from ancestral lands and traditional practices leading to unprecedented challenges in food security and, subsequently, dietary identity (Richmond et al., 2020). Richmond et al. (2020) note the pronounced discrepancy between food sovereignty for Indigenous peoples residing on reserves and those who reside in urban locales, illustrating the tensions that arise when community values are confronted with urban food environments. This conflict can lead to the gradual erosion of traditional food taboos, as urban Indigenous populations adapt to more Westernized diets filled with convenience foods that do not necessarily resonate with their cultural identities.

Moreover, the phenomenon of migration and the diaspora contributes to an evolution of food taboos as individuals navigate hybrid identities that amalgamate culinary practices from their homeland with new cultural influences. This dynamic pushes forth a landscape where food choices become not only functional but also symbolic of cultural retention and adaptation (Tesfa et al., 2023). In diasporic communities, food becomes a crucial vessel for the transmission of traditions, yet the need to assimilate can lead to conflicts over dietary practices, which may significantly differ from those of the original homeland. Syncretism and hybrid dietary practices emerge as common themes in these changing food landscapes. Many communities exhibit a blend of culturally relevant foods and practices that are curated from both their Indigenous and new backgrounds selectively adhering to, modifying, or even discarding certain tabooed foods (Arzoaquoi et al., 2015; Tripura, 2025). Such adaptations highlight resilience in the face of cultural pressures and the capacity for communities to redefine their food identities through intercultural exchange. For example, researchers have reported that while traditional taboos may remain significant, modern influences often lead individuals to create a balance between honoring their heritage and embracing new culinary traditions (Gebregziabher et al., 2023).

Further research illustrates that syncretism is not merely a passive acceptance but is in fact an active negotiation of cultural identity, where individuals display agency in their food choices in response to modernity. The role of community figures, such as elders or religious leaders, also comes into play as they influence dietary practices, either in support of traditional taboos or in adjusting these to accommodate contemporary realities (Acire et al., 2023; Mohammed et al., 2019). Nowhere is this more visible than in the context of pregnancy, where dietary restrictions remain firmly entrenched, yet there is a growing awareness of the health implications brought on by enforced food taboos (Chakona & Shackleton, 2019; Ekwochi et al., 2016). The challenge remains for communities to find ways to preserve essential aspects of their culinary heritage while ensuring the dietary needs of future generations are met.

Studies further demonstrate that traditional taboos are heavily imbued with societal significance, often reflecting broader social, economic, and environmental contexts (Tela et al., 2020). For example, in examining the communal beliefs surrounding food practices, socioeconomic status can dictate adherence levels to taboos; thus, individuals from lower-income backgrounds exhibit differing dietary restrictions compared to those who possess more financial resources, highlighting a clear connection between economic factors and culinary choices (Amare et al., 2022). This sprawling tapestry of influences calls for a multifaceted approach in understanding how food taboos evolve, intertwining traditional narratives, contemporary challenges, communal identities, and individual agency. Communities face the dual challenge of preserving their unique food identities while also adapting to changing circumstances that shape consumer habits. As food taboos shift in diaspora and urbanized Indigenous communities, understanding these dynamics is crucial not only for addressing health and nutrition but also for fostering cultural resilience in a rapidly changing world (Kesa et al., 2024).

14.9 Policy and Public Health Implications of Food Taboos

Food taboos during pregnancy represent a deeply rooted cultural phenomenon across the globe, significantly influencing nutritional practices and maternal health outcomes. These taboos often emerge from traditional beliefs that associate certain foods with adverse effects on pregnancy and childbirth, leading to various dietary restrictions among pregnant women. In many cultures, pregnant women are subjected to various food taboos that dictate what can and cannot be consumed. For example, Tofik et al. (2025) identified that certain foods are avoided by pregnant women in Deder, Eastern Ethiopia, due to fears related to birth weight and potential complications like obstructed labor. They suggested implementing community-based education campaigns and encouraging male partner involvement as vital strategies to address these taboos through culturally appropriate mechanisms (Tofik et al., 2025). Similarly, Dalaba et al. (2021) emphasized that food taboos in rural northern Ghana stem from beliefs that noncompliance could upset ancestors and lead to harmful consequences. These findings underscore the importance of incorporating cultural beliefs into nutrition education to ensure that interventions resonate with the target population.

The nature of food taboos often reflects broader social dynamics and beliefs around health and nutrition. Food taboos are typically not arbitrary; they are imbued with symbolic meanings. Wulandari et al. found that some pregnant women avoided specific foods such as durian and young pineapple due to fears of miscarriage, while still consuming other foods seen as less risky. This selective avoidance illustrates the complex interplay between food perception and nutritional intake (Wulandari et al., 2023). Such taboos may also be reinforced by community elders, as narrated in studies by Chakona and Shackleton (2019), where knowledge about food taboos is often passed down through generations. The reliance on traditional knowledge underscores healthcare providers' need to engage with local frameworks when designing interventions.

Culturally sensitive health education is crucial in dismantling misconceptions surrounding food taboos. Research conducted by Oluleke et al. (2024) highlighted that the notion of "big babies" among women directly correlates with the avoidance of foods believed to contribute to difficult labor or the need for cesarean sections. Such perceptions are prevalent in various cultures, suggesting that educating women on what constitutes a nutritious and balanced diet during pregnancy, free of stigma associated with food choices, is critical for overcoming taboos. Additionally, Zerfu et al. (2016) emphasized how socioeconomic factors influence food restrictions during pregnancy, revealing a complex layer of causality where poverty and education intertwine with cultural beliefs. Engaging traditional knowledge holders in food policy is essential for effective public health strategies. Policies that overlook local beliefs may falter or fail to achieve their intended outcomes. Maggiulli et al. (2022) emphasized the cross-cultural differences in food taboos during pregnancy, advocating for a greater understanding of local practices in shaping dietary recommendations. The engagement of community leaders and traditional healers can facilitate the acceptance of nutrition interventions, as they often serve as trusted sources of information and advice (de et al., 2012). This involvement creates opportunities for dialogue where misconceptions can be addressed collaboratively, fostering adherence to beneficial nutritional practices.

Given the multifaceted nature of food taboos, integrative approaches that combine nutrition-sensitive and nutrition-specific interventions are necessary. As Mwangome et al. (2010) pointed out, changing underlying social determinants such as poverty and education is crucial to mitigating the impacts of dietary restrictions. Nutrition-sensitive interventions can address the broader context within which food taboos exist, employing multisectoral collaboration to improve healthcare access and elevate educational efforts on nutrition.

Habtu et al. (2022) further reiterated that direct nutrition interventions alone are insufficient to alleviate malnutrition unless they are coupled with broader strategies aimed at empowering women and enhancing the economic conditions of families.

Furthermore, nutrition interventions must consider the unique cultural contexts within which different communities operate. The systematic reviews by Gwynn et al. (2019) and Browne et al. (2020) demonstrated that communities with culturally relevant nutrition policies showed marked improvements in dietary practices and health outcomes. These findings suggest personalized, context-specific strategies are essential for crafting effective food policies. Furthermore, ensuring that public health initiatives respect local customs can enhance the acceptance of such programs, ultimately leading to improved maternal and child health outcomes. It is also necessary to promote interdisciplinary collaborations to address the complexities of food taboos better. Frumence et al. (2024) advocated involving multiple stakeholders, such as nutritionists, community health workers, and local leaders, to create comprehensive strategies addressing the direct and indirect causes of undernutrition. Within these frameworks, practitioners can integrate traditional dietary practices with modern nutritional sciences to facilitate healthier pregnancy outcomes while respecting cultural values.

In the wake of these considerations, measuring the success of these culturally sensitive interventions is paramount. Implementing robust monitoring and evaluation frameworks, as discussed by Andersen et al. (2022), can help track the efficacy of nutrition interventions and ensure accountability within policy initiatives. Continuous data collection and analysis can provide insights into how food taboos evolve and how effectively community engagement strategies work to alleviate their adverse health impacts. Ultimately, realizing effective nutrition interventions in food taboos during pregnancy requires a profound understanding of cultural beliefs, engagement with traditional knowledge holders, and the application of culturally sensitive health education. This multifaceted approach recognizes the importance of integrating local customs with nutritional guidelines to foster compliance and promote healthier maternal and child outcomes. Policymakers and health practitioners must navigate these complexities with sensitivity, ensuring interventions are viable, respectful, and empowering to the communities they aim to serve. By fostering partnerships with local stakeholders and educators, it is possible to pave the way toward achieving improved nutritional health outcomes amid the many layers of cultural beliefs surrounding food consumption.

14.10 Ethical and Methodological Considerations in Food Taboos

In examining the ethical and methodological considerations in addressing food taboos, especially concerning Indigenous knowledge systems, it is crucial to navigate the challenges of insider versus outsider perspectives. Insiders in specific cultural contexts typically hold valuable knowledge regarding food practices and taboos. At the same time, outsiders may lack understanding or awareness of the deeper meanings behind the cultural significance of food. This epistemic difference is critical, as outsider researchers might unintentionally perpetuate stereotypes or disseminate oversimplified knowledge, contributing to cultural appropriation and disempowerment (Jernigan et al., 2023; Tauri, 2017). Observing how food taboos shape diets among Indigenous populations can illuminate broader issues of health equity and food sovereignty; however, it demands careful engagement with cultural ontologies, particularly given the colonial histories of exploitation and marginalization that Indigenous communities have faced (David-Chavez et al., 2024; Sinclaire et al., 2021).

Ethical engagement with Indigenous knowledge systems entails recognizing their rightful ownership over food-related knowledge and practices. The necessity for collaborative and respectful partnerships emerges as a central theme in Indigenous research ethics (David-Chavez et al., 2024; Nesterova, 2019). Academics have suggested adopting relational ethics frameworks that prioritize community input and agency, thus ensuring research practices contribute to community empowerment rather than exploitation (David-Chavez, 2019; Dieter et al., 2018). This relational accountability requires non-Indigenous researchers to disclose their personal positionalities, motivations, and interests in the research outcome to build transparent relationships with Indigenous populations (Howard, 2017). By fostering such collaborative environments, the risk of reinforcing colonial paradigms diminishes, allowing for genuine co-created knowledge that respects traditional practices surrounding food and health (David-Chavez & Gavin, 2018; Keane, 2021). Simultaneously, researchers must navigate the tension between promoting health equity and avoiding cultural relativism. While the advocacy for health equity emphasizes the need to address disparities caused by systemic inequalities across various cultural contexts, cultural relativism risks undermining universal health standards that could enhance the well-being for all groups, including those with food taboos. The disjunction between health promotion practices and cultural food preferences can lead to conflicts where health messages might not resonate with individuals whose diets are dictated by cultural beliefs (Ring et al., 2023; Villalona et al., 2021). Thus, balancing involves integrating culturally relevant approaches that honor food traditions while promoting practices that facilitate better health outcomes. This approach is crucial in public health settings where diverse dietary practices need careful consideration in health education messages (Jernigan et al., 2023; Smith & Blumenthal, 2012).

Moreover, methodological frameworks such as community-based participatory research (CBPR) effectively bridge these divergent needs (Jull et al., 2017, 2018). CBPR emphasizes the importance of community involvement in the research process, ensuring that the knowledge produced is relevant and actionable within the community context. This participatory approach directly addresses inequalities by engaging marginalized populations, fostering trust, and addressing historical grievances stemming from hierarchical research practices (David-Chavez & Gavin, 2018; Nikolakis & Hotte, 2021). Researchers can codevelop interventions that align more closely with Indigenous knowledge and contemporary health needs by allowing communities to identify their priorities and food-related challenges (Dieter et al., 2018; Smith & Blumenthal, 2012). In Indigenous food taboos, food sovereignty is vital in empowering communities to reclaim their traditional food systems disrupted by colonial practices (Thompson et al., 2012). This concept advocates for Indigenous peoples' right to self-determination in their food systems and acknowledges the connections between food practices and health outcomes. Establishing food sovereignty is essential for mitigating food insecurity and revitalizing cultural identities deeply intertwined with culinary traditions and associated taboos (Jernigan et al., 2023; Thompson et al., 2012). The challenges of food sovereignty further intersect with the need for cultural safety in healthcare, which ensures that Indigenous clients feel respected and understood within health systems that are often influenced predominantly by non-Indigenous perspectives (Greenwood et al., 2017; Sinclaire et al., 2021).

The primary research ethics concerns within these contexts extend to informed consent and reciprocal knowledge-sharing. Disempowerment can result from traditional research ethics frameworks that fail to accommodate Indigenous ways of knowing, leading to misrepresentation and distrust among Indigenous populations (David-Chavez, 2019; Tauri, 2017). Therefore, engaging with ethical spaces—mutual spaces where diverse worldviews are shared and respected—is necessary for fostering authentic dialogues between researchers and Indigenous communities (Greenwood et al., 2017; Nikolakis & Hotte, 2021). This approach aligns closely with the "Two-Eyed Seeing" concept, which incorporates Indigenous and Western scientific knowledge systems (Sinclaire et al., 2021). Integrating Indigenous knowledge with broader health policies necessitates understanding food taboos from an insider perspective and an ethical framework that promotes shared ownership of knowledge. The challenge remains to create an academic environment wherein Indigenous peoples are respected as coresearchers and their cultural insights are viewed as valuable contributions rather than mere subjects of study. This process includes ongoing actions that uphold the principles of respect, reciprocity, and relevance in designing health interventions that honor the complexities of cultural practices while striving for health equity (David-Chavez et al., 2024; Dieter et al., 2018; Keane, 2021).

Furthermore, one must consider the implications of academic and institutional structures that often uphold Western epistemological dominance, which can stigmatize non-Western dietary practices as inferior or unhealthy (Ring et al., 2023; Villalona et al., 2021). Thus, a radical shift toward equity-focused research methodologies is required—ones that adjust the power dynamics inherent in research and lead to actions grounded in mutual respect for cultural traditions. This involves redefining success based on the communities' well-being rather than traditional academic metrics, instead valuing the lived experiences and knowledge within Indigenous populations (Dieter et al., 2018; Howard, 2017). As researchers work to balance promoting health equity while ensuring that cultural integrity and autonomy are respected, a transformative approach that prioritizes Indigenous voices will ultimately contribute to healthier, more resilient communities. Therefore, the pathways toward effective and inclusive practices must continue to be explored and refined as the dialogue between divergent worldviews unfolds.

14.11 Conclusion

Food taboos embody a significant cultural phenomenon influencing Indigenous people's dietary practices and health. Recognizing their complex roots and effects is imperative for developing culturally sensitive approaches that promote nutritional well-being and harness the potential of traditional knowledge in addressing modern health challenges. They are deeply rooted in Indigenous communities' social, spiritual, and ecological fabric. They function as dietary rules and mechanisms for maintaining cultural identity, regulating resource use, and reinforcing social norms and hierarchies. This chapter has shown that food taboos, while often adaptive and symbolically meaningful, can have complex nutritional consequences, especially when they limit access to essential nutrients among vulnerable groups such as pregnant women, lactating mothers, and children. The intersection of food taboos with gender roles, ritual practices, and environmental knowledge underscores the multifaceted nature of these beliefs. While often serving to reinforce patriarchal structures and impact women's health negatively, they also open avenues for empowerment and community identity. The interplay between adherence to these taboos, women's health, and agency underscores the necessity for nuanced approaches in addressing dietary practices prioritizing gender equality and women's health outcomes. As Indigenous societies increasingly encounter the pressures of globalization, religious conversion, migration, and climate change, traditional food taboos are being reshaped, reinterpreted, or, in some cases, abandoned. These shifts bring opportunities and challenges: While some outdated or harmful restrictions may fade, cultural erosion and loss of ecological knowledge may also occur. Understanding food taboos requires a nuanced, interdisciplinary perspective that values Indigenous knowledge systems while addressing public health concerns. This chapter calls for culturally sensitive and participatory approaches to health promotion and nutrition policy that engage respectfully with community elders, knowledge holders, and local institutions. By doing so, interventions can support culturally meaningful foodways that promote social cohesion and nutritional well-being.

Chapter Reflection

1. How do different disciplines (sociology, anthropology, ecology, and evolution) explain the origins and purposes of food taboos?
2. In what ways do gender-based food taboos impact maternal and child nutrition?
3. How can food taboos support ecological sustainability and create nutritional risks? Give one cultural example.
4. How are food taboos changing due to globalization, religion, urbanization, and education?
5. How do food taboos shape cultural identity and power, particularly for women?
6. What ethical principles should guide researchers studying food taboos in Indigenous communities?
7. Propose a culturally sensitive health intervention to address harmful food taboos without undermining cultural beliefs.

8. How does food sovereignty help balance cultural preservation with nutritional health?
9. How can the "Two-Eyed Seeing" approach bridge Indigenous knowledge and modern nutrition science?
10. How has this chapter changed your view of the link between culture, nutrition, and health?

REFERENCES

Abdullah, S. (2022). Menstrual food restrictions and taboos: A qualitative study on rural, resettlement, and urban Indigenous Temiar of Malaysia. *PLOS One*, *17*(12), e0279629. https://doi.org/10.1371/journal.pone.0279629

Acire, P., Bagonza, A., & Opiri, N. (2023). The misbeliefs and food taboos during pregnancy and early infancy: A pitfall to attaining adequate maternal and child nutrition outcomes among the rural Acholi communities in northern Uganda. https://doi.org/10.21203/rs.3.rs-3069493/v1

Akiso, D., Mosisa, M., & Erkalo, D. (2024). Food taboos practice and associated factors among pregnant women attending antenatal care at Doctor Bogalech Gebre Memorial General Hospital, Durame Town, southern Ethiopia, 2022. *Journal of Family Medicine and Primary Care*, *13*(7), 2632–2638. https://doi.org/10.4103/jfmpc.jfmpc_1308_23

Alamirew, S., Lemke, S., Freyer, B., & Stadlmayr, B. (2024). Dietary behaviour of pregnant women in Ethiopia: The missing aspect of care. *Nutrients*, *16*(19), 3227. https://doi.org/10.3390/nu16193227

Amare, W., Tura, A., Semahegn, A., & Roba, K. (2022). Food taboos among pregnant women and associated factors in eastern Ethiopia: A community-based cross-sectional study. *Sage Open Medicine*, *10*. https://doi.org/10.1177/20503121221133935

Andersen, C., Cain, J., Chaudhery, D., Ghimire, M., Higashi, H., & Tandon, A. (2022). Assessing public financing for nutrition in Bhutan, Nepal, and Sri Lanka. *Maternal and Child Nutrition*, *18*(3), e13320. https://doi.org/10.1111/mcn.13320

Angsongna, A., Armah, F. A., Boamah, S., Hambati, H., Luginaah, I., Chuenpagdee, R., & Campbell, G. (2016). A systematic review of resource habitat taboos and human health outcomes in the context of global environmental change. *Global Bioethics*, *27*(2–4), 91–111. https://doi.org/10.1080/11287462.2016.1212608

Arzoaquoi, S., Essuman, E., Gbagbo, F., Tenkorang, E., Soyiri, I., & Laar, A. (2015). Motivations for food prohibitions during pregnancy and their enforcement mechanisms in a rural Ghanaian district. *Journal of Ethnobiology and Ethnomedicine*, *11*(1). https://doi.org/10.1186/s13002-015-0044-0

Babu, M., & Patnaik, R. (2021). Environment and food taboos: Breastfeeding, antenatal, and postnatal practices among Koya Dora of East Godavari District, Andhra Pradesh. *The Oriental Anthropologist a Bi-Annual International Journal of the Science of Man*, *21*(1), 164–174. https://doi.org/10.1177/0972558x211004711

Begossi, A., Hanazaki, N., & Ramos, R. (2004). Food chain and the reasons for fish food taboos among Amazonian and Atlantic forest fishers (Brazil). *Ecological Applications*, *14*(5), 1334–1343. https://doi.org/10.1890/03-5072

Blumer, H. (1969). *Symbolic interactionism: Perspective and method.* University of California Press.

Blythe, J. (2013). Preference organization driving structuration: Evidence from Australian Aboriginal interaction for pragmatically motivated grammaticalization. *Language*, *89*(4), 883–919. https://doi.org/10.1353/lan.2013.0057

Browne, J., Lock, M., Walker, T., Egan, M., & Backholer, K. (2020). Effects of food policy actions on Indigenous peoples' nutrition-related outcomes: A systematic review. *BMJ Global Health*, *5*(8), e002442. https://doi.org/10.1136/bmjgh-2020-002442

Bukachi, S. A., Ngutu, M., Muthiru, A. W., Lépine, A., Kadiyala, S., & Dominguez-Salas, P. (2022). Gender and sociocultural factors in animal source foods (ASFs) access and consumption in lower-income households in urban informal settings of Nairobi, Kenya. *Journal of Health Population and Nutrition*, *41*(1), 30. https://doi.org/10.1186/s41043-022-00307-9

Chakona, G., & Shackleton, C. (2019). Food taboos and cultural beliefs influence pregnant women's food choices and dietary preferences in the Eastern Cape, South Africa. *Nutrients*, *11*(11), 2668. https://doi.org/10.3390/nu11112668

Cook, E. (2023). Anthropological and sociological perspectives on food allergy. *Clinical and Experimental Allergy*, *53*(10), 989–1003. https://doi.org/10.1111/cea.14387

Costa, S., Zepeda, L., & Sirieix, L. (2014). Exploring the social value of organic food: A qualitative study in France. *International Journal of Consumer Studies*, *38*(3), 228–237. https://doi.org/10.1111/ijcs.12100

Dalaba, M. A., Nonterah, E. A., Chatio, S. T., Adoctor, J. K., Watson, D., Barker, M., Ward, K. A., & Debpuur, C. (2021). Culture and community perceptions on diet for maternal and child health: A qualitative study in rural northern Ghana. *BMC Nutrition*, *7*(1). https://doi.org/10.1186/s40795-021-00439-x

David-Chavez, D. (2019). A guiding model for decolonizing environmental science research and restoring relational accountability with Indigenous communities. https://doi.org/10.31237/osf.io/ec9s5

David-Chavez, D., & Gavin, M. (2018). A global assessment of Indigenous community engagement in climate research. *Environmental Research Letters*, *13*(12), 123005. https://doi.org/10.1088/1748-9326/aaf300

David-Chavez, D., Gavin, M., Ortiz, N., Valdez, S., & Carroll, S. (2024). A values-centered relational science model: Supporting Indigenous rights and reconciliation in research. *Ecology and Society*, *29*(2). https://doi.org/10.5751/es-14768-290211

de, J., Bouttasing, N., Sampson, L., Perks, C., Osrin, D., & Prost, A. (2012). Identifying priorities to improve maternal and child nutrition among the Khmu ethnic group, Laos: A formative study. *Maternal and Child Nutrition*, *9*(4), 452–466. https://doi.org/10.1111/j.1740-8709.2012.00406.x

Dieter, J., McKim, L., Tickell, J., Bourassa, C., Lavallee, J., & Boehme, G. (2018). The path of creating co-researchers in the File Hills Qu'appelle Tribal Council. *International Indigenous Policy Journal*, *9*(4). https://doi.org/10.18584/iipj.2018.9.4.1

Domínguez-Salas, P., Alarcón, P., Häsler, B., Dohoo, I. R., Colverson, K., Kimani-Murage, E. W., Alonso, S., Ferguson, E., Fèvre, E. M., Rushton, J., & Grace, D. (2016). Nutritional characterisation of low-income households of Nairobi: Socioeconomic, livestock and gender considerations and predictors of malnutrition from a cross-sectional survey. *BMC Nutrition*, *2*(1), 47. https://doi.org/10.1186/s40795-016-0086-2

Douglas, M. (1966). *Purity and danger: An analysis of concepts of pollution and taboo*. Routledge.

Dumbili, E. (2016). Gendered sexual uses of alcohol and associated risks: A qualitative study of Nigerian university students. *BMC Public Health*, *16*(1). https://doi.org/10.1186/s12889-016-3163-1

Ekwochi, U., Osuorah, C., Ndu, I., Ifediora, C., Asinobi, I., & Eke, C. (2016). Food taboos and myths in southeastern Nigeria: The belief and practice of mothers in the region. *Journal of Ethnobiology and Ethnomedicine*, *12*(1). https://doi.org/10.1186/s13002-016-0079-x

Esimai, O., Ojofeitimi, E., & Oyebowale, O. (2001). Sociocultural practices influencing under five nutritional status in an urban community in Osun State, Nigeria. *Nutrition and Health*, *15*(1), 41–46. https://doi.org/10.1177/026010600101500105

Ezeifeka, C. (2019). Patriarchal legitimization strategies in Igbo gender-related taboos: A case for critical discourse analysis. *Advances in Social Sciences Research Journal*, *6*(3), 383–400. https://doi.org/10.14738/assrj.63.6229

Fessler, D. M. T., & Navarrete, C. D. (2003). Domain-specific variation in disgust sensitivity across the menstrual cycle. *Evolution and Human Behavior*, *24*(6), 406–417.

Frumence, G., Jin, Y., Kasangala, A., Bakar, S., Mahiti, G., & Ochieng, B. (2024). A systems approach in the prevention of undernutrition among children under five in Tanzania: Perspectives from key stakeholders. *Nutrients*, *16*(11), 1551. https://doi.org/10.3390/nu16111551

Gebregziabher, H., Kahsay, A., Gebrearegay, F., Berhe, K., Gebremariam, A., & Gebretsadik, G. (2023). Food taboos and their perceived reasons among pregnant women in Ethiopia: A systematic review, 2022. *BMC Pregnancy and Childbirth*, *23*(1). https://doi.org/10.1186/s12884-023-05437-4

Geta, T., Gebremedhin, S., & Omigbodun, A. (2022). Dietary diversity among pregnant women in Gurage Zone, South-Central Ethiopia: Assessment based on longitudinal repeated measurement. *International Journal of Women S Health*, *14*, 599–615. https://doi.org/10.2147/ijwh.s354536

Getnet, W., Aycheh, W., & Tessema, T. (2018). Determinants of food taboos in the pregnant women of the Awabel District, East Gojjam Zone, Amhara Regional State in Ethiopia. *Advances in Public Health*, *2018*, 1–6. https://doi.org/10.1155/2018/9198076

Gibson, E., Stacey, N., Sunderland, T. C. H., & Adhuri, D. S. (2021). Coping or adapting? experiences of food and nutrition insecurity in specialised fishing households in Komodo district, eastern Indonesia. *BMC Public Health*, *21*(1). https://doi.org/10.1186/s12889-021-10248-3

Girma, A., Ejigu, A., Ayalew, E., & Getachew, D. (2022). Determinants of dietary practice among pregnant women at the public hospitals in Bench-Sheko and Kaffa Zones, Southwest Ethiopia. *BMC Nutrition*, *8*(1), 88. https://doi.org/10.1186/s40795-022-00588-7

Gitungwa, H., Gustafson, C. R., Jimenez, E. Y., Peterson, E. W., Mwanzalila, M., Makweta, A., Komba, E., Kazwala, R. R., Mazet, J. A. K., & VanWormer, E. (2021). Female and male-controlled livestock holdings impact pastoralist food security and women's dietary diversity. *One Health Outlook*, *3*(1), 3. https://doi.org/10.1186/s42522-020-00032-5

Golden, C., & Comaroff, J. (2015). The human health and conservation relevance of food taboos in northeastern Madagascar. *Ecology and Society*, *20*(2), 42. https://doi.org/10.5751/es-07590-200242

Golden, C., Gupta, A., Vaitla, B., & Myers, S. (2016). Ecosystem services and food security: Assessing inequality at community, household and individual scales. *Environmental Conservation*, *43*(4), 381–388. https://doi.org/10.1017/s0376892916000163

Gowder, S. (2024). Social aspects of food and nutrition: An overview. *Journal of Ecohumanism*, *3*(7), 2953–2961. https://doi.org/10.62754/joe.v3i7.4431

Greenwood, M., Lindsay, N., King, J., & Loewen, D. (2017). Ethical spaces and places: Indigenous cultural safety in British Columbia health care. *Alternative an International Journal of Indigenous Peoples*, *13*(3), 179–189. https://doi.org/10.1177/1177180117714411

Gulema, H., Demissie, M., Worku, A., Yadeta, T., Fasil, N., & Berhane, Y. (2023). The association between intrahousehold food allocation social norms and thinness among young adolescent girls: A community-based study. *Ethiopian Journal of Health Sciences*, *33*(6). https://doi.org/10.4314/ejhs.v33i6.4

Gwynn, J., Sim, K., Searle, T., Senior, A., Lee, A., & Brimblecombe, J. (2019). Effect of nutrition interventions on diet-related and health outcomes of Aboriginal and Torres Strait Islander Australians: A systematic review. *BMJ Open*, *9*(4), e025291. https://doi.org/10.1136/bmjopen-2018-025291

Habtu, M., Gebremariam, A., Umugwaneza, M., Mochama, M., & Munyan-shongore, C. (2022). Effect of integrated nutrition-sensitive and nutrition-specific intervention package on maternal malnutrition among pregnant women in Rwanda. *Maternal and Child Nutrition, 18*(3), e13367. https://doi.org/10.1111/mcn.13367

Harris-Fry, H., Shrestha, N., Costello, A., & Saville, N. (2017). Determinants of intra-household food allocation between adults in South Asia – A systematic review. *International Journal for Equity in Health, 16*(1), 107. https://doi.org/10.1186/s12939-017-0603-1

Harvey, D. (2017). "Gimme a pigfoot and a bottle of beer": Food as cultural performance in the aftermath of hurricane Katrina. *Symbolic Interaction, 40*(4), 498–522. https://doi.org/10.1002/symb.318

Henryks, J., & Brimblecombe, J. (2016). Mapping point-of-purchase influencers of food choice in Australian remote Indigenous communities. *Sage Open, 6*(1). https://doi.org/10.1177/2158244016629183

Howard, H. (2017). Co-producing community and knowledge: Indigenous epistemologies of engaged, ethical research in an urban context. *Engaged Scholar Journal Community-Engaged Research Teaching and Learning, 2*(1), 205–224. https://doi.org/10.15402/esj.v2i1.207

Ikhajiagbe, B., Ogwu, M. C., Ogochukwu, O. F., Odozi, E. B., Adekunle, I. J., & Omage, Z. E. (2021). The place of neglected and underutilized legumes in human nutrition and protein security in Nigeria. *Critical Reviews in Food Science and Nutrition, 62*(14), 3930–3938. https://doi.org/10.1080/10408398.2020.1871319

Ikwuka, U., Galbraith, N., & Nyatanga, L. (2013). Causal attribution of mental illness in south-eastern Nigeria. *International Journal of Social Psychiatry, 60*(3), 274–279. https://doi.org/10.1177/0020764013485331

Imarhiagbe, O., & Ogwu, M. C. (2022). Sacred groves in the Global South: A panacea for sustainable biodiversity conservation. In S. C. Izah (Ed.), *Biodiversity in Africa: Potentials, threats and conservation.* Sustainable Development and Biodiversity (Vol. 29, pp. 525–546). Springer. http://doi.org/10.1007/978-981-19-3326-4_20

Jernigan, V. B. B., Nguyen, C. J., Maudrie, T. L., Demientieff, L. X., Black, J. C., Mortenson, R., Wilbur, R. E., Clyma, K. R., Lewis, M., & Lopez, S. (2023). Food sovereignty and health: A conceptual framework to advance research and practice. *Health Promotion Practice*, *24*(6), 1070–1074. https://doi.org/10.1177/15248399231190367

Jull, J., Giles, A., & Graham, I. (2017). Community-based participatory research and integrated knowledge translation: Advancing the co-creation of knowledge. *Implementation Science*, *12*(1), 150. https://doi.org/10.1186/s13012-017-0696

Jull, J., Ninomiya, M., Compton, I., & Picard, A. (2018). Fostering the conduct of ethical and equitable research practices: The imperative for integrated knowledge translation in research conducted by and with Indigenous community members. *Research Involvement and Engagement*, *4*(1), 45. https://doi.org/10.1186/s40900-018-0131-1

Kahsay, Z., Birhanu, Z., Chaka, M., & Gebreyesus, H. (2017). Foods tabooed for pregnant women in Ambala District of Afar Region, Ethiopia: An inductive qualitative study. *BMC Nutrition*, *3*(1), 40. https://doi.org/10.1186/s40795-017-0159-x

Keane, M. (2021). Research ethics and diversity of worldviews: Integrated worlds and ubuntu. *Scholarship of Teaching and Learning in the South*, *5*(2), 22–35. https://doi.org/10.36615/sotls.v5i2.194

Kesa, H., Tchuenchieu, A., Zuma, M., & Mbhenyane, X. (2024). Availability and accessibility of Indigenous foods in Gauteng region, South Africa. *Frontiers in Sustainable Food Systems*, *8*. https://doi.org/10.3389/fsufs.2024.1385230

Köhler, R., Sae-tan, S., Lambert, C., & Biesalski, H. (2018). Plant-based food taboos in pregnancy and the postpartum period in southeast Asia – A systematic review of literature. *Nutrition and Food Science*, *48*(6), 949–961. https://doi.org/10.1108/nfs-02-2018-0059

Landim, A., Souza, J., Santos, L., Lins-Neto, E., Silva, D., & Ferreira, F. (2023). Food taboos and animal conservation: A systematic review on how cultural expressions influence interaction with wildlife species. *Journal of Ethnobiology and Ethnomedicine*, *19*(1), 31. https://doi.org/10.1186/s13002-023-00600-9

Lekey, A., Masumo, R. M., Jumbe, T., Ezekiel, M., Daudi, Z., Mchome, N. J., David, G., Onesmo, W., & Leyna, G. H. (2024). Food taboos and preferences among adolescent girls, pregnant women, breastfeeding mothers, and children aged 6–23 months in mainland Tanzania: A qualitative study. *Plos Global Public Health*, *4*(8), e0003598. https://doi.org/10.1371/journal.pgph.0003598

Lévi-Strauss, C. (1969). *The raw and the cooked*. Harper and Row.

Maggiulli, O., Rufo, F., Johns, S., & Wells, J. (2022). Food taboos during pregnancy: Meta-analysis on cross cultural differences suggests specific, diet-related pressures on childbirth among agriculturalists. *PeerJ, 10*, e13633. https://doi.org/10.7717/peerj.13633

Malinowski, B. (1944). *A scientific theory of culture and other essays*. University of North Carolina Press.

Maria, K., Josephine, A., & Florence, E. (2023). *Gender and spirituality in anti-colonial struggles in Uganda: A case of Nyabingi movement in Kigezi region (1900–1945)*. RHSS. https://doi.org/10.7176/rhss/13-18-02

Martínez, A. (2013). Reconsidering acculturation in dietary change research among Latino immigrants: Challenging the preconditions of U.S. migration. *Ethnicity and Health, 18*(2), 115–135. https://doi.org/10.1080/13557858.2012.698254

McNamara, K., & Wood, E. (2019). Food taboos, health beliefs, and gender: Understanding household food choice and nutrition in rural Tajikistan. *Journal of Health Population and Nutrition, 38*(1), 17. https://doi.org/10.1186/s41043-019-0170-8

Meyer-Rochow, V. B. (2009). Food taboos: Their origins and purposes. *Journal of Ethnobiology and Ethnomedicine, 5*, 18. https://doi.org/10.1186/1746-4269-5-18

Mohammed, S., Hailu, T., Larijani, B., & Esmaillzadeh, A. (2019). Food taboo among pregnant Ethiopian women: Magnitude, drivers, and association with anemia. *Nutrition Journal, 18*(1), 19. https://doi.org/10.1186/s12937-019-0444-4

Mwangome, M., Prentice, A., Plugge, E., & Nweneka, C. (2010). Determinants of appropriate child health and nutrition practices among women in rural Gambia. *Journal of Health Population and Nutrition, 28*(2), 167–172. https://doi.org/10.3329/jhpn.v28i2.4887

Nesterova, Y. (2019). Teaching Indigenous children in Taiwan: Tensions, complexities and opportunities. *Global Studies of Childhood, 9*(2), 156–166. https://doi.org/10.1177/2043610619846349

Nguyen, P. H., Kachwaha, S., Tran, L. M., Sanghvi, T., Ghosh, S., Kulkarni, B., Beesabathuni, K., Menon, P., & Sethi, V. (2021). Maternal diets in India: Gaps, barriers, and opportunities. *Nutrients, 13*(10), 3534. https://doi.org/10.3390/nu13103534

Nikolakis, W., & Hotte, N. (2021). Implementing "ethical space": An explor-atory study of Indigenous-conservation partnerships. *Conservation Science and Practice, 4*(1), e580. https://doi.org/10.1111/csp2.580

Nwankwo, E. A., & Itanyi, E. I. (2019). Heritage studies and challenges: Im-plications on research results from Igboland, Nigeria. *Heliyon, 5*(12), e02962. https://doi.org/10.1016/j.heliyon.2019.e02962

Nweke, K., & Okpaleke, I. (2019). The re-emergence of African spiritualities: Prospects and challenges. *Transformation an International Journal of Holistic Mission Studies, 36*(4), 246–265. https://doi.org/10.1177/0265378819866215

Ogwu, M. C. (2023). Local food crops in Africa: Sustainable utilization, threats, and traditional storage strategies. In S. C. Izah & M. C. Ogwu (Eds.), *Sustainable utilization and conservation of Africa's biological resources and environment.* Sustainable Development and Biodiversity (Vol. 888, pp. 353–374). Springer. https://doi.org/10.1007/978-981-19-6974-4_13

Ogwu, M. C., Ojo, A. O., & Osawaru, M. E. (2025). Quantitative ethnobotany of Afenmai people of Southern Nigeria: An assessment of their crop utilization, and preservation methods. *Genetic Resources and Crop Evolution, 72*, 5807–5829. https://doi.org/10.1007/s10722-024-02302-x

Ogwu, M. C., & Osawaru, M. E. (2022). Traditional methods of plant conservation for sustainable utilization and development. In S. C. Izah (Ed.), *Biodiversity in Africa: Potentials, threats and conservation.* Sustainable Development and Biodiversity (Vol. 29, pp. 451–472). Springer. http://doi.org/10.1007/978-981-19-3326-4_17

Ogwu, M. C., Osawaru, M. E., Amodu, E., & Osamo, F. (2023). Comparative morphology, anatomy and chemotaxonomy of two *Cissus* Linn. species. *Brazilian Journal of Botany, 46*, 397–412. https://doi.org/10.1007/s40415-023-00881-0

Oluleke, M., Ogunwale, A., Arulogun, O., & Adelekan, A. (2024). Dietary intake knowledge and reasons for food restriction during pregnancy among pregnant women attending primary health care centers in Ile-Ife, Nigeria. *International Journal of Population Studies, 2*(1), 103. https://doi.org/10.18063/ijps.2016.01.006

Ọnụkawa, E. (2021). Leftwardness. An aspect of prohibitions in the Igbo cul-ture. *Anthropos, 116*(2), 379–384. https://doi.org/10.5771/0257-9774-2021-2-379

Placek, C., Madhivanan, P., & Hagen, E. (2017). Innate food aversions and culturally transmitted food taboos in pregnant women in rural southwest India:

Separate systems to protect the fetus? *Evolution and Human Behavior*, *38*(6), 714–728. https://doi.org/10.1016/j.evolhumbehav.2017.08.001

Radcliffe-Brown, A. R. (1952). *Structure and function in primitive society*. Free Press.

Ramulondi, M., de Wet, H., & Ntuli, N. R. (2021). Traditional food taboos and practices during pregnancy, postpartum recovery, and infant care of Zulu women in northern KwaZulu-Natal. *Journal of Ethnobiology and Ethnomedicine*, *17*(1). https://doi.org/10.1186/s13002-021-00451-2

Rayna, S. E., Khan, F. A., Samin, S., Siraj, S., Nizam, S., Islam, S. A., & Khalequzzaman, Md. (2024). A qualitative study on food taboos among rural pregnant women in Bangladesh: Motivators for adherence and influencers of taboo-breaking behavior. https://doi.org/10.1101/2024.09.09.24313362

Richmond, C., Steckley, M., Neufeld, H., Kerr, R., Wilson, K., & Dokis, B. (2020). First Nations food environments: Exploring the role of place, income, and social connection. *Current Developments in Nutrition*, *4*(8), nzaa108. https://doi.org/10.1093/cdn/nzaa108

Ring, M., Ai, D., Maker-Clark, G., & Sarazen, R. (2023). Cooking up change: DEIB principles as key ingredients in nutrition and culinary medicine education. *Nutrients*, *15*(19), 4257. https://doi.org/10.3390/nu15194257

Rozin, P., & Fallon, A. E. (1987). A perspective on disgust. *Psychological Review*, *94*(1), 23–41.

Simoons, F. J. (1994). *Eat not this flesh: Food avoidances from prehistory to the present*. University of Wisconsin Press.

Sinclaire, M., Schultz, A., Linton, J., & McGibbon, E. (2021). Etuaptmumk (two-eyed seeing) and ethical space: Ways to disrupt health researchers' colonial attraction to a singular biomedical worldview. *Witness the Canadian Journal of Critical Nursing Discourse*, *3*(1), 57–72. https://doi.org/10.25071/2291-5796.94

Smith, M. (2023). Historical and social science perspectives on food allergy. *Clinical and Experimental Allergy*, *53*(9), 902–910. https://doi.org/10.1111/cea.14360

Smith, S., & Blumenthal, D. (2012). Community health workers support community-based participatory research ethics: Lessons learned along the research-to-practice-to-community continuum. *Journal of Health Care for the Poor and Underserved*, *23*(4a), 77–87. https://doi.org/10.1353/hpu.2012.0156

Tauri, J. (2017). Research ethics, informed consent and the disempowerment of First Nation peoples. *Research Ethics*, *14*(3), 1–14. https://doi.org/10.1177/1747016117739935

Tela, F., Gebremariam, L., & Beyene, S. (2020). Food taboos and related misperceptions during pregnancy in Mekelle City, Tigray, Northern Ethiopia. *PLOS One*, *15*(10), e0239451. https://doi.org/10.1371/journal.pone.0239451

Tesfa, A., Begna, Z., Sisay, D., Muche, T., Kabthymer, R. H., Tesfaye, T., & Ewune, H. A. (2023). A systematic review and thematic synthesis of qualitative research studies on food taboos related to pregnancy in Ethiopia. *Journal of BioMed Research and Reports*, *2*(4). https://doi.org/10.59657/2837-4681.brs.23.022

Thomas, N., & Emond, R. (2017). Living alone but eating together: Exploring lunch clubs as a dining out experience. *Appetite*, *119*, 34–40. https://doi.org/10.1016/j.appet.2017.03.003

Thompson, S., Kamal, A., Alam, M., & Wiebe, J. (2012). Community development to feed the family in northern Manitoba communities: Evaluating food activities based on their food sovereignty, food security, and sustainable livelihood outcomes. *Canadian Journal of Nonprofit and Social Economy Research*, *3*(2). https://doi.org/10.22230/cjnser.2012v3n2a121

Tofik, A., Gobena, T., Eyeberu, A., Debella, A., Gebremichael, B., Gamachu, M., Deressa, A., Ayana, G. M., Birhanu, A., Zakaria, H. F., Jibro, U., & Mussa, I. (2025). Food taboo practices among pregnant women in Deder Town, eastern Ethiopia, 2024. *PLOS One*, *20*(4), e0321773. https://doi.org/10.1371/journal.pone.0321773

Tripura, L. (2025). Food taboos among Indigenous pregnant women of Khagrachari District, Bangladesh. *Sage Open Medicine*, *13*. https://doi.org/10.1177/20503121251342979

Tsegaye, D., Tamiru, D., & Belachew, T. (2021). Food-related taboos and misconceptions during pregnancy among rural communities of Illu Aba Bor zone, southwest Ethiopia. A community-based qualitative cross-sectional study. *BMC Pregnancy and Childbirth*, *21*(1). https://doi.org/10.1186/s12884-021-03778-6

Tugume, P., Mustafa, A., Walusansa, A., Ojelel, S., Nyachwo, E., Muhumuza, E., Nampeera, M., Kabbale, F., & Ssenku, J. (2024). Unravelling taboos and cultural beliefs associated with hidden hunger among pregnant and breastfeeding women in Buyende District Eastern Uganda. *Journal of Ethnobiology and Ethnomedicine*, *20*(1), 46. https://doi.org/10.1186/s13002-024-00682-z

Vasilevski, V., & Carolan, M. (2016). Food taboos and nutrition-related pregnancy concerns among Ethiopian women. *Journal of Clinical Nursing*, *25*(19–20), 3069–3075. https://doi.org/10.1111/jocn.13319

Villalona, S., Ortiz, V., Castillo, W., & Laumbach, S. (2021). Cultural relevancy of culinary and nutritional medicine interventions: A scoping review. *American Journal of Lifestyle Medicine*, *16*(6), 663–671. https://doi.org/10.1177/15598276211006342

Wamoyi, J., Mshana, G., Mongi, A., Neke, N., Kapiga, S., & Changalucha, J. (2014). A review of interventions addressing structural drivers of adolescents' sexual and reproductive health vulnerability in sub-Saharan Africa: Implications for sexual health programming. *Reproductive Health*, *11*(1). https://doi.org/10.1186/1742-4755-11-88

Wulandari, C., Kardina, R., & Wijaya, S. (2023). Analysis of food taboo culture with protein intake in pregnant women. *Medical Technology and Public Health Journal*, *6*(2), 207–212. https://doi.org/10.33086/mtphj.v6i2.3434

Yen, A., Bilney, C., Shackleton, M., & Lawler, S. (2017). Current issues involved with the identification and nutritional value of wood grubs consumed by Australian aborigines. *Insect Science*, *25*(2), 199–210. https://doi.org/10.1111/1744-7917.12430

Zerfu, T., Umeta, M., & Baye, K. (2016). Dietary habits, food taboos, and perceptions towards weight gain during pregnancy in Arsi, rural central Ethiopia: A qualitative cross-sectional study. *Journal of Health Population and Nutrition*, *35*(1), 22. https://doi.org/10.1186/s41043-016-0059-8

Zinyemba, L. (2023). The role played by taboos in exposing women to maternal mortality in Binga, Zimbabwe. *Journal of Development Administration*, *8*(4), 113–120. https://doi.org/10.4314/jda.v8i4.3

CHAPTER 15

Food, Identity, and Symbolism in Indigenous Societies

Abstract

This chapter explores the relationship between food, identity, and symbolism in Indigenous societies, emphasizing how food operates as a powerful medium of cultural expression, social organization, and spiritual significance. Moving beyond its biological role, food is examined as a living archive of ancestral knowledge, a marker of group identity, and a tool for maintaining ecological and cosmological balance. Drawing on symbolic anthropology, identity theory, and ethnographic case studies, the chapter investigates how food is embedded in rituals, taboos, oral traditions, and gendered practices that shape community life. It examines the symbolic roles of food in rites of passage, seasonal festivals, funerals, and sacred ceremonies, illustrating how culinary practices reinforce belonging, continuity, and cultural resilience. Attention is given to the impacts of colonization and globalization on Indigenous food systems and the emergence of revitalization efforts through food sovereignty movements. These initiatives reclaim traditional knowledge and promote autonomy, particularly in diasporic and urban contexts where Indigenous communities creatively adapt foodways to sustain cultural identity. The chapter also discusses the intersections of food with gender, class, and ecological knowledge, highlighting how culinary practices embody systems of knowledge, resistance, and social values. Through this lens, Indigenous food is revealed as sustenance and a symbolic and political force—a conduit of memory, healing, and intergenerational solidarity. This chapter contributes to ongoing conversations in food studies, Indigenous knowledge systems, and cultural sustainability by situating food as central to the lived experiences and enduring legacies of Indigenous peoples.

Keywords: food symbolism, Indigenous identity, cultural rituals, food taboos, oral traditions, biocultural diversity, food and gender

15.1 Introduction

Food plays a multifaceted role within Indigenous societies, extending far beyond mere sustenance. It serves as a fundamental component in expressing cultural identity, heritage, and social cohesion. Indigenous food practices embody the cultural beliefs, ancestral traditions, and community values of these groups. The complex intersections of food, identity, and symbolism create a framework for analyzing the ways in which Indigenous peoples assert their cultural values in the face of modern challenges. In Indigenous contexts, food acquisition methods and dietary preferences are deeply intertwined with socioeconomic conditions, historical injustices, and cultural practices. Various studies highlight the shift in the diets of Indigenous populations, particularly as traditional foodways have been overshadowed by the influx of processed and non-Indigenous foods. For instance, research indicates a significant erosion of traditional dietary practices among Aboriginal communities in Australia due to colonial impacts such as the establishment of community stores (Patel et al., 2021). This signifies that food is not merely a dietary concern; it also highlights a cultural loss and a disconnect from ancestral ways of living. Additionally, the literature discusses the broader implications of these dietary shifts on community identity and resilience (Patel et al., 2021).

One of the foundational concepts in understanding Indigenous food systems is food sovereignty, which advocates for the right of Indigenous peoples to control their food systems, which are inherently linked to their cultural integrity and autonomy. Indigenous food sovereignty emphasizes the importance of revitalizing traditional food practices, as seen among the Syilx Okanagan community's initiative to reintroduce cultural foods like sockeye salmon into their diets (Blanchet et al., 2022). This not only fosters food security but also serves to reconnect community members with cultural traditions, thereby reinforcing social bonds and fostering resilience and self-determination. Moreover, through frameworks of cultural food security, Indigenous communities can navigate the challenges posed by colonial legacies that have disrupted their traditional food systems. An example is highlighted in the analysis of food security in First Nations households in Canada, which illustrates the intricate relationship between food access and cultural identity (Batal et al., 2021). The structural barriers impacting food acquisition perpetuate food insecurity and hinder cultural expression and the transmission of traditional knowledge through food practices.

Additionally, the significance of food in rituals and cultural events cannot be overstated. Indigenous food is crucial in community ceremonies, reinforcing identity and collective memory. Traditional foods consumed during important cultural events reflect their symbolic meanings and the values they represent within Indigenous narratives. For instance, food is often a central element in community gatherings and celebrations, where it is utilized for nourishment, storytelling, and the affirmation of cultural practices (Utami et al., 2021). Such affirmative acts cultivate a shared understanding and collective identity among community members, aligning with broader cultural preservation and revitalization goals. In examining the symbolic use of food, various studies reveal how it serves as a communication tool in Indigenous contexts. Foods are often used in rituals that symbolize cultural teachings and worldviews (Nyarota et al., 2022). Traditional meals may reflect the interrelationship between humans and nature, highlighting a worldview that sees food as sacred. This component is essential to preserving Indigenous identities, particularly in modern settings where globalization threatens traditional ways of life and cultural expressions. For instance, Indigenous peoples often advocate teaching younger generations about local food sources and traditional dietary practices as a form of cultural resilience (Poirier et al., 2024).

The nutritional aspect of Indigenous food must also be considered, as traditional diets are frequently depicted as healthier options compared to processed alternatives widely available today. Studies have shown that traditional foods consumed among Indigenous communities are often nutrient-dense, providing significant health benefits that align with cultural practices (Blanchet et al., 2020). This intersection between health and food traditions showcases how traditional foods can serve both as a means of sustaining health and celebrating cultural identity. The impact of colonization on Indigenous diets is profound; research indicates that the displacement and dispossession of Indigenous peoples from their lands have had deleterious effects on their ability to access traditional foods, leading to increased food insecurity and health disparities (Rondoni, 2022). As such, Indigenous food systems play an integral role in combating food insecurity while simultaneously fostering cultural identity and community resilience. For example, initiatives to engage communities in local food production have proven effective in revitalizing traditional agricultural practices and promoting cultural education among younger generations (Skinner et al., 2013; Young et al., 2024).

In contemporary discussions, there is a growing recognition of the importance of integrating Indigenous perspectives into food security policies. The Truth and Reconciliation Commission's Calls to Action emphasize the need for policies that reflect Indigenous worldviews and prioritize traditional knowledge (Sumner et al., 2023). Implementing such frameworks can facilitate a more significant appreciation for the food systems that Indigenous peoples have maintained and adapted over generations. Food within Indigenous societies serves as a critical lens through which to understand identity, culture, and community resilience. It transcends mere caloric intake, embodying the essence of cultural traditions, challenges posed by colonialism, and the ongoing efforts for food sovereignty. Through a comprehensive investigation of the symbolic meanings associated with Indigenous foods, we can appreciate how they facilitate cultural continuity, social cohesion, and a connection to the environment. The importance of revitalizing and maintaining traditional food practices cannot be overstated, as they are vital for physical nourishment and sustaining the cultural identity and agency of Indigenous communities in a rapidly changing world.

This chapter aims to illuminate the profound ways in which food functions as a vessel of identity, memory, and symbolic meaning in Indigenous societies. By analyzing food beyond its nutritional value, the chapter contributes to understanding how Indigenous communities embed social values, cosmological beliefs, and ecological ethics into their culinary practices. It also addresses how Indigenous food systems adapt in diasporic and contemporary contexts, asserting identity and sovereignty amid globalization. This chapter contributes to food studies, Indigenous knowledge research, and cultural sustainability by positioning food as a central axis of sociobiological and symbolic continuity.

15.2 Food as Cultural Identity

Food is a powerful medium through which cultural identities are expressed, negotiated, and maintained (Table 15.1). The intersection of food and identity has been the focus of scholarly inquiry across various disciplines, notably symbolic anthropology. Theories proposed by thinkers such as Mary Douglas and Claude Lévi-Strauss have laid the groundwork for understanding food as a cultural symbol. In her exploration of food taboos, Douglas emphasized how certain foods function as boundary markers within a culture, delineating which groups belong together and which do not, thereby shaping communal identities and societal structures (Mintz & Bois, 2002). Similarly, Lévi-Strauss's structuralist approach dissects the underlying rules of food preparation and consumption, arguing that these rules reveal cultural logic essential to understanding human behavior within specific social contexts (Mintz & Bois, 2002).

Table 15.1: Some Foods Used as Cultural Identity in Indigenous Societies

Cultural Group	Cultural Identity Food	Cultural Significance	Symbolic Functions
Maya (Central America)	Maize (corn)	Considered the sacred origin of humanity in the Popol Vuh, a daily staple and ritual food	Links people to ancestral origins and cosmology
Igbo (Nigeria)	Yam	Symbol of wealth, masculinity, and agricultural skill; central to the New Yam Festival	Used in marriage negotiations, community rituals, and social hierarchy
Inuit (Arctic Circle)	Seal and whale meat	Reflects adaptation to the Arctic environment and communal hunting traditions	Reinforces survival ethics and collective identity in harsh environments
Quechua (Andes, Peru)	Quinoa and potatoes	Domesticated over centuries, tied to traditional farming and ecological knowledge	Emblematic of Andean resilience and spiritual relationship with Pachamama (Earth Mother)
Zulu (South Africa)	Sour milk (amasi) and maize porridge	Daily dietary foundation; served in ceremonies and ancestral offerings	Reinforces continuity, respect for elders, and communal life
Ainu (Japan)	Salmon and wild plants	Tied to rivers, seasonal cycles, and animist beliefs	Represents harmony with nature and cyclical time
Ashanti (Ghana)	Fufu and light soup	Shared during family gatherings, funerals, and royal ceremonies	Symbolizes unity, respect, and communal bonds
Ojibwe (North America)	Wild rice (manoomin)	Sacred gift from the Creator; central in ceremonies and storytelling	Embodies cultural survival, spirituality, and ecological stewardship

Furthermore, identity theory has gained traction in cultural and food studies, illuminating how individuals and communities forge their identities concerning their food practices. For instance, Robinson and Getz (2013) point out that food enthusiasts often seek to express their food-related identity through both consumption and the social experiences intertwined, suggesting that food choices are not merely personal preferences but are also indicators of a wider social identity. This dynamic emphasizes how food consumption reflects one's social context and individual identities, merging to create a collective cultural identity expressed through shared culinary practices. Commencing with the notion of commensality, the act of sharing meals holds significant social meaning within many cultures. Commensality signifies belonging and communal ties, reinforcing social hierarchies and group memberships. Rosenblum (2010) notes that the cultural customs surrounding eating—who we eat with—contribute to the delineation of "us" versus "them." Eating patterns often exclude others based on food taboos or dietary restrictions, reinforcing boundaries and delineating group identity (Rosenblum, 2010). Additionally, the social activities surrounding food preparation and sharing, such as communal feasts or family dinners, solidify bonds and reinforce collective memories associated with specific cultural practices.

Food can be a defining factor for group membership and heritage, carrying the weight of historical and cultural narratives. Multiple case studies highlight the cultural significance of specific foods among various groups. For example, maize plays a foundational role in Mesoamerican societies, not only as a staple food but also as a symbol of life, culture, and identity (Perry, 2017). Enset among the Gurage in Ethiopia illustrates a similar phenomenon, serving as a critical food source that encapsulates the identity and heritage of the Gurage people, thus reinforcing their social cohesion and cultural continuity (Perry, 2017). In Central Africa, cassava serves a comparable function, being deeply integrated into the culinary traditions and social fabric of the communities that rely on it, representing resilience and adaptability despite historical challenges (Quintero-Ángel et al., 2019). The transmission of identity through food practices—especially in the context of preparation and consumption—grants individuals the tools for cultural expression and acts as a mechanism for intergenerational continuity. Plastow et al. highlight that traditional food practices among aging populations can act as a means of reclaiming and maintaining cultural identities, asserting that these interactions with traditional cuisine can invigorate the sense of belonging across generations (Plastow et al., 2014). This assertion elucidates the vital role food plays in sustaining ongoing dialogue about identity within family units and community structures.

Moreover, the role of food in identity formation is further complicated by modern globalization, where culinary practices from different cultures converge, leading to hybrid identities. Many individuals adopt cosmopolitan identities incorporating diverse culinary preferences while negotiating their original cultural roots. Miočević et al. (2021) demonstrate that expatriates often seek out host country foods as a form of cultural exploration, attempting to balance their culinary heritage with newfound experiences. This complex negotiation illustrates how food transcends mere sustenance and serves as an emblem of cultural dialogism, blending the old with the new. The social context in which food is consumed can substantially influence perceptions and evaluations, as outlined by Hackel et al. (2015). They argue that social identities can dynamically affect people's preferences and choices regarding food, underscoring the social nature of food evaluation and its connection to identity. The implications of this phenomenon extend to political and ethical considerations regarding food consumption, where certain dietary practices become a form of sociopolitical identity, as noted by Chuck et al. (2016). This politicized approach to dietary choices reveals how food can embody resistance, belonging, or ideological standings within broader societal narratives. The symbolism imbued in culinary practices—rooted in cultural heritage and social dynamics—plays a crucial role in shaping how communities define themselves and interact with others. Through commensality, food taboos, and selective consumption, groups delineate their boundaries while reinforcing heritage through traditional food practices. This discourse reveals how food can signify and encapsulate cultural identity, thereby advocating for further exploration of food's role as a vehicle of social meaning within anthropological and sociological studies.

15.3 Symbolism and Ritual Use of Food

The symbolism and ritual use of food within various cultural and religious contexts deeply intertwine with human social practices. It plays a crucial role in rites of passage, where food serves as a medium for communal bonding, spiritual connection, and cultural identity reinforcement (Table 15.2). For instance, in Native American cultures, food items such as maize, beans, and squash are not only staples but also embody significant spiritual meanings. They facilitate connection during ceremonies related to birth, initiation, marriage, and funerals. These rites serve as transitions in life, where food acts as a communal symbol representing life's continuity and the sustenance provided by nature. During marriage ceremonies, specific foods may symbolize fertility and prosperity, while during funerals, foods might be offered to honor the deceased and celebrate their journey into the afterlife, encapsulating beliefs regarding mortality and the legacy left behind within that community (Colby et al., 2012).

Turning toward sacred and ceremonial foods, specific items such as kola nuts in West Africa and peyote in Native American traditions hold profound spiritual significance. Kola nuts are used in various ceremonies to signify hospitality, respect, and unity among participants, often pivotal in marriage ceremonies and community assemblies. Sharing kola nuts can symbolize the collective spirit and bonding among participants, fostering relationships and reinforcing social harmony (Bonvillain & Porter, 2014). Meanwhile, peyote, a hallucinogenic cactus, is integral to the rituals of the Native American Church, where its consumption is seen as a conduit for communion with the divine. This sacred food embodies not just a means of spiritual healing but also represents a cultural identity, intertwining with concepts of health, recovery, and resistance against historical trauma (Davis et al., 2022; Rokach, 2020). The ritual use of peyote often aligns with a broader understanding of food as a medium to create community, draw connections with the spiritual realm, and foster personal healing.

Harvest festivals represent another critical domain where food resonates with cosmological meanings (Table 15.2). These festivals, celebrated across cultures, embody gratitude for the bounty provided by the earth while also signifying the cyclical nature of life and seasons. In various Indigenous practices, foods harvested during these occasions are often used for ceremonial exchanges marked by feasting, rituals, and community gatherings. Such festivals are laden with symbolic meanings, reflecting the interdependence between humans and nature while reaffirming communal ties. The preparation and sharing of specific dishes during these gatherings act as a ritualistic acknowledgment of the earth's gifts, transforming agricultural cycles into opportunities for social cohesion and cultural expression (Colby et al., 2012; Jones, 2007). Like rituals surrounding life events, harvest festivals encapsulate the profound significance of food that transcends mere sustenance; they embody homage to ancestors and the beliefs underpinning agricultural practices in their respective cultures.

Moreover, the implications of food symbolism extend into the realm of psychological and communal health. For instance, Figure 15.1 depicts members of a host family flaying an ox head as part of a ritual offering to the gods. In many Indigenous and ancestral food systems, meat preparation is not only a culinary act but also a sacred gesture of devotion (Riccio, 2022). Such rituals highlight the deep-rooted connection between food, sacrifice, and ceremonial respect in

Table 15.2: Some Symbolism and Ritual Use of Food in Indigenous Societies

Ritual Context	Role of Food	Examples of Foods Used	Examples of Indigenous Cultures
Rites of Passage	Food marks key life transitions (birth, puberty, marriage, and death), reinforcing social status and communal identity.	Millet porridge, goat meat, yam, palm wine, rice, and milk	Zulu (South Africa), Yoruba (Nigeria), Navajo (the United States), and Balinese (Indonesia)
Initiation Ceremonies	Foods are used to symbolize strength, endurance, and maturity during rites of passage into adulthood.	Roasted meat, fermented cereals, and honey beer	Maasai (Kenya/ Tanzania), Senufo (Côte d'Ivoire), and Ainu (Japan)
Marriage Rituals	Food exchanges symbolize union, fertility, and family alliance; feasting signifies community blessing.	Yams, kola nuts, cooked chicken, and wedding cakes	Igbo (Nigeria), Quechua (Peru), and Sámi (Scandinavia)
Funerary Rituals	Offerings and feasts honor the dead, assist in the transition to the ancestral realm, and reaffirm community ties.	Cooked rice, millet beer, tubers, and dried fish	Akan (Ghana), Ifugao (Philippines), and Hopi (the United States)

Ritual Context	Role of Food	Examples of Foods Used	Examples of Indigenous Cultures
Sacred/ Ceremonial Use	Foods serve as offerings or conduits to deities, ancestors, or spirits, often consumed communally or ritually.	Kola nuts, peyote, sacred mushrooms, and maize cakes	Yoruba (Nigeria), Huichol (Mexico), and Maori (New Zealand)
Harvest Festivals	Celebrates agricultural abundance; rituals renew ecological balance and give thanks to spirits or deities.	First yams, maize, sorghum, cassava, and plantains	Ashanti (Ghana— Homowo Festival), Igbo (New Yam Festival, Nigeria), and Aymara (Bolivia)
Healing and Protection	Foods used in spiritual or physical healing, protective rituals, or purification ceremonies.	Herbal infusions, bitter roots, sacred soup, and medicinal broths	Dagara (Burkina Faso), Mapuche (Chile), and Tibetans (Himalayas)

traditional communities. Also, Figure 15.2 shows the *junku*, a ceremonial tray used by the Eastern Minyag people of China (Punzi, 2025). It contains ritual items such as corn, eggs, fermented liquor, incense, herbs, and deity images. Food items on the tray, especially corn and spirits, are central to spiritual offerings, symbolizing abundance and serving as a medium to connect with ancestors and deities. The *junku* reflects how food and ritual deeply intertwine in Minyag culture, where nourishment and spirituality are expressed through shared ceremonial practices (Punzi, 2025). Studies indicate that participating in food rituals fosters a sense of belonging and strengthens interpersonal relationships, serving as a therapeutic avenue for individuals grappling with historical or contemporary traumas. By partaking in communal meals, individuals

Figure 15.1: The Flaying of the Ox Head by Members of the Host Family to Make the Meat an Offering to the Gods. (Photo: Thomas Riccio)

Source: Riccio (2022)

Figure 15.2: The *Junku*–Eastern Minyag Ritual Objects Incorporating Food and Drinks as Symbolic Items

Source: Punzi (2025)

engage in practices that nurture the soul, as outlined by various researchers who highlight the transformative nature of these experiences (Bassett et al., 2012; Bonvillain & Porter, 2014). This is particularly evident within Native American communities where rituals, including the use of peyote, provide not just spiritual guidance but also avenues for collective healing, reinforcing resilience against the backdrop of historical injustices (Calabrese, 2008). Food is vital for cultural expression, community connectivity, and spiritual reflection across various rites of passage, sacred practices, and seasonal celebrations (Imarhiagbe & Ogwu, 2022; Obongodot & Ogwu, 2023). The intertwining of food within these cultural narratives underscores its prominence as more than sustenance; it embodies collective identity, continuity, and resilience, reflective of the intricate relationships societies hold with both their natural environments and cultural traditions.

15.4 Food Taboos and Social Order

Food taboos serve as a critical framework within which societies define concepts of purity, gender roles, and social status. The nuances of these taboos can often be traced back to historical traditions, social structures, and cultural beliefs that inform practices related to food consumption (Table 15.3). The complexity of food taboos is illustrated in the context of adolescent dietary practices, demonstrating the interdependence between familial influence and societal norms. In a study by Chitakunye and Maclaran (2008), the authors highlight how young people's food consumption is shaped by formal and informal rules, reflecting evolving social dynamics wherein intergenerational conflicts play a significant role. The enforcement of food rules across generations reinforces cultural values and social identity, wherein familial practices serve as vectors of traditional knowledge. Research by Kharuhayothin and Kerrane (2018) elucidates how childhood experiences contribute to adult food preferences and behaviors, supporting the idea that intergenerational influences persist into adulthood regarding familial food socialization practices. When parents instill specific dietary norms, they essentially transfer symbolic meanings and cultural identities to their children, laying the groundwork for future consumption practices. Such intergenerational linkages provide a compelling narrative on how food preferences evolve through familial channels, leading to the understanding that food taboos are not merely personal choices but are also reflective of complex social networks (Ogwu, 2023; Ogwu & Osawaru, 2022).

Table 15.3: Examples of Food Taboos in Indigenous Societies

Type of Food Taboo	Prohibited Food Item(s)	Symbolic or Social Reason	Examples of Indigenous Cultures
Gender-Based Taboos	Certain meats (e.g., bushmeat, eggs, or organs)	The belief that women may become infertile, impure, or overpower male energies	Gikuyu (Kenya), Tikar (Cameroon), Inuit (Canada, etc.)
Age-Based Taboos	Alcohol or strong spices	Young people are seen as spiritually or physically unready to consume adult foods	Ashaninka (Peru), Maasai (Kenya/Tanzania, etc.)
Pregnancy-Related Taboos	Twin bananas, snails, pineapple, etc.	Thought to cause congenital disabilities, difficult labor, or spiritual contamination	Igbo (Nigeria), Hmong (Vietnam/Laos), Hausa (Nigeria, etc.)
Ritual Purity Taboos	Fish, pork, or certain birds	Considered unclean or spiritually defiling in sacred contexts	Dinka (South Sudan), Batak (Indonesia), Brahmin Hindus (India, etc.)
Caste/Status-Based Taboos	Foods handled or cooked by lower-status individuals	Maintains social hierarchy and spiritual cleanliness	Caste-based Hindu groups (India), and some Andean societies
Period of Mourning	Meat, salt, or spicy foods	Symbolizes humility, detachment, and spiritual reflection	Akan (Ghana), Igbo (Nigeria), Balinese (Indonesia, etc.)
Hunting-Related Taboos	Killing/eating totem animals or endangered species	The animal may be viewed as an ancestor, spirit guide, or clan protector	Ojibwe (Canada/the United States), Aboriginal Australians, Shona (Zimbabwe, etc.)

Type of Food Taboo	Prohibited Food Item(s)	Symbolic or Social Reason	Examples of Indigenous Cultures
Seasonal/ Temporal Taboos	Fresh yams or fruits before the harvest ritual	Ensures community-level permission and spiritual gratitude before consumption	Igbo (New Yam Festival), Ashanti (Ghana), Aymara (Bolivia, etc.)

A significant case study that illustrates the enforcement of food taboos is that of menstrual food taboos, which exemplify the intersection of gender roles with dietary restrictions. Menstrual taboos often involve strict dietary rules that seek to regulate the behavior and social interactions of menstruating individuals, emphasizing purity and social separation from others. Abdullah's (2022) qualitative study on the Temiar people in Malaysia elucidates how these taboos function to maintain cultural harmony and balance. Menstrual taboos often dictate prohibitions, such as the avoidance of sour foods, believed to disrupt physical processes related to menstruation and highlight a belief system that ties food, bodily experience, and social order together (Ogwu & Osawaru, 2023; Osawaru & Ogwu, 2023). The implications of these taboos extend beyond individual dietary choices, shaping social expectations and interactions within various community contexts (Kumar & Srivastava, 2011).

Moreover, menstrual taboos contribute to a broader symbolic separation, reinforcing gender roles within societies. This separation results in specific expectations of behavior and dietary practices assigned to menstruating women, steeped in historical and cultural significance. The prohibition of certain foods during menstruation can be understood as a mechanism to maintain societal cohesion and symbolize purity during a time viewed as a moment of physical vulnerability (Akin, 2003). Such mechanisms highlight how food rules can transcend personal health issues and represent societal norms and moral frameworks dictating gender-specific behaviors (Ogwu et al., 2023). Cross-cultural comparisons reveal varied expressions of these taboos, demonstrating how food practices communicate moral values and sociocultural identities. For example, in regions of India, certain foods are avoided by menstruating women due to beliefs about their influence on menstrual flow, showcasing a connection between dietary choices and maintaining social order during critical life stages such as menstruation (Abdullah, 2022).

The emotional and symbolic significance of foods involved in taboos cannot be underestimated. Fessler and Navarrete discuss the psychological underpinnings of food taboos, arguing that they arise from a complex interplay of cultural conditioning and fear of pathogens. The aversion to certain food items during sensitive times, such as menstruation, can also be viewed through this lens, where avoidance reflects ingrained societal beliefs about hygiene and moral conduct (Fessler & Navarrete, 2003). Intergenerational practices thus play a pivotal role in perpetuating the meanings intrinsic to food taboos. The reinforcement of dietary restrictions across generations signifies that such taboos are not static but evolve and adapt, reflecting changing dynamics within families and communities, as explored in Liu's (2017) work, highlighting how family-based food practices adapt across generations in response to social factors. Emphasis on familial food rituals underscores how the social context shapes how taboos circulate and transform while maintaining continuity and identity (Ogwu et al., 2023, 2025).

Additionally, societal structures influence the acceptance and enforcement of food taboos, especially against economic and health-related policies. The work of Vereecken et al. discusses how socioeconomic status can shape access to food and resultant practices, emphasizing the interaction between societal values and individual choices (Vereecken et al., 2004). As adolescents navigate these structures, they position themselves within a network of inherited food practices, contributing to their understanding of identity, social norms, and familial expectations. As communities evolve, the interlinked dimensions of food consumption and social behavior indicate that taboos can mediate broader social change. Shukla's (2024) study emphasizes how intergenerational relations within food systems can illuminate shifts in practices across diverse contexts, moving beyond food security to interrogate social welfare. This understanding underscores the importance of holistic frameworks examining food systems encapsulating both traditional and contemporary interactions (Ikhajiagbe et al., 2021). The reinforcement of food taboos thus becomes a focal point for examining the interplay between cultural identity, social interactions, and familial relationships.

15.5 Language, Storytelling, and Food Symbolism

The interconnectedness of language, storytelling, and food symbolism is fundamental to cultural expression, informing identity, wisdom, and communal ties across diverse societies. Food-related metaphors are pervasive in language, serving as vehicles for deeper meanings. For example, phrases like "milk of the land" symbolize fertility, sustenance, and the nurturing qualities of a region or culture, reinforcing the bond between people and their environment (Singer, 2019). This symbolic resonance is crucial as it encapsulates the emotional significance of food, showcasing its role in shaping cultural narratives and identities. The importance of oral traditions surrounding food connects communities to their history and heritage. Mythological tales often explore the origins of food, illustrating moral values and societal norms embedded within culinary practices. Many cultures have creation myths that involve food, symbolizing the sustenance provided by nature or divine entities. These stories effectively encode wisdom regarding agricultural practices, ethical treatment of nature, and communal sharing, which are essential for survival and cultural continuity (Long, 2025; Orea-Giner et al., 2024). Furthermore, folklore studies emphasize that food is integral to cultural identity, representing local community customs, rituals, and economic relations (Long, 2025).

Proverbs and folk sayings embody culinary wisdom, reflecting communal knowledge and lived experiences. These aphorisms often encapsulate practical life lessons from agricultural practices or culinary traditions, providing a cultural and ethical framework for behavior. Expressions about the value of sharing food, such as "you are what you eat," convey societal values regarding health, identity, and community (Long, 2025; Pamantung et al., 2021). Such proverbs are mnemonic devices, ensuring the transmission of cultural values to new generations and encapsulating complex societal norms related to food consumption and solidarity (Long, 2025).

In recent years, the merger of food and digital storytelling has gained prominence, particularly through social media platforms where visual and narrative representations enhance culinary marketing. Social media influencers utilize narrative paradigms to engage audiences emotionally, drawing from cultural narratives to enrich viewers' understanding of food products. The storytelling strategies employed by food vloggers not only convey information about culinary products but also invoke a sense of nostalgia, enhancing consumer experience by fostering connections to heritage and place (Wahyudi et al., 2024). This digital landscape leverages the symbolism of food, showcasing its role as an element of cultural identity and allowing for complex storytelling that aligns consumer preferences with traditional practices. The strategic use of storytelling within food marketing is fundamental in promoting local cuisines, especially in the face of global culinary trends. Successful culinary marketing involves integrating local narratives within promotional strategies, allowing consumers to engage with the cultural significance and authenticity of regional dishes. By emphasizing local origin stories, culinary businesses can reinforce the unique characteristics of their offerings and enhance consumer loyalty through shared cultural heritage (Nasution, 2025). This illustrates how storytelling acts as a marketing strategy and a form of cultural preservation, effectively bridging past and present culinary traditions.

Moreover, food festivals have emerged as vital spaces for food storytelling, where the intersection of community, identity, and gastronomy can be celebrated. These events foster experiences tied to food authenticity and local traditions, helping communities maintain cultural narratives while engaging with broader audiences (Orea-Giner et al., 2024). Storytelling during food festivals positions local products within larger discourses of heritage and innovation, offering insights into the unique relationships between food, place, and identity. Research exploring food-related storytelling consistently highlights its role in articulating cultural identity. Studies investigating food portrayal in folklore reveal that culinary traditions are dynamic expressions of cultural identity that reflect sociohistorical transformations (Long, 2025). The narratives surrounding food shape individuals' perceptions and relationships within their communities, offering frameworks for interpreting their experiences and identities in a culturally significant manner (McGovern et al., 2021).

In addition to reflection and identity, food symbolism often intersects with broader sustainability and ethical consumption themes. The narrative aspects of food in contemporary discussions encompass moral dimensions associated with sourcing, production, and culinary practices. When consumers engage with food narratives, they grapple with the ethical implications of their choices, drawing upon traditional wisdom to guide them toward more sustainable practices (Rivera-Toapanta et al., 2021). This convergence of food, storytelling, and ethics serves as a reminder of the interconnectedness of environment, culture, and individual choices in shaping future food systems. The ongoing dialogue about food, origins, and meanings enables deeper engagement with food practices, resulting in collective understanding and respect for cultural diversity. As culinary narratives evolve, they inform how societies negotiate cultural heritage and adapt to global influences, ensuring local identities remain vibrant and relevant in an interconnected world. Celebrating unique, local food narratives is essential for cultural preservation and identity affirmation within the broader discourse of globalization (Castelló, 2020). The multifaceted relationship between language, storytelling, and food symbolism becomes evident through the lenses of folklore, cultural sociology, and marketing strategies. Each element enriches human experiences, embodying the complex interplay of tradition, innovation, and community. Language is a powerful medium for expressing cultural narratives that shape our understanding of food and its significance.

15.6 Diasporic and Contemporary Reconfigurations of Food Identity and Symbolism

Exploring Indigenous foodways within diaspora contexts entails a complex interplay of cultural resilience, transformation, and the reimagining of identity. As Indigenous communities migrate or find themselves in diaspora, their food practices combine traditional knowledge with contemporary adaptations. In these settings, food operates not merely as sustenance but in fact as an embodiment of cultural memory and identity. The movement toward food sovereignty has been instrumental in this ongoing process, positing food as a fundamental human right intertwined with cultural heritage and governance (Coté, 2016; Elliott et al., 2021; Pimbert & Claeys, 2024). In many instances, the food sovereignty movement transcends simple nourishment, calling for the right of communities to define their food systems while reinforcing cultural traditions. This is especially salient for Indigenous populations, whose culinary practices are forcibly interrupted by colonial histories and contemporary globalization mechanisms (Coté, 2016; Elliott et al., 2021; Ruelle, 2017). The revitalization of traditional foodways within diaspora communities asserts cultural identity and resists the erasure of Indigenous knowledge and practices (Hoover, 2021; Maudrie et al., 2023). Indeed, by reestablishing connections to their food systems, Indigenous peoples in diaspora reclaim agency over their culinary heritage, fostering a sense of pride and continuity in urban environments (Babcock & Budowle, 2022; Clendenning et al., 2015).

Urban spaces often act as crucibles for cultural exchange, where food becomes a site of resistance and resilience (Ogwu, 2019). Indigenous foodways serve as a counter-narrative to the dominant food systems characterized by industrialization and commodification. As urban Indigenous populations articulate their food identities, they engage with movements like Slow Food and other local food initiatives that advocate for sustainable, ethical, and culturally relevant food practices (Babcock & Budowle, 2022; Clendenning et al., 2015; Figueroa-Helland et al., 2018). These initiatives highlight the importance of localized food systems that respect Indigenous land practices and knowledge systems, positioning them as essential for cultural survival and resistance against colonial legacies (Elliott et al., 2021; Maudrie et al., 2023).

Furthermore, the role of Indigenous chefs and culinary practitioners has become increasingly prominent. They act as cultural stewards, invoking ancestral knowledge through their food, which helps revitalize traditional culinary practices and address community health issues. These chefs, often involved in food sovereignty movements, promote culturally significant ingredients, leading to a renaissance of precolonial food practices (Hoover, 2021; Sandoval & Wathne, 2022). This ongoing culinary renaissance enhances community health and revitalizes an economic ecosystem that supports Indigenous food systems in urban settings, creating channels for traditional knowledge to thrive (Maudrie et al., 2023; Ruelle, 2017). In urban contexts, Indigenous food sovereignty movements encapsulate various social and cultural continuity aspects. Such movements are about food production, reestablishing social bonds and communal ties, and the reclamation of land-based knowledge. Initiatives that emphasize relationality and reciprocity—central tenets of Indigenous knowledge systems—are increasingly gaining traction as necessary paradigms for sustainable food practices in urban areas (Babcock & Budowle, 2022; Sandoval & Wathne, 2022; Young et al., 2024). These initiatives reflect an understanding of food systems not just as resources for consumption but as integral to social health and community resilience, thereby challenging traditional notions of food security that tend to overlook cultural aspects (Clendenning et al., 2015; Shattuck et al., 2015).

Therefore, Indigenous culinary claims function within a broad spectrum of resistance and resilience strategies encompassing slow-food movements, local agriculture, and community gardens. These efforts can challenge food insecurity and the socioeconomic and political marginalization of Indigenous communities in urbanized landscapes (Elliott et al., 2021; Nyarota et al., 2022). Through these efforts, Indigenous populations in urban settings successfully assert their right to food sovereignty while simultaneously crafting a narrative that fosters cultural pride and community empowerment (Babcock & Budowle, 2022; Maudrie et al., 2023; Pimbert & Claeys, 2024). The intricacies of food identity in diaspora also highlight the potential for transformative practices through collective action. As Indigenous groups engage in revitalization movements, culinary arts become a form of activism that promotes not only cultural significance but also the restoration of ecological health (Coté, 2019; Figueroa-Helland et al., 2018; Oldham et al., 2024). This reinforces the view that food sovereignty is not merely an agricultural principle but in fact a holistic framework that nurtures cultural, environmental, and social justice, creating pathways for reclaiming rights and narratives in the face of adversity (Coté, 2016; Noll & Murdock, 2019; Sandoval & Wathne, 2022).

As urban Indigenous populations navigate the challenges of diaspora, the symbolic significance of food emerges as central to the negotiation of identity, autonomy, and belonging. By reclaiming their food systems, they assert a cultural territory that strengthens community ties and fosters intergenerational knowledge transfer (Carlile et al., 2021; Hoover, 2021; Pimbert & Claeys, 2024). This cultural and culinary hybridity process enables Indigenous groups to cultivate resilience and pride in diverse urban landscapes, challenging dominant food narratives that often marginalize Indigenous claims on land and food (Carlile et al., 2021; Kim, 2018).

Moreover, the intersection of food sovereignty and ecological stewardship is increasingly recognized as vital to achieving comprehensive Indigenous self-determination. Food systems rooted in traditional ecological knowledge sustain Indigenous peoples' cultural practices and contribute to the overarching environmental sustainability narratives. The focus on agroecology and community-led agriculture signifies a vital step toward reclaiming Indigenous ways of knowing and being within contemporary sociopolitical frameworks (Figueroa-Helland et al., 2018; Oldham et al., 2024; Ruelle, 2017). The work of Indigenous chefs and culinary practitioners is also paramount in this reclamation process. They serve as facilitators of cultural resurgence, using food as a medium to educate others about the importance of Indigenous food systems and health advocacy. The relationships built through culinary practices reflect broader cultural narratives that emphasize healing, community engagement, and the intergenerational sharing of knowledge (Nyarota et al., 2022; Oldham et al., 2024). Their contributions underscore the intricate relationship between identity, agency, and the environment in urban food experiences.

As we delve further into the dynamics of food identity and symbolism, it becomes clear that Indigenous perspectives on food critically assess broader social contexts—deconstructing the industrial food systems that often strive to homogenize disparate food cultures in the pursuit of globalized convenience. In contrast, the reinvigoration of local practices presents an opportunity for preservation and innovation that honors the past while forging pathways to a culturally vibrant future (Young et al., 2024). The narratives woven through foodways articulate a profound sense of place and belonging for Indigenous peoples, reaffirming cultural identities that may otherwise dissipate in diasporic contexts. The intertwining of traditional knowledge with contemporary urban experiences propels movements surrounding food sovereignty, facilitating a reimagining of cultural landscapes. As Indigenous communities assert their rights to food and self-determination, the role of food emerges as a critical dimension of cultural pride and resistance in urban contexts. Through the culinary arts and organic practices grounded in ancestral knowledge, these communities honor their heritage and actively shape the food systems that nourish them.

15.7 Food Identity Intersections with Gender, Class, and Ecology

The intersection of food identity with gender, class, and ecological considerations reflects the deep-rooted symbolism that food holds within cultural contexts. Gender roles are fundamental in shaping food symbolism and identity, as societal expectations often influence the relationships between individuals and their food practices. For instance, specific food rituals are frequently associated with gendered behaviors, where women traditionally take on roles as primary caregivers and food preparers. This is pivotal in ceremonies and celebrations, positioning women at the center of family traditions and cultural identity formation. Food can bind individual identities with collective and national identities, thus underscoring the social conventions that govern eating practices (Ibrahim & Howarth, 2016; Moscato & Ozanne, 2019).

Furthermore, class-based distinctions in ceremonial and everyday foods reinforce socioeconomic hierarchies. Different social classes often have access to varying qualities and types of food, which transform eating practices into markers of status and identity. Media narratives repeatedly frame food consumption, often reinforcing social perceptions and judgments associated with food choices (Ibrahim & Howarth, 2016). For instance, luxurious foods might be celebrated in higher social circles during ceremonial occasions, reflecting affluence and social power. Conversely, foods consumed by lower socioeconomic classes might be depicted as inadequate or culturally inferior, despite their nutritional value and cultural significance. This highlights the complex interplay between food choices, social identity, and class dynamics, suggesting that food practices have symbolic meanings beyond sustenance (Gamble, 2017).

In addition to socioeconomic influences, ecological knowledge is an embedded aspect of food rituals and practices. Indigenous communities illustrate the sustainable integration of ecological understanding into their culinary traditions, exemplifying the powerful connection between culture and the environment. Traditional ecological knowledge serves as a framework through which various communities adapt to their environments, ensuring the sustainability of their food sources (Ghosh-Jerath et al., 2020; Qodim, 2023; Sulistiyowati et al., 2022). Such practices are often ritualized, reflecting deep-rooted cultural beliefs about maintaining harmony with nature while supporting community resilience. Food, therefore, transcends its immediate material function to embody cultural narratives that articulate humanity's relationship with its environment

(Gómez-Baggethun et al., 2013; McGraw & Krátký, 2017). Healthy practices grounded in local ecological knowledge often incorporate biodiversity, leading to dietary traditions that preserve cultural and environmental contexts (Siahaya et al., 2016; Thayyib, 2021).

Moreover, food plays a crucial role in enacting and preserving cultural identity within communities, especially in multicultural settings. The establishment and maintenance of cultural identity through food can be seen in various ethnic groups, where traditional foods signify social codes and collective memory. For example, within the Sabah community in Malaysia, traditional foods are intertwined with the layers of cultural identity, allowing different ethnicities to express their unique narratives through culinary practices (Lucas, 2022; Yakin et al., 2022). The cultural symbolism attached to food communicates intricate social dynamics, where preparing and sharing food becomes a significant ritual for fostering community bonds and affirming identities (Chelum et al., 2023). Exploring ecological knowledge revealed through food rituals further emphasizes the interconnectedness of cultural identity and environmental stewardship. Communities often imbue their food practices with meanings highlighting conservation efforts rooted in traditional beliefs (Geng et al., 2017; Sulistiyowati et al., 2022; Thayyib, 2021). Recognizing particular food sources as sacred or significant promotes respect for the environment and ensures the longevity of traditional practices amid global pressures. The reverence for food in community rituals can thus serve as a vehicle for environmental education, fostering awareness and action toward sustainability and ecological integrity (Prasetyo, 2023).

15.8 Case Studies in Food Identity and Symbolism

The examination of food identity and symbolism reveals a rich tapestry of cultural practices across diverse regions, highlighting how particular foods serve as nourishment and critical markers of identity and social coherence. In Aboriginal Australia, wild yams (*Dioscorea* spp.) are integral to cultural identity and spiritual connection to the land. Wild yams hold a dual significance for Indigenous Australians: They are a food source and a symbol of identity connected to ancestral narratives and traditions. The foraging of wild yams embodies a practice of bush gathering that reinforces community bonds and cultural heritage. This practice is steeped in ecological knowledge passed down through generations, reflecting a deep understanding of biodiversity and sustainable management of wild resources (Mintz & Bois, 2002). Harvesting these yams often coincides with specific cultural rituals, reinforcing their role as a communal activity that underscores identity (Mintz & Bois, 2002). Moreover, in some regions, the popularity of wild yams has spurred a revival of interest in traditional foodways, allowing contemporary Aboriginal communities to reclaim their cultural narratives and assert their identities in modern society.

In Southeast Asia, fermented fish, including shrimp paste and fish sauce varieties, epitomize dietary staples and cultural heritage. Fermented fish is prevalent in various regional cuisines, such as Thailand, Vietnam, and Laos, exemplifying how food practices encode identity and community values. Fermented fish products are often associated with traditional cooking methods and rituals, deeply embedding them in the social fabric of these cultures (Everett & Aitchison, 2008; Yapici, 2018). Furthermore, fermentation is symbolic of Indigenous knowledge systems that involve careful attention to environmental resources, thus reinforcing a connection to place and identity. These practices serve as markers of cultural authenticity, allowing communities to navigate modernity while maintaining ties to their traditional heritage.

Millet beer, particularly in the Sahelian region of Africa, is a vital cultural emblem tied to social rituals and agricultural practices. The brewing and consumption of millet beer are deeply embedded in various rites of passage and communal gatherings, facilitating social interaction and cohesion (Morell-Hart, 2012). In many Sahelian communities, millet beer is essential in celebrations, such as weddings and harvest festivals, reinforcing social bonds and cultural identity through shared experiences (Morell-Hart, 2012). The production of millet beer is also indicative of the agricultural calendars of these communities, symbolizing resistance and resilience against climatic challenges faced in the arid Sahel. As communities adapt to external pressures, the ritualistic consumption of millet beer symbolizes continuity and the reaffirmation of identity amid change.

These ethnographic examples illustrate that food choices and practices cannot be disentangled from identity, tradition, and cultural sovereignty issues. In Aboriginal Australia, wild yams are emblematic of a profound connection to the land and ancestral knowledge, while in Southeast Asia, fermented fish encapsulates culinary heritage and community resilience. Similarly, in the Sahel, millet beer serves as a vital cultural marker that embodies social rituals and agricultural practices that shape community identity. Through these cases, it becomes evident that food is not merely a means of survival; it is a vital medium through which identities are constructed, expressed, and negotiated. The case studies outlined above demonstrate that foodways are essential in comprehending the dynamic interplay between culture and identity. They illuminate how foods serve as conduits for expressing and negotiating social identities, anchoring individuals within their cultural contexts while also allowing for adaptability and resilience in the face of changing circumstances. Ethnographic studies like these underscore the importance of food practices in understanding human sociocultural dynamics, offering valuable insights into how communities navigate their identities within a complex global landscape (Mintz & Bois, 2002). Food identity reinforces the notion that food is a powerful symbol of culture, a medium through which identities are articulated, and a resource for fostering community solidarity. The roles played by Indigenous foods are emblematic of their respective cultures and illustrative of the broader narratives surrounding food anthropology, signaling a need for continued scholarship in this vital field of study (Everett & Aitchison, 2008; Morell-Hart, 2012).

15.9 Conclusion

Food in Indigenous societies transcends its biological function to become a central medium through which identity, spirituality, and social order are expressed, negotiated, and preserved. This chapter has demonstrated that food is not merely consumed for sustenance but in fact serves as a potent symbol of cultural continuity, belonging, and meaning. Through rituals, taboos, storytelling, and communal practices, food operates as a language that conveys moral values, gender roles, spiritual beliefs, and ecological knowledge. Whether through the sacred use of ingredients in rites of passage, structuring social boundaries via food taboos, or encoding ancestral wisdom in proverbs and myths, food remains an essential axis of Indigenous cosmologies and social structures. Moreover, the dynamic nature of Indigenous food identities—especially in diasporic and urban contexts—reveals food's capacity to serve as a medium of resilience, adaptation, and political assertion. Amid the forces of globalization, colonization, and ecological change, Indigenous communities continue to mobilize their food traditions to reclaim autonomy, resist cultural erosion, and reimagine their place in contemporary societies. This chapter calls attention to the importance of supporting Indigenous food sovereignty, revitalizing traditional food knowledge, and promoting inclusive food policies that respect cultural diversity. Ultimately, preserving the symbolic and identity-bearing roles of food in Indigenous societies is crucial for sustaining both cultural integrity and broader ecological and social well-being.

Chapter Reflection

1. How does food function as both a biological necessity and a cultural symbol in Indigenous societies?
2. In what ways do Indigenous food practices serve as living archives of ancestral knowledge and identity?
3. Explain the role of food in rites of passage, sacred ceremonies, and harvest festivals within Indigenous cultures.
4. How do food taboos reflect and reinforce gender roles, purity concepts, and social structures in Indigenous societies?
5. Discuss the significance of storytelling, oral traditions, and proverbs in preserving food-related cultural knowledge.
6. How do Indigenous communities in diaspora use food to maintain cultural identity and resist cultural erosion?
7. In what ways do food sovereignty movements empower Indigenous peoples to reclaim autonomy over their food systems?
8. Analyze the intersection of food identity with gender, class, and ecological knowledge. How do these factors shape food practices?
9. Using examples from the chapter, explain how specific Indigenous foods (e.g., wild yams, fermented fish, millet beer) symbolize cultural resilience.
10. Reflect on how globalization, colonization, and urbanization have impacted Indigenous food identities, and suggest ways to support Indigenous food sovereignty.

REFERENCES

Abdullah, S. (2022). Menstrual food restrictions and taboos: A qualitative study on rural, resettlement and urban Indigenous Temiar of Malaysia. *PLOS One*, *17*(12), e0279629. https://doi.org/10.1371/journal.pone.0279629

Akin, D. (2003). Concealment, confession, and innovation in Kwaio women's taboos. *American Ethnologist*, *30*(3), 381–400. https://doi.org/10.1525/ae.2003.30.3.381

Babcock, A., & Budowle, R. (2022). An appreciative inquiry and inventory of Indigenous food sovereignty initiatives within the Western U.S. *Journal of*

Agriculture Food Systems and Community Development, 11(2), 1–21. https:// doi.org/10.5304/jafscd.2022.112.016

Bassett, D., Tsosie, U., & Nannauck, S. (2012). "Our culture is medicine": Perspectives of native healers on posttrauma recovery among American Indian and Alaska Native patients. *The Permanente Journal, 16*(1), 19–27. https://doi. org/10.7812/tpp/11-123

Batal, M., Chan, H. M., Fediuk, K., Ing, A., Berti, P. R., Mercille, G., Sadik, T., & Johnson-Down, L. (2021). First Nations households living on-reserve experience food insecurity: Prevalence and predictors among ninety-two First Nations communities across Canada. *Canadian Journal of Public Health, 112*(S1), 52–63. https://doi.org/10.17269/s41997-021-00491-x

Blanchet, R., Willows, N., Johnson, S., & Batal, M. (2022). Enhancing cultural food security among the Syilx Okanagan adults with the reintroduction of Okanagan sockeye salmon. *Applied Physiology Nutrition and Metabolism, 47*(2), 124–133. https://doi.org/10.1139/apnm-2021-0321

Blanchet, R., Willows, N., Johnson, S., Initiatives, O., & Batal, M. (2020). Traditional food, health, and diet quality in Syilx Okanagan adults in British Columbia, Canada. *Nutrients, 12*(4), 927. https://doi.org/10.3390/nu12040927

Bonvillain, N., & Porter, F. (2014). Native American religion. In M. Issitt & C. Main (Eds.), *Hidden religion: The greatest mysteries and symbols of the world's religious beliefs* (pp. 393–423). Bloomsbury Publishing Plc. https://doi. org/10.5040/9798400663277.0031

Calabrese, J. (2008). Clinical paradigm clashes: Ethnocentric and political barriers to Native American efforts at self-healing. *Ethos, 36*(3), 334–353. https:// doi.org/10.1111/j.1548-1352.2008.00018.x

Carlile, R., Kessler, M., & Garnett, T. (2021). What is food sovereignty?. https://doi.org/10.56661/f07b52cc

Castelló, E. (2020). Storytelling in applications for the EU quality schemes for agricultural products and foodstuffs: Place, origin and tradition. *Spanish Journal of Agricultural Research, 18*(2), e0105. https://doi.org/10.5424/sjar/2020182-16192

Chelum, A., Magiman, M., Chan, S., & Kundat, F. (2023). Food dishes in the Nyangahant ritual as symbols of nonverbal communication in the Salako community of Pueh Village, Lundu, Sarawak. *International Journal of Academic*

Research in Business and Social Sciences, *13*(8), 103–115. https://doi.org/10.6007/ijarbss/v13-i8/18011

Chitakunye, D., & Maclaran, P. (2008). The everyday practices surrounding young people's food consumption. *Young Consumers Insight and Ideas for Responsible Marketers*, *9*(3), 215–227. https://doi.org/10.1108/17473610810901642

Chuck, C., Fernandes, S. A., & Hyers, L. L. (2016). Awakening to the politics of food: Politicized diet as social identity. *Appetite*, *107*, 425–436. https://doi.org/10.1016/j.appet.2016.08.106

Clendenning, J., Dressler, W., & Richards, C. (2015). Food justice or food sovereignty? Understanding the rise of urban food movements in the US. *Agriculture and Human Values*, *33*(1), 165–177. https://doi.org/10.1007/s10460-015-9625-8

Colby, S., McDonald, L., & Adkison, G. (2012). Traditional Native American foods: Stories from northern plains elders. *Journal of Ecological Anthropology*, *15*(1), 65–73. https://doi.org/10.5038/2162-4593.15.1.5

Coté, C. (2016). "Indigenizing" food sovereignty. Revitalizing Indigenous food practices and ecological knowledges in Canada and the United States. *Humanities*, *5*(3), 57. https://doi.org/10.3390/h5030057

Coté, C. (2019). Hishuk'ish Tsawalk—everything is one. Revitalizing place-based Indigenous food systems through the enactment of food sovereignty. *Journal of Agriculture Food Systems and Community Development*, *9*, 37–48. https://doi.org/10.5304/jafscd.2019.09a.003

Davis, A., Arterberry, B., Xin, Y., Agin-Liebes, G., Schwarting, C., & Williams, M. (2022). Race, ethnic, and sex differences in prevalence of and trends in hallucinogen consumption among lifetime users in the United States between 2015 and 2019. *Frontiers in Epidemiology*, *2*. https://doi.org/10.3389/fepid.2022.876706

Elliott, H., Mulrennan, M., & Cuerrier, A. (2021). Resurgence, refusal, and reconciliation through food movement organizations: A case study of food secure Canada's 2018 assembly. *Journal of Agriculture Food Systems and Community Development*, *10*(3), 1–21. https://doi.org/10.5304/jafscd.2021.103.009

Everett, S., & Aitchison, C. (2008). The role of food tourism in sustaining regional identity: A case study of Cornwall, southwest England. *Journal of Sustainable Tourism*, *16*(2), 150–167. https://doi.org/10.2167/jost696.0

Fessler, D., & Navarrete, C. (2003). Meat is good to taboo: Dietary proscriptions as a product of the interaction of psychological mechanisms and social processes. *Journal of Cognition and Culture*, *3*(1), 1–40. https://doi.org/10.1163/156853703321598563

Figueroa-Helland, L., Thomas, C., & Aguilera, A. (2018). Decolonizing food systems: Food sovereignty, Indigenous revitalization, and agroecology as counter-hegemonic movements. *Perspectives on Global Development and Technology*, *17*(1–2), 173–201. https://doi.org/10.1163/15691497-12341473

Gamble, L. (2017). Feasting, ritual practices, social memory, and persistent places: New interpretations of shell mounds in southern California. *American Antiquity*, *82*(3), 427–451. https://doi.org/10.1017/aaq.2017.5

Geng, Y., Hu, G., Ranjitkar, S., Shi, Y., Zhang, Y., & Wang, Y. (2017). The implications of ritual practices and ritual plant uses on nature conservation: A case study among the Naxi in Yunnan province, southwest China. *Journal of Ethnobiology and Ethnomedicine*, *13*(1). https://doi.org/10.1186/s13002-017-0186-3

Ghosh-Jerath, S., Kapoor, R., Singh, A., Downs, S., Barman, S., & Fanzo, J. (2020). Leveraging traditional ecological knowledge and access to nutrient-rich Indigenous foods to help achieve SDG 2: An analysis of the Indigenous foods of Sauria Paharias, a vulnerable tribal community in Jharkhand, India. *Frontiers in Nutrition*, *7*. https://doi.org/10.3389/fnut.2020.00061

Gómez-Baggethun, E., Corbera, E., & Reyes-García, V. (2013). Traditional ecological knowledge and global environmental change: Research findings and policy implications. *Ecology and Society*, *18*(4), 72. https://doi.org/10.5751/es-06288-180472

Hackel, L., Coppin, G., Wohl, M., & Bavel, J. (2015). From groups to grits: Social identity shapes evaluations of food pleasantness. *SSRN Electronic Journal*. https://doi.org/10.2139/ssrn.2662835

Hoover, E. (2021). "Our own foods as a healing": The role of health in the Native American food sovereignty movement. *Journal for the Anthropology of North America*, *24*(2), 89–97. https://doi.org/10.1002/nad.12154

Ibrahim, Y., & Howarth, A. (2016). Contamination, deception and "othering": The media framing of the horsemeat scandal. *Social Identities*, *23*(2), 212–231. https://doi.org/10.1080/13504630.2016.1207512

Ikhajiagbe, B., Ogwu, M. C., Ogochukwu, O. F., Odozi, E. B., Adekunle, I. J., & Omage, Z. E. (2021). The place of neglected and underutilized legumes in human nutrition and protein security in Nigeria. *Critical Reviews in Food Science and Nutrition, 62*(14), 3930–3938. https://doi.org/10.1080/10408398.2020.1871319

Imarhiagbe, O., & Ogwu, M. C. (2022). Sacred groves in the Global South: A Panacea for sustainable biodiversity conservation. In S. C. Izah (Ed.), *Biodiversity in Africa: Potentials, threats and conservation.* Sustainable Development and Biodiversity (Vol. 29, pp. 525–546). Springer. http://doi.org/10.1007/978-981-19-3326-4_20

Jones, P. (2007). The Native American church, peyote, and health: Expanding consciousness for healing purposes. *Contemporary Justice Review, 10*(4), 411–425. https://doi.org/10.1080/10282580701677477

Kharuhayothin, T., & Kerrane, B. (2018). Learning from the past? An exploratory study of familial food socialization processes using the lens of emotional reflexivity. *European Journal of Marketing, 52*(12), 2312–2333. https://doi.org/10.1108/ejm-10-2017-0694

Kim, H. (2018). Contending visions of local agriculture in a non-agricultural state: Food sovereignty in Singapore. *Journal of Globalization Studies, 9*(2), 92–106. https://doi.org/10.30884/jogs/2018.02.06

Kumar, A., & Srivastava, K. (2011). Cultural and social practices regarding menstruation among adolescent girls. *Social Work in Public Health, 26*(6), 594–604. https://doi.org/10.1080/19371918.2010.525144

Liu, C. (2017). Family-based food practices and their intergenerational geographies in contemporary Guangzhou, china. *Transactions of the Institute of British Geographers, 42*(4), 572–583. https://doi.org/10.1111/tran.12178

Long, L. (2025). *Food and folklore.* Bloomsbury Academic. https://doi.org/10.1093/obo/9780197764381-0024

Lucas, L. (2022). Food, diversity, and cultural identity. *Contingent Horizons the York University Student Journal of Anthropology, 6*(1), 1–12. https://doi.org/10.25071/2292-6739.112

Maudrie, T. L., Nguyen, C. J., Wilbur, R. E., Mucioki, M., Clyma, K. R., Ferguson, G. L., & Jernigan, V. B. B. (2023). Food security and food sovereignty:

The difference- between surviving and thriving. *Health Promotion Practice*, *24*(6), 1075–1079. https://doi.org/10.1177/15248399231190366

McGovern, J., Burt, K., & Schwittek, D. (2021). Food for thought: Culturally diverse older adults' views on food and meals captured by student-led digital storytelling in the Bronx. *Urban Social Work*, *5*(1), 60–75. https://doi.org/10.1891/usw-d-18-00005

McGraw, J., & Krátký, J. (2017). Ritual ecology. *Journal of Material Culture*, *22*(2), 237–257. https://doi.org/10.1177/1359183517704881

Mintz, S., & Bois, C. (2002). The anthropology of food and eating. *Annual Review of Anthropology*, *31*(1), 99–119. https://doi.org/10.1146/annurev.anthro.32.032702.131011

Miočević, D., Kvasina, A., & Crnjak-Karanović, B. (2021). Cosmopolitanism and expatriate's preference for host country food: The conditional effects of experiential capital and retail development. *International Journal of Consumer Studies*, *46*(2), 676–688. https://doi.org/10.1111/ijcs.12719

Morell-Hart, S. (2012). Foodways and resilience under apocalyptic conditions. *Culture Agriculture Food and Environment*, *34*(2), 161–171. https://doi.org/10.1111/j.2153-9561.2012.01075.x

Moscato, E., & Ozanne, J. (2019). Rebellious eating: Older women misbehaving through indulgence. *Qualitative Market Research an International Journal*, *22*(4), 582–594. https://doi.org/10.1108/qmr-07-2018-0082

Nasution, Z. (2025). Traditional culinary marketing communication strategies in the face of foreign culinary trends. *Literacy International Scientific Journals of Social Education Humanities*, *4*(1), 292–300. https://doi.org/10.56910/literacy.v4i1.2148

Noll, S., & Murdock, E. (2019). Whose justice is it anyway? Mitigating the tensions between food security and food sovereignty. *Journal of Agricultural and Environmental Ethics*, *33*(1), 1–14. https://doi.org/10.1007/s10806-019-09809-9

Nyarota, M., Chikuta, O., Musundire, R., & Kazembe, C. (2022). Towards cultural heritage preservation through Indigenous culinary claims: A viewpoint. *Journal of African Cultural Heritage Studies*, *3*(1), 136–150. https://doi.org/10.22599/jachs.114

Obongodot, N. U., & Ogwu, M. C. (2023). Plant food for human health: Case study of Indigenous vegetables in Akwa Ibom State, Nigeria. In S. C. Izah, M. C. Ogwu, & M. Akram (Eds.), *Herbal medicine phytochemistry*. Reference Series in Phytochemistry (pp. 1–38). Springer. https://doi. org/10.1007/978-3-031-21973-3_2-1

Ogwu, M. C. (2019). Towards sustainable development in Africa: The challenge of urbanization and climate change adaptation. In P. B. Cobbinah, & M. Addaney (Eds.), *The geography of climate change adaptation in urban Africa* (pp. 29–55). Palgrave Macmillan. https://doi.org/10.1007/978-3-030-04873-0_2

Ogwu, M. C. (2023). Local food crops in Africa: Sustainable utilization, threats, and traditional storage strategies. In S. C. Izah & M. C. Ogwu (Eds.), *Sustainable utilization and conservation of Africa's biological resources and environment.* Sustainable Development and Biodiversity (Vol. 888, pp. 353–374). Springer. https://doi.org/10.1007/978-981-19-6974-4_13

Ogwu, M. C., Ogwu, H. I., & Kosoe, E. A. (2023). Plants used in the management and treatment of cardiovascular diseases: Case Study of the Benin People of Southern Nigeria. In S. C. Izah, M. C. Ogwu, & M. Akram (Eds.), *Herbal medicine phytochemistry*. Reference Series in Phytochemistry (pp. 1–31). Springer. https://doi.org/10.1007/978-3-031-21973-3_4-1

Ogwu, M. C., Ojo, A. O., & Osawaru, M. E. (2025). Quantitative ethnobotany of Afenmai people of Southern Nigeria: An assessment of their crop utilization, and preservation methods. *Genetic Resources and Crop Evolution, 72*, 5807–5829. https://doi.org/10.1007/s10722-024-02302-x

Ogwu, M. C., & Osawaru, M. E. (2022). Traditional methods of plant conservation for sustainable utilization and development. In S. C. Izah (Ed.), *Biodiversity in Africa: Potentials, threats and conservation.* Sustainable Development and Biodiversity (Vol. 29, pp. 451–472). Springer. http://doi.org/10.1007/978-981-19-3326-4_17

Ogwu, M. C., & Osawaru, M. E. (2023). Plants used in the management and treatment of male reproductive health issues: Case Study of Benin People of Southern Nigeria. In S. C. Izah, M. C. Ogwu, & M. Akram (Eds.), *Herbal medicine phytochemistry*. Reference Series in Phytochemistry (pp. 1–39). Springer. https://doi.org/10.1007/978-3-031-21973-3_56-1

Ogwu, M. C., Osawaru, M. E., Amodu, E., & Osamo, F. (2023). Comparative morphology, anatomy and chemotaxonomy of two *Cissus* Linn. species. *Brazilian Journal of Botany, 46*, 397–412. https://doi.org/10.1007/s40415-023-00881-0

Oldham, O., Newton, P., & Short, N. (2024). Land-based resistance: Enacting Indigenous self-determination and Kai sovereignty. *Elementa Science of the Anthropocene, 12*(1). https://doi.org/10.1525/elementa.2023.00118

Orea-Giner, A., Fusté-Forné, F., & Todd, L. (2024). The origin story: Behind the scenes of food festivals. *Event Management, 28*(4), 585–598. https://doi.org/10.3727/152599523x16957834460312

Osawaru, M. E., & Ogwu, M. C. (2023). Plants used in the management and treatment of female reproductive health issues: Case Study from Southern Nigeria. In S. C. Izah, M. C. Ogwu, & M. Akram (Eds.), *Herbal medicine phytochemistry*. Reference Series in Phytochemistry (pp. 1–37). Springer. https://doi.org/10.1007/978-3-031-21973-3_5-1

Pamantung, R., Mantau, M., Sahetapy, J., & Manangkot, V. (2021). Abstraction of Minahasan folklore in food tradition. *Linguistics and Culture Review, 5*(S2), 1172–1183. https://doi.org/10.21744/lingcure.v5ns2.1726

Patel, J., Durey, A., Naoum, S., Kruger, E., & Slack-Smith, L. (2021). Oral health education and prevention strategies among remote aboriginal communities: A qualitative study. *Australian Dental Journal, 67*(1), 83–93. https://doi.org/10.1111/adj.12890

Perry, M. (2017). Feasting on culture and identity: Food functions in a multicultural and transcultural Malaysia. *3l the Southeast Asian Journal of English Language Studies, 23*(4), 184–199. https://doi.org/10.17576/3l-2017-2304-14

Pimbert, M., & Claeys, P. (2024). *Food sovereignty*. Oxford Research Encyclopedia of Anthropology. https://doi.org/10.1093/acrefore/9780190854584.013.297

Plastow, N., Atwal, A., & Gilhooly, M. (2014). Food activities and identity maintenance in old age: A systematic review and meta-synthesis. *Aging and Mental Health, 19*(8), 667–678. https://doi.org/10.1080/13607863.2014.971707

Poirier, B., Soares, G., Neufeld, H., Hedges, J., Sethi, S., & Jamieson, L. (2024). Conceptualising the relationships between food sovereignty, food security, and oral health among global Indigenous communities: A scoping review. *Public Health Nutrition, 27*(1). https://doi.org/10.1017/s1368980024001198

Prasetyo, S. (2023). Harmony of nature and culture: Symbolism and environmental education in ritual. *Journal of Contemporary Rituals and Traditions, 1*(2), 67–76. https://doi.org/10.15575/jcrt.361

Punzi, V. (2025). Fresh twigs, drying blood, and popped corn: The ephemeral materiality of eastern minyag ritual objects. *Religions, 16*(5), 539. https://doi.org/10.3390/rel16050539

Qodim, H. (2023). Nature harmony and local wisdom: Exploring Tri Hita Karana and traditional ecological knowledge of the Bali Aga community in environmental protection. *Religious Jurnal Studi Agama-Agama Dan Lintas Budaya, 7*(1), 1–10. https://doi.org/10.15575/rjsalb.v7i1.24250

Quintero-Ángel, M., Mendoza, D., & Quintero-Angel, D. (2019). The cultural transmission of food habits, identity, and social cohesion: A case study in the rural zone of Cali-Colombia. *Appetite, 139*, 75–83. https://doi.org/10.1016/j.appet.2019.04.011

Riccio, T. (2022). Zhuiniu water buffalo ritual of the miao: Cultural narrative performed. *Religions, 13*(4), 303. https://doi.org/10.3390/rel13040303

Rivera-Toapanta, E., Kallas, Z., Čandek-Potokar, M., Gonzàlez, J., Gil, M., Varela, E., Faure, J., Cerjak, M., Urška, T., & Roig, J. (2021). Marketing strategies to self-sustainability of autochthonous swine breeds from different EU regions: A mixed approach using the world café technique and the analytical hierarchy process. *Renewable Agriculture and Food Systems, 37*(1), 92–102. https://doi.org/10.1017/s1742170521000363

Robinson, R., & Getz, D. (2013). Food enthusiasts and tourism. *Journal of Hospitality and Tourism Research, 40*(4), 432–455. https://doi.org/10.1177/1096348013503994

Rokach, A. (2020). Belonging, togetherness, and food rituals. *Open Journal of Depression, 9*(4), 77–85. https://doi.org/10.4236/ojd.2020.94007

Rondoni, C. (2022). Extractivism and unjust food insecurity for Peru's Loreto Indigenous communities. *Sustainability, 14*(12), 6954. https://doi.org/10.3390/su14126954

Rosenblum, J. (2010). *Food and identity in early rabbinic Judaism.* Cambridge University Press. https://doi.org/10.1017/cbo9780511730375

Ruelle, M. (2017). Ecological relations and Indigenous food sovereignty in Standing Rock. *American Indian Culture and Research Journal, 41*(3), 113–125. https://doi.org/10.17953/aicrj.41.3.ruelle

Sandoval, L., & Wathne, S. (2022). Food sovereignty movements. In D. A. Snow, D. della Porta, D. McAdam, & B. Klandermans (Eds.), *The Wiley-Blackwell encyclopedia of social and political movements* (pp. 1–4). John Wiley & Sons. https://doi.org/10.1002/9780470674871.wbespm561

Shattuck, A., Schiavoni, C., & VanGelder, Z. (2015). Translating the politics of food sovereignty: Digging into contradictions, uncovering new dimensions. *Globalizations, 12*(4), 421–433. https://doi.org/10.1080/14747731.2015.1041243

Shukla, A. (2024). Examining the role of intergenerational relations in food systems: Evidence from western India. *Progress in Development Studies, 24*(3), 234–251. https://doi.org/10.1177/14649934231215116

Siahaya, M., Hutauruk, T., Aponno, H., Hatulesila, J., & Mardhanie, A. (2016). Traditional ecological knowledge on shifting cultivation and forest management in east Borneo, Indonesia. *International Journal of Biodiversity Science Ecosystems Services and Management, 12*(1–2), 14–23. https://doi.org/10.1080/21513732.2016.1169559

Singer, A. (2019). The strategic, shifting work of market devices: Selective stories and oriented knowledge. *Cultural Sociology, 13*(2), 198–216. https://doi.org/10.1177/1749975519838597

Skinner, K., Hanning, R., Desjardins, E., & Tsuji, L. (2013). Giving voice to food insecurity in a remote Indigenous community in subarctic Ontario, Canada: Traditional ways, ways to cope, ways forward. *BMC Public Health, 13*(1). https://doi.org/10.1186/1471-2458-13-427

Sulistiyowati, E., Setiadi, S., & Haryono, E. (2022). Food traditions and biodiversity conservation of the Javanese community in Gunungkidul Karst, Yogyakarta Province, Indonesia. *Biodiversitas Journal of Biological Diversity, 23*(4). https://doi.org/10.13057/biodiv/d230443

Sumner, J., McMurtry, J., & Tarhan, D. (2023). Growing community sustenance: The social economy as a route to Indigenous food sovereignty. *Canadian Journal of Nonprofit and Social Economy Research, 14*(S1). https://doi.org/10.29173/cjnser535

Thayyib, M. (2021). The ecological insight of the Bunga' Lalang rice farming tradition in Luwu Society, South Sulawesi, Indonesia. *Journal of Ethnology and Folkloristics, 15*(1), 140–153. https://doi.org/10.2478/jef-2021-0008

Utami, N., Sayuti, S., & Jailani, J. (2021). Indigenous artifacts from remote areas, were used to design a lesson plan for preservice math teachers regarding sustainable education. *Heliyon*, *7*(3), e06417. https://doi.org/10.1016/j.heliyon.2021.e06417

Vereecken, C., Bobelijn, K., & Maes, L. (2004). School food policy at primary and secondary schools in Belgium-Flanders: Does it influence young people's food habits? *European Journal of Clinical Nutrition*, *59*(2), 271–277. https://doi.org/10.1038/sj.ejcn.1602068

Wahyudi, S., Rahmanto, A., & Naini, A. (2024). Kisarasa YouTube channel's gastronomic storytelling: A narrative paradigm study. *Jurnal Multidisiplin Madani*, *4*(7), 988–1000. https://doi.org/10.55927/mudima.v4i7.10370

Yakin, H., Totu, A., Lokin, S., Sintang, S., & Mahmood, N. (2022). Tamu: Its roles as a medium of cultural identity preservation among Sabah ethnic in the era of information technology and industrial revolution 4.0. *E-Bangi Journal of Social Science and Humanities*, *19*(5). https://doi.org/10.17576/ebangi.2022.1905.10

Yapici, S. (2018). Food and identity in central Asia. *Central Asian Survey*, *38*(3), 434–436. https://doi.org/10.1080/02634937.2018.1549398

Young, L., Shukla, S., & Wilson, T. (2024). Indigenous values and perspectives for strengthening food security and sovereignty: Learning from a community-based case study of *Misko-ziibiing* (Bloodvein River First Nation), Manitoba, Canada. *Frontiers in Sustainable Food Systems*, *8*. https://doi.org/10.3389/fsufs.2024.1321231

CHAPTER 16

Indigenous Foods in Urban and Diaspora Communities

Abstract

This chapter critically examines the preservation, transformation, and sociocultural significance of Indigenous foods within urban and diaspora communities. As people migrate and settle in new urban environments through voluntary relocation, forced displacement, or generational diaspora, Indigenous food systems often travel with them, evolving in response to new geographies, markets, and sociopolitical conditions. The chapter explores how traditional food knowledge, ingredients, and culinary practices are retained, adapted, or reimagined in these contexts, emphasizing resilience and innovation. It analyzes the role of food in maintaining cultural identity, fostering intergenerational continuity, and asserting belonging in multicultural and often marginalizing urban settings. Key themes include the accessibility and affordability of traditional ingredients, the rise of diaspora-owned food businesses, and how Indigenous cuisines are commodified or appropriated in global cities. The chapter also explores how dietary transitions, which are often prompted by acculturation, limited access, or economic pressures, may impact nutritional health and cultural cohesion among migrant populations. Drawing on interdisciplinary case studies from West African, Indigenous Mexican, Caribbean, and Pacific Islander communities, it illustrates the diversity of Indigenous food expressions and the challenges they face in urban and diasporic spaces. Notably, the chapter discusses community-led efforts such as urban gardening, cultural food festivals, and food sovereignty movements that aim to reclaim and revitalize Indigenous foodways. It concludes with policy and advocacy recommendations to protect cultural food heritage and promote sustainable, inclusive urban food systems that honor Indigenous contributions.

Keywords: Indigenous foods, diaspora communities, urban food systems, cultural resilience, culinary adaptation, food sovereignty, migration, identity, traditional knowledge, nutritional transition

16.1 Introduction

Urban and diaspora communities consist of individuals who have migrated from their native geographical locations to urban centers while often maintaining connections to their cultural and historical backgrounds. These communities may include Indigenous peoples who have relocated from their traditional lands due to sociopolitical and economic factors such as colonization, urbanization, and globalization. Consequently, these populations experience significant challenges accessing traditional foods and maintaining cultural practices. Indigenous food systems are critical for sustaining Indigenous peoples' identities, health, and community cohesion in urban and diaspora settings. Food systems rooted in Indigenous traditions provide nutritional sustenance and avenues for cultural expression, social connectivity, and political advocacy. Scholars have noted that the essence of Indigenous food sovereignty extends beyond food security; it encapsulates the right of Indigenous peoples to define their food systems and practices, which are deeply connected to their heritage and land (Levkoe et al., 2021; Sumner et al., 2023; Timler et al., 2019). This chapter posits that even with disconnection from traditional landscapes, Indigenous communities in urban environments are actively mobilizing to reclaim and preserve their food systems, often utilizing innovative strategies that blend cultural identity with food practices (Nikolaus et al., 2022; Shafiee et al., 2023).

The interdependence of food sovereignty and Indigenous identity is particularly pronounced in urban settings, where traditional practices may be challenged. For many Indigenous peoples, particularly those in the diaspora, the connection to food is intertwined with their historical narratives and cultural identities. Food serves not merely as a means of sustenance but as a bridge that links individuals to their lineage, community, and ancestral lands (Jernigan et al., 2023; Poirier et al., 2024). Research indicates that many members of urban Indigenous communities desire to reconnect with traditional foods; however, they often encounter barriers such as limited access and availability (Nikolaus et al., 2022; Richmond et al., 2020). This disconnection can contribute to dietary shifts that increase food insecurity and associated health issues within urban Indigenous populations (Nikolaus et al., 2022; Shafiee et al., 2023). Indigenous food sovereignty initiatives in urban areas, including community gardens and food-sharing programs, are crucial to enhancing cultural pride and community solidarity (Kuhnlein, 2014; Levkoe et al., 2021). For example, Timler et al. (2019) emphasize that food sovereignty requires empowering Indigenous communities to navigate their food systems and practices, regardless of urban pressures. These initiatives address food insecurity and function as platforms for cultural education and social cohesion, essential for the resilience of Indigenous communities confronting modern challenges (Judge et al., 2022; Phillipps et al., 2022).

Moreover, the role of women in these food sovereignty movements cannot be understated. Women frequently act as vital transmitters of traditional food knowledge and practices. As Phillipps et al. (2022) noted, urban Indigenous women are reclaiming their roles in food culture, thus fostering identity justice and advocating for environmental sustainability. Highlighting women's leadership in food initiatives underscores the necessity to integrate gender perspectives in discussions on Indigenous food sovereignty, as they are integral to nurturing community resilience through food practices. Urbanization significantly impacts the food practices of Indigenous peoples, presenting multiple challenges that impede their capacity to sustain food sovereignty. The geographical discon-nection from traditional lands restricts access to culturally significant foods, and dietary alterations may lead to adverse health consequences, particularly increasing vulnerability to chronic diseases (Jernigan et al., 2023; Nikolaus et al., 2022). Decolonizing food systems remains crucial for fostering greater equity in urban spaces. Indigenous communities have called for an intersectional policy approach to food access, recognizing the diverse cultural, economic, and social factors that shape their experiences (Phillipps et al., 2022). Such policies should promote access to traditional foods and incorporate Indigenous knowledge and practices into broader urban food systems, thereby acknowledging food's central role in fostering health equity and social justice (Maudrie et al., 2021; Nikolaus et al., 2022).

This chapter aims to explore how Indigenous food systems are preserved, adapted, and transformed within urban and diaspora communities, focusing on the interplay between cultural identity, migration, and food practices. It critically examines the sociocultural, economic, and political factors that influence access to traditional ingredients, the continuity of culinary knowledge, and the resilience of Indigenous dietary patterns in nontraditional settings. The chapter investigates the nutritional transitions and health implications associated with migration and urbanization while highlighting community-driven strategies such as urban agriculture, cultural food festivals, and food justice movements supporting the revitalization of Indigenous foodways. Drawing on global case studies, it evaluates how Indigenous food practices are maintained, reinterpreted, or commodified in diverse urban contexts and offers practical insights for policy and advocacy aimed at fostering inclusive and sustainable food systems. In contributing to the broader sociobiology of Indigenous foods, this chapter emphasizes the coevolution of human social structures and dietary behaviors. It expands sociobiological discourse beyond ancestral environments by situating food within contemporary diasporic and urbanized landscapes, where biocultural adaptation, identity negotiation, and sociopolitical dynamics converge. By analyzing food as both a biological necessity and a cultural-symbolic system, the chapter illustrates how Indigenous foodways serve as resilient, cohesive, and resistance mechanisms. Ultimately, it reinforces the importance of understanding food not merely as sustenance but in fact as a living archive of evolutionary, ecological, and social processes embedded in human history and ongoing adaptation.

16.2 Migration, Urbanization, and the Displacement of Indigenous Food Systems

The intricate relationships between migration, urbanization, and the displacement of Indigenous food systems reveal profound historical and contemporary dynamics influencing food traditions. These relationships operate on multiple levels, ranging from socioeconomic constraints to cultural resilience, shaping dietary practices and community identities. Historically, migration has significantly altered Indigenous food systems. Indigenous peoples often relocated due to colonial pressures, industrial developments, or forced assimilation policies, leading to an erosion of their traditional sources of food. Shafiee et al. (2024a, 2024b) highlight that urban Indigenous populations face various barriers to maintaining their dietary practices in urban settings, citing economic constraints and limited access to traditional foods as primary hindrances. Richmond et al. (2020) elaborate on this point by discussing how food environments reflect individual choices and are deeply embedded in political and social contexts that dictate access to traditional food sources. The authors argue that Indigenous food systems are fundamentally connected to place, income, and social ties, which are often disrupted when communities migrate to urban areas.

Contemporary patterns of urbanization exacerbate these historical challenges. Many Indigenous peoples find themselves disconnected from their traditional food sources, leading to dietary shifts characterized by reliance on processed foods that are readily available but nutritionally inadequate. Urban developments can also encroach upon traditional lands where food is gathered, further distancing Indigenous communities from their ancestral practices. Moreover, the psychological impacts of historical trauma compound these issues, as highlighted by

Shafiee et al. (2024a), who note the internal and community dynamics affecting dietary choices among urban Indigenous persons in Saskatchewan (Canada). Transformation rather than absolute loss of food traditions can also be observed in diasporic contexts. The diaspora can be a site of cultural dislocation and a space for cultural resilience. For instance, the work of Chariyev et al. (2025) reflects how traditional food can transform, blending with local ingredients and styles to adapt to new environments while maintaining core cultural elements. In diasporic communities globally, culinary practices often evolve, underscoring a balance between retention of identity and adaptation to new cultural landscapes. Mehta (2005) promotes this idea, indicating that the semiotics of food can illuminate how African and Caribbean diasporas use culinary practices as links to cultural identity amid their exilic experiences. Urbanization has influenced the structural conditions under which Indigenous dietary practices are sustained and prompted new frameworks of solidarity and identity. Askew (2021) examines how thoughtful engagement with food practices can bridge gaps created by urban settings, creating a sense of community and belonging even in unfamiliar environments. The interplay between cuisine and identity becomes increasingly complex in urban contexts, where food signifies a connection to heritage while engaging with contemporary food environments.

The tensions between loss and transformation of Indigenous food systems in urban and diasporic landscapes necessitate a nuanced understanding of culinary practices as sites of negotiation. For instance, the Warlpiri matriarchs (Nangala in the Warlpiri traditional kinship system [Australia]) adapt their practices while retaining deep-rooted connections to their cultural identity despite being uprooted from their traditional homelands. This adaptation serves as a survival mechanism and becomes critical to their identity formation in a new context. The preservation and transformation of traditional diets also hint at broader sociopolitical narratives. Income levels, social connections, and cultural heritage impact how Indigenous peoples negotiate their identities through food as they navigate urban environments (Richmond et al., 2020). Also, food production and consumption become acts of resistance and affirmation of cultural identity amid modern constraints. Abdallah et al. (2019) present a case study of Lebanese food traditions in London, illustrating how diasporic identity remains deeply linked to homeland cuisine, thus fostering a sense of community among migrants while adapting to new culinary landscapes.

Nonetheless, not all transformations are beneficial or empowering. Urbanization often creates food deserts where affordable, nutritious, and culturally relevant foods become scarce, exacerbating health disparities among Indigenous populations.

Davies et al. (2023) advocate for developing validated dietary assessment methods tailored for Aboriginal and Torres Strait Islander Australians, emphasizing that these communities experience a disproportionate burden of health issues related to nutrition. The rapid change in dietary patterns among urban Indigenous populations can lead to adverse health outcomes such as obesity and diabetes, underscoring the link between urbanization, dietary practices, and health. From the perspective of the global diaspora, food emerges not just as sustenance but as a form of cultural capital that endures despite geographical displacements. Research suggests that cultural expressions through cuisine fortify communal bonds among diasporic populations. For example, Thomas's (2004) examination of Vietnamese diaspora food practices underscores the shifts in dietary habits and social structure in contexts both at home and abroad, capturing the essence of transitioning tastes amid ongoing globalization. The interactions between migration, urbanization, and the displacement of Indigenous food systems reveal a complex tapestry of loss, adaptation, and resilience. These processes shape not only dietary practices but also societal and cultural identities, underscoring the need for careful consideration of how urban environments and diasporic contexts can foster or hinder the survival of Indigenous culinary traditions. Ultimately, the legacy of these shifts continues to challenge Indigenous communities as they navigate their identities in an increasingly urbanized world.

16.3 Retention and Adaptation of Indigenous Food Practices in Urban and Diaspora Communities

The retention and adaptation of Indigenous food practices in urban and diaspora communities through oral practices, communal cooking, food festivals, and so on reflect a vibrant interplay of cultural preservation mechanisms that are critical in maintaining cultural identities, especially as urban environments pose unique challenges and opportunities for Indigenous peoples (Table 16.1). Oral history plays a vital role in retaining Indigenous food practices, promoting communal bonds, and facilitating the transmission of traditional knowledge across generations. The Indigenous Foods Knowledges Network emphasizes that storytelling is essential for recreating culinary practices and that it enhances understanding of food sovereignty initiatives within urban settings (Jäger et al., 2025). Miltenburg et al. (2023) highlight how urban Indigenous communities proactively revitalize these practices, enabling them to sustain cultural heritage despite modern challenges.

Table 16.1: Mechanisms of Retention and Adaptation of Indigenous Food Practices in Urban and Diaspora Communities

Practice	Function	Examples
Oral History and Storytelling	Transmits cultural knowledge and culinary traditions intergenerationally	Narratives of food origins, ceremonial recipes, and storytelling circles
Communal Cooking	Strengthens community ties and intergenerational learning	Cooking workshops, youth-led kitchen programs, and shared meals
Food Festivals	Celebrates identity and adapts traditions for broader visibility	Events showcasing fusion dishes and traditional meals, often in urban markets
Culinary Syncretism	Fosters hybrid culinary identities and ensures cultural relevance	Mixing Indigenous ingredients with global dishes (e.g., taro burgers, bannock tacos, etc.)
Intergenerational Transmission	Preserves cultural memory and sustains practices through lived experience	Elders mentoring youth, recipe documentation, inter-family food projects
Community-Led Food Projects	Reclaims food sovereignty and addresses food insecurity	Community gardens, Indigenous farmers' markets, and seed-saving initiatives
Nutritional Revitalization	Links traditional diets to modern health and sustainability goals	Reintroducing ancestral superfoods, youth health campaigns using traditional recipes
Cultural Relearning and Innovation	Engages urban Indigenous populations in reimagining food heritage	Educational programs, culinary incubators, and integration of digital platforms for food education

Communal cooking is another significant method facilitating the retention and adaptation of Indigenous food practices. This practice reinforces community ties and serves as a platform for intergenerational learning. Shafiee et al. (2024) note that communal cooking events create opportunities for reviving traditional foods and adapting them to contemporary preferences, ensuring that urban Indigenous diets retain cultural richness. These events often engage younger community members, blending traditional culinary techniques with modern tastes and promoting a culturally vibrant exchange. Food festivals also serve as expressions of Indigenous identity and are crucial for adapting food practices. Community gatherings around food reinforce cultural narratives and embrace modern influences. This dynamic is evident in events showcasing traditional dishes alongside fusion foods, allowing a dialogue between heritage and innovation that responds to evolving tastes and nutritional needs (Jernigan et al., 2023). Culinary syncretism reflects the creation of hybrid recipes that merge Indigenous culinary practices with those of other cultures in urban settings. Jäger et al. (2025) argue that adapting traditional foods is essential for maintaining cultural relevance and resilience in changing socioeconomic environments. For example, Urban 'Āina (Indigenous Hawaiian) practices demonstrate how traditional ingredients can be integrated into contemporary dishes, rejuvenating urban food landscapes and addressing modern issues like food security (Deluze et al., 2023).

Intergenerational transmission of food practices is integral to preserving cultural memory among Indigenous communities in urban environments. Richmond et al. (2020) stress that the complexities of urban food environments necessitate innovative pedagogical approaches that align traditional practices with current socioeconomic landscapes. This exchange between generations fosters cultural preservation while meeting contemporary health and wellness considerations. Urban Indigenous populations face numerous barriers, including limited access to traditional food sources and historical trauma's psychological impacts. However, proactive efforts to sustain these practices underscore their resilience (Jernigan et al., 2023). Navigating the influences of modern urban life alongside traditional sustenance methods allows food to be a form of cultural expression that is both adaptive and resistant (Wilson et al., 2021).

Community-led projects on food sovereignty address contemporary food insecurity while revitalizing Indigenous diets by reintroducing traditional ingredients and methods (Miltenburg et al., 2022). This reclamation is crucial given the increasing interest in ethnic foods, which risk commodifying traditional practices and necessitate balancing cultural visualization with authenticity (Ganesan et al., 2020; Shafiee et al., 2024). As Indigenous cultures intersect with urban environments, a new culinary identity emerges responsive to historical legacies and contemporary realities. The process of unlearning and relearning reflects a critical engagement with urban food systems, emphasizing the importance of addressing both struggles and innovations in food practices (Overend & Rai, 2024). Urban Indigenous communities leverage their heritage to create spaces where traditional food practices can flourish, such as community gardens, farmers' markets, and food cooperatives, which strengthen community ties and support collaborative efforts combining traditional and modern agricultural techniques (Howard-Bobiwash et al., 2021; Miltenburg et al., 2022).

Furthermore, the renewed interest in the nutritional benefits of traditional foods plays a significant role in reengaging Indigenous youth, who often led initiatives to revive these practices. Through such ventures, they gain practical skills and cultivate pride in their cultural identities while aligning traditional food practices with contemporary health dialogues focused on nutrition and sustainability (Shafiee et al., 2024). The processes of retention and adaptation of Indigenous food practices in urban and diaspora communities highlight a dynamic interplay of oral history, communal cooking experiences, celebrations, and intergenerational learning. Each element is critical in enhancing cultural vitality and resilience amid urban challenges. As these communities continue to innovate while honoring their past, they weave a unique tapestry of Indigenous culinary identity that enriches urban life and fosters collective healing and empowerment (Obongodot & Ogwu, 2023).

16.4 Socioeconomic and Political Dimensions of Indigenous Foods in Urban and Diaspora Communities

The socioeconomic and political dimensions of Indigenous foods in urban and diaspora communities are complex, encompassing many factors ranging from access to traditional ingredients to legal and policy challenges.

16.4.1 Access to Traditional Ingredients in Urban Markets

Urban markets often present significant challenges and opportunities for Indigenous communities seeking traditional foods. The availability of Indigenous ingredients can vary widely, often hindered by urban planning and public health policies that do not prioritize culturally specific food practices. For instance, Shafiee et al. (2024) emphasize the importance of developing supportive environments for traditional food practices, advocating for culturally specific food markets and community gathering spaces to enhance access to traditional ingredients in urban settings. Such spaces facilitate access to foods integral to Indigenous diets and foster community relationships, which are crucial for the intergenerational transfer of traditional knowledge. Additionally, the accessibility of traditional foods is also impacted by vendors operating within urban landscapes. Traditional food markets serve as critical nodes for food access among urban populations, yet they often face threats from urban development and changing consumer preferences that favor industrialized food systems (Ogwu, 2023; Ogwu & Osawaru, 2022; Osawaru & Ogwu, 2020). The competition posed by supermarkets and formal food retail sectors can diminish the visibility and viability of traditional food vendors, making it imperative for policies to support the integration of Indigenous foods into urban food systems (Cook et al., 2024). Moreover, the socioeconomic context of urban Indigenous peoples significantly affects their access to traditional foods. According to Wanyama et al. (2019), low-income households in urban slums often encounter barriers that limit their dietary diversity and access to fresh produce, further complicating the food security challenges these communities face. Thus, while there is a distinct demand for Indigenous foods, structural inequalities manifesting as economic disadvantages can inhibit access to these crucial food sources.

16.4.2 Economic Role of Indigenous Food Businesses and Vendors

Indigenous food businesses play a pivotal role in the economic landscape of urban environments. They enhance food choices available to Indigenous communities and contribute to local economies by creating jobs and promoting entrepreneurship among Indigenous peoples. A study by Aryeetey et al. (2016) found that traditional markets can foster livelihoods, especially in contexts with limited formal employment options. This entrepreneurial aspect is crucial as many Indigenous vendors capitalize on their cultural heritage to offer unique food products that connect communities to their roots (Blennerhassett et al., 2021; Egboduku et al., 2024). Moreover, traditional food markets are vital for food sovereignty—a core principle among Indigenous communities aimed at ensuring access to traditional food systems. Urban Indigenous food businesses provide economic opportunities and serve as platforms for cultural expression and the preservation of traditional diets, facilitating food security and community resilience (Gendron et al., 2016). These businesses embody a dual role: They function economically while reinforcing cultural identities and practices.

Despite these economic benefits, Indigenous food businesses and vendors often confront challenges ranging from regulatory barriers to limited access to capital (Obahiagbon et al., 2023a, 2023b, 2024a, 2024b). For instance, legal restrictions and urban food policies may impose operational hurdles that disproportionately impact Indigenous vendors, limiting their growth potential and accessibility to larger markets (Phillipps et al., 2022). The intersectionality of race, gender, and socioeconomic status further complicates the ability of urban Indigenous women, who are often tasked with managing food procurement for their families, to maintain access to traditional foods (Judge et al., 2022).

16.4.3 Legal and Policy Challenges: Import Restrictions and Urban Food Regulations

Legal and policy challenges present significant obstacles to accessing traditional Indigenous foods in urban areas, underscoring the need for policymakers to understand the implications of existing regulations on food access for Indigenous communities. One profound barrier is the regulatory framework governing wild foods, which is frequently designed without considering the unique needs and practices of Indigenous peoples (Kosoe et al., 2023). This misalignment between regulations and cultural practices can exacerbate food insecurity among urban Indigenous populations, who often rely on these foods for nutrition and cultural identity. Compounding these challenges are issues surrounding import restrictions that limit the availability of traditional ingredients from rural areas or Indigenous lands. Legal frameworks prioritizing industrial and agricultural outputs while marginalizing Indigenous food systems can diminish the economic viability of traditional food practices in urban centers (Gendron et al., 2016). Furthermore, food insecurity among Indigenous peoples that impose restrictions can severely affect accessibility and contribute to a decline in traditional knowledge, necessitating community-led efforts to reclaim and revitalize Indigenous food systems (Elliott et al., 2012). Additionally, sociopolitical landscapes influence regulatory practices, particularly in urban environments where gentrification and neoliberal policies often reshape food access. Elliott et al. (2012) underscore how these policies fall short in addressing the nuanced realities of Indigenous food insecurity, contributing to a cycle of disenfranchisement for urban Indigenous peoples. Policy analysis must consider the historical context of dispossession and marginalization experienced by Indigenous populations, proposing interventions that support the reintegration of traditional food systems into contemporary urban life.

16.5 Health, Nutrition, and Food Sovereignty in Diaspora Settings

The health, nutrition, and food sovereignty in diaspora settings represent multi-dimensional discussions that encompass historical, sociopolitical, and cultural narratives. These issues become increasingly significant when considering Indigenous populations in urban and diasporic contexts. This discourse is particularly relevant in studying how Indigenous diets influence health outcomes, highlighting the ongoing challenges tied to nutritional resilience and transitions within these communities. The dietary patterns of Indigenous populations significantly affect their health outcomes. Several studies suggest that these populations often experience dietary transitions that align with broader trends in urban integration and globalization, leading to maladaptive health outcomes such as obesity and diabetes. Notably, a study conducted by Hodge et al. (2011) indicated that Indigenous Australians face health disparities linked to inadequate diets tied to low socioeconomic status and access to healthy food options. Similarly, a comparative analysis by Roldán Amaro et al. (2017) highlighted significant health service coverage gaps in marginalized Indigenous populations compared to their urban counterparts, suggesting that these disparities are starkly pronounced in dietary and nutritional contexts.

Furthermore, nutritional resilience emerges prominently in discussions about urban Indigenous communities. While urbanization and globalization lead to rapid dietary changes, many Indigenous individuals and groups strive for nutritional resilience by reclaiming traditional food cultivation and consumption practices. The efforts are underscored by movements advocating for food sovereignty, reflecting a desire to halt the detrimental shift toward a diet based predominantly on processed and nutrient-poor foods. For instance, the work by Polanco and Rodríguez-Cruz (2019) emphasizes the reclamation of Taíno foodways as a method of cultural preservation and health improvement through traditional diets like yuca and taro. This raises critical discourse about how Indigenous dietary practices can reassert cultural identity while promoting health in diasporic settings. Food sovereignty emerges at the intersection of diet, culture, and political activism. It embodies the struggles of Indigenous peoples to regain control over food systems historically undermined by colonial practices. Thus, groups advocating for food sovereignty are seen as fighting for better nutrition, cultural recognition, and self-determination. Gentles-Gibbs and Murphy (2025) address this intersection by discussing the importance of community-centered practices rooted in decolonization for establishing sustainable, health-promoting systems. Such frameworks allow urban Indigenous populations to assert their identities through localized food systems that honor ancestral practices and knowledge.

Alongside these efforts, decolonizing diets is integral to reshaping food narratives within diasporic communities. The reevaluation of traditional food systems as acts of resistance against colonialism has gained momentum. For example, Olivella and Tittler (2021) showcase how reclaiming traditional diets mirrors broader decolonization efforts, encapsulating the collective yearning for cultural resurgence and health equity. This decolonization process is about returning to traditional foods and acknowledging and reversing the impacts of colonial rule on food systems. Therefore, the movements for food justice emphasize ethical and sustainable consumption that aligns with Indigenous identities and histories. As diaspora communities navigate the complexities of identity, the intersectionality of food, culture, and health outcomes becomes even more apparent. The marginalized experiences of these populations highlight how diaspora settings often exacerbate issues surrounding health due to limited access to traditional food sources and culturally appropriate health services. Research indicates that Indigenous people living in urban environments face many challenges, such as cultural alienation from food sources aligned with their heritage (Roldán Amaro et al., 2017). Such challenges are compounded by the high costs of traditional foods, often leading to reliance on unhealthy alternatives that worsen health disparities, as reflected in the study by Hodge et al. (2011).

Nutritional transitions in diasporic settings can also reflect shifts in cultural practices and identity among Indigenous peoples. While urban environments might offer opportunities for integrating new foods and dietary practices, they can dilute longstanding culinary traditions simultaneously. Aydın's (2024) analysis of the political implications of diaspora suggests that reestablishing cultural identity often necessitates reclaiming nutritional practices that reflect Indigenous heritage, thereby fostering a stronger community connection and health resilience. Thus, nutrition transitions in diasporic contexts underscore the importance of integrating cultural identity into health practices to navigate contemporary health challenges effectively. The collective actions for food justice, alongside decolonizing diets, illustrate a significant response to colonial frameworks that have historically marginalized Indigenous peoples in discussions about health and nutrition. Mishra (2017) notes that identity crises in diasporic communities often stem from alienation, affecting their dietary choices and health outcomes. In this complex interplay of identity, culture, and nutrition, initiatives for food justice gain their urgency. Many Indigenous groups champion the food sovereignty movement, advocating for the right to healthy, culturally appropriate food produced through ecologically sound and sustainable methods (Polanco & Rodríguez-Cruz, 2019).

Moreover, decolonizing education further enhances the discourse surrounding food sovereignty and health in diaspora settings. By integrating Indigenous knowledge systems into educational frameworks, communities can empower newer generations to acknowledge and value their culinary traditions as part of their identity (Lewis, 2018). This reawakening fosters resilience in dietary habits and ensures the continuity of Indigenous knowledge systems that are critical for future generations. In discussing the role of collective memory and cultural practices, revitalizing Indigenous culinary traditions among diasporic populations can emerge as a pivotal strategy for promoting health and identity. The analysis by Suzuki (2019) underscores the importance of rituals and collective memory in food practices, which is essential for maintaining a sense of belonging and heritage among diasporic individuals. By engaging in traditional food practices, urban Indigenous populations fortify their health through access to nutritious food and reclaim agency over their culinary histories.

16.6 Media, Identity, and Culinary Representation

In the modern context of culinary representation, various platforms such as social media, cookbooks, and food documentaries serve as vital mediums for Indigenous communities to express and reinforce their cultural identities. Social media, in particular, has emerged as a powerful tool for Indigenous youths, allowing them to explore their identities and connect with others with similar backgrounds and experiences. This connectivity is critical in a world where cultural disconnection can often lead to a dilution of identity, especially among young Indigenous people facing challenges associated with modernity and globalization. According to Rice et al. (2016), social media facilitates creative expression and fosters a sense of community among Indigenous individuals, highlighting their shared experiences and cultural practices. It acts as a platform wherein traditional culinary practices can be shared widely and appreciated, enhancing visibility for Indigenous foods.

Furthermore, food is an essential vector through which cultural identity can be articulated and celebrated. Errmann et al. (2023) argue that food choices can reflect ethical and cultural identity, suggesting that culinary practices reflect and shape moral frameworks surrounding consumption. This idea underscores the importance of Indigenous foods in social media discussions, where these foods are not only shared for their taste but also for their significance as cultural symbols that convey deeper narratives of history, ethics, and belonging. Thus, showcasing Indigenous dishes online contributes to a broader understanding of Indigenous identity while challenging preconceived notions surrounding culinary traditions.

In examining diaspora chefs and cultural advocates, the case studies reveal how these individuals utilize traditional foods to bridge cultural gaps and advocate for broader social issues. For instance, in various communities, chefs of Indigenous descent leverage their platforms to educate others about the significance of their culinary heritage. According to Gichunge et al. (2016), traditional foods are perceived as a core component of cultural identity, and discussions about them reflect the persistent connection of individuals from diaspora communities to their roots. These diaspora chefs often navigate the complexities of identity within multicultural contexts, reinforcing their cultural narratives through the foods they prepare and present. Moreover, the interplay between food, identity, and community can significantly impact social cohesion among Indigenous peoples. As Scott et al. (2019) note, the structural influences of marketing and media play a vital role in defining food-related behaviors among youth, which can either support or undermine cultural identity. When culinary representations are dominated by commercial interests that redefine Indigenous foods solely as exotic or niche, it can create a divide that undermines the authentic representation of these cultural practices. This highlights the crucial role of authentic portrayal by diaspora chefs who strive to bring cultural significance lost in mainstream representations back into focus.

Additionally, issues of cultural food security come to the forefront when discussing Indigenous identity and culinary practices. Wright et al. emphasize that access to cultural foods is integrally tied to mental well-being and identity affirmation, particularly among second-generation immigrants (Wright et al., 2021). These access issues pertain not only to physical availability but also to the socioeconomic frameworks that enable or restrict the ability of individuals and communities to partake in their culinary traditions. Interventions highlighting the importance of cultural foods and their connection to identity can enhance community resilience and foster a stronger sense of self among Indigenous populations. The role of food as a symbol of cultural identity is critical in reinforcing community ties and continuity. İflazoğlu and Aksoy (2023) highlight that iconic foods within a region can encapsulate the historical, geographical, and social dimensions integral to a community's identity. This encapsulation illustrates how foods carry narratives encompassing traditions, values, and shared histories. Within these narratives, diasporic individuals and communities find their stories reflected, offering a constructive space to negotiate their identities in a transnational context.

In culinary representation, documents and cookbooks serve as archives of cultural heritage that can influence individual and community identity. By compiling traditional recipes and practices, food authors accentuate the importance of these culinary traditions in constructing a shared cultural identity. Culinary texts often intertwine personal narratives with historical contexts, which reinvigorate and preserve cultural practices for future generations. This selective preservation inherently shapes communal identity by reinforcing what is considered valuable and representative of a particular culture.

Therefore, social media, cookbooks, and food documentaries significantly intertwine with culinary identity and representation. By utilizing these platforms, Indigenous peoples can reclaim their narratives against the backdrop of pervasive stereotypes and commodification. For instance, the increased portrayal of Indigenous foods in popular media acts not only as a celebration of these culinary practices but also as a platform for discourse on contemporary issues pertaining to identity politics, colonization, and cultural validity. The multicultural lens through which many diaspora chefs approach their culinary landscapes helps to build a dialogue surrounding cultural preservation and food politics, which provides insight into the lived experiences of Indigenous peoples in a globalized world. Consequently, these representations invite broader societal engagement and understanding, fostering an appreciation for the rich culinary histories embedded within Indigenous foods. This process also illuminates the ongoing struggles for recognition and respect among Indigenous communities, necessitating critically examining the narratives told and shared across social media and other platforms. Illustrating these narratives encourages a more profound respect for how food encapsulates culture, solidarity, and the profound human connections fostered through shared meals and culinary traditions.

16.7 Case Analysis of Indigenous Foods in Urban and Diaspora Communities

Indigenous foodways are powerful carriers of cultural memory, resilience, and adaptive strategies in global urban and diasporic contexts. This section analyzes six key case studies where Indigenous food traditions rooted in specific bioregions, spiritual beliefs, and intergenerational knowledge systems are preserved, reinterpreted, and institutionalized in cities worldwide. These culinary landscapes reveal the sociobiological importance of food as a medium through which communities maintain biocultural continuity, assert collective identities, and respond to displacement and globalization.

16.7.1 West African Foods in New York, London, and Accra

In metropolitan centers like New York and London, West African culinary traditions have created diasporic spaces where food is a sensory link to ancestral homelands. In the Bronx, Accra Restaurant (Figure 16.1) and Papaye are emblematic of Ghanaian diasporic identity, serving dishes like jollof rice, fufu, and light soup—each rooted in Indigenous ecological knowledge and traditional cooking techniques (Adjaye, 2017). These restaurants nourish immigrant communities and function as sites of cultural education and economic agency. In London, neighborhoods like Peckham—often called "Little Lagos"—are home to popular establishments like 805 Restaurant and Enish, which serve Nigerian classics such as egusi soup and amala. Meanwhile, in Accra, a rapidly urbanizing capital, restaurants such as Buka and Santoku are redefining local foods through fine dining models, appealing to middle-class and international clientele. This circular exchange illustrates a feedback loop of culinary globalization, where diasporic reinterpretations influence food aesthetics and consumption patterns in their places of origin (Obeng-Odoom, 2015).

Figure 16.1: The Front of Accra Restaurant in Bronx (New York), United States

16.7.2 Indigenous Mexican Foods in U.S. Cities

Mexican Indigenous foodways, deeply rooted in Mesoamerican civilizations such as the Nahua, Zapotec, and Mixtec, continue to thrive and adapt in U.S. urban centers. In Los Angeles, Guelaguetza has become a cultural institution by preserving Oaxacan traditions through dishes like mole negro, tlayudas, and memelas, grounded in locally grown heirloom corn varieties (Figure 16.2). These dishes maintain spiritual and seasonal dimensions linked to Indigenous agricultural calendars and rituals (Pilcher, 2012). In the Bronx, Tlaxcalli explicitly references the Indigenous state of Tlaxcala and uses hand-ground masa to create authentic tortillas and tamales, reflecting an effort to resist the standardization of Mexican cuisine. Beyond commerce, these food spaces empower Indigenous migrants—many undocumented and marginalized—through employment, language preservation, and culinary storytelling. They also serve broader educational roles, introducing non-Indigenous consumers to the complexity and sustainability of traditional Mesoamerican diets.

Figure 16.2: Entrance of the Guelaguetza Restaurant (Los Angeles), United States

16.7.3 Pacific Islander Food Traditions in Urban Oceania and the United States

Pacific Islander cuisines—including Native Hawaiian, Samoan, Tongan, and Fijian traditions—exemplify resilience in maintaining Indigenous ecological relationships, despite profound colonial disruptions. In Honolulu, Highway Inn (Figure 16.3), established in 1947, continues to serve traditional Hawaiian meals such as laulau (steamed pork wrapped in taro leaves), poi (fermented taro paste), and lomi salmon—each drawing on ahupua'a land division systems that integrated mountains, rivers, and sea (Kanahele, 1986). These foods are not merely nostalgic but are actively embedded in sovereignty movements that challenge land dispossession and promote Native Hawaiian food self-determination. In Auckland, Kai Pasifika showcases a regional mix of Samoan oka (raw fish salad), Tongan lu pulu (taro leaves in coconut cream), and Fijian curries, celebrating oceanic biodiversity and traditional food preservation methods. Pacific Islander restaurants are often integrated into broader efforts in urban agriculture, seed saving, and youth culinary training, linking cultural revival with food sovereignty.

Figure 16.3: The Front of the Highway Inn in Honolulu (Hawaii, United States)

16.7.4 Afro-Caribbean Foodscapes in Canadian Urban Centers

Afro-Caribbean food traditions in Canadian cities like Toronto and Montreal are vibrant expressions of diasporic identity, memory, and resistance. In Toronto's Little Jamaica, establishments such as Randy's (Figure 16.4), The Real Jerk and Rasta Pasta serve traditional Jamaican dishes like ital stew, curried goat, and ackee and saltfish. These foods reflect Rastafarian and Maroon culinary systems, which emphasize self-sufficiency, veganism, and spiritual purity. Their presence in the urban foodscape resists the erasure of Caribbean histories and fosters solidarity across Black diasporas (Cummings, 2020). In Montreal, Boom J's Cuisine brings Grenadian and Trinidadian food cultures to the fore, offering doubles, oil down, and pelau, alongside cultural programming that includes poetry readings, drumming, and language workshops. These spaces act as "third places"—neither home nor work—where community resilience, intergenerational knowledge, and political solidarity are cultivated through food.

Figure 16.4: The Entrance to Randy's in Little Jamaica (Toronto, Canada)

16.7.5 Indian Foods and Culture in Europe and North America

Indian food culture has become one of the most globally embedded Indigenous culinary systems, owing to centuries of migration and colonial contact. In the United Kingdom, Dishoom restaurants evoke the atmosphere of 1940s Bombay cafés while offering intensely regional dishes such as pau bhaji, keema pav, and masala chai. By curating historical narratives and evocative décor, these restaurants transform dining into cultural memory experiences (Banerji, 2016). In North America, Adda Indian Canteen in New York (Figure 16.5) and Pippali in Chicago are cultural ambassadorships for regional Indian cuisines, including Bengali, Gujarati, and Hyderabadi traditions. These restaurants satisfy diasporic cravings and decolonize food narratives by challenging Western misconceptions of Indian food as monolithic or overly spicy. Many also partner with local farms and spice cooperatives to ethically source ingredients, further bridging ecological and cultural sustainability.

Figure 16.5: Entrance to the Adda Indian Canteen in New York

16.7.6 Chinese Foods and Culture in Europe and North America

Chinese diasporic cuisines have shaped urban foodscapes globally, evolving from survival-based street foods to celebrated culinary arts. In San Francisco's Chinatown, Z and Y Restaurant are renowned for its authentic Szechuan cuisine, including mapo tofu and spicy boiled fish, which showcase regional biodiversity and traditional fermentation techniques (Figure 16.6; Anderson, 1988). In Paris, Chez Vong blends Cantonese culinary heritage with fine dining, serving delicacies such as Peking duck and shark fin soup (now increasingly substituted due to sustainability concerns). These restaurants navigate complex terrain: While serving as custodians of culinary heritage, they must also contend with cultural appropriation, exoticization, and shifting regulatory environments. Still, Chinese food establishments remain intergenerationally run, often doubling as sites for language retention, kinship support, and informal economies—echoing the sociobiological function of food in sustaining familial and ethnic continuity.

Figure 16.6: Z and Y Chinese Restaurant in San Francisco's Chinatown

16.8 Challenges and Opportunities Associated with Indigenous Foods in Urban and Diaspora Communities

One critical challenge facing Indigenous foods in urban environments is the cultural erasure and commodification of traditional food practices. Traditional foods are not merely sources of nutrition; they embody cultural heritage and identity, reinforcing community connections through practices of sharing and intergenerational knowledge transmission (Miltenburg et al., 2023; Shafiee et al., 2024). The significance of these foods as cultural touchstones is underlined by the ongoing commodification processes, whereby Indigenous foods may be misrepresented or stripped of their cultural context to fit capitalist frameworks. This phenomenon is accompanied by regulatory barriers that often limit access to traditional foods in urbanized settings, as explored in comparative policy analyses conducted across various Canadian territories (Judge et al., 2022). Such regulations can perpetuate systemic inequalities, undermining the food sovereignty of urban Indigenous populations by restricting their ability to access the foods integral to their cultural identities (Phillipps et al., 2022; Ray et al., 2019).

Compounding these issues is the challenge of gentrification within urban food spaces, which exacerbates the obstacles to community access to Indigenous foods. Urban neighborhoods heavily influenced by gentrification undergo rapid transformations that can marginalize local Indigenous populations. Research indicates that gentrification displaces Indigenous communities from traditional food sources and spaces, thereby diminishing their ability to engage in cultural culinary practices (Whyte, 2018). This displacement not only affects the availability of traditional foods but also connects to broader issues of social justice and equity (Sugimoto, 2018). Urban Indigenous individuals often find themselves navigating a food environment largely dominated by commercialized food systems, which do not cater to their dietary needs or cultural preferences (Brant et al., 2023; Wilson et al., 2021).

In response to these formidable challenges, urban gardening and farming initiatives are emerging as crucial revitalization tools, enhancing food sovereignty and promoting healthier food environments among Indigenous populations. Community gardens and urban farming projects provide platforms for restoring traditional food practices and rebuilding relationships with the land (Miltenburg et al., 2023; Sumner et al., 2023). These initiatives foster a sense of community, resilience, and empowerment by allowing individuals to grow traditional foods that resonate with their cultural identities (McEachern et al., 2022). Ecological and cultural sustainability is achieved through the cultivation of Indigenous plants and practices that have defined community identities for generations, offering solutions to contemporary food security issues while also combatting socioeconomic inequalities (Cousins & Witkowski, 2015; Hansell, 2025).

Furthermore, recent studies underscore the importance of community-led efforts that reflect Indigenous methodologies in addressing food security and sovereignty. These initiatives prioritize Indigenous knowledge systems and actively involve community members in decision-making processes concerning their food systems (Miltenburg et al., 2022; Shafiee et al., 2024). For instance, learning circles and adaptive strategies have emerged as effective means to facilitate knowledge transfer and capacity building among Indigenous youth, thereby linking cultural preservation with food sovereignty goals (McEachern et al., 2022; Neufeld & Wilson, 2020). Moreover, institutional partnerships that prioritize culturally safe and context-specific food programs can contribute to more robust food sovereignty initiatives within urban centers (Jernigan et al., 2023; Phillipps et al., 2022).

The European-centric frameworks that often dominate food policy discussions neglect the unique contexts and needs of Indigenous populations, necessitating an urgent reevaluation and integration of Indigenous food sovereignty principles into broader food systems discussions (Nguyen et al., 2023). Notably, the connection between food sovereignty and health emerges as a critical aspect of exploring food systems; rejuvenating traditional diets has potential for enhanced health outcomes among Indigenous populations residing in urban areas (Nguyen et al., 2023; Ray et al., 2019). Holistic approaches that connect food practices with cultural identity and community well-being are essential for fostering sustainable food systems that honor Indigenous rights to self-determination.

Addressing the complex challenges associated with urban Indigenous foods requires an informed understanding of both the sociopolitical landscape and the historical contexts that have shaped these experiences. As urban Indigenous communities continue to advocate for their rights to engage with traditional food systems, further scholarly attention will be vital in illustrating how effective community responses can combat food injustice while reinforcing the resilience of cultural identities amid ongoing external pressures. Through collaborative efforts that empower Indigenous voices and integrate traditional knowledge into urban food practices, the potential for revitalization and sustainability within food systems expands significantly.

16.9 Sustainability of Indigenous Foods in Urban and Diaspora Communities

The sustainability of Indigenous foods within urban and diaspora communities is a multifaceted issue involving various stakeholders, including local governments and nongovernmental organizations (NGOs). These entities play a crucial role in supporting Indigenous food practices through policy development, advocacy, and community engagement. Local governments increasingly recognize the importance of integrating Indigenous food systems into broader urban planning and food security initiatives. This integration is critical in addressing historical injustices and fostering food sovereignty, a pressing concern in many marginalized Indigenous communities (Levkoe et al., 2019; Robin, 2019). Food policy councils, for instance, have emerged as platforms for promoting coordinated policies that ensure healthy, equitable, and sustainable food systems in communities that often lack adequate representation in decision-making (James et al., 2021; Levkoe et al., 2019).

NGOs significantly contribute by mobilizing resources and networks to bolster Indigenous food sovereignty. They forge partnerships with Indigenous communities to develop programs aimed at preserving traditional ecological knowledge and revitalizing Indigenous food practices. Such collaborations are essential for moving away from capitalist food production models that often disregard the cultural significance of Indigenous foods, as outlined by Levkoe et al. (2021) and Robin (2019). NGOs also facilitate educational initiatives, equipping Indigenous communities with skills to reclaim and sustain their food systems in urban settings (Ray et al., 2019). For example, urban Indigenous food initiatives that integrate traditional food narratives with modern nutritional practices can enhance food sovereignty while addressing health equity concerns (Ray et al., 2019; Weiler et al., 2014).

Educational programs are pivotal in promoting Indigenous food practices and can take various forms. School curricula incorporating Indigenous knowledge about food systems can bridge the gap between traditional practices and contemporary education. This can foster a greater understanding and appreciation of Indigenous foods among non-Indigenous students while revitalizing cultural practices among Indigenous youth. Researchers such as Michnik et al. (2021) and Donatuto et al. (2020) emphasize that integrating elders' knowledge and traditional practices into educational frameworks supports cultural transmission and strengthens community ties through shared learning experiences.

Furthermore, these educational programs can highlight the health benefits of Indigenous foods, addressing malnutrition and food insecurity prevalent in many Indigenous populations (Lemke & Delormier, 2017; Liduo et al., 2024). Moreover, educational institutions can catalyze sustainable Indigenous food systems by fostering partnerships with urban Indigenous communities. Developing curricula that incorporate Indigenous knowledge regarding food sovereignty can significantly influence students' perceptions of nutrition and environmental stewardship.

Additionally, local governments and NGOs can safeguard Indigenous cultural heritage through intellectual property rights, which protect traditional knowledge and practices. This is especially vital in the context of globalization, where Indigenous foods and practices may be appropriated without acknowledgment or compensation (Ogwu et al., 2023, 2025). Establishing clear guidelines and policies for recognizing Indigenous intellectual property rights is necessary to ensure communities retain control over their cultural heritage (Farfán et al., 2021; Maritz & Foley, 2018). Such protection preserves the integrity of Indigenous food practices and empowers communities economically, fostering sustainable development. Policy and advocacy efforts are fundamental for creating an environment conducive to revitalizing Indigenous food systems. There is an increasing recognition of the need for policies that genuinely reflect Indigenous communities' interests rather than merely including them in existing frameworks that might be inherently colonial. Rotz et al. (2023) note the tendency in settler colonial contexts to impose frameworks that do not effectively serve Indigenous needs. Consequently, Indigenous communities should lead the design of food policies and address their unique challenges, fostering self-determination. Innovative advocacy strategies can leverage frameworks established by the food sovereignty movement, emphasizing communities' rights to define their food systems and practices. The principles within the food sovereignty discourse, such as decolonization, respect, and responsibility, are crucial for guiding policy reforms that support Indigenous food systems (James et al., 2021). By enshrining these principles into legislation, governments can create a supportive environment nurturing the revival of Indigenous food practices and the health of Indigenous communities.

In urban and diaspora contexts, Indigenous food practices face the challenge of maintaining cultural relevance while adapting to modern lifestyles. This duality necessitates innovative approaches that respect traditional methods while accommodating contemporary realities. Exploring Indigenous food practices through an ecological lens, as emphasized by Green and Chenarides (2020) and Grenz and Armstrong (2023), can facilitate programs that bolster both environmental sustainability and cultural resilience. Urban gardens and community kitchens can serve as platforms for cultivating Indigenous foods and providing community gathering and cultural expression spaces.

16.10 Conclusion

The importance of Indigenous foods within urban and diaspora communities is profound. These foods serve as vital links to heritage, identity, and health, significantly influencing the overall well-being of community members. However, persistent challenges necessitate ongoing advocacy and action to promote food sovereignty within urban environments. Indigenous foods in urban and diaspora communities represent a profound continuity of cultural heritage, adaptation, and resistance in the face of displacement, globalization, and urban homogenization. This chapter has shown that, far from being lost in the migratory process, Indigenous food systems are actively preserved, reinterpreted, and revitalized in new social and geographic contexts. Through everyday culinary practices, food entrepreneurs, intergenerational knowledge transfer, and community-based initiatives, Indigenous communities maintain vital links to their ancestral roots while navigating the complexities of modern urban life. Despite significant challenges such as restricted access to traditional ingredients, cultural commodification, and the health implications of dietary transitions, Indigenous foodways continue to serve as tools for identity formation, social cohesion, and cultural sovereignty. The chapter's case studies illuminate the creativity and resilience with which diaspora and urban communities engage their food traditions, often blending innovation with preservation to sustain meaningful connections to their heritage. From a sociobiological perspective, the endurance of Indigenous food practices in diasporic and urban settings underscores the adaptive nature of human cultural systems. Food, as both a biological necessity and a symbolic medium, reflects the coevolution of human societies with their environments, even as those environments shift. The ongoing presence of Indigenous foods in global cities challenges dominant narratives of cultural assimilation and highlights the importance of biocultural diversity in achieving sustainable and equitable food futures. As such, supporting Indigenous food systems in urban and diaspora contexts is not only a matter of cultural preservation but also a matter of ecological integrity, public health, and social justice.

Chapter Reflection

1. How do Indigenous communities preserve and adapt food traditions in urban and diaspora settings?
2. What does food sovereignty mean for Indigenous peoples in cities compared to their homelands?
3. What socioeconomic challenges limit access to traditional foods in urban areas?
4. How has migration shaped both the loss and transformation of Indigenous food systems?
5. How do legal and policy frameworks impact Indigenous food access in urban contexts?
6. What health challenges arise from dietary shifts in urban Indigenous populations, and how can traditional diets help?
7. How do media platforms (social media, cookbooks, documentaries) shape Indigenous food identities?
8. Choose one case study: How does it show both preservation and change in diaspora foodways?
9. How does gentrification affect Indigenous food spaces in cities?
10. What roles do governments, NGOs, and schools play in sustaining Indigenous food systems?

REFERENCES

Abdallah, A., Fletcher, T., & Hannam, K. (2019). Lebanese food, "Lebaneseness" and the Lebanese diaspora in London. *Hospitality and Society, 9*(2), 145–160. https://doi.org/10.1386/hosp.9.2.145_1

Adjaye, J. K. (2017). Diaspora foodways: Culinary cultures in African immigrant communities. *African Studies Review, 60*(3), 45–65. https://doi.org/10.1017/asr.2017.92

Anderson, E. N. (1988). *The food of China.* Yale University Press.

Aryeetey, R., Oltmans, S., & Owusu, F. (2016). Food retail assessment and family food purchase behavior in Ashongman estates, Ghana. *African Journal of Food Agriculture Nutrition and Development, 16*(4), 11386–11403. https://doi.org/10.18697/ajfand.76.15430

Askew, A. (2021). Mindful eating: Reinvigorating American culinary diaspora in a low-income community. *Tourism Culture and Communication, 21*(1), 49–53. https://doi.org/10.3727/109830421x16135685359956

Aydın, F. (2024). Book review: Diaspora as translation and decolonization Ipek Demir. *Memory Studies, 17*(4), 987–990. https://doi.org/10.1177/17506980241255083

Banerji, C. (2016). *Eating India: Exploring a nation's cuisine.* Bloomsbury.

Blennerhassett, C., Moore-Cherry, N., & Bonnin, C. (2021). Street markets, urban development, and immigrant entrepreneurship: Unpacking precarity in Moore Street, Dublin. *Urban Studies, 59*(13), 2739–2755. https://doi.org/10.1177/00420980211040928

Brant, S., Williams, K., Andrews, J., Hammelmann, C., & Levkoe, C. (2023). Indigenous food systems and food sovereignty: A collaborative conversation from the American Association of Geographers 2022 annual meeting. *Journal of Agriculture Food Systems and Community Development, 12*(3), 1–14. https://doi.org/10.5304/jafscd.2023.123.012

Chariyev, K., Abdullayev, A., Khamdamova, L., Sapaev, I. B., Kholnazarova, L., Khudayberganov, K., Tursunov, M., & Zokirov, K. (2025). Traditional food processing methods in Uzbekistan and their role in sustainable agriculture. *Natural and Engineering Sciences, 10*(2), 459–468. https://doi.org/10.28978/nesciences.1757008

Cook, B., Trevenen-Jones, A., & Sivasubramanian, B. (2024). Nutritional, economic, social, and governance implications of traditional food markets for vulnerable populations in sub-Saharan Africa: A systematic narrative review. *Frontiers in Sustainable Food Systems, 8.* https://doi.org/10.3389/fsufs.2024.1382383

Cousins, S., & Witkowski, E. (2015). Indigenous plants: Key role players in community horticulture initiatives. *Human Ecology Review, 21*(1). https://doi.org/10.22459/her.21.01.2015.03

Cummings, M. (2020). Caribbean food in Canada: Identity, diaspora, and urban culture. *Canadian Journal of Cultural Studies, 35*(1), 88–105. https://doi.org/10.3138/cjcs.2020.35.1.88

Davies, A., Coombes, J., Wallace, J., Glover, K., Porykali, B., Allman-Farinelli, M., Kunzli-Rix, T., & Rangan, A. (2023). Yarning about diet: The applicability of dietary assessment methods in Aboriginal and Torres Strait islander Australians—A scoping review. *Nutrients*, *15*(3), 787. https://doi.org/10.3390/nu15030787

Deluze, A. K., Enos, K., Mossman, K., Gunasekera, I., Espiritu, D., Jay, C., Jackson, P., Connelly, S., Han, M. H., Giardina, C. P., McMillen, H., & Meyer, M. (2023). Urban 'āina: An Indigenous, biocultural pathway to transforming urban spaces. *Sustainability*, *15*(13), 9937. https://doi.org/10.3390/su15139937

Donatuto, J., Campbell, L., LeCompte, J., Rohlman, D., & Tadlock, S. (2020). The story of 13 Moons: Developing an environmental health and sustainability curriculum founded on Indigenous first foods and technologies. *Sustainability*, *12*(21), 8913. https://doi.org/10.3390/su12218913

Egboduku, W. O., Egboduku, T., Golohor, O. M., Imarhiagbe, O., & Ogwu, M. C. (2024). Cassava as raw material for sustainable bioeconomy development. In M. C. Ogwu, S. C. Izah, A. A. Cunha Alves, & S. C. Babu (Eds.), *Plant biology, sustainability and climate change, sustainable cassava* (pp. 57–73). Academic Press. https://doi.org/10.1016/B978-0-443-21747-0.00022-9

Elliott, B., Jayatilaka, D., Brown, C., Varley, L., & Corbett, K. K. (2012). "We are not being heard": Aboriginal perspectives on traditional foods access and food security. *Journal of Environmental and Public Health*, *2012*, 1–9. https://doi.org/10.1155/2012/130945

Errmann, A., Conroy, D., & Young, J. (2023). The lab, land, and longing: Discursive constructions of Australian identities in "future" food consumption. *Journal of Consumer Culture*, *24*(1), 193–210. https://doi.org/10.1177/14695405231207602

Farfán, J., Chaux, J., & Torres, D. (2021). Food autonomy: Decolonial perspectives for Indigenous health and buen vivir. *Global Health Promotion*, *28*(3), 50–58. https://doi.org/10.1177/1757975920984206

Ganesan, K., Govindasamy, A. R., Wong, J. K. L., Rahman, S. A., Aguol, K. A., Hashim, J., & Bala, B. (2020). Environmental challenges and traditional food practices: The Indigenous Lundayeh of Long Pasia, Sabah, Borneo. *Etropic Electronic Journal of Studies in the Tropics*, *19*(1). https://doi.org/10.25120/etropic.19.1.2020.3734

Gendron, F., Hancherow, A., & Norton, A. (2016). Exploring and revitalizing Indigenous food networks in Saskatchewan, Canada, as a way to improve food security. *Health Promotion International*, *32*(5), 808–817. https://doi.org/10.1093/heapro/daw013

Gentles-Gibbs, N., & Murphy, K. (2025). Decolonizing social work: Lessons for social work practice and education from the Jamaican diaspora transnational movement. *International Social Work*, *68*(4), 566–576. https://doi.org/10.1177/00208728251319491

Gichunge, C., Somerset, S., & Harris, N. (2016). Using a household food inventory to assess the availability of traditional vegetables among resettled African refugees. *International Journal of Environmental Research and Public Health*, *13*(1), 137. https://doi.org/10.3390/ijerph13010137

Green, K., & Chenarides, L. (2020). Using a sensory learning framework to design effective curricula: Evidence from Indigenous nutrition education programs. *Sustainability*, *12*(17), 7077. https://doi.org/10.3390/su12177077

Grenz, J., & Armstrong, C. (2023). Pop-up restoration in colonial contexts: Applying an Indigenous food systems lens to ecological restoration. *Frontiers in Sustainable Food Systems*, *7*. https://doi.org/10.3389/fsufs.2023.1244790

Hansell, R. (2025). Indigenous food sovereignty: Literature review. *Fourth World Journal*, *24*(2), 147–184. https://doi.org/10.63428/7zjhvq86

Hodge, A., Cunningham, J., Maple-Brown, L., Dunbar, T., & O'Dea, K. (2011). Plasma carotenoids are associated with socioeconomic status in an urban Indigenous population: An observational study. *BMC Public Health*, *11*(1). https://doi.org/10.1186/1471-2458-11-76

Howard-Bobiwash, H., Joe, J., & Lobo, S. (2021). Concrete lessons: Policies and practices affecting the impact of COVID-19 for urban Indigenous communities in the United States and Canada. *Frontiers in Sociology*, *6*. https://doi.org/10.3389/fsoc.2021.612029

İflazoğlu, N., & Aksoy, M. (2023). Exploring iconic foods of Hatay cuisine as a cultural identity. *Journal of Tourism and Gastronomy Studies*, *11*(4), 3262–3279. https://doi.org/10.21325/jotags.2024.1341

Jäger, M. B., Johnson, N., Burk, E., Christensen, N., Ferguson, D. B., Honani, S., Huntington, O., Larson, S., Jennings, L., Johnson, M. K., Juan, A., Strawhacker, C., Taylor, M., Todd, W. F., Walker, A., & Carroll, S. (2025). Our lands tell our stories: Supporting Indigenous co-led research through the Indigenous Foods Knowledges Network. *Arctic Science*, *11*, 1–11. https://doi.org/10.1139/as-2024-0063

James, D., Bowness, E., Robin, T., McIntyre, A., Dring, C., Desmarais, A. A., & Wittman, H. (2021). Dismantling and rebuilding the food system after

COVID-19: Ten principles for redistribution and regeneration. *Journal of Agriculture Food Systems and Community Development, 10*(2), 1–23. https://doi.org/10.5304/jafscd.2021.102.019

Jernigan, V. B. B., Taniguchi, T., Nguyen, C. J., London, S. M., Henderson, A., Maudrie, T. L., Blair, S., Clyma, K. R., Lopez, S. V., & Jacob, T. (2023). Food systems, food sovereignty, and health: Conference shares linkages to support Indigenous community health. *Health Promotion Practice, 24*(6), 1109–1116. https://doi.org/10.1177/15248399231190360

Judge, C., Spring, A., & Skinner, K. (2022). A comparative policy analysis of wild food policies across Ontario, northwest territories, and Yukon Territory, Canada. *Frontiers in Communication, 7*. https://doi.org/10.3389/fcomm.2022.780391

Kanahele, G. S. (1986). *Ku Kanaka: Stand tall—A search for Hawaiian values*. University of Hawai'i Press.

Kosoe, E. A., Achana, G. T. W., & Ogwu, M. C. (2023). Regulations and policies for herbal medicine and practitioners. In S. C. Izah, M. C. Ogwu, & M. Akram (Eds.), *Herbal medicine phytochemistry. reference series in phytochemistry*. Springer. https://doi.org/10.1007/978-3-031-21973-3_33-1

Kuhnlein, H. (2014). Food system sustainability for health and well-being of Indigenous peoples. *Public Health Nutrition, 18*(13), 2415–2424. https://doi.org/10.1017/s1368980014002961

Lemke, S., & Delormier, T. (2017). Indigenous peoples' food systems, nutrition, and gender: Conceptual and methodological considerations. *Maternal and Child Nutrition, 13*(S3). https://doi.org/10.1111/mcn.12499

Levkoe, C., McLaughlin, J., & Strutt, C. (2021). Mobilizing networks and relationships through Indigenous food sovereignty: The Indigenous food circle's response to the COVID-19 pandemic in northwestern Ontario. *Frontiers in Communication, 6*. https://doi.org/10.3389/fcomm.2021.672458

Levkoe, C., Ray, L., & McLaughlin, J. (2019). The Indigenous food circle: Reconciliation and resurgence through food in northwestern Ontario. *Journal of Agriculture Food Systems and Community Development, 9*, 1–14. https://doi.org/10.5304/jafscd.2019.09b.008

Lewis, J. (2018). Releasing a tradition. *The Cambridge Journal of Anthropology, 36*(2), 21–33. https://doi.org/10.3167/cja.2018.360204

Liduo, L., Zhangshuirui, Z., & Wook, T. (2024). The impact of Malaysia's feeding program on Orang Asli families. *International Journal of Social Science and Human Research*, *7*(6). https://doi.org/10.47191/ijsshr/v7-i06-56

Maritz, A., & Foley, D. (2018). Expanding Australian Indigenous entrepreneurship education ecosystems. *Administrative Sciences*, *8*(2), 20. https://doi.org/10.3390/admsci8020020

Maudrie, T., Colón-Ramos, U., Harper, K., Jock, B., & Gittelsohn, J. (2021). A scoping review of the use of Indigenous food sovereignty principles for intervention and future directions. *Current Developments in Nutrition*, *5*(7), nzab093. https://doi.org/10.1093/cdn/nzab093

McEachern, L., Yessis, J., Zupko, B., Yovanovich, J., Valaitis, R., & Hanning, R. (2022). Learning circles: An adaptive strategy to support food sovereignty among First Nations communities in Canada. *Applied Physiology Nutrition and Metabolism*, *47*(8), 813–825. https://doi.org/10.1139/apnm-2021-0776

Mehta, B. J. (2005). Culinary diasporas: Identity and the language of food in gisèle pineau's un papillon dans la cité and l'exil selon julia. *International Journal of Francophone Studies*, *8*(1), 23–51. https://doi.org/10.1386/ijfs.8.1.23/1

Michnik, K., Thompson, S., & Beardy, B. (2021). Moving your body, soul, and heart to share and harvest food. *Canadian Food Studies / La Revue Canadienne Des Études Sur L Alimentation*, *8*(2). https://doi.org/10.15353/cfs-rcea.v8i2.446

Miltenburg, E., Neufeld, H., & Anderson, K. (2022). Relationality, responsibility and reciprocity: Cultivating Indigenous food sovereignty within urban environments. *Nutrients*, *14*(9), 1737. https://doi.org/10.3390/nu14091737

Miltenburg, E., Neufeld, H., Perchak, S., & Skene, D. (2023). "Where creator has my feet, there i will be responsible": Place-making in urban environments through Indigenous food sovereignty initiatives. *International Journal of Environmental Research and Public Health*, *20*(11), 5970. https://doi.org/10.3390/ijerph20115970

Mishra, S. (2017). Location of diaspora in vs. Naipaul's a bend in the river. *KMC Research Journal*, *1*(1), 1–8. https://doi.org/10.3126/kmcrj.v1i1.28238

Neufeld, H., & Wilson, H. (2020). Exploring undergraduate Indigenous students' experiences with institutional and community food systems in a Canadian urban setting. *Current Developments in Nutrition*, *4*, nzaa043_103. https://doi.org/10.1093/cdn/nzaa043_103

Nguyen, C. J., Wilbur, R. E., Henderson, A., Sowerwine, J., Mucioki, M., Sarna-Wojcicki, D., Ferguson, G. L., Maudrie, T. L., Moore-Wilson, H., Wark, K., & Jernigan, V. B. B. (2023). Framing an Indigenous food sovereignty research agenda. *Health Promotion Practice, 24*(6), 1117–1123. https://doi.org/10.1177/15248399231190362

Nikolaus, C. J., Johnson, S., Benally, T., Maudrie, T., Henderson, A., Nelson, K., Lane, T., Segrest, V., Ferguson, G. L., Buchwald, D., Jernigan, V. B. B., & Sinclair, K. (2022). Food insecurity among American Indian and Alaska Native people: A scoping review to inform future research and policy needs. *Advances in Nutrition, 13*(5), 1566–1583. https://doi.org/10.1093/advances/nmac008

Obahiagbon, E. G., & Ogwu, M. C. (2023a). Consumer perception and demand for sustainable herbal medicine products and market. In S. C. Izah, M. C. Ogwu, & M. Akram (Eds.), *Herbal medicine phytochemistry.* Reference Series in Phytochemistry (pp. 1–34). Springer. https://doi.org/10.1007/978-3-031-21973-3_65-1

Obahiagbon, E. G., & Ogwu, M. C. (2023b). Sustainable supply chain management in the herbal medicine industry. In S. C. Izah, M. C. Ogwu, & M. Akram (Eds.), *Herbal medicine phytochemistry.* Reference Series in Phytochemistry (pp. 1–29). Springer. https://doi.org/10.1007/978-3-031-21973-3_64-1

Obahiagbon, E. G., & Ogwu, M. C. (2024a). Organic food preservatives: The shift towards natural alternatives and sustainability in the Global South's Markets. In M. C. Ogwu, S. C. Izah, & N. R. Ntuli (Eds.), *Food safety and quality in the Global South* (pp. 299–329). Springer. https://doi.org/10.1007/978-981-97-2428-4_10

Obahiagbon, E. G., & Ogwu, M. C. (2024b). The nexus of business, sustainability, and herbal medicine. In S. C. Izah, M. C. Ogwu, & M. Akram (Eds.), *Herbal medicine phytochemistry.* Reference Series in Phytochemistry (pp. 1–42). Springer. https://doi.org/10.1007/978-3-031-21973-3_67-1

Obeng-Odoom, F. (2015). *The urban political economy of West African cities: Accra, Ghana.* Palgrave Macmillan.

Obongodot, N. U., & Ogwu, M. C. (2023). Plant food for human health: Case study of Indigenous vegetables in Akwa Ibom State, Nigeria. In S. C. Izah, M. C. Ogwu, & M. Akram (Eds.), *Herbal medicine phytochemistry.* Reference Series in Phytochemistry (pp. 1–38). Springer. https://doi.org/10.1007/978-3-031-21973-3_2-1

Ogwu, M. C. (2023). Local food crops in Africa: Sustainable utilization, threats, and traditional storage strategies. In S. C. Izah & M. C. Ogwu (Eds.), *Sustainable utilization and conservation of Africa's biological resources and environment.*

Sustainable Development and Biodiversity (Vol. 888, pp. 353–374). Springer. https://doi.org/10.1007/978-981-19-6974-4_13

Ogwu, M. C., Ojo, A. O., & Osawaru, M. E. (2025). Quantitative ethnobotany of Afenmai people of Southern Nigeria: An assessment of their crop utilization, and preservation methods. *Genetic Resources and Crop Evolution, 72*, 5807–5829. https://doi.org/10.1007/s10722-024-02302-x

Ogwu, M. C., & Osawaru, M. E. (2022). Traditional methods of plant conservation for sustainable utilization and development. In S. C. Izah (Ed.), *Biodiversity in Africa: Potentials, threats and conservation.* Sustainable Development and Biodiversity (Vol. 29, pp. 451–472). Springer. http://doi.org/10.1007/978-981-19-3326-4_17

Ogwu, M. C., Osawaru, M. E., Amodu, E., & Osamo, F. (2023). Comparative morphology, anatomy and chemotaxonomy of two *Cissus* Linn. species. *Brazilian Journal of Botany, 46*, 397–412. https://doi.org/10.1007/s40415-023-00881-0

Olivella, M., & Tittler, J. (2021). *Changó, decolonizing the African diaspora.* Routledge. https://doi.org/10.4324/9781003163428

Osawaru, M. E., & Ogwu, M. C. (2021). Plants and plant products in local markets within Benin city and environs. In N. Oguge, D. Ayal, L. Adeleke, & I. da Silva (Eds.), African handbook of climate change adaptation. Springer. https://doi.org/10.1007/978-3-030-45106-6_159

Overend, A., & Rai, R. (2024). Un-learning and re-learning: Reflections on relationality, urban berry foraging, and settler research uncertainties. *Canadian Food Studies/La Revue Canadienne Des Études Sur L Alimentation, 11*(2), 40–57. https://doi.org/10.15353/cjds.v11i2.649

Phillipps, B., Skinner, K., Parker, B., & Neufeld, H. (2022). An intersectionality-based policy analysis examining the complexities of access to wild game and fish for urban Indigenous women in northwestern Ontario. *Frontiers in Communication, 6*. https://doi.org/10.3389/fcomm.2021.762083

Pilcher, J. M. (2012). *Planet Taco: A global history of Mexican food.* Oxford University Press.

Poirier, B., Soares, G., Neufeld, H., Hedges, J., Sethi, S., & Jamieson, L. (2024). Conceptualising the relationships between food sovereignty, food security, and oral health among global Indigenous communities: A scoping review. *Public Health Nutrition, 27*(1), e147. https://doi.org/10.1017/s1368980024001198

Polanco, V., & Rodríguez-Cruz, L. (2019). Decolonizing the Caribbean diet: Two perspectives on possibilities and challenges. *Journal of Agriculture Food Systems and Community Development, 9*, 1–6. https://doi.org/10.5304/jafscd.2019.09b.004

Ray, L., Burnett, K., Cameron, A., Joseph, S., LeBlanc, J., Parker, B., Recollet, A., & Sergerie, C. (2019). Examining Indigenous food sovereignty as a conceptual framework for health in two urban communities in northern Ontario, Canada. *Global Health Promotion, 26*(3_suppl), 54–63. https://doi.org/10.1177/1757975919831639

Rice, E., Haynes, E., Royce, P., & Thompson, S. (2016). Social media and digital technology use among Indigenous young people in Australia: A literature review. *International Journal for Equity in Health, 15*(1). https://doi.org/10.1186/s12939-016-0366-0

Richmond, C., Steckley, M., Neufeld, H., Kerr, R., Wilson, K., & Dokis, B. (2020). First Nations food environments: Exploring the role of place, income, and social connection. *Current Developments in Nutrition, 4*(8), nzaa108. https://doi.org/10.1093/cdn/nzaa108

Robin, T. (2019). Our hands at work: Indigenous food sovereignty in Western Canada. *Journal of Agriculture Food Systems and Community Development, 9*, 1–15. https://doi.org/10.5304/jafscd.2019.09b.007

Roldán Amaro, J., Álvarez Izazaga, M., del Refugio Carrasco Quintero, M., Guarneros, N., Ledesma-Solano, J. A., Cuchillo-Hilario, M., & Chávez, A. (2017). Marginalization and health service coverage among Indigenous, rural, and urban populations: A public health problem in Mexico. *Rural and Remote Health, 17*, 3948. https://doi.org/10.22605/rrh3948

Rotz, S., Xavier, A., & Robin, T. (2023). "It wasn't built for us": The possibility of Indigenous food sovereignty in settler colonial food bureaucracies. *Journal of Agriculture Food Systems and Community Development, 12*(3), 1–18. https://doi.org/10.5304/jafscd.2023.123.009

Scott, S., Elamin, W., Giles, E., Hillier-Brown, F., Byrnes, K., Connor, N., Newbury-Birch, D., & Ells, L. (2019). Socio-ecological influences on adolescent (aged 10–17) alcohol use and unhealthy eating behaviours: A systematic review and synthesis of qualitative studies. *Nutrients, 11*(8), 1914. https://doi.org/10.3390/nu11081914

Shafiee, M., Al-Bazz, S., Lane, G., Szafron, M., & Vatanparast, H. (2024a). Exploring healthy eating perceptions, barriers, and facilitators among urban Indigenous peoples in Saskatchewan. *Nutrients*, *16*(13), 2006. https://doi. org/10.3390/nu16132006

Shafiee, M., Al-Bazz, S., Szafron, M., Lane, G., & Vatanparast, H. (2024b). "I haven't had moose meat in a long time": Exploring urban Indigenous perspectives on traditional foods in Saskatchewan. *Nutrients*, *16*(15), 2432. https://doi. org/10.3390/nu16152432

Shafiee, M., Lane, G., Szafron, M., Hillier, K., Pahwa, P., & Vatanparast, H. (2023). Exploring the implications of COVID-19 on food security and coping strategies among urban Indigenous peoples in Saskatchewan, Canada. *Nutrients*, *15*(19), 4278. https://doi.org/10.3390/nu15194278

Sugimoto, T. (2018). "Someone else's land is our garden!" Risky labor in Taipei's Indigenous food boom. *Gastronomica*, *18*(2), 46–58. https://doi.org/10.1525/ gfc.2018.18.2.46

Sumner, J., McMurtry, J., & Tarhan, D. (2023). Growing community sustenance: The social economy as a route to Indigenous food sovereignty. *Canadian Journal of Nonprofit and Social Economy Research*, *14*(S1). https://doi. org/10.29173/cjnser535

Suzuki, T. (2019). Diasporic identity and mourning: Commemorative practices among Okinawan repatriates from colonial Micronesia. *Portal Journal of Multidisciplinary International Studies*, *16*(1–2), 29–45. https://doi.org/10.5130/ pjmis.v16i1-2.6276

Thomas, M. (2004). Transitions in taste in Vietnam and the diaspora. *The Australian Journal of Anthropology*, *15*(1), 54–67. https://doi.org/10.1111/j.1835-9310.2004. tb00365.x

Timler, K., Varcoe, C., & Brown, H. (2019). Growing beyond nutrition :: How a prison garden program highlights the potential of shifting from food security to food sovereignty for Indigenous peoples. *International Journal of Indigenous Health*, *14*(2), 95–114. https://doi.org/10.32799/ijih.v14i2.31938

Wanyama, R., Gödecke, T., & Qaim, M. (2019). Food security and dietary quality in African slums. *Sustainability*, *11*(21), 5999. https://doi.org/10.3390/ su11215999

Weiler, A., Hergesheimer, C., Brisbois, B., Wittman, H., Yassi, A., & Spiegel, J. (2014). Food sovereignty, food security, and health equity: A meta-narrative mapping exercise. *Health Policy and Planning*, *30*(8), 1078–1092. https://doi.org/10.1093/heapol/czu109

Whyte, K. (2018). Food sovereignty, justice, and Indigenous peoples: An essay on settler colonialism and collective continuance. In A. Barnhill, M. Budolfson, & T. Doggett (Eds.), *The Oxford handbook of food ethics* (pp. 345–366). Oxford University Press. https://doi.org/10.1093/oxfordhb/9780199372263.013.34

Wilson, H., Neufeld, H., Anderson, K., Wehkamp, C., & Khoury, D. (2021). Exploring Indigenous undergraduate students' experiences within urban and institutional food environments. *Sustainability*, *13*(18), 10268. https://doi.org/10.3390/su131810268

Wright, K. E., Lucero, J. E., Ferguson, J. K., Granner, M. L., Devereux, P. G., Pearson, J. L., & Crosbie, E. (2021). The impact that cultural food security has on identity and well-being in the second-generation U.S. American minority college students. *Food Security*, *13*(3), 701–715. https://doi.org/10.1007/s12571-020-01140-w

CHAPTER 17

Revitalizing Indigenous Food Systems for Sustainability

Abstract

This chapter highlights a growing global movement to revitalize Indigenous food systems as a pathway toward ecological sustainability, cultural resilience, and food sovereignty. Indigenous foodways are rooted in deep relationships with land, water, biodiversity, and community and have long embodied sustainable principles of reciprocity, seasonality, and regeneration. However, colonization, land dispossession, industrial agriculture, and climate change have severely disrupted these systems. The chapter outlines the concept of food system revitalization as an active process of reclaiming, restoring, and reimagining traditional practices in contemporary contexts. It explores key drivers of revitalization, including youth-led initiatives, cultural resurgence movements, and increased recognition of Indigenous ecological knowledge in the face of environmental crises. Drawing from case studies across diverse regions, including the Americas, Oceania, Africa, and Asia, the chapter highlights community-based innovations such as seed saving networks, traditional farming methods, language revitalization through culinary practices, and Indigenous-led food sovereignty campaigns. It also analyzes how Indigenous food systems intersect with public policy, education, and governance frameworks, highlighting enabling conditions and systemic barriers. Ethical considerations such as cultural appropriation, commercialization, and power imbalances in partnerships are addressed, emphasizing the importance of Indigenous leadership and consent. By presenting Indigenous food systems as dynamic, adaptive, and inherently sustainable, the chapter argues for their centrality in addressing the interconnected challenges of biodiversity loss, climate change, and nutritional insecurity. It concludes with a call for sustained, respectful engagement with Indigenous food movements as part of broader efforts toward just and sustainable futures.

Keywords: food sovereignty, ecological knowledge, seed saving, cultural revitalization, traditional agriculture, community resilience, decolonization, biocultural diversity

17.1 Introduction

Indigenous food systems represent a profound connection between communities and their ecological and cultural landscapes. These systems are crucial for sustenance and embody resilience, knowledge, and sustainability tailored to the specific contexts of Indigenous peoples. Importantly, they are grounded in the principles of biocultural heritage, encapsulating traditional practices, ecological knowledge, and community governance. Evidence indicates that these food systems are integral to the health and well-being of Indigenous communities, serving as a foundation for identity, cohesion, and resilience in the face of external pressures such as modernization and climate change (Robin, 2019; Swiderska et al., 2022). The historical relationship between Indigenous peoples and their food systems has profound interdependence and spiritual significance. Traditional food systems were inherently designed to ensure ecological balance and communal well-being, governed by Indigenous knowledge systems emphasizing sustainability and reciprocity (Robin & Hart, 2025). However, the advent of colonization introduced significant disruptions. Indigenous land dispossession and the imposition of non-Indigenous food systems led to a disconnection from traditional food practices, contributing to an alarming increase in food insecurity and noncommunicable diseases, such as diabetes and hypertension, among Indigenous populations (Bagelman et al., 2016; Timler et al., 2019).

Research indicates that these disruptions have not only diminished access to traditional food but have also driven the adoption of processed foods that are nutritionally inadequate (Robin et al., 2020). As articulated by various scholars, this transformation underscores the need to critically examine food systems, emphasizing the importance of cultural and historical factors in the discourse on nutrition and health (Ahmed et al., 2023; Jernigan et al., 2023). Furthermore, ongoing social injustices, environmental degradation, and climate change threaten Indigenous food sovereignty, necessitating concerted efforts for revitalization that address these multifaceted challenges (Delormier et al., 2017; Swiderska et al., 2022). The detrimental legacy of colonial practices persists, as many Indigenous communities encounter structural inequities that limit their autonomy over food production and distribution. Activists and scholars alike argue that reclaiming control over food systems is integral to restoring dignity, health, and identity among Indigenous peoples (Gone & Kirmayer, 2020; Robin, 2019). Therefore, understanding this historical context is critical to framing efforts to revitalize Indigenous food systems within broader social justice movements. Such policies must be grounded in the principles of self-determination and acknowledge the unique cultural contexts of Indigenous peoples.

Food sovereignty is vital for understanding the revival of Indigenous food systems. It emphasizes the right of people to define their food systems, which incorporates cultural traditions, ecological sustainability, and community governance (Coté, 2016; Jernigan et al., 2023). Indigenous food sovereignty shifts the focus from mere food security, which often reflects a supply-oriented approach, to a more holistic understanding that encompasses cultural and political dimensions (Neufeld et al., 2017; Obahiagbon & Ogwu, 2023a, 2023b, 2024a, 2024b). This framework addresses immediate dietary needs and heralds a resurgence of Indigenous identity and cultural practices deeply rooted in traditional foodways (Ghosh-Jerath et al., 2019; Michnik et al., 2021). Research highlights that successful food sovereignty initiatives are characterized by intercultural collaborations engaging Indigenous knowledge and scientific methodologies. While addressing food sovereignty, it is critical to acknowledge the intersectional complexities impacting Indigenous communities. Socioeconomic status, geographic location, and historical trauma significantly affect food access and cultural practices (Phillipps et al., 2022; Sherriff et al., 2022). Thus, policy interventions aimed at supporting Indigenous food systems must employ an intersectionality lens to understand and address these multifaceted challenges, ensuring the inclusion of diverse voices in the dialogue concerning food sovereignty.

Efforts to revitalize Indigenous food systems can take many forms, from community-based initiatives to policy-level actions. One promising approach involves fostering community gardens and urban agriculture projects that provide fresh foods and serve as educational platforms for passing down traditional food knowledge to younger generations (Bagelman et al., 2016; Skinner et al., 2013). These initiatives underscore the agency of communities to reclaim their food narrative, empowering them to adapt to modern contexts while honoring their ancestral traditions (Ghosh-Jerath et al., 2019; Ullrich, 2019). Equally important is the role of Indigenous women in leading food revitalization efforts. Women have historically been the custodians of traditional food practices, and their leadership is crucial in fostering food sovereignty and community well-being (Phillipps et al., 2022; Robin & Hart, 2025). Engaging women in food-related decision-making processes strengthens the cultural fabric of communities and reinforces intergenerational connections. Moreover, the recognition of women's roles can catalyze broader social and environmental justice movements, as they advocate for both the preservation of traditional practices and the need for systemic change (Jernigan et al., 2023; Michnik et al., 2021). Additionally, collaborating with universities and research institutions can provide vital resources and knowledge to Indigenous communities seeking to revitalize their food systems. This collaboration should prioritize Indigenous methodologies and ethics in research practices, ensuring that studies reflect the needs and aspirations of the communities involved (Ahmed et al., 2023; Jernigan et al., 2023).

This chapter critically explores the processes, motivations, and implications of revitalizing Indigenous food systems within a global context increasingly shaped by ecological crisis, cultural erosion, and food insecurity. It examines how Indigenous communities reclaim ancestral foodways, integrate traditional environmental knowledge (TEK) with contemporary innovations, and foster local food sovereignty movements that promote cultural resilience and sustainability. Through a transdisciplinary and comparative lens, the chapter investigates diverse case studies across continents, highlighting grassroots actions such as seed saving, traditional farming, land stewardship, and intergenerational knowledge transfer. It aims to identify the political, infrastructural, and ethical challenges these movements face, particularly in ongoing colonization, land dispossession, and climate change, while offering policy insights for inclusive and equitable food futures. This chapter contributes to the sociobiological understanding of Indigenous food systems by framing food revitalization as a coevolutionary process involving biocultural adaptation, environmental feedback, and social organization. It extends sociobiology beyond evolutionary dietary patterns to encompass the lived, relational, and symbolic dimensions of food as an ecological and cultural strategy. The chapter underscores how Indigenous knowledge systems serve survival and sustainability functions by emphasizing the adaptive intelligence embedded in traditional food practices such as rotational farming, foraging, and seasonal observance.

17.2 Defining Revitalization in the Context of Indigenous Foodways

Revitalization within the context of Indigenous foodways encapsulates a multifaceted approach that integrates cultural, ecological, and political dimensions. At its core, revitalization is predicated on the concepts of food sovereignty, which asserts the rights of Indigenous peoples to control their food systems, including food production, quality, and distribution (Babcock & Budowle, 2022; Johnson-Jennings et al., 2020). This control is crucial for restoring relationships between people and food and with the land itself, fostering a sense of cultural identity deeply intertwined with traditional food practices (Blanchet et al., 2021; Young et al., 2024). Food sovereignty is differentiated from mere preservation efforts or commodification of food systems by focusing on self-determination and autonomy. It transcends merely restoring Indigenous foodways by ensuring that these practices are not simply treated as commodities within larger economic systems that can exploit or misrepresent them (Coté, 2016). Instead, revitalization includes a reclamation of Indigenous knowledge regarding food production, preparation, and consumption, leading to the nurturing of community ties and cultural practices that have been disrupted over generations of colonial oppression (Elliott et al., 2021; Ruelle et al., 2022).

The revitalization of Indigenous foodways can be viewed as a means of cultural rejuvenation alongside environmental sustainability. Engaging with traditional food practices cultivates ecological knowledge that empowers communities to make decisions about land and resources in a manner that honors Indigenous traditions and worldviews (Lewis, 2021; Maudrie et al., 2021). The relationships maintained through food practices illustrate how cultural identity and ecological stewardship depend on and thrive within dynamic community interactions often rooted in historical contexts (Sumner et al., 2023). Importantly, revitalization distinguishes itself from preservation as it emphasizes adaptability and ongoing engagement with food systems in changing environmental and sociopolitical conditions. Indigenous foodways have repeatedly demonstrated resilience by linking traditional ecological knowledge to modern nutritional practices, thus positioning them as living cultural expressions rather than relics to be preserved (Gordon et al., 2018; Noe, 2023). In contemporary contexts, including traditional foods in diets is often framed as political sovereignty, celebrating local customs and asserting rights over historically marginalized ancestral lands (Elliott et al., 2021; Lewis, 2021).

The cultural dimension of revitalization is particularly highlighted in contemporary Indigenous food sovereignty initiatives that integrate learning about traditional practices into educational frameworks, particularly through land-based learning (Bowra et al., 2020). This academic approach not only transmits knowledge but also reinforces communal bonds by fostering a collective understanding that connects members of these communities to their cultural heritage and each other. Moreover, practical activities like hunting, gathering, and preparing traditional foods facilitate a physical and emotional connection to the land that has significant implications for identity and wellness (Michnik et al., 2021; Sumner et al., 2023). Ecologically, revitalization involves reviving traditional Indigenous food practices and reevaluating agricultural methods that favor biodiversity, sustainability, and ecological balance over industrial farming practices that often compromise them (Elliott et al., 2021). For example, restoring native bison populations in Northern Great Plains communities embodies a return to sustainable food practices by integrating ecological restoration with Indigenous knowledge systems, thereby enhancing tribal food sovereignty (Shamon et al., 2022). At a political level, revitalization serves as an assertion of rights amidst ongoing struggles against marginalization. The Indigenous food sovereignty movement encourages political action that demands participatory governance of food systems, focusing on reclaiming community decision-making power (Skinner et al., 2013). Furthermore, it acts as a platform for social justice, drawing attention to the inequalities perpetuated by colonial histories and advocating for reinstating relationships with food and broader social, political, and environmental systems (Riley & Donaldson, 2024). These movements are intertwined with contemporary calls for reconciliation and recognition of Indigenous rights.

Therefore, revitalizing Indigenous foodways is a purposeful and holistic endeavor beyond simple preservation or economic commodification. It celebrates and restores Indigenous identities and ecological practices that have withstood external pressures while fostering community resilience and promoting health outcomes among Indigenous populations (Burnette et al., 2020; Rowe et al., 2024). The objective is not merely to resurrect past practices but to adapt and thrive within a contemporary context where Indigenous knowledge is regarded as central to sustainability, cultural integrity, and political agency.

17.3 Drivers of the Revitalization of Indigenous Food Systems

The revitalization of Indigenous food systems is a multifaceted process driven by several interrelated factors, including cultural resurgence movements, youth-led initiatives, climate change awareness, and global Indigenous rights movements.

Cultural resurgence movements serve as a fundamental driver in revitalizing Indigenous food systems (Table 17.1). These movements emphasize the necessity of reconnecting Indigenous peoples with their ancestral lands and traditional food practices. As Klopotek et al. (2022) described, the food sovereignty movement

Table 17.1: Some Key Drivers of the Revitalization of Indigenous Food Systems

Driver	Contribution	Examples/Applications
Cultural Resurgence Movements	Reconnecting Indigenous peoples with ancestral lands, food practices, and spiritual relationships.	Food sovereignty movements; cultural revitalization through traditional diets and ceremonies.
Youth-Led Initiatives	Engaging youth in gardening, harvesting, cooking, and storytelling to foster identity and health.	The Ode'imin Giizis project in Minnesota integrates cultural teachings with practical food skills.
Climate Change Awareness	Recognition of the threat that climate change poses to traditional foods, prompting adaptive responses using Indigenous ecological knowledge.	Use of TEK in sustainable land and food management practices.
Global Indigenous Rights Movements	Advocacy for land, food, and cultural sovereignty through international frameworks like UNDRIP.	Legal claims to food and territory and increased visibility of Indigenous governance models.
Intergenerational Knowledge Transfer	Elders teach youth about traditional foods, farming, and values, bridging cultural gaps created by colonial histories.	Story circles, community kitchens, and land-based education initiatives.
Food Insecurity Response	Revitalization as a response to chronic food insecurity in Indigenous communities caused by colonization and structural inequities.	Restoring access to nutrient-rich traditional foods to combat diet-related diseases and malnutrition.
Innovative Methodologies	Use participatory research, digital storytelling, and localized health models to empower communities in food planning and education.	Community-designed nutrition programs and digital archiving of traditional food knowledge.

asserts Indigenous peoples' rights to control their food systems, fostering a cultural renaissance that reinforces identity, and ecological relationships tied to specific traditional practices. In Canadian contexts, conventional food practices not only provide nutritional benefits but also uphold cultural identity, as indicated by Marushka et al. (2021), who highlight how traditional food contributes significantly to cultural, spiritual, and community well-being among Indigenous populations. This reclamation is essential for addressing food insecurity, a condition that disproportionately affects Indigenous communities due to historical injustices resulting from colonization (Jernigan et al., 2021).

Youth-led initiatives are critical in this cultural resurgence by facilitating intergenerational knowledge transfer. Various community-driven projects show that engaging youth in traditional food practices enhances their understanding of Indigenous heritage and promotes healthier lifestyle choices (Lopes et al., 2024; Sarkar et al., 2019). Programs such as the Ode'imin Giizis initiative in Minnesota exemplify this dual approach—combining cultural teachings with practical skills like gardening and cooking to foster a sense of pride and health awareness among Indigenous youth (Ahmed et al., 2022). This interconnection between cultural knowledge and community health underscores the importance of education in revitalization efforts.

Moreover, the growing awareness of climate change and its adverse effects on biodiversity and food security galvanizes Indigenous communities to adapt and strengthen their food systems. Ahmed et al. (2023) discusses how incorporating Indigenous knowledge systems into land management practices can lead to more sustainable ecological outcomes. This incorporation illustrates that TEK is vital for addressing contemporary environmental challenges such as climate change (Ahmed et al., 2023). This awareness not only reinforces the need for Indigenous food systems but also highlights their adaptability and resilience in the face of external pressures threatening their traditional ways of life. The rise of global Indigenous rights movements, notably the adoption of frameworks such as the United Nations Declaration on the Rights of Indigenous Peoples, further empowers communities to advocate for their rights to land and food sovereignty. As Hudson et al. articulate, these movements encapsulate a broader call for self-determination and equitable access to resources necessary for sustaining Indigenous cultures and food practices (Yongabi, 2023). This framework legitimizes Indigenous claims to land and food resources, fostering a renewed commitment to restoring traditional practices as part of a decolonization strategy aimed at reversing the damage wrought by colonial practices.

Correlatively, initiatives aimed at intergenerational knowledge transfer underscore the importance of cultural continuity in food systems. Hudson et al. (2023) explored the role of Indigenous elders in fostering intergenerational solidarity, revealing how their contributions to knowledge transfer are pivotal for maintaining community wellness. Engaging elders in educational settings fosters a deeper understanding of traditional practices among younger generations, effectively bridging the gap created by colonial disruptions (Viscogliosi et al., 2020). As a result, knowledge transfer becomes not just an act of retention but a powerful tool for revitalization and empowerment within communities. Also, innovative approaches, such as participatory action research and digital tools, offer unique frameworks for enhancing community engagement and resilience. As Churchill et al. (2020) noted, leveraging local contexts and Indigenous perspectives in health and nutrition initiatives is crucial for developing culturally relevant interventions that promote systemic change. By embracing such innovative methodologies, communities can solidify their efforts to revitalize their food systems while addressing the nutritional challenges historically marginalized populations face.

17.4 Case Studies of Revitalization Efforts for Indigenous Food Systems

Revitalizing Indigenous food systems requires place-based approaches that honor cultural traditions, ecological knowledge, and the right to self-determination. Across the globe, Indigenous communities are engaging in efforts to reclaim, adapt, and sustain food practices that were historically disrupted by colonialism, industrial agriculture, land dispossession, and environmental degradation. These revitalization initiatives are about restoring food production and reasserting cultural identity, community resilience, and ecological balance. The following case studies offer regional perspectives on how Indigenous foodways are being reimagined to meet contemporary social, ecological, and political challenges.

17.4.1 North America: Reestablishing Seed Sovereignty and Traditional Farming

In North America, revitalization efforts are rooted in a return to seed sovereignty and traditional agricultural practices such as Three Sisters farming, which involves the intercropping of corn, beans, and squash—a method known for enhancing soil fertility, reducing pests, and supporting nutritional diversity (Figure 17.1;

Hoover, 2017). These practices are deeply embedded in the cosmologies and seasonal rhythms of Haudenosaunee, Anishinaabe, and other Indigenous nations. Organizations like the Indigenous Seed Keepers Network and the Native American Food Sovereignty Alliance have emerged as leaders in promoting seed exchanges, seed-saving education, and legal advocacy to protect heirloom varieties from corporate patents and genetic erosion (Kellogg, 2019). The work of tribal seed banks, such as the one led by Rowen White (Mohawk), exemplifies how seed sovereignty reconnects communities to ancestral foodways, facilitates intergenerational learning, and strengthens biocultural resilience. These initiatives also intersect with movements to decolonize diets and resist health inequities caused by industrial food systems.

Figure 17.1: Woman Tending a Traditional Three Sisters Garden (Corn, Beans, and Squash)

In Mesoamerica, this milpa system, a centuries-old practice among Maya communities—is being revitalized as an ecological and cultural counterpoint to industrial monoculture. Milpa involves poly cropping maize, beans, squash, chili, and root crops within agroforestry systems that mimic forest dynamics, conserve soil, and protect biodiversity (Altieri & Toledo, 2011). This system integrates agriculture with sacred cycles and cosmologies, particularly the Mayan calendar and traditional land tenure systems. Amid pressures from logging, GMOs, and agribusiness, communities in Mexico, Belize, and Guatemala are restoring milpa as a form of resistance and renewal. Networks like Cultura Milpa and Red de Guardianes de Semillas promote seed sovereignty, soil restoration, and youth engagement through culturally grounded agroecological education (Nigh & Diemont, 2013). These efforts also challenge the narrative that Indigenous systems are outdated, instead demonstrating their sophistication in climate adaptation, biodiversity maintenance, and food autonomy.

17.4.2 Oceania: Reviving Oceanic Food Systems and Traditional Fishing Practices

Across Oceania, Indigenous communities are restoring marine-based food systems that align with customary tenure and spiritual relationships to the sea. For example, in Polynesian and Melanesian societies, traditional marine governance practices such as rau'i (Cook Islands) and tabu (Fiji and Vanuatu) involve seasonal closures of fishing grounds to allow species regeneration (Jupiter et al., 2014). These customary systems are forms of adaptive management that predate modern conservation science but are increasingly recognized for their effectiveness in sustaining fisheries and protecting biodiversity. In Hawai'i, the grassroots network Kua'āina Ulu 'Auamo (KUA) has been instrumental in reviving 'āina-based stewardship models that combine Indigenous knowledge ('ike kūpuna) with scientific tools to manage coastal ecosystems (McMillen et al., 2017). Efforts include community fishpond restoration (loko i'a), limu (seaweed) cultivation, and educational exchanges that link elders with youth. These projects reassert Indigenous governance over ocean resources and reflect a holistic approach to food systems where cultural well-being and ecosystem health are inseparable (Montgomery & Vaughan, 2020).

17.4.3 Africa: Restoring Indigenous Grains and Agroecological Knowledge

In sub-Saharan Africa, communities are revaluing neglected and underutilized crops such as fonio (*Digitaria exilis*), a drought-tolerant, nutrient-rich grain in West Africa, and teff (*Eragrostis tef*), a staple crop in Ethiopia with deep historical and ceremonial significance. Both grains are resilient to climate variability and require fewer inputs than cash crops like maize or wheat (Padulosi et al., 2013). The African Orphan Crops Consortium has championed the genomic research and policy support necessary to increase the visibility and market viability of these crops. At the same time, local initiatives focus on farmer-to-farmer knowledge exchange, seed sharing, and the preservation of cultivation techniques (Demissie & Arega, 2017). In Ethiopia, the Ethiopian Biodiversity Institute supports community seed banks and promotes teff-based diets as alternatives to ultra-processed imports. Revitalization efforts are about restoring lost grains and sustaining agroecological knowledge systems that promote soil conservation, intercropping, and spiritual connections to land (Ogwu, 2023; Ogwu & Osawaru, 2022). These initiatives resist dependency on external inputs and reinforce Indigenous sovereignty over agricultural systems.

17.4.4 Asia: Community-Based Wild Food Foraging and Sacred Forest Stewardship

In Asia, many Indigenous communities maintain a close relationship with forest ecosystems, relying on wild food foraging and sacred forest stewardship for nutrition, medicine, and spiritual practices (Figure 17.2). The Adi, Apatani, and other Indigenous groups in Northeast India manage agricultural forest mosaics that combine wet-rice cultivation with traditional aquaculture and seasonal foraging (Roy et al., 2018). Sacred groves, maintained through customary taboos and rituals, function as biodiversity hotspots and spiritual sanctuaries that ensure the regeneration of medicinal and edible species (Tiwari et al., 1998). These systems are now being revitalized through partnerships with ethnobotanists, NGOs, and Indigenous researchers who document traditional knowledge and advocate for land tenure recognition. Similar practices persist in regions of the Philippines, Indonesia, and Vietnam, where community forests are comanaged using traditional ecological calendars that guide harvesting, hunting, and planting cycles. Such revitalization initiatives affirm Indigenous worldviews where food, ecology, and cosmology are inseparable and highlight the role of cultural institutions in ecological governance.

Figure 17.2: Shintō's Sacred Forest in Japan Helps Protect Biodiversity Used as Food

Source: Toya (2017)

These regional efforts collectively demonstrate that revitalizing Indigenous food systems is a multidimensional endeavor involving ecological restoration, political autonomy, and cultural survival. They also illustrate the deep-rooted sociobiological relationship between human communities and their environments, where food serves as sustenance and a medium for transmitting knowledge, asserting rights, and shaping adaptive futures. The continued success of these initiatives depends on sustained support, respectful collaboration, and policy environments that recognize Indigenous leadership, land rights, and knowledge systems as central to global sustainability.

17.5 Indigenous Food Knowledge Systems and Sustainability

Indigenous Food Knowledge Systems (IFKS) are pivotal in fostering sustainable food practices and ecological management across diverse ecosystems. Just like TEK, IFKS is often ingrained in the cultural identities of Indigenous peoples and offers a comprehensive understanding of local environments, aligning human activities with natural processes. This knowledge is crucial in maintaining social-ecological resilience, particularly for communities reliant on environmental resources (Obongodot & Ogwu, 2023). The intricate link between traditional food knowledge and ecosystem management supports biodiversity and integrates Indigenous practices into contemporary governance frameworks. This integration is necessary for addressing the dual crises of biodiversity loss and climate change, ensuring that local practices inform wider sustainable development strategies (Mugabirwe & Turyamureeba, 2025; Oskal et al., 2023).

The knowledge embedded in Indigenous communities is fundamentally linked to their unique systems of timekeeping, phenology, and climate responsiveness. Indigenous peoples often possess an acute understanding of seasonal changes and natural cycles, guiding their agricultural practices and resource management. For example, practices like selective harvesting and sustainable hunting cater to immediate food needs and uphold long-term ecological balance. The Suku Anak Dalam community in Indonesia exemplifies how Indigenous knowledge utilizes ecological theories of sustainability, with community practices rooted in spiritual connections to the land serving as cultural and environmental touchstones (Ginting, 2025). Additionally, communities display remarkable adaptability to climate crises through localized strategies that recognize and integrate natural rhythms, supporting crop diversity and resilience against climate variability (Mugabirwe & Turyamureeba, 2025).

Sustainable harvesting practices in Indigenous cultures are underpinned by a holistic understanding of ecosystems and their interdependencies. This often involves rotational grazing systems and biocultural indicators that have evolved over generations. For instance, rotational grazing can prevent overgrazing and support soil health and plant biodiversity, as illustrated in New Zealand's agricultural practices (Rowarth et al., 2021). These practices emphasize soil carbon retention and minimizing environmental impacts, which allow for a more sustainable approach to livestock management. Moreover, studies have shown that Indigenous practices foster biodiversity by integrating traditional ecological knowledge into resource management, helping maintain species' health and their habitats (Atlas et al., 2020; Rankoana, 2016). Indigenous knowledge systems also contribute to wildlife management, ensuring that traditional harvesting practices are conducted sustainably. Indigenous communities effectively manage fish and wildlife populations by utilizing place-based knowledge and selective harvesting methods. For example, the management of Pacific salmon fisheries showcases how Indigenous practices can serve as models for sustainable fishing, reinforcing ecosystem resilience (Atlas et al., 2020). Furthermore, community-led initiatives for species management, such as those observed with muskoxen in Greenland, highlight the importance of local ecological knowledge in establishing sustainable harvest limits (Cuyler et al., 2020).

Integrating Indigenous practices with modern science can lead to innovative solutions addressing environmental challenges. Community empowerment through applying Indigenous practices, such as developing an environmental health and sustainability curriculum based on First Foods, embodies how knowledge transfer and education can enhance awareness and foster sustainable practices within the community (Donatuto et al., 2020). Such efforts enable a fusion of traditional knowledge with contemporary scientific understandings, nurturing a comprehensive outlook on sustainability (Donatuto et al., 2020). Adapting to rapid environmental changes necessitates reinforcing Indigenous knowledge systems, particularly in the face of climate change. Indigenous peoples have historically developed strategies such as drought prediction and resource management that prioritize ecosystem health while meeting community needs. The ability to forecast climatic events through traditional practices enhances food security and fosters resilience against adverse conditions, vital for sustainable livelihoods in rural settings (Mugabirwe & Turyamureeba, 2025). Therefore, integrating these strategies into formal climate adaptation policies offers a pathway to strengthening the socioecological fabric of Indigenous communities and enhancing their responsiveness to climate variability.

Moreover, the application of Indigenous knowledge in environmental governance extends to water conservation practices, which have been shown to effectively combat water scarcity through techniques such as rainwater harvesting and wetland management (Sahani et al., 2025). These practices, grounded in cultural traditions, address ecological needs while empowering communities by ensuring equitable access to vital resources and the continuity of local knowledge systems across generations. Collectively, these practices emphasize the significant overlap between cultural and environmental sustainability, fostering a broader understanding of the interconnectedness of all life forms and their habitats. As we explore the nuanced intersections of Indigenous food knowledge systems and sustainability, it becomes increasingly clear how these traditional practices can inform global efforts toward ecological balance and food sovereignty. Integrating TEK into modern governance and sustainability frameworks is beneficial and essential for developing a comprehensive understanding of environmental management in diverse communities. By valuing and prioritizing Indigenous knowledge systems—encompassing cultural practices as vital aspects of sustainability—we can cultivate an ecosystem management paradigm that honors the ecological wisdom of Indigenous peoples while addressing the pressing challenges of the tweny-first century (Budiyoko & Verrysaputro, 2025; Negi et al., 2021).

17.6 Infrastructure and Innovation for the Revitalization of Indigenous Food Systems

The revitalization of Indigenous food systems represents a critical intersection of health, culture, and ecological sustainability. Community gardens, seed banks, and Indigenous-run food hubs are pivotal components in this endeavor. In regions such as Northern Canada, initiatives like hoop house gardening in Wapekeka First Nation illustrate the potential of localized food production to enhance food security by mitigating reliance on expensive market food (Thompson et al., 2018). These initiatives build on traditional land-based food practices while fostering community resilience through increased capacity for nutrition (Mucioki et al., 2024). Moreover, community-specific approaches, informed by TEK, underscore the importance of historical and cultural legacies in shaping contemporary gardening practices (Timler et al., 2019). The role of technology in revitalizing Indigenous food systems is equally significant. Geographic information systems and other mapping technologies enable the documentation of traditional food practices and the establishment of community gardens, thus preserving invaluable Indigenous knowledge while providing a platform for storytelling (Timler et al., 2019). Through digital documentation and storytelling platforms, Indigenous communities can assert their narratives, promoting awareness and appreciation of their cultural practices (Emmanuel et al., 2023).

Furthermore, integrating the revitalization of Indigenous languages with food systems presents an essential avenue for cultural preservation and education. Including Indigenous food names and culinary vocabulary in community gardens and educational materials nurtures linguistic diversity and strengthens cultural identities (Meissner, 2023). This linguistic reclamation fosters a connection to food sources. It affirms the knowledge systems underpinning Indigenous food practices, reinforcing ties between language, culture, and ecology while cultivating a deeper understanding of biodiversity and ecological stewardship among Indigenous and non-Indigenous community members (Turpin & Si, 2017). As communities engage in initiatives like seed banks and gardening projects, they cultivate physical spaces and relationships with the land that reflect a profound sense of reciprocity and interconnectedness. The historical context of Indigenous land-use practices highlights the ecological benefits arising from traditional approaches to land and resource management (Armstrong et al., 2021). For instance, studies reveal how Indigenous management practices contribute positively to biodiversity and ecosystem health (Kumar, 2021). Reclaiming their agricultural legacies, Indigenous communities can revive their food sovereignty and cultural identities (ápi, 2011).

Cultivating community gardens often emerges as a vehicle for addressing various public health challenges Indigenous peoples face. Evidence suggests that gardening interventions significantly enhance community engagement and well-being by allowing for culturally centered practices that resonate with Indigenous populations (Emmanuel et al., 2023). The substantial health benefits derived from active participation in gardening encompass nutritional improvements and the promotion of social cohesion and mental health (Moscou, 2022). The collaborative nature of community gardening fosters mutual aid, emphasizing collective responsibility while addressing food insecurity in urban Indigenous communities (Tebulo, 2025). Integrating medicinal plants into community gardening initiatives further enriches the landscape of Indigenous health. Medicinal plants, long utilized by Indigenous healers, can be incorporated within community gardens to promote preventative health measures and enhance local wellness strategies (Tebulo, 2025). By integrating traditional healing practices into gardening efforts, communities preserve invaluable knowledge and provide holistic health pathways that honor Indigenous healing traditions. Current frameworks that address food sovereignty and security underscore the necessity of adapting to colonial histories and acknowledging the ongoing impacts of such legacies on Indigenous health and food systems (Timler et al., 2019). Incorporating aspects of colonial history allows for a more nuanced comprehension of food system challenges, directing attention to the crucial need for policy changes that genuinely reflect Indigenous priorities and aspirations for self-determination (Timler et al., 2019). Initiatives stemming from Indigenous leadership and community input are essential for developing sustainable, culturally appropriate solutions to food access issues (Moscou, 2022).

Additionally, the role of Indigenous-run food hubs is critical, serving as points for food distribution and as centers for education and community engagement. They facilitate the exchange of traditional knowledge while serving as platforms for outreach and networking among Indigenous food producers (Thompson et al., 2018). By promoting local food systems and supporting Indigenous agriculture, food hubs contribute to resilient communities by prioritizing sustainability and cultural integrity (Mucioki et al., 2024). Ultimately, revitalizing Indigenous food systems through community gardens, seed banks, and technology is a multifaceted endeavor beyond mere food production. It encapsulates a broader movement toward cultural reclamation, ecological sustainability, and health equity. The revitalization of these systems is intrinsically tied to the recognition and celebration of Indigenous knowledge, values, and practices, encompassing a holistic approach prioritizing community well-being and environmental stewardship (Tebulo, 2025). As Indigenous communities continue to lead these initiatives, the transformation within food systems is a powerful testament to their resilience, creativity, and commitment to a sustainable future.

17.7 The Role of Policy, Education, and Governance in the Revitalization of Indigenous Food Systems

The revitalization of Indigenous food systems is a critical endeavor that intersects with public policy, education, and governance. Recognizing the importance of integrating Indigenous food systems into these domains is vital for enhancing food sovereignty, preserving cultural heritage, and fostering economic resilience within Indigenous communities. Integrating Indigenous food systems into public policy is essential for ensuring these systems receive the recognition and support they need to thrive. According to Swiderska et al. (2022), the urgent need to protect and revitalize Indigenous food systems arises from various factors, including climate instability and unsustainable agrifood systems. This protection can be achieved by formulating policies that uphold Indigenous rights and promote the sustainable use of local biological resources. Additionally, Hunter et al. (2019) emphasize the viability of integrating neglected and underutilized species into production systems to enhance food security that respects Indigenous agricultural practices.

Furthermore, legal frameworks that protect Indigenous foodways and land rights are essential for maintaining the integrity and sustainability of these systems. These frameworks serve as critical tools for addressing past injustices and offering pathways for community governance. As highlighted by Michnik et al. (2021), community-led initiatives that emphasize traditional ecological knowledge and cultural beliefs play an integral role in fostering food sovereignty. Policymakers must engage with Indigenous communities to understand their historical contexts and current challenges, ensuring that public policies address food access, cultural restoration, and environmental health. Tribal sovereignty, embodying the right of Indigenous communities to self-govern and control their food systems, is paramount in legislative efforts aimed at revitalizing traditional food practices. Johnson-Jennings et al. (2020) assert that exercising food sovereignty grants Indigenous nations the authority to manage their food production, ensuring that it aligns with their cultural values and ecological knowledge. Acknowledging Indigenous peoples' barriers in navigating contemporary food systems is crucial, as identified by Phillipps et al. (2022), where access to traditional food sources is complicated by urbanization and colonial policies.

Moreover, education is critical in revitalizing Indigenous food systems by fostering an understanding of traditional food practices among younger generations. Shafiee et al. (2024) discuss how programs that connect youth with Elders facilitate intergenerational knowledge transfer and reinforce community ties, thereby preserving cultural heritage and enhancing social structures. Education systems must, therefore, be designed to incorporate Indigenous perspectives and methodologies, recognizing the profound connections between cultural identity and traditional foodways. Community-based governance models are instrumental in facilitating the revitalization of Indigenous food systems. Levkoe et al. (2019) illustrate how community initiatives, like those championed by the Indigenous Food Circle, respond to food insecurity by advocating for self-determination in food practices. These models prioritize relationships among community members, emphasizing collective action and resilience in addressing food sovereignty challenges. The importance of support networks within these communities underscores the necessity of collaborative frameworks that allow individuals to share knowledge and resources across generations and cultural lines.

As Indigenous food systems are reemphasized within governance frameworks, a decolonial lens becomes essential for evaluating existing policies. Coté (2016) discusses the "indigenizing" of food sovereignty as a movement emphasizing the cultural responsibilities that Indigenous peoples maintain towards their environments, thus advancing a holistic understanding of food sovereignty that addresses food security and cultural and ecological sustainability. Efforts to integrate Indigenous food systems must also incorporate findings from public health research highlighting the links between diet, health outcomes, and nutritional equity in Indigenous populations. Merz and Steinberg (2014) note that understanding these implications is vital in countering the chronic health disparities faced by these communities. Policies that embrace traditional foodways significantly improve Indigenous peoples' health indicators.

However, a critical barrier remains in the context of urban Indigenous populations, where access to traditional food resources is often limited due to urban policies that fail to account for Indigenous needs. Judge et al. (2022) analyze the regulation of traditional food access in urban settings and highlight the need for inclusive policies to facilitate access to traditional foods that accommodate the cultural practices of urban Indigenous communities. Also, fostering resilience in food systems requires a comprehensive approach that combines modern methodologies with traditional Indigenous knowledge. Isumonah (2024) emphasizes the importance of integrated strategies to combat food insecurity effectively, suggesting that local practices and traditions must inform policies designed to enhance food security. Communities can promote self-determination and agency over their food systems by establishing educational initiatives emphasizing Indigenous knowledge and food practices. Ruelle (2017) addresses the importance of reestablishing relationships between Indigenous peoples and their traditional food sources through initiatives that enhance health outcomes and cultural practices, solidifying the connection between health, identity, and environmental stewardship.

17.8 Challenges in the Revitalization of Indigenous Food Systems

The revitalization of Indigenous food systems presents several challenges, particularly concerning cultural appropriation and commercialization, access to vital resources, the balance of traditional and modern practices, and existing power dynamics in collaborations with external entities such as NGOs, academia, and state bodies (Table 17.2). Each of these issues is deeply interwoven with the history of colonialism, resource exploitation, and the systemic inequities that Indigenous communities face today. A significant risk in revitalizing Indigenous food systems is the potential for cultural appropriation and commercialization. This issue revolves around the appropriation of Indigenous knowledge and practices by non-Indigenous entities for profit, often without proper acknowledgment or benefit to the Indigenous communities themselves. Lopes et al. illustrate the need for inclusive, culturally appropriate methodologies that engage Aboriginal communities meaningfully in all project phases, emphasizing that such approaches are necessary to avoid misrepresenting and commodifying Indigenous food systems (Lopes et al., 2024).

Access to essential resources such as land, water, and markets is another critical challenge. Jernigan et al. (2021) demonstrate that food insecurity prevalent among Indigenous populations largely stems from historical injustices, including forced relocations and disruption of traditional subsistence patterns. These communities

Table 17.2: Challenges in the Revitalization of Indigenous Food Systems

Challenge	Resulting Issue	Implications	Potential Strategies
Cultural Appropriation	Misuse or commodification of Indigenous food knowledge without consent or benefit-sharing.	Undermines Indigenous sovereignty, erases cultural context, and leads to exploitation.	Implement ethical guidelines, require Free Prior Informed Consent, and promote community-led initiatives.
Land Inaccessibility	Loss of ancestral territories due to colonization, urbanization, or extractive industries.	It limits traditional food practices, restricts cultivation and foraging, and exacerbates food insecurity.	Advocate for land restitution, enforce Indigenous land rights, and support land trust mechanisms.
Policy Exclusion	Lack of supportive legislation and representation in food and agricultural policy frameworks.	Marginalizes Indigenous priorities and prevents systemic support for food sovereignty.	Codevelop policy with Indigenous stakeholders; include traditional foodways in national food strategies.
Market Dependency	Overreliance on commercial food systems due to disrupted traditional supply chains.	Increases exposure to processed, nutrient-poor foods and undermines food autonomy.	Promote local food hubs, seed exchanges, and Indigenous-led value chains.

Erosion of Traditional Knowledge	Intergenerational disconnect and loss of oral knowledge and practices.	Weakens cultural identity, farming techniques, and ecological memory.	Support land-based education, elder-youth mentorship, and language revitalization linked to food.
Climate Change	Disrupts seasonal cycles, damages ecosystems and alters crop viability.	Threatens food availability and traditional ecological knowledge systems.	Integrate Indigenous knowledge into climate adaptation planning and strengthen agroecological practices.
Power Imbalances in Collaborations	Dominance of non-Indigenous researchers, NGOs, or funders in food projects.	This results in top-down programs, a lack of community ownership, and research extraction.	Foster equitable partnerships, use participatory action research, and center Indigenous leadership.
Urbanization and Displacement	The movement away from traditional lands due to economic or social pressures.	Hinders access to culturally significant foods and weakens community food networks.	Develop urban Indigenous gardens and create culturally inclusive food access programs.
Commercialization Pressures	Pressure to commodify Indigenous foods for tourism or niche markets.	Risks homogenization of cultural practices and disempowerment of communities.	Encourage ethical entrepreneurship and community-controlled food enterprises.

Lack of Institutional Recognition	Educational and scientific systems undervalue Indigenous ecological and nutritional knowledge.	Marginalizes TEK and hinders its integration into sustainable development models.	Embed Indigenous food knowledge in school curricula and sustainability research.
Health Disparities	High rates of diet-related illnesses due to disconnection from traditional diets.	Increases vulnerability and healthcare burdens in Indigenous communities.	Reintroduce traditional diets via health interventions and culturally relevant nutrition education.

frequently encounter systemic barriers to accessing land and water, exacerbated by policies prioritizing economic profit over community welfare. This trend can lead to reliance on external food sources that do not adequately fulfill the cultural and nutritional needs of Indigenous people (Heaney et al., 2024). Furthermore, land management practices often favor the interests of the state and corporate entities, sidelining Indigenous perspectives on ecological stewardship and sustainability. The revitalization efforts also navigate the competing pressures of preserving traditional values while embracing modern innovations. Incorporating technology, such as improved food processing methods, can enhance food security and sustainability within Indigenous food systems. However, it risks diluting traditional practices and knowledge (Heaney et al., 2024). A balance must be struck where traditional Indigenous food practices are honored while adapting to contemporary dietary and environmental challenges. Such balancing acts require community discussions and consensus, as noted by Robin, who stresses the importance of understanding Indigenous perspectives when considering food systems and cultural revitalization (Robin, 2018).

Power dynamics in collaborations with NGOs, academia, and state agencies complicate the revitalization process, often leading to conflicts over priorities and methodologies. The predominance of Western epistemologies can overshadow Indigenous knowledge systems, resulting in projects that fail to reflect the needs or aspirations of Indigenous communities. Jennings et al. argue that Indigenous knowledge should guide food and land practices to address diet-related diseases, advocating for equitable partnerships instead of top-down approaches (Jennings et al., 2020). Collaboration models prioritizing Indigenous voices are essential for fostering genuine empowerment and self-determination among these communities.

Additionally, Indigenous food sovereignty must be reinstated to address food insecurity and promote health outcomes. The right to access culturally appropriate foods underlines physical health and social and spiritual well-being; Detwiler highlights that this relationship to food nurtures cultural identity and communal ties (Detwiler, 2020). Oloko and Shukla (2018) affirm that traditional food gathering provides a holistic view of health and community sustenance while embodying values such as reciprocity, which is essential for maintaining cultural heritage. As Indigenous communities strive toward revitalization, it remains crucial to document and protect their food knowledge and practices. Communities jeopardize invaluable cultural heritage without frameworks that incorporate traditional knowledge into food systems (Derashri et al., 2025). Emphasizing the role of digital repositories, as Johnson-Jennings et al. (2020) discuss, can sustain Indigenous food wisdom while enhancing information accessibility. Creating platforms for intergenerational teaching fosters learning and reinforces cultural practices that have endured through adversity.

17.9 Sustainability Indicators of a Revitalized Indigenous Food Systems

The revitalization of Indigenous food systems is intricately tied to multiple aspects of sustainability, encompassing improved health outcomes, increased biodiversity, cultural identity reinforcement, and the promotion of food sovereignty. These factors synergistically contribute to an overall paradigm shift towards sustainable practices embedded within Indigenous communities. The health and nutrition outcomes correlated with revived Indigenous food systems underscore an urgent need for holistic approaches to tackling prevalent health crises among Indigenous populations. Research indicates a global pattern of obesity and undernutrition exacerbated by disconnections from traditional diets and food systems, spotlighting the critical health implications of revitalizing Indigenous food practices (Reynolds et al., 2023; Tremblay et al., 2020). For instance, the documentation of traditional diets can enhance food and nutrition security in Indigenous communities, as these food systems possess extensive biodiversity that can contribute to improved dietary diversity and nutritional outcomes (Kuhnlein, 2014; Kuhnlein & Chotiboriboon, 2022). Moreover, local food systems can integrate ethnobiological knowledge that highlights food plants and fosters community resilience against chronic illnesses related to poor diet (Lemke & Delormier, 2017).

Furthermore, incorporating active living education alongside traditional knowledge and food systems can significantly bolster mental health among Indigenous youth, further reinforcing the health benefits of this cultural reintegration (Ferguson et al., 2017; Walker et al., 2023). The increase in biodiversity through enriched Indigenous food systems is paramount, as these systems often represent complex ecosystems where traditional practices promote biodiversity conservation (Good et al., 2020; O'Bryan et al., 2020). Indigenous agricultural methods, such as companion planting, exemplify the capacity for cultural practices to restore and maintain ecological balance while ensuring food security (Vijayan et al., 2022). The interdependence between species and the land cultivated within Indigenous frameworks contributes to broader ecological restorations, addressing the biodiversity crisis by integrating Indigenous Knowledge with scientific insights for sustainable ecosystem governance (Constant & Tshisikhawe, 2018; Klopotek et al., 2022). This reciprocal relationship not only aids in the preservation of flora and fauna but also enhances the resilience of landscapes in the face of climate change (Tengö et al., 2014).

Food sovereignty emerges as a central theme in relation to strengthened cultural identity and youth engagement. The revitalization of Indigenous food systems is linked to reaffirming cultural identity as communities reconnect with ancestral practices and knowledge (Grenz & Armstrong, 2023; Kassam & Bernardo, 2022). Programs aimed at revitalizing traditional diets and practices resonate particularly with Indigenous youth, allowing for integrating cultural heritage into everyday life, which is vital for fostering a sense of belonging and identity (Coté, 2016; Fa & Luiselli, 2025). The intergenerational transmission of food-related knowledge is essential for preserving cultural practices, sustaining community bonds, and enhancing youth engagement, thereby cultivating new stewards of cultural identity (Bagelman, 2018; Reynolds et al., 2023).

Moreover, food sovereignty and economic empowerment for Indigenous communities are increasingly recognized as foundational to achieving sustainable and equitable food systems. Indigenous food sovereignty movements advocate for the right to self-determination in food production, emphasizing the importance of local and traditional agricultural practices that empower communities economically (Dawson et al., 2021). As Indigenous communities reclaim control over their food sources, they assert their rights to healthy and culturally appropriate food and to the ecosystems that support them (Brondízio et al., 2021; Lemke & Delormier, 2017). This control leads to economic opportunities not previously accessible, allowing for holistic community development when integrated with knowledge-sharing initiatives and collaborative projects with external partners that respect Indigenous rights and knowledge (Saxena, 2020). The inclusive approach fostered through collaboration between Indigenous peoples and external groups can stimulate innovative strategies for revitalizing Indigenous food systems, benefiting community health and ecological integrity. This collaborative knowledge coproduction model supports the notion of embedding Indigenous perspectives into broader sustainable practices, ensuring that policies reflect the needs and aspirations of Indigenous communities rather than imposing top-down solutions that undermine their traditional systems (O'Bryan et al., 2020; Vijayan et al., 2022).

17.10 Conclusion

Revitalizing Indigenous food systems is a transformative act that intersects cultural resilience, ecological restoration, and food sovereignty. Far from being static traditions of the past, these foodways represent dynamic systems of knowledge and practice that are highly adaptive to contemporary challenges, including climate change, biodiversity loss, nutritional inequities, and sociopolitical marginalization. This chapter has illustrated how Indigenous communities are reimagining food futures anchored in ancestral wisdom and local autonomy through a rich array of strategies. These revitalization efforts are inherently multidimensional, integrating cultural resurgence, intergenerational knowledge transmission, ecological stewardship, and political self-determination. They also offer powerful alternatives to industrial agrifood regimes by demonstrating how reciprocal, place-based relationships with land and food foster ecological sustainability, social cohesion, and cultural identity. However, sustaining these movements requires addressing structural challenges ranging from land dispossession and policy exclusion to cultural appropriation and extractive research. Meaningful support must be grounded in Indigenous-led governance, ethical collaboration, and legal recognition of food sovereignty and land rights. Importantly, revitalization must not be approached as a one-size-fits-all intervention but as a community-driven process informed by local cosmologies, seasonal knowledge, and spiritual values. From a sociobiological perspective, Indigenous food systems exemplify the coadaptation of human societies and ecological systems over millennia. They are living embodiments of biocultural diversity, where food serves as a conduit for transmitting memory, enacting identity, and sustaining collective life. As global communities confront intersecting crises, the revitalization of Indigenous foodways offers a pathway toward sustainability and a compelling model for relational living grounded in reciprocity, resilience, and justice. Supporting these systems is thus an act of restoration and a moral and ecological imperative for our shared future.

Chapter Reflection

1. How does the revitalization of Indigenous food systems contribute to sustainability and food sovereignty?
2. In what ways does colonization continue to affect Indigenous foodways today?
3. What role do youth and intergenerational knowledge transfer play in Indigenous food system revitalization?
4. How does traditional ecological knowledge (TEK) complement modern environmental science?
5. What are some of the ethical concerns, such as cultural appropriation, involved in Indigenous food system revitalization?
6. How do case studies from North America, Oceania, Africa, and Asia illustrate diverse strategies for food system revitalization?
7. What role do policy, governance, and legal frameworks play in enabling or hindering food sovereignty?
8. How do community gardens, seed banks, and food hubs serve as tools for revitalizing Indigenous food systems?
9. What are the main challenges Indigenous communities face in restoring food sovereignty, and how might these be addressed?
10. Why is it important to view Indigenous food systems as dynamic, living systems rather than static traditions?

REFERENCES

Ahmed, F., Liberda, E., Solomon, A., Davey, R., Sutherland, B., & Tsuji, L. (2023). Indigenous land-based approaches to well-being: The niska (goose) harvesting program in subarctic Ontario, Canada. *International Journal of Environmental Research and Public Health, 20*(4), 3686. https://doi.org/10.3390/ijerph20043686

Altieri, M. A., & Toledo, V. M. (2011). The agroecological revolution in Latin America: Rescuing nature, ensuring food sovereignty and empowering peasants. *Journal of Peasant Studies*, *38*(3), 587–612. https://doi.org/10.1080/03066150.2011.582947

ápi, S. (2011). Indigenous nation rebuilding through gardening. *International Journal of Critical Indigenous Studies*, *4*(1), 41–48. https://doi.org/10.5204/ijcis.v4i1.70

Armstrong, C., Miller, J., McAlvay, A., Ritchie, P., & Lepofsky, D. (2021). Historical Indigenous land use explains plant functional trait diversity. *Ecology and Society*, *26*(2). https://doi.org/10.5751/es-12322-260206

Atlas, W. I., Ban, N. C., Moore, J. W., Tuohy, A. M., Greening, S., Reid, A. J., Morven, N., White, E., Housty, W. G., Housty, J. A., Service, C. N., Greba, L., Harrison, S., Sharpe, C., Butts, K. I. R., Shepert, W. M., Sweeney-Bergen, E., Macintyre, D., Sloat, M. R., & Connors, K. (2020). Indigenous systems of management for culturally and ecologically resilient Pacific salmon (*Oncorhynchus* spp.) fisheries. *Bioscience*, *71*(2), 186–204. https://doi.org/10.1093/biosci/biaa144

Babcock, A., & Budowle, R. (2022). An appreciative inquiry and inventory of Indigenous food sovereignty initiatives within the Western U.S. *Journal of Agriculture Food Systems and Community Development*, *11*(2), 1–21. https://doi.org/10.5304/jafscd.2022.112.016

Bagelman, C. (2018). Unsettling food security: The role of young people in Indigenous food system revitalisation. *Children and Society*, *32*(3), 219–232. https://doi.org/10.1111/chso.12268

Bagelman, J., Deveraux, F., & Hartley, R. (2016). Feasting for change: Reconnecting with food, place, and culture. *International Journal of Indigenous Health*, *11*(1), 6–17. https://doi.org/10.18357/ijih111201616016

Blanchet, R., Batal, M., Johnson-Down, L., Johnson, S., Okanagan Nation Salmon Reintroduction Initiatives, & Willows, N. (2021). An Indigenous food sovereignty initiative is positively associated with well-being and cultural connectedness, as shown by a survey of Syilx Okanagan adults in British Columbia, Canada. *BMC Public Health*, *21*(1). https://doi.org/10.1186/s12889-021-11229-2

Bowra, A., Mashford-Pringle, A., & Poland, B. (2020). Indigenous learning on Turtle Island: A review of the literature on land-based learning. *Canadian Geographer / Le Géographe Canadien*, *65*(2), 132–140. https://doi.org/10.1111/cag.12659

Brondízio, E. S., Aumeeruddy-Thomas, Y., Bates, P., Cariño, J., Fernández-Llamazares, Á., Ferrari, M. F., Galvin, K., Reyes-García, V., McElwee, P., Molnár, Z., Samakov, A., & Shrestha, U. (2021). Locally based, regionally manifested, and globally relevant: Indigenous and local knowledge, values, and practices for nature. *Annual Review of Environment and Resources*, *46*(1), 481–509. https://doi.org/10.1146/annurev-environ-012220-012127

Budiyoko, B., & Verrysaputro, E. (2025). Exploring agricultural local wisdom in the Indigenous Bonokeling Community of Banyumas District in Central Java, Indonesia. *IOP Conference Series Earth and Environmental Science*, *1441*(1), 012025. https://doi.org/10.1088/1755-1315/1441/1/012025

Burnette, C., Lesesne, R., Temple, C., & Rodning, C. (2020). Family as the conduit to promote Indigenous women and men's enculturation and wellness: "I wish I had learned earlier". *Journal of Evidence-Based Social Work*, *17*(1), 1–23. https://doi.org/10.1080/26408066.2019.1617213

Churchill, M., Smylie, J., Wolfe, S., Bourgeois, C., Moeller, H., & Firestone, M. (2020). Conceptualising cultural safety at an Indigenous-focused midwifery practice in Toronto, Canada: Qualitative interviews with Indigenous and non-Indigenous clients. *BMJ Open*, *10*(9), e038168. https://doi.org/10.1136/bmjopen-2020-038168

Constant, N. L., & Tshisikhawe, M. P. (2018). Hierarchies of knowledge: Ethnobotanical knowledge, practices and beliefs of the Vhavenda in South Africa for biodiversity conservation. *Journal of Ethnobiology and Ethnomedicine*, *14*(1), 56. https://doi.org/10.1186/s13002-018-0255-2

Coté, C. (2016). "Indigenizing" food sovereignty. Revitalizing Indigenous food practices and ecological knowledges in Canada and the United States. *Humanities*, *5*(3), 57. https://doi.org/10.3390/h5030057

Cuyler, C., Daniel, C. J., Enghoff, M., Levermann, N., Møller-Lund, N., Hansen, P. N., Damhus, D., & Danielsen, F. (2020). Using local ecological knowledge as evidence to guide management: A community-led harvest calculator for muskoxen in Greenland. *Conservation Science and Practice*, *2*(3), e159. https://doi.org/10.1111/csp2.159

Dawson, N. M., Coolsaet, B., Sterling, E. J., Loveridge, R., Gross-Camp, N. D., Wongbusarakum, S., Sangha, K. K., Scherl, L. M., Phan, H. P., Zafra-Calvo, N., Lavey, W. G., Byakagaba, P., Idrobo, C. J., Chenet, A., Bennett, N. J., Mansourian, S., & Rosado-May, F. (2021). The role of Indigenous peoples and local communities in effective and equitable conservation. *Ecology and Society*, *26*(3), 19. https://doi.org/10.5751/es-12625-260319

Delormier, T., Horn-Miller, K., McComber, A., & Marquis, K. (2017). Reclaiming food security in the Mohawk community of Kahnawà:ke through Haudenosaunee responsibilities. *Maternal and Child Nutrition, 13*(S3), e12556. https://doi.org/10.1111/mcn.12556

Demissie, A., & Arega, M. (2017). Conservation and sustainable use of tef genetic resources in Ethiopia. *Plant Genetic Resources, 15*(6), 478–486. https://doi.org/10.1017/S1479262117000274

Derashri, A., Sharma, D., Dwivedi, A., & Rajput, D. (2025). Future directions and innovations. In S. K. Jain, R. Gupta, S. Vengurlekar, & N. Bais (Eds.), *Quality assurance of ethno-herbals: Cultivating confidence in alternative medicine* (pp. 121–136). Bentham Science Publishers. https://doi.org/10.2174/97898152 74554125010011

Detwiler, D. (2020). Food deserts and food insecurity: In tribal lands and from coast to coast. In *Building the future of food safety technology: Blockchain and beyond* (pp. 121–125). Academic Press. https://doi.org/10.1016/b978-0-12-818956-6.00011-7

Donatuto, J., Campbell, L., LeCompte, J., Rohlman, D., & Tadlock, S. (2020). The story of 13 Moons: Developing an environmental health and sustainability curriculum founded on Indigenous first foods and technologies. *Sustainability, 12*(21), 8913. https://doi.org/10.3390/su12218913

Elliott, H., Mulrennan, M., & Cuerrier, A. (2021). Resurgence, refusal, and reconciliation through food movement organizations: A case study of food secure Canada's 2018 assembly. *Journal of Agriculture Food Systems and Community Development, 10*(3), 1–21. https://doi.org/10.5304/jafscd.2021.103.009

Emmanuel, R., Read, U. M., Grande, A. J., & Harding, S. (2023). Acceptability and feasibility of community gardening interventions for the prevention of non-communicable diseases among Indigenous populations: A scoping review. *Nutrients, 15*(3), 791. https://doi.org/10.3390/nu15030791

Fa, J., & Luiselli, L. (2025). Weaving the middle spaces between Indigenous and scientific knowledge for biodiversity conservation and ecology. *African Journal of Ecology, 63*(2), e70030. https://doi.org/10.1111/aje.70030

Ferguson, M., Brown, C., Georga, C., Miles, E., Wilson, A., & Brimblecombe, J. (2017). Traditional food availability and consumption in remote aboriginal communities in the northern territory, Australia. *Australian and New Zealand Journal of Public Health, 41*(3), 294–298. https://doi.org/10.1111/1753-6405.12664

Ghosh-Jerath, S., Downs, S., Singh, A., Paramanik, S., Goldberg, G., & Fanzo, J. (2019). Innovative matrix for applying a food systems approach for developing interventions to address nutrient deficiencies in Indigenous communities in India: A study protocol. *BMC Public Health, 19*(1), 944. https://doi.org/10.1186/s12889-019-6963-2

Ginting, S. (2025). Indonesia's Indigenous Suku Anak Dalam: Knowledge for food and environmental sustainability. *Frontiers in Sustainable Food Systems, 9.* https://doi.org/10.3389/fsufs.2025.1587094

Gone, J., & Kirmayer, L. (2020). Advancing Indigenous mental health research. *Transcultural Psychiatry, 57*(2), 235–249. https://doi.org/10.1177/1363461520923151

Good, A., Sims, L., Clarke, K., & Russo, F. (2020). Indigenous youth reconnect with cultural identity: The evaluation of a community- and school-based traditional music program. *Journal of Community Psychology, 49*(2), 588–604. https://doi.org/10.1002/jcop.22481

Gordon, K., Xavier, A., & Neufeld, H. (2018). Healthy roots: Building capacity through shared stories rooted in Haudenosaunee knowledge to promote Indigenous foodways and well-being. *Canadian Food Studies / La Revue Canadienne Des Études Sur L Alimentation, 5*(2), 180–195. https://doi.org/10.15353/cfs-rcea.v5i2.210

Grenz, J., & Armstrong, C. (2023). Pop-up restoration in colonial contexts: Applying an Indigenous food systems lens to ecological restoration. *Frontiers in Sustainable Food Systems, 7.* https://doi.org/10.3389/fsufs.2023.1244790

Heaney, D., Padilla-Zakour, O., & Chen, C. (2024). Processing and preservation technologies to enhance Indigenous food sovereignty, nutrition security, and health equity in north America. *Frontiers in Nutrition, 11.* https://doi.org/10.3389/fnut.2024.1395962

Hoover, E. (2017). "You can't say you're sovereign if you can't feed yourself": Defending food sovereignty in the United States. *American Indian Culture and Research Journal, 41*(3), 31–70. https://doi.org/10.17953/aicrj.41.3.hoover

Hudson, M., Carroll, S. R., Anderson, J., Blackwater, D., Cordova-Marks, F. M., Cummins, J., David-Chavez, D., Fernandez, A., Garba, I., Hiraldo, D., Jäger, M. B., Jennings, L. L., Martinez, A., Sterling, R., Walker, J. D., & Rowe, R. K. (2023). Indigenous peoples' rights in data: A contribution toward Indigenous research sovereignty. *Frontiers in Research Metrics and Analytics, 8.* https://doi.org/10.3389/frma.2023.1173805

Hunter, D., Borelli, T., Beltrame, D. M. O., Oliveira, C. N. S., Coradin, L., Wasike, V. W., Wasilwa, L., Mwai, J., Manjella, A., Samarasinghe, G. W. L., Madhujith, T., Nadeeshani, H. V. H., Tan, A., Ay, S. T., Güzelsoy, N., Lauridsen, N., Gee, E., & Tartanac, F. (2019). The potential of neglected and underutilized species for improving diets and nutrition. *Planta*, *250*(3), 709–729. https://doi.org/10.1007/s00425-019-03169-4

Isumonah, K. (2024). An examination of food insecurity among Canadian Aboriginal people. *Journal of Global Health Economics and Policy*, *4*, e2024009. https://doi.org/10.52872/001c.126467

Jennings, D., Johnson-Jennings, M., & Little, M. (2020). Utilizing Webs to share ancestral and intergenerational teachings: The process of co-building an online digital repository in partnership with Indigenous communities. *Genealogy*, *4*(3), 70. https://doi.org/10.3390/genealogy4030070

Jernigan, V. B. B., Maudrie, T. L., Nikolaus, C. J., Benally, T., Johnson, S., Teague, T., Mayes, M., Jacob, T., & Taniguchi, T. (2021). Food sovereignty indicators for Indigenous community capacity building and health. *Frontiers in Sustainable Food Systems*, *5*. https://doi.org/10.3389/fsufs.2021.704750

Jernigan, V. B. B., Nguyen, C. J., Maudrie, T. L., Demientieff, L. X., Black, J. C., Mortenson, R., Wilbur, R. E., Clyma, K. R., Lewis, M., & Lopez, S. (2023). Food sovereignty and health: A conceptual framework to advance research and practice. *Health Promotion Practice*, *24*(6), 1070–1074. https://doi.org/10.1177/15248399231190367

Johnson-Jennings, M., Jennings, D., Paul, K., & Little, M. (2020). Identifying needs and uses of digital Indigenous food knowledge and practices for an Indigenous food wisdom repository. *Alternative an International Journal of Indigenous Peoples*, *16*(4), 290–299. https://doi.org/10.1177/1177180120954446

Judge, C., Spring, A., & Skinner, K. (2022). A comparative policy analysis of wild food policies across Ontario, northwest territories, and Yukon territory, Canada. *Frontiers in Communication*, *7*. https://doi.org/10.3389/fcomm.2022.780391

Jupiter, S. D., Cohen, P. J., Weeks, R., Tawake, A., & Govan, H. (2014). Locally-managed marine areas: Multiple objectives and diverse strategies. *Pacific Conservation Biology*, *20*(2), 165–179. https://doi.org/10.1071/PC140165

Kassam, K., & Bernardo, J. (2022). Role of biodiversity in ecological calendars and its implications for food sovereignty: Empirical assessment of the resilience of indicator species to anthropogenic climate change. *GeoHealth*, *6*(10). https://doi.org/10.1029/2022gh000614

Kellogg, J. (2019). Reviving seeds, reclaiming food: Indigenous seed sovereignty in the United States. *Agriculture and Human Values*, *36*(4), 677–689. https://doi.org/10.1007/s10460-019-09942-4

Klopotek, B., Claybrook, T., & Scott, J. (2022). Indigenous companion planting in the great churn: Three sisters in Kalapuya ilihi. *Environment and Planning E Nature and Space*, *6*(3), 1889–1904. https://doi.org/10.1177/25148486221126618

Kuhnlein, H. (2014). How ethnobiology can contribute to food security. *Journal of Ethnobiology*, *34*(1), 12–27. https://doi.org/10.2993/0278-0771-34.1.12

Kuhnlein, H., & Chotiboriboon, S. (2022). Why and how to strengthen Indigenous peoples' food systems with examples from two unique Indigenous communities. *Frontiers in Sustainable Food Systems*, *6*. https://doi.org/10.3389/fsufs.2022.808670

Kumar, A. (2021). How can India leverage its botanic gardens for the conservation and sustainable utilization of wild food plant resources through the implementation of a global strategy for plant conservation? *Journal of Zoological and Botanical Gardens*, *2*(4), 586–599. https://doi.org/10.3390/jzbg2040042

Lemke, S., & Delormier, T. (2017). Indigenous peoples' food systems, nutrition, and gender: Conceptual and methodological considerations. *Maternal and Child Nutrition*, *13*(S3). https://doi.org/10.1111/mcn.12499

Levkoe, C., Ray, L., & McLaughlin, J. (2019). The Indigenous food circle: Reconciliation and resurgence through food in northwestern Ontario. *Journal of Agriculture Food Systems and Community Development*, *9*, 1–14. https://doi.org/10.5304/jafscd.2019.09b.008

Lewis, C. (2021). Brief research commentary: The US Indigenous food sovereignty movement's impact on understandings of COVID-19 in Indian country. *Culture Agriculture Food and Environment*, *43*(2), 107–113. https://doi.org/10.1111/cuag.12280

Lopes, C. V. A., de Sousa Alves Neri, J. L., Hunter, J., Ronto, R., & Mihrshahi, S. (2024). Interventions and programs using native foods to promote health: A scoping review. *Nutrients*, *16*(23), 4222. https://doi.org/10.3390/nu16234222

Lopes, C., Mihrshahi, S., Hunter, J., Ronto, R., & Cawthorne, R. (2024). Co-designing research for sustainable food systems and diets with aboriginal communities: A study protocol. *International Journal of Environmental Research and Public Health, 21*(3), 298. https://doi.org/10.3390/ijerph21030298

Marushka, L., Batal, M., Tikhonov, C., Sadik, T., Schwartz, H., Ing, A., Fediuk, K., & Chan, H. (2021). Importance of fish for food and nutrition security among First Nations in Canada. *Canadian Journal of Public Health, 112*(S1), 64–80. https://doi.org/10.17269/s41997-021-00481-z

Maudrie, T., Colón-Ramos, U., Harper, K., Jock, B., & Gittelsohn, J. (2021). A scoping review of the use of Indigenous food sovereignty principles for intervention and future directions. *Current Developments in Nutrition, 5*(7), nzab093. https://doi.org/10.1093/cdn/nzab093

McMillen, H., Ticktin, T. & Springer, H. K. (2017). The future is behind us: Traditional ecological knowledge and resilience over time on Hawai'i Island. *Regional Environmental Change 17*, 579–592. https://doi.org/10.1007/s10113-016-1032-1

Meissner, S. (2023). "World"-traveling in Tule Canoes: Indigenous philosophies of language and an ethic of incommensurability. *Hypatia, 38*(4), 849–870. https://doi.org/10.1017/hyp.2023.82

Merz, C., & Steinberg, M. (2014). Applying a political economy of health standpoint to traditional food acquisition practices and the inequitable prevalence of obesity and diabetes amongst First Nations peoples in British Columbia. *Environmental Health Review, 57*(3), 65–70. https://doi.org/10.5864/d2014-028

Michnik, K., Thompson, S., & Beardy, B. (2021). Moving your body, soul, and heart to share and harvest food. *Canadian Food Studies / La Revue Canadienne Des Études Sur L Alimentation, 8*(2). https://doi.org/10.15353/cfs-rcea.v8i2.446

Montgomery, M., & Vaughan, M. (2020). Kīpuka Kuleana: Restoring reciprocity to coastal land tenure and resource use in Hawai'i. In N. Turner (Ed.), *Plants, people and places: The roles of ethnobotany and ethnoecology in Indigenous peoples' land rights in Canada and beyond* (pp. 96, 238). McGill-Queen's University Press.

Moscou, K. (2022). Planting seeds of change: Voices of Indigenous youth on wholistic health. *International Indigenous Policy Journal, 13*(2). https://doi.org/10.18584/iipj.2022.13.2.10977

Mucioki, M., Kelly, S., Holen, D., Powell, B., Galbreath, T., Paterno, S., Mixon, R., & Chi, G. (2024). Gardening practices in Alaska build on traditional food system foundations. *Agriculture and Human Values*, *42*(2), 965–981. https://doi.org/10.1007/s10460-024-10652-6

Mugabirwe, O., & Turyamureeba, R. (2025). Indigenous knowledge for climate change adaptation and resilience in rural communities of Ankole sub-region, Uganda. African *Journal of Climate Change and Resource Sustainability*, *4*(1), 239–251. https://doi.org/10.37284/ajccrs.4.1.2925

Sarkar, D., Walker-Swaney, J., & Shetty, K. (2019). Food diversity and Indigenous food systems to combat diet-linked chronic diseases. *Current Developments in Nutrition, 4*(Suppl. 1), 3–11. https://doi.org/10.1093/cdn/nzz099

Negi, V., Pathak, R., Thakur, S., Joshi, R., Bhatt, I., & Rawal, R. (2021). Scoping the need of mainstreaming Indigenous knowledge for sustainable use of bioresources in the Indian Himalayan region. *Environmental Management*, *72*(1), 135–146. https://doi.org/10.1007/s00267-021-01510-w

Neufeld, H., Richmond, C., & Centre, S. (2017). Impacts of place and social spaces on traditional food systems in southwestern Ontario. *International Journal of Indigenous Health*, *12*(1), 93–115. https://doi.org/10.18357/ijih112201716903

Nigh, R., & Diemont, S. A. W. (2013). The Maya milpa: Fire and the legacy of living soil. *Frontiers in Ecology and the Environment*, *11*(S1), e45–e54. https://doi.org/10.1890/120344

Noe, S. (2023). Indigenous foodways as persistence in the Alta California mission system. *American Antiquity*, *88*(4), 451–475. https://doi.org/10.1017/aaq.2023.53

O'Bryan, C. J., Garnett, S. T., Fa, J. E., Leiper, I., Rehbein, J. A., Fernández-Llamazares, Á., Jackson, M. V., Jonas, H. D., Brondizio, E. S., Burgess, N. D., Robinson, C. J., Zander, K. K., Molnár, Z., Venter, O., & Watson, J. (2020). The importance of Indigenous peoples' lands for the conservation of terrestrial mammals. *Conservation Biology*, *35*(3), 1002–1008. https://doi.org/10.1111/cobi.13620

Obahiagbon, E. G., & Ogwu, M. C. (2023a). Sustainable supply chain management in the herbal medicine industry. In S. C. Izah, M. C. Ogwu, & M. Akram (Eds.), *Herbal medicine phytochemistry*. Reference Series in Phytochemistry (pp. 1–29). Springer. https://doi.org/10.1007/978-3-031-21973-3_64-1

Obahiagbon, E. G., & Ogwu, M. C. (2023b). Consumer perception and demand for sustainable herbal medicine products and market. In S. C. Izah, M. C. Ogwu,

Note: Please confirm is this correct

& M. Akram (Eds.), *Herbal medicine phytochemistry.* Reference Series in Phyto-chemistry (pp. 1–34). Springer. https://doi.org/10.1007/978-3-031-21973-3_65-1

Obahiagbon, E. G., & Ogwu, M. C. (2024a). Organic food preservatives: The shift towards natural alternatives and sustainability in the Global South's Markets. In M. C. Ogwu, S. C. Izah, & N. R. Ntuli (Eds.), *Food safety and quality in the Global South* (pp. 299–329). Springer. https://doi.org/10.1007/978-981-97-2428-4_10

Obahiagbon, E. G., & Ogwu, M. C. (2024b). The nexus of business, sustain-ability, and herbal medicine. In S. C. Izah, M. C. Ogwu, & M. Akram (Eds.), *Herbal medicine phytochemistry.* Reference Series in Phytochemistry (pp. 1–42). Springer. https://doi.org/10.1007/978-3-031-21973-3_67-1

Obongodot, N. U., & Ogwu, M. C. (2023). Plant food for human health: Case study of Indigenous vegetables in Akwa Ibom State, Nigeria. In S. C. Izah, M. C. Ogwu, & M. Akram (Eds.), *Herbal medicine phytochemistry.* Reference Series in Phytochemistry (pp. 1–38). Springer. https://doi.org/10.1007/978-3-031-21973-3_2-1

Ogwu, M. C. (2023). Local food crops in Africa: Sustainable utilization, threats, and traditional storage strategies. In S. C. Izah & M. C. Ogwu (Eds.), *Sustainable utilization and conservation of Africa's biological resources and environment.* Sustainable Development and Biodiversity (Vol. 888, pp. 353–374). Springer. https://doi.org/10.1007/978-981-19-6974-4_13

Ogwu, M. C., & Osawaru, M. E. (2022). Traditional methods of plant conservation for sustainable utilization and development. In S. C. Izah (Ed.), *Biodiversity in Africa: Potentials, threats and conservation.* Sustainable Development and Biodiversity (Vol. 29, pp. 451–472). Springer. http://doi.org/10.1007/978-981-19-3326-4_17

Oloko, M., & Shukla, S. (2018). "Nya Anghuwa Che" (our food gives us life): Exploring Indigenous perspectives on traditional food gathering and foraging in an Irigwe community from Nigeria. *Ab-Original, 2*(1), 1–22. https://doi.org/10.5325/aboriginal.2.1.0001

Oskal, A., Sara, R., Krarup-Hansen, K., Smuk, I., & Mathiesen, S. (2023). Reindeer herders' food knowledge systems. In S. D. Mathiesen, I. M. G. Eira, E. I. Turi, A. Oskal, M. Pogodaev, & M. Tonkopeeva (Eds.), *Reindeer husbandry.* Springer Polar Sciences (pp. 139–168). Springer. https://doi.org/10.1007/978-3-031-42289-8_6

Padulosi, S., Thompson, J., & Rudebjer, P. (2013). *Fighting poverty, hunger and malnutrition with neglected and underutilized species: Needs, challenges and the way forward.* Bioversity International. https://hdl.handle.net/10568/97014

Phillipps, B., Skinner, K., Parker, B., & Neufeld, H. (2022). An intersectionality-based policy analysis examining the complexities of access to wild game and fish for urban Indigenous women in northwestern Ontario. *Frontiers in Communication*, *6*. https://doi.org/10.3389/fcomm.2021.762083

Rankoana, S. (2016). Sustainable use and management of Indigenous plant resources: A case of Mantheding Community in Limpopo Province, South Africa. *Sustainability*, *8*(3), 221. https://doi.org/10.3390/su8030221

Reynolds, A., Keough, M. T., Blacklock, A., Tootoosis, C., Whelan, J., Bomfim, E., Mushquash, C., Wendt, D. C., O'Connor, R. M., Burack, J. A., & Burack, J. (2023). The impact of cultural identity, parental communication, and peer influence on substance use among Indigenous youth in Canada. *Transcultural Psychiatry*, *61*(3), 351–360. https://doi.org/10.1177/13634615231191999

Riley, K., & Donaldson, E. (2024). Food (inter)activism in the Marquesas, French Polynesia. *American Anthropologist*, *126*(4), 707–711. https://doi.org/10.1111/aman.28014

Robin, T. (2018). Responsibilities and reflections: Indigenous food, culture, and relationships. *Canadian Food Studies / La Revue Canadienne Des Études Sur L Alimentation*, *5*(2), 9–12. https://doi.org/10.15353/cfs-rcea.v5i2.216

Robin, T. (2019). Our hands at work: Indigenous food sovereignty in western Canada. *Journal of Agriculture Food Systems and Community Development*, *9*, 1–15. https://doi.org/10.5304/jafscd.2019.09b.007

Robin, T., Dennis, M., & Hart, M. (2020). Feeding Indigenous people in Canada. *International Social Work*, *65*(4), 652–662. https://doi.org/10.1177/0020872820916218

Robin, T., & Hart, M. (2025). Cree food knowledge and being well. *International Journal of Environmental Research and Public Health*, *22*(2), 181. https://doi.org/10.3390/ijerph22020181

Rowarth, J., Manning, M., Roberts, A., & King, W. (2021). New-generative agriculture - based on science, informed by research and honed by New Zealand farmers. *Journal of New Zealand Grasslands*, *82*, 221–229. https://doi.org/10.33584/jnzg.2020.82.430

Rowe, S., Brady, C., Sarang, R., Wiipongwii, T., Leu, M., Jennings, L., Peterson, T., Boston, J., Roach, B., Phillips, J., & Conrad, Z. (2024). Improving Indigenous food sovereignty through sustainable food production: A narrative review. *Frontiers in Sustainable Food Systems*, *8*. https://doi.org/10.3389/fsufs.2024.1341146

Roy, P. S., Singh, S., & Roy, A. (2018). *Sacred groves of India: A biodiversity treasure*. Indian Council of Forestry Research and Education.

Ruelle, M. (2017). Ecological relations and Indigenous food sovereignty in Standing Rock. *American Indian Culture and Research Journal*, *41*(3), 113–125. https://doi.org/10.17953/aicrj.41.3.ruelle

Ruelle, M., Skye, A., Collins, E., & Kassam, K. (2022). Ecological calendars, food sovereignty, and climate adaptation in standing rock. *GeoHealth*, *6*(12). https://doi.org/10.1029/2022gh000621

Sahani, A., Gupta, G., Anand, S., Sharma, V., Singh, R., Kamil, A., Kumar, H., Gagan, A., Mayala, V. K., Mishra, A. P., Jaiswal, S., & Anand, J. (2025). Indigenous knowledge and water conservation practices in South Africa: A systematic literature review. *Journal of Environmental and Earth Sciences*, *7*(2). https://doi.org/10.30564/jees.v7i2.7988

Saxena, L. (2020). Community self-organisation from a social-ecological perspective: "Burlang Yatra" and revival of millets in Odisha (India). *Sustainability*, *12*(5), 1867. https://doi.org/10.3390/su12051867

Shafiee, M., Al-Bazz, S., Szafron, M., Lane, G., & Vatanparast, H. (2024). "I haven't had moose meat in a long time": Exploring urban Indigenous perspectives on traditional foods in Saskatchewan. *Nutrients*, *16*(15), 2432. https://doi.org/10.3390/nu16152432

Shamon, H., Cosby, O. G., Andersen, C. L., Augare, H., Stiffarm, J. B., Bresnan, C. E., Brock, B. L., Carlson, E., Deichmann, J. L., Epps, A., Guernsey, N., Hartway, C., Jørgensen, D., Kipp, W., Kinsey, D., Komatsu, K. J., Kunkel, K., Magnan, R., Martin, J. M., ... Akre, T. (2022). The potential of bison restoration as an ecological approach to future tribal food sovereignty on the Northern Great Plains. *Frontiers in Ecology and Evolution*, *10*. https://doi.org/10.3389/fevo.2022.826282

Sherriff, S., Kalucy, D., Tong, A., Naqvi, N., Nixon, J., Eades, S., Ingram, T., Slater, K., Dickson, M., Lee, A., & Muthayya, S. (2022). Murradambirra dhangaang (make food secure): Aboriginal community and stakeholder perspectives on food insecurity in urban and regional Australia. *BMC Public Health*, *22*(1). https://doi.org/10.1186/s12889-022-13202-z

Skinner, K., Hanning, R., Desjardins, E., & Tsuji, L. (2013). Giving voice to food insecurity in a remote Indigenous community in subarctic Ontario, Canada:

Traditional ways, ways to cope, ways forward. *BMC Public Health*, *13*(1), 427. https://doi.org/10.1186/1471-2458-13-427

Sumner, J., McMurtry, J., & Tarhan, D. (2023). Growing community sustenance: The social economy as a route to Indigenous food sovereignty. *Canadian Journal of Nonprofit and Social Economy Research*, *14*(S1). https://doi.org/10.29173/cjnser535

Swiderska, K., Argumedo, A., Wekesa, C., Ndalilo, L., Song, Y., Rastogi, A., & Ryan, P. (2022). Indigenous peoples' food systems and biocultural heritage: Addressing Indigenous priorities using decolonial and interdisciplinary research approaches. *Sustainability*, *14*(18), 11311. https://doi.org/10.3390/su141811311

Tebulo, K. (2025). Community gardens: Cultivating medicinal plants for local disease prevention. *Research Output Journal of Biological and Applied Science*, *5*(2), 1–4. https://doi.org/10.59298/rojbas/2025/521400

Tengö, M., Brondízio, E., Elmqvist, T., Malmer, P., & Spierenburg, M. (2014). Connecting diverse knowledge systems for enhanced ecosystem governance: The multiple evidence base approach. *Ambio*, *43*(5), 579–591. https://doi.org/10.1007/s13280-014-0501-3

Thompson, H., Mason, C., & Robidoux, M. (2018). Hoop house gardening in the Wapekeka First Nation as an extension of land-based food practices. *Arctic*, *71*(4), 407–421. https://doi.org/10.14430/arctic4746

Timler, K., Varcoe, C., & Brown, H. (2019). Growing beyond nutrition:: How a prison garden program highlights the potential of shifting from food security to food sovereignty for Indigenous peoples. *International Journal of Indigenous Health*, *14*(2), 95–114. https://doi.org/10.32799/ijih.v14i2.31938

Tiwari, B. K., Barik, S. K., & Tripathi, R. S. (1998). Sacred groves of Meghalaya. In P. S. Ramakrishnan, K. G. Saxena, & U. M. Chandrashekara (Eds.), *Conserving the sacred for biodiversity management* (pp. 253–262). UNESCO and Oxford-IBH Publishing.

Toya, M. (2017, January 6). *Shintō's sacred forests and Japanese environmentalism*. Nippon.com. https://www.nippon.com/en/views/b05214/

Tremblay, R., Landry-Cuerrier, M., & Humphries, M. (2020). Culture and the social-ecology of local food use by Indigenous communities in northern North America. *Ecology and Society*, *25*(2). https://doi.org/10.5751/es-11542-250208

Turpin, M., & Si, A. (2017). Edible insect larvae in Kaytetye: Their nomenclature and significance. *Journal of Ethnobiology*, *37*(1), 120–140. https://doi.org/10.2993/0278-0771-37.1.120

Ullrich, J. (2019). For the love of our children: An Indigenous connectedness framework. *Alternative an International Journal of Indigenous Peoples*, *15*(2), 121–130. https://doi.org/10.1177/1177180119828114

Vijayan, D., Ludwig, D., Rybak, C., Kaechele, H., Hoffmann, H., Schönfeldt, H. C., Mbwana, H. A., Rivero, C. V., & Löhr, K. (2022). Indigenous knowledge in food system transformations. *Communications Earth and Environment*, *3*(1), 213. https://doi.org/10.1038/s43247-022-00543-1

Viscogliosi, C., Asselin, H., Basile, S., Borwick, K., Couturier, Y., Drolet, M., Gagnon, D., Obradovic, N., Torrie, J., Zhou, D., & Levasseur, M. (2020). Importance of Indigenous elders' contributions to individual and community wellness: Results from a scoping review on social participation and intergenerational solidarity. *Canadian Journal of Public Health*, *111*(5), 667–681. https://doi.org/10.17269/s41997-019-00292-3

Walker, S., Kannan, P., Bhawra, J., & Katapally, T. (2023). Evaluation of a longitudinal digital citizen science initiative to understand the impact of culture on Indigenous youth mental health: Findings from a quasi-experimental qualitative study. *PLOS One*, *18*(12), e0294234. https://doi.org/10.1371/journal.pone.0294234

Yongabi, K. (2023). Perspectives on Indigenous knowledge in mitigating climate change. *Research and Reviews on Healthcare Open Access Journal*, *8*(4). https://doi.org/10.32474/rrhoaj.2023.08.000293

Young, L., Shukla, S., & Wilson, T. (2024). Indigenous values and perspectives for strengthening food security and sovereignty: Learning from a community-based case study of Misko-Ziibiing (Bloodvein River First Nation), Manitoba, Canada. *Frontiers in Sustainable Food Systems*, *8*. https://doi.org/10.3389/fsufs.2024.1321231

www.ingramcontent.com/pod-product-compliance
Lightning Source LLC
Chambersburg PA
CBHW041345210526
45162CB00015B/8